THE ISLE OF WIGHT FLORA

Following page
The western tip of the Island, showing the lowland heath of
Headon Warren in close proximity to the calcareous grassland
of the Needles headland. The range of habitats to be found
within a close distance, a consequence of the varied geology,
epitomises the diverse nature of the Island's flora. (KM)

The
ISLE OF WIGHT FLORA

Colin Pope, Lorna Snow, David Allen

WITH CONTRIBUTIONS FROM
Allan Insole, Francis Rose, Rob Scaife, Denis Simmons,
Richard Smout and Nick Stewart

EDITOR: Anne Marston

THE DOVECOTE PRESS
in association with
THE ISLE OF WIGHT NATURAL HISTORY
AND ARCHAEOLOGICAL SOCIETY

First published in 2003 by The Dovecote Press Ltd
Stanbridge, Wimborne, Dorset BH21 4JD

in association with
The Isle of Wight Natural History & Archaeological Society
Salisbury Gardens, Dudley Road, Ventnor, Isle of Wight PO38 1EJ

ISBN 1 904349 28 5

© The Isle of Wight Natural History & Archaeological Society 2003
Paintings © Dolly Norledge
Photographs © retained by photographers

The Isle of Wight Natural History & Archaeological Society
has asserted its rights under the Copyright, Designs
and Patent Act 1988 to be identified as author of this work

Designed by The Dovecote Press Ltd

Typeset in Monotype Sabon
Printed and bound by KHL Printing in Singapore

A CIP catalogue record for this book is available
from the British Library

All rights reserved

Contents

Acknowledgements 6

The Contributors 7

Introduction 9
RICHARD SMOUT

Geology & Physical Features 12
ALLAN N. INSOLE

The Island Climate 17
DENIS J. SIMMONS

The Palaeoecological Background 19
ROB SCAIFE

Island Distinctiveness 32
FRANCIS ROSE

*A History of Botanical Recording
in the Isle of Wight* 36
DAVID E. ALLEN

Island Habitats – Past & Present 49
COLIN POPE

Vascular Plants 64
COLIN POPE

Bryophytes 199
(Liverworts & Mosses)
LORNA SNOW

Lichens 218
COLIN POPE

Stoneworts 234
NICK STEWART

Appendices 237
Recorders
Extinctions

Index 239

Acknowledgements

The project to publish this Flora for the Isle of Wight has been guided throughout by a small committee of the Isle of Wight Natural History and Archaeological Society and we have been most fortunate in having the willing co-operation from many individuals with skills to offer.

The book has been enriched by the contributions of national and local experts, so that it presents a rounded picture of the Island's flora in context, and we wish to record our thanks to those who have given of their time to write the specialist chapters. The illustrations demonstrate the local distinctiveness of the Flora and we are grateful to the artist and photographers for permission to use their work. Mr A. Brent-Good kindly allowed us to reproduce illustrations from his family flora.

A list of some of the individuals who have contributed towards recording the Island's flora is given at Appendix 1. Where space permits, observations have been credited to named recorders. We are grateful to the many specialist referees who have confirmed the identification of material sent to them.

Many people have assisted by supplying information, and by commenting critically on various sections. In particular we would like to record our thanks to Jean Paton (Liverworts), Howard Matcham and Rod Stern (Mosses), and Francis Rose and Mark Seaward (Lichens). We have also received valuable assistance from Bill Shepard, Rebecca Loader, David Pearman, Pete Selby, Alex Lockton and Mike Wood. We are most grateful to some of the good Island folk who have contributed local information, including Bill Shepard, Brian Warne, Jack Lavers, Reuben Abbott, Richard Lightbown, Jimmy Winter, Dorothy Pope, John Heal, Gwen Bunce and the late Clifford Matthews. In the twenty-first century, the compilation of records is routinely handled by the use of a computer database. Recorder 2000 was used for processing the Bryophyte records and Aditsite software for the vascular plants and lichen records. Particular thanks must be given to Paul Griffiths from Aditsite, who has given great assistance.

We are grateful to the Ordnance Survey for permission to reproduce extracts of Ordnance Survey based mapping (© Crown Copyright NC/03/91/9179). The geology maps are reproduced by permission of the British Geological Survey. (© NERC. All rights reserved. IPR/42-29C.) We acknowledge the assistance of the Isle of Wight Council in the production of these maps.

Proofreaders have not only looked for obvious errors, but have in many instances commented helpfully on the text. We would like to thank David Biggs, Sheila and David Burch, Margaret Burnhill, Ann Campbell, Beth Dollery, Jackie Hart, Cedric Harrald, Margaret Jackson, Keith Marston, Jillie Pope, Margaret Savory, Richard Smout, Lorna Snow, Les and Sheila Street and Maureen Whitaker for undertaking this task. Sue Telfer has been responsible for the business and finance aspects, a role vital to the management of the venture.

We must thank our Publisher, David Burnett, of The Dovecote Press for his wise advice at all stages of the process, and the design and layout of the book.

The Society extends its thanks to the following organisations for financial support:

The Botanical Society of the British Isles: organises plant distribution surveys, publishes handbooks and offers a panel of referees to assist with identification. Worldwide membership for professional and amateur botanists. Dept. of Botany, The Natural History Museum, Cromwell Road, London SW7 5BD.

The British Bryological Society: Dept. of Botany, National Museum of Wales, Cardiff CF1 3NE.

The Wild Flower Society: for the conservation of wild flowers and the countryside, including promoting understanding of wildflowers by children. For membership, contact: 82A, High Street, Sawston, Cambridge CB2 4HJ.

Leader +. This project was part-financed by the European Community IW Rural Action Zone Leader + 2000-2006 Programme.

The Daisy Rich Trust.

The Contributors

DR COLIN POPE has a botany degree and doctorate in plant ecology from the University of London. He has been the vice-county recorder for vascular plants since 1995, and is the author of a lichen flora published in 1983. He is co-author of *The Nature of Hampshire and the Isle of Wight*, published in 1986. A native of the Island with a comprehensive knowledge of its natural history, he is employed as the Ecology Officer for the Isle of Wight Council. He is President Elect of the Isle of Wight Natural History and Archaeological Society.

LORNA SNOW has a long association with the Island having spent six months of each year at the family house near Sandown, before becoming a resident in the 1970s. She first joined the Isle of Wight Natural History and Archaeological Society in 1956 and became interested in botany during the survey work for the previous flora. A developing interest in mosses and liverworts led to her taking on the recording of bryophytes in 1977.

DR DAVID E. ALLEN has made forays since 1974 from his home in Winchester to investigate the Island's brambles, the particularly challenging group in which he is a leading European specialist. A past President of the Botanical Society of the British Isles and the Society for the History of Natural History, he is the author of a similar work on the flora of the Isle of Man as well as several books and many articles on British naturalists since the seventeenth century. Formerly a research administrator by profession, he is an associate of the Wellcome Trust Centre for the History of Medicine at University College London and also of London's Natural History Museum.

DR ALLAN N. INSOLE received his geology degree and doctorate, related to the Tertiary geology of the Isle of Wight, from the University of Bristol. He lived on the Island for number of years, and was Museums Officer for the Isle of Wight County Council. He is a past President of the Isle of Wight Natural History and Archaeological Society. He is a co-author of the Geologists' Association *Field Guide to the Isle of Wight*. He is currently a freelance lecturer and spends most of his time teaching on a Shell training programme in Nigeria.

DR FRANCIS ROSE MBE was until his retirement in 1981 Reader in Biogeography at King's College, University of London. He is an experienced botanist who has worked in the field in every county of Britain, except Orkney, Shetland and the Outer Isles. Records from his collection of nearly 200 notebooks are being extracted and deposited at the National Museum of Wales. He has done much survey work in Kent, Sussex and Hampshire, and is a co-author of the *Flora of Hampshire* published in 1996 He has also written lichen floras of Sussex and Somerset, and researched the information that epiphytic lichens can give on the continuity and age of old woodlands throughout Britain and Western Europe.

DR ROB SCAIFE is a consultant Quaternary paleoecologist and archaeobotanist, and is a visiting research fellow in the Department of Geography, University of Southampton. He specialises in the palynology of archaeological sites in southern and eastern England, and the Italian Alps. Having undertaken his doctoral research on the Isle of Wight, he has continued his study of the vegetation history of the Island. He is resident on the Island, and is a keen field botanist.

NICK STEWART is a freelance botanist with a particular interest in aquatic botany and ecology. He is the national referee for stoneworts (charophytes) and co-author of the *Red Data Book* on the group and of the stoneworts section of the *British Freshwater Algal Flora*.

DENIS J. SIMMONS was born and brought up on the Isle of Wight, and observing the weather was his main interest from an early age. Since the 1940s he has kept records from a variety of locations on the Island, and since 1969, from his weather station on the outskirts of Newport. He is a member of Climatological Observers Link, a nationwide amateur weather recorders' organisation. He counts himself greatly blessed to have lived on the Island, with its equitable climate and clean air.

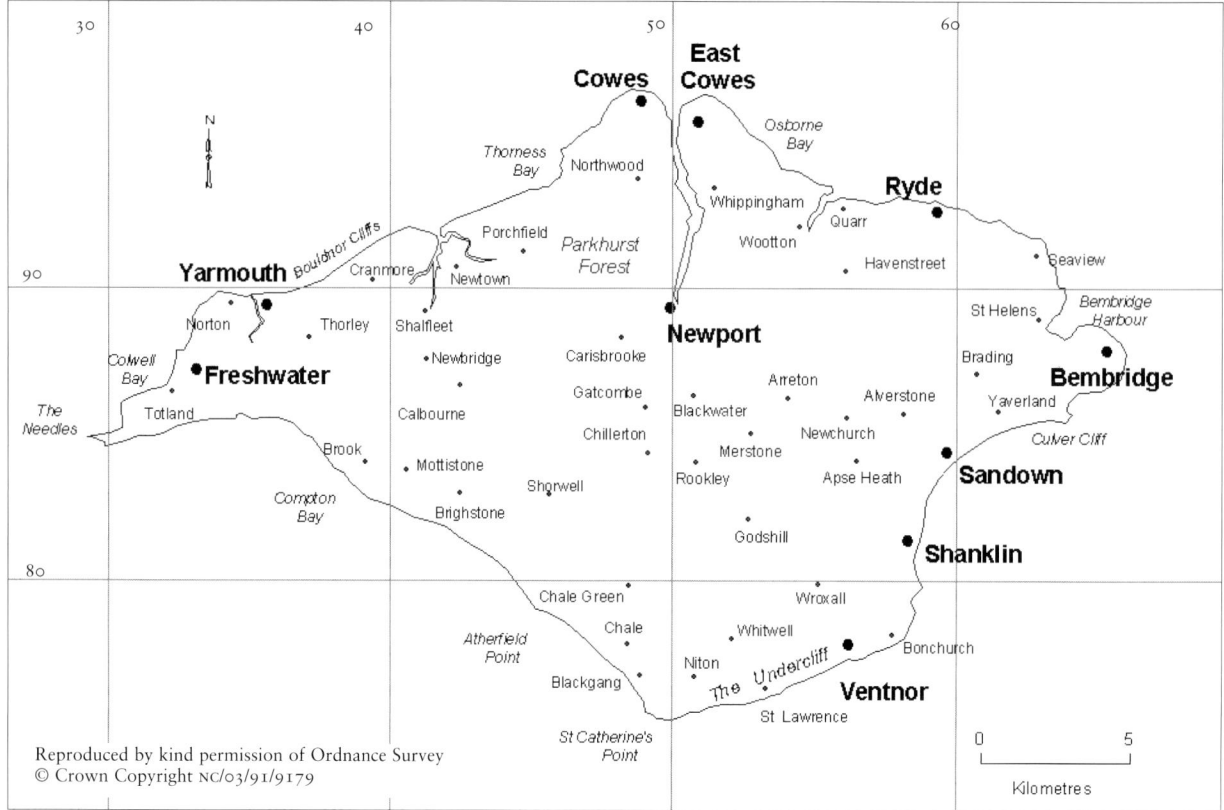

RICHARD SMOUT was born and brought up in Huntingdonshire. He worked for thirteen years as an archivist in Gloucestershire before moving to the Isle of Wight to take up the post of County Archivist in 1995. His main interests are ornithology, entomology and the changes in our attitude to natural history over the centuries. He has co-authored an article on the history of the Wydcombe estate. He is currently President of the Isle of Wight Natural History and Archaeological Society.

ANNE MARSTON is the leader of the Botanical Section of the Isle of Wight Natural History and Archaeological Society. She has a wide-ranging interest in natural history and has been involved in communicating this to others through working in environmental education for a number of years. She is the assistant Ecology Officer for the Isle of Wight Council.

DOLLY NORLEDGE is a founder member of the Leicester Society of Botanical Illustration and a member of the Southampton Flower Painting Society. She studied at Southfield College of Art, Leicester and Goldsmiths College, University of London. She is a visiting lecturer at various Universities and Colleges, and teaches the subject to local groups. She has exhibited widely both in England and abroad, and she has been awarded medals from the Royal Horticultural Society for Botanical Illustration. Her paintings have been reproduced in *Botanical Paintings* by Evans and Evans, and in *Drawing Flowers* by Margaret Sevens.

Each photograph has been credited with the photographers initials, as follows: AB Andy Butler, CNP Connie Pelham, DC Daphne Carter, DDa Dave Dana, JDS Jonathan Sleath, KM Keith Marston, KR Ken Richards, RDP Ron Porley, SB Sue Blackwell, ST Sylvia Taylor

Introduction

RICHARD SMOUT

'It is the botanist . . . who will find in the Isle of Wight that he has been very specially provided for. You get wild flowers, dale flowers, hedge flowers, forest flowers, sand flowers, creepers innumerable and ivy beyond convenience. . . . Does the reader devote himself or herself to Botany? Then let him seek the neighbourhood of Rookley, Freshwater, Alverstone, Thorley, Shanklin, Quarr, and Carisbrooke.'

The Isle of Wight, J. REDDING WARE (1869)

There is little doubt that the wealth of plant life on the Isle of Wight is one of the key elements that publicists for the Island have sought to highlight. Since the nineteenth century, the principal image that has been used to promote the Island to the outside world has been that of 'The Garden Isle'. For generations, prospective visitors have been hooked by the lure of a phrase that implied not just a green and pleasant land, but also all the implicit advantages of the sunny, mild climate, which would encourage such a richness of growth.

In 1988, Oliver Frazer, a notable past president of the Isle of Wight Natural History and Archaeological Society, wrote an article entitled 'Nature's Garden Isle' for the Ordnance Survey's Leisure Guide. He pointed out that, of all the terms used to describe the Island, 'none has been more long-lasting and appropriate than the 'Garden Isle' in recognition of its natural beauty'. [1]

His decision to stress the importance of the role of nature was a welcome corrective. In recent years the concept has been seized upon by some who would favour the reduction of the Island's publicly-owned spaces, and its highway verges, to a series of well-tended lawns, interspersed with a few choice bulbs. In fairness, this attitude has been with us for some time. A sale particular, dated 1900, offering land for sale at the Sea Copse Hill estate, west of Wootton advised that, 'the wood could be used as pleasure grounds, and would form gardens of no ordinary beauty, embellished here and there with forest trees, here and there by primitive bits of jungle, here and there by newly-formed flower beds and shrubberies, and everywhere would be heard the melodies of song birds and (unless uprooted) natural carpets of wild flowers of varied hues.' [2]

The use of the word 'Garden' was not, however, originally just a reference to the prolific growth of some introduced species (although the presence of certain half-hardy plants attracted the admiration of many 19th century visitors). Instead, it alluded to the richness of vegetation and the fertility of the soil to be found here. Some of these elements are described in the breathless prose of Hill's *Directory* of 1871. Writing of the Undercliff, that most admired of all habitats, the guide says: 'The myrtle blooms in this favoured spot; the geranium flourishes even in the chill autumnal months; an Italian atmosphere seems to breathe its balm around; leafiness makes a very bower of each sequestered knoll; even to the marge of the lower cliff slopes the luxuriant verdure . . .' [3]

Similarly Roscoe, in 1843, talks of the wealth of species to be found, both native and the 'green-house tribes': 'In regard to its plants and flowers . . . the island boasts an envious distinction, for nowhere are they surpassed in size, fragrance and beauty. They are equally varied in regard to bloom and species . . .' [4]

The term 'Garden Isle' appears to have risen to prominence in the middle of the nineteenth century. It is certainly in use by the time of White's *Directory* of 1859, where it is stated that the Island, 'may be said with justice to be the realization of the Poet's dream of a Calypso's Isle . . . all the phases of natural beauty adorn the Garden Isle' [5]

Literary analogies are also to be found in the preface to Bromfield's *Flora Vectensis*, which appeared in 1856 and where the publisher chose significantly to quote from Spenser's 'The Faerie Queen' :

> *It was a chosen plott of fertile land*
> *Emongst wide waves sett like a little nest,*
> *As if it had by Nature's cunning hand*
> *Bene choycely picked out from all the rest*
> *And laid forth for ensample of the best* [6]

Until the 1850s, the overwhelming bulk of references

are not to the Garden Isle, but to the Island as 'The Garden of England', making it one of a number of places to have claimed that accolade over the years. Its other epithet 'The British Madeira' was used rather less frequently. It is hard to trace the earliest use of the term. Sir Richard Worsley in his *History of the Isle of Wight* refers to the phrase in 1781. It was clearly an analogy that was already in vogue: 'In general, such is the purity of the air, the fertility of the soil, and the beauty and variety of the landscapes, that this island has often been styled the Garden of England.' [7]

The concept is expanded upon by Edward Wedlake Brayley and John Britton in 1805. As well as referring to some of the influences cited by Worsley, they add: '. . . the vegetation is so abundant, that this Island has often been styled the Garden of England: an appellation, perhaps, that is partly suggested to the mind by the innumerable plants and flowers which grow everywhere in wild luxuriance: among them are ophrys apifera, or bee-orchis; the digitalis, or foxglove and the crithmum maritimum, or rock-samphire.' [8]

Others were less impressed by the concept. William Gilpin, on his visit in 1798, clearly felt that the Island had been spoilt by the inroads of agriculture. 'The Isle of Wight is, in fact, a large garden, or rather a field, which in every part has been disfigured by the spade, the coulter, and the harrow . . .Yet these manufactured scenes are commonly thought to be picturesque . . .' [9]

Interestingly, some of the agricultural experts were included in the voices that praised the wealth of aspects of the Island's flora. In 1813, Charles Vancouver, talking of the downs and their 'high-land pasture' stated that it is 'found to consist of a prodigious variety of plants and grasses'. [10]

His contemporary, William Marshall, commenting on Vancouver's report, identified it as one of the most powerful pleas for the conservation of unimproved grasslands. He said that the author showed, 'the propriety, or necessity, of preserving them in their natural state; and arraigning the officious meddlings of theory, with well-grounded practice'. [11]

As with all gardens, some of the contents were viewed with more favour than others. Agriculturalists may have approved of keeping downland pasture in its natural state, but, in a ploughed field, the flowers were definitely weeds. At least one court case, from the 1830s, refers in passing to 'charlick' and charlock-pickers in the fields around Alvington, in the parish of Carisbrooke. [12] Most famously, field cow-wheat was, in the early nineteenth century, sufficient of a pest to have earned itself the name of 'poverty weed'. This was due to the fact that it tainted the flour made from the wheat with which it had been harvested, to the extent that it made it unmarketable. Nowadays field cow-wheat is one of the Island's rarest species and considerable work is undertaken each year by local naturalists to ensure its survival on the Isle of Wight. [13]

On the other hand, some species were continually being singled out for special attention. This was due, in part, to the shameless plagiarism that took place within successive generations of guidebooks. It is surely no coincidence that White's *Directory* of 1859 highlights the bee-orchis, foxglove and rock-samphire, exactly the same three species identified by Brayley and Britton over half a century before. [14] Even ivy, where not 'beyond convenience' was welcome in far more places than nowadays. Late nineteenth century views of Carisbrooke Castle show portions of the ruins invisible under the weight of foliage. [15]

David Allen's essay on Historical Botanists, which follows, demonstrates how successive generations of botanists have sought to provide a more precise description of the contents of the 'garden'. Of all the Floras, the most famous is William Arnold Bromfield's *Flora Vectensis*. His was not the first description of the Island's flora, but the comment in the preface to his book was, nevertheless, a fair analysis of the situation: 'Of all the districts into which England is divided . . . there is perhaps no one that offers a more interesting or promising field for botanical research than the Isle of Wight; yet, singular as it may appear, hardly any spot of equal extent, within the same distance of the metropolis, has received so small a share of attention . . .' [16]

Once Bromfield's findings had permeated through into the consciousness of the wider publishing world, guide books in the 1860s and 1870s gave far more hard facts about the highlights of the Island's Flora. The enthusiastic amateur was guided to the prime sites, and enabled to study the outstanding aspects of the Island's plant life, and all this during a comparatively brief stay. From now on the Garden Isle image was confirmed by a detailed and readily accessible analysis of its constituent parts. [17,18]

1919 saw the establishment of the Isle of Wight Natural History Society, later to become the Isle of Wight Natural History and Archaeological Society. The stated objects of the Society are 'the promotion and advancement of the study of the flora, fauna, geology, and archaeology of the county'. With these aims in mind, it is hardly surprising that the Society has, for many years, been seeking to update Frank Morey's monumental *Guide to the Natural History of the Isle of Wight*, published in 1909 and now almost a century old. This goal seems particularly appropriate when so many of the contributors to Morey's *Guide*, not least Frank Morey himself, played a pivotal role in the establishment of the Society. To create such a detailed panorama would be difficult to achieve nowadays, but it is hoped that this *Flora* may prove to be the start of such a process and lead to further volumes on other aspects of the Island's natural history. Providing up-to-date information about our flora and fauna also contributes to a wider picture. Globally, there is a growing concern about the rate of change of the natural world and our attention is frequently drawn to species in danger of extinction.

This awareness of the importance of variation in the status of species is nothing new. Chris Preston has pointed out that Bromfield was very conscious of the rapid pace of change on the Island. The upsurge in building, and increase in the area under cultivation posed a significant threat of local extinctions even in his time.[19]

Twenty-five years ago, Jim Bevis, Reg Kettell, and Bill Shepard produced an updated *Flora*, which was published by the Society. The introduction to this work referred to the dangers of our slipping into the twenty-first century, 'with no comprehensive assessment of our flora having been written for more than a hundred years. Such a lapse would reduce future botanists to mere conjecture regarding the changes that were taking place'.[20]

If anything the pace of change in the last quarter century has been even greater, with habitat loss, climatic change, and the problems attached to invasive species. It is, therefore, important to leave behind a further 'accurate yardstick' with which to mark the early years of a new millennium. The new work also includes accounts of bryophytes, lichens and stoneworts, which did not come within the scope of the 1978 *Flora*. Many details have been covered in past *Proceedings* of the Isle of Wight Natural History and Archaeological Society but the account of the bryophytes of the Island is the first full account to be published since Livens's contribution to Morey's *Guide*.

This new book would not have been possible without the outstanding contributions of its authors, the editor, a dedicated body of volunteers from the Society who have worked behind the scenes, and above all the very large number of individuals who have taken the trouble to record, and send in their field observations. Some are members of the Society; many are visitors to the Island willing to share their observations with others. To all who have helped we are very grateful.

Over two centuries ago, in 1795, John Albin observed that:

'We should be highly remiss in the discharge of that duty to which we have pledged ourselves in the public estimation, if we were to neglect the notice of the numerous plants and herbs in this island. Almost every species, which are to be found in any other part of England, are met with here; a circumstance which must be extremely agreeable to the philosophic mind, and grateful to the botanist and man of science. They abound in quantity, as well as in variety, so that persons are annually employed in the summer season to collect those of a medical nature, by professional and other gentlemen who visit the island for that purpose.'[21]

This *Flora* is a prime example of the way in which the Society continues, in Albin's tradition, to 'discharge that duty to which we have pledged ourselves'.

REFERENCES

1. Frazer, O. (1988) *Isle of Wight* Leisure Guide Automobile Association and the Ordnance Survey
2. Isle of Wight County Record Office (IWCRO) (1900) Particulars, Plans, View and Conditions of Sale of the Freehold Sea-Copse Hill Estate.
3. Hill, J.W. and Co. (1871) *An Historical and Commercial Directory of the Isle of Wight.*
4. Roscoe, T. (1843) *Summer Tour to the Isle of Wight.*
5. White, W. (1859) *History, Gazetteer and Directory of Hampshire and the Isle of Wight.*
6. Bromfield, W.A. (1856) *Flora Vectensis: being a systematic description of the phaenogamous or flowering plants and ferns indigenous to the Isle of Wight*, eds. Hooker W.J. and Salter T.B.
7. Worsley, R. (1781) *The History of the Isle of Wight.*
8. Brayley, E.W. and Britton J. (1805) *The Beauties of England and Wales; or Delineations, Topographical, Historical, and Descriptive, of each county*, VI.
9. Gilpin, W. (1798) *Observations on the Western Parts of England, relative chiefly to Picturesque Beauty*, London.
10. Vancouver, C. (1813) *General View of the Agriculture of Hampshire, including the Isle of Wight.*
11. Marshall, W. (1817) *The Review and Abstract of the County Reports to the Board of Agriculture, Volume 5*, (reprinted 1969, David and Charles.)
12. IWCRO CPS12 1832 County Petty Sessions, minutes of evidence.
13. Bevis J.H., Kettell R.E. and Shepard B. (1978) *Flora of the Isle of Wight.*
14. White, W. (*op.cit.*).
15. IWCRO CAR 192, car 205, c.1890-1900 Photographs of Carisbrooke Castle
16. Bromfield, W. (*op. cit.*).
17. Venables, Rev. E. (1860) *A Guide to the Isle of Wight, its approaches and places of resort*, London.
18. White W. (1878) *History, Gazetteer and Directory of the County of Hampshire, including the Isle of Wight.*
19. Preston, C.D. (2003) Perceptions of change in English county Floras, 1660-1960, *Watsonia* 24: 287-304.
20. Bevis J.H., Kettell R.E. and Shepard B. (*op. cit.*).
21. Albin, J. (1795), *A New, Correct and Much Improved History of the Isle of Wight . . . Comprehending whatever is curious or worthy of Attention in Natural History.*

Geology & Physical Features

ALLAN N. INSOLE

GEOLOGY

The Isle of Wight is the smallest English county, but within its boundaries it exhibits a remarkably varied geological succession. The various geological horizons and the soils that are derived from them are important factors controlling the physical features, climate and distribution of flora and fauna on the Island.

The Island's surface geology consists of a succession of sedimentary rocks, laid down under water or by winds. They range from early Cretaceous to early Oligocene, exposed beneath more recent Quaternary deposits (less than 2 million years old). For a modern detailed account of the Island's geology and geological history, the reader is referred to Insole, Daley and Gale (1998).

CRETACEOUS

The oldest rocks are the Wealden Group, found in a small area in Sandown Bay and a larger one along the shores of Brighstone Bay. In both areas, the Wealden Group can be subdivided into two units: the Wessex Formation and the Vectis Formation. The Wessex Formation is a non-marine sequence comprising varicoloured, but mainly red, mudstones. This unit represents the accumulated deposits of floods on an alluvial plain. The Vectis Formation consists mainly of dark grey siltstones and mudstones, deposited in shallow lagoons. A phase of river delta building deposited a unit of sandstone within the formation, seen at Barnes High on the south-west coast near Brighstone.

The Lower Greensand Group succession was deposited under mainly marine conditions when the Wessex Group coastal plain was submerged by the sea about 112 million years ago. It underlies the greater part of the southern half of the Island. The Lower Greensand of the Isle of Wight is divided into four formations, which can be recognised in all the available exposures: Atherfield Clay, Ferruginous Sands, Sandrock and Carstone in ascending order. The Atherfield Clay Formation comprises a sequence of brown-grey silty muds, silts and fine sands. It is overlain without any apparent break by a thick sequence of alternating dark silty clays and muddy glauconitic sands, which form the Ferruginous Sands Formation. The Sandrock Formation succession has alternating silty clays and muddy sands. The Carstone Formation comprises ferruginous, medium- to coarse-grained sands and sandstones with occasional thin pebbly bands.

In common with other areas in southern England, the Gault Clay and Upper Greensand form an unbroken sequence. The Gault clay consists of dark blue-grey silty muds, and is involved in various forms of mass movement, especially along the southern coast, which has earned it the local name of the 'Blue Slipper'. However, this term is often used on the Island for any blue, green or grey mudrock that is liable to lose its coherence when wet, regardless of its origin. The Upper Greensand is a glauconitic siltstone or fine-grained sand and sandstone, with bands of calcareous and siliceous concretions. The uppermost part of the formation, the greyish 'Chert Beds', is the most conspicuous part of the succession.

The last part of the Cretaceous story is the Chalk Group, whose deposition began about 99 million years ago. It is composed mainly of the skeletal remains of minute planktonic algae, and comprises a thick sequence of grey to white limestones. The lower part of the group contains up to 20% clay, which gives the rock the grey appearance.

PALAEOGENE

About 65 million years ago, sea level fell and the Cretaceous rocks were gently buckled by earth movements. Subsequently, erosion removed the uppermost part of the Chalk Group. About 56 million years ago, a new geography developed in the area, as sea level began to rise again and the eroded Chalk surface was gradually submerged. For the next 25 million years, from the late Palaeocene to earliest Oligocene, the shoreline of the sea constantly shifted. Sea levels rose and fell, the land intermittently subsided and a river built out a delta eastwards across the area. Consequently, the rock sequence laid down during this period was deposited in many different environments:

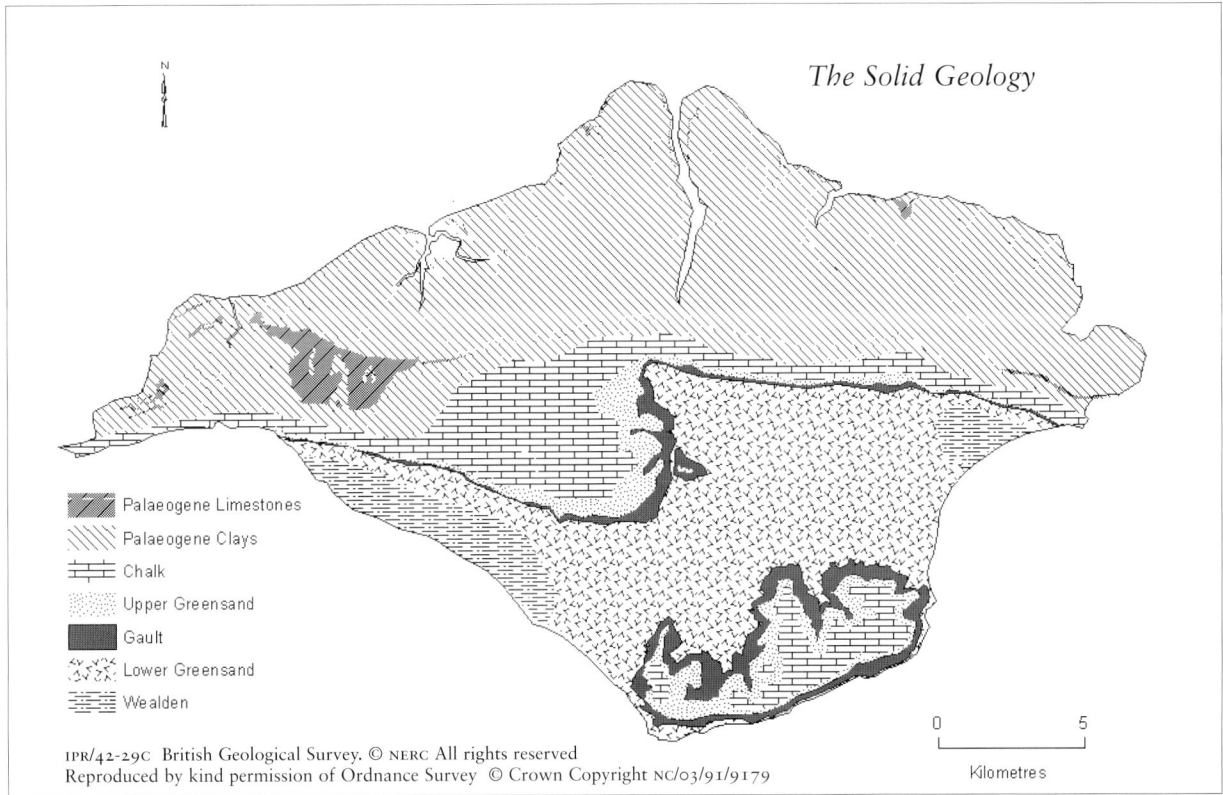

The Solid Geology

Palaeogene Limestones
Palaeogene Clays
Chalk
Upper Greensand
Gault
Lower Greensand
Wealden

IPR/42-29C British Geological Survey. © NERC All rights reserved
Reproduced by kind permission of Ordnance Survey © Crown Copyright NC/03/91/9179

shallow marine shelf, beaches, tidal flats, coastal marshes, lagoons, estuaries, rivers and lakes. There are also fossil soils, which represent periods of exposure and weathering.

This Palaeogene succession has been assigned to fourteen formations and it underlies the northern half of the Island. The Reading Formation has at its base a thin conglomerate that rests on an irregular surface at the top of the Chalk. This unit is overlain by a sequence comprising mainly red and purple colour-mottled muds, which are considered to represent non-marine sediments that were subsequently subjected to weathering and soil formation. The London Clay is not uniform as its name suggests. It comprises a series of units, each consisting of a base, sometimes marked by pebbles. Above the base come mainly silty muds; the upper part of the units is characterised by two lithologies: one consists of laminated silty sands and muds which are interpreted as tidal flat deposits, and the other comprises well-sorted, cross-bedded sands which are considered to have been deposited in tidal channels. These sands were formerly called the 'Bagshot Sands'.

The Bracklesham Group is a complex unit consisting of five formations, and it shows considerable vertical and lateral variation. Like the London Clay, it comprises a rhythmic sequence. In the east of the Island the lower part of each unit comprises green sands and sandy muds. The upper parts of the units consist of interbedded muds, silts, sands and lignites. At the western end of the Island, the character of the Group is somewhat different as it was laid down under more marginally marine conditions than in the east. The lower parts of each unit comprise sands, within which is a mixture of beach, fluvial and aeolian (wind-blown) dune deposits. The upper parts of the units comprise interbedded muds, silts and sands deposited in tidal lagoons and channels. The repeated occurrence of lignites and palaeosols (fossil soils) within the upper parts of the units probably represents coastal marshes.

The Barton Group comprises four formations: Boscombe Sand, Barton Clay, Chama Sand and Becton Sand (formerly Barton Sand), in ascending order. The Boscombe Sand Formation only occurs at the westernmost part of the Island, and comprises well-sorted sands. The succeeding Barton Clay consists of more or less silty muds, and the Chama Sand has thin sandy muds. At the top of the group come the mainly clean quartz sands of the Becton Sand Formation.

The Solent Group comprises two formations, the lower Headon Hill Formation and the upper Bouldnor Formation, which are composed of sands, silts and muds, separated by a primarily carbonate unit, the Bembridge Limestone. Deposition took place in a complex low-lying coastal environment. Salinities would have varied from place to place, and through time, depending on a number of factors, including subsidence, global sea level changes, proximity to river channels and climate.

About 32 million years ago, the sea retreated completely from this area. The whole rock sequence was buckled by major earth movements, which were associated with the formation of the Alps. In Alum Bay and Whitecliff Bay, some of the originally horizontal strata can be seen to be steeply inclined or even vertical as a result of this folding, which probably commenced in the Eocene but did not reach its culmination until the Miocene, about 15 million years ago. The area then underwent erosion to form the landscape as we see it today.

QUATERNARY

This final phase of the geological history of the Island is very poorly known. The only available evidence comprises relatively thin, disconnected patches of unconsolidated sediments. Few of these Drift or Superficial deposits can be dated, making their correlation and interpretation largely speculative.

Thin, unbedded deposits of angular flint pebbles in a muddy sand matrix, termed the Angular Flint Gravels of the Downs, occur in isolated patches at high levels on the Chalk Downs of St. Boniface and Bowcombe. They are considered to be residual deposits formed by *in situ* dissolution of the White Chalk Formation.

In the late Pliocene, about 2 million years ago, the Earth's climate began to experience a series of fluctuations. Periodically, the average annual temperature fell to such an extent that glaciers and ice sheets developed in high latitudes. During these 'ice ages' or glacials, ice sheets spread southwards across the British Isles. While they never reached as far south as the Solent, the climate locally would have been similar to that of Arctic Canada today. At the same time, global sea levels fell by as much as 300 m so that the English Channel became dry land. As a result of the lowered sea level, the local rivers deepened their valleys to as much as 48m below current sea level.

The glacials were separated by warmer periods called interglacials, when the average annual temperatures were as high, or even higher, than they are today. As the temperature rose at the start of an interglacial, the ice sheets melted and sea level rose, sometimes above present day levels. The over-deepened river valleys were inundated to form broad estuaries.

The prevailing extremely cold conditions during the last glacial (Devensian) resulted in the development locally of Head and Brickearth. The former are deposits produced by downslope movement of weathered bedrock and drift by a mixture of solifluction and downwash, while the latter appears to represent either primary or reworked wind-blown silt, known as loess.

Several different types of deposit were formed locally during the interglacials. The Older River Gravels (formerly called Plateau Gravels) and the River Terrace Deposits are the most difficult to place chronologically; these were laid down as valley fills, and their origin may be partly marine or fluvial.

About 10,000 years ago at the end of the Pleistocene, the current interglacial phase, the Holocene, began. The climate improved, and sea level began to rise. By about 8000 years ago, the sea had flooded all the local river valleys and had broken through the Chalk ridge, which had formerly extended westwards to Dorset. The Island had finally become an island. As sea level rose, muds and peat accumulated in the newly formed estuaries, sedimentation usually keeping pace with the rising sea level.

PHYSICAL FEATURES AND SOILS

The Island divides naturally into four distinct regions: the southern bowl, the southern downs, the central ridge and the northern dissected plateau. A glance at the geological map shows that this physical division reflects the underlying geology. This geological foundation has been etched by recent events (ice ages, sea level changes, weathering and erosion) to produce the landscape that we see today. The disposition of the different rock types and their character not only controls the physical features of the Island but also the nature of the soil and, through this, the type of vegetation that it supports. So, there is a close association between the underlying geology and the geographical distribution of many of the Island's plants.

The southern bowl of the Island, lying between the central ridge and the southern downs, is underlain mainly by the Wealden and Lower Greensand Groups. These have given rise to a gently rolling landscape with isolated knolls and hogback ridges formed where sandstones occur. At the coast, the Wealden Group is subject to rapid erosion and produces relatively low terraced cliffs. The terraces provide locally sheltered conditions and some of them contain temporary ponds. In contrast, where the Lower Greensand Group reaches the coast it tends to form high perpendicular or stepped cliffs, as at Blackgang and Red Cliff at Sandown.

The Wealden Group, being composed mainly of mudstones, gives rise to heavy clay soils, prone to waterlogging. In general, its outcrop is given over to

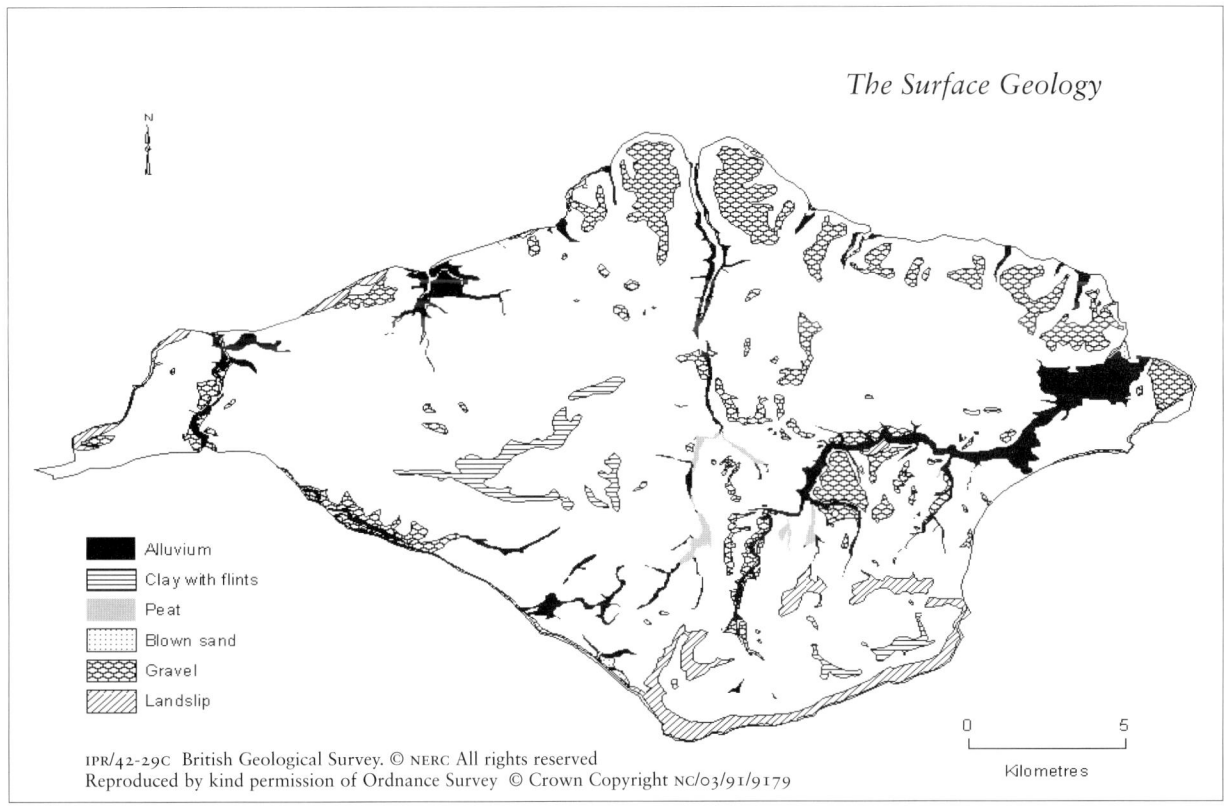

The Surface Geology

- Alluvium
- Clay with flints
- Peat
- Blown sand
- Gravel
- Landslip

IPR/42-29C British Geological Survey. © NERC All rights reserved
Reproduced by kind permission of Ordnance Survey © Crown Copyright NC/03/91/9179

rough pasture, with a heathy vegetation, although some small patches of ancient woodland also occur. The Lower Greensand Group generally produces a light sandy, rapidly draining soil, which provides some of the best arable land on the Island.

Along the southernmost part of the Island rise the southern downs, deeply indented on their inland side and steeply precipitous facing the English Channel. They are underlain by Gault Clay and Upper Greensand and capped by the Chalk Group. The Gault Clay occurs at the base of the downs, while the Upper Greensand forms the steep, sometimes vertical, slopes above. All around the southern downs, the combination of the massive Upper Greensand overlying the plastic Gault Clay has resulted in huge landslips in the past. The most spectacular landslips occur between Luccombe and Blackgang, where massive blocks containing Upper Greensand and sometimes the overlying Chalk Group have moved seawards. This has produced a series of irregular sloping terraces, with large blocks of sandstone and limestone scattered along them, an area which is referred to as the Undercliff. Altogether this landslipped area stretches for nearly 10 km along the coast and affects a zone between 0.4 and 1.2 km wide, making this the largest area of inhabited coastal landslip in north-west Europe. Other, less obvious, ancient landslips occur around the northern rim of the southern downs.

Most of the higher parts of the southern downs are underlain by the Chalk Group, which yields light, thin calcareous soils rich in flints. The soils on the lower slopes are more complex. Only relatively small areas of Gault Clay and Upper Greensand occur without either a veneer of Chalk downwash or disturbance by landslips. However, where they do occur, as might be expected, they produce heavy clay and light silty soils respectively. Where the Gault and Upper Greensand are covered by Chalk downwash, calcareous soils occur. Where they are involved in, or covered by landslips, the picture can be very complex since the soil formed at any particular spot will depend on the nature of the rocks involved in the landslip. It is quite possible to have small areas of alkaline calcareous, acid silty and heavy clay soils adjacent to each other. Thus, the Undercliff possesses a series of microhabitats and is a rich site for a diverse range of species with an extreme southern distribution.

The central east-west 'backbone' of the Island is underlain mainly by limestones of the Chalk Group. Except in the centre of the Island between Chillerton and Calbourne, where it broadens considerably, this is a very narrow hogback ridge. Where it reaches the sea at Culver Cliff and west of Compton Bay, it forms high vertical cliffs. One of the prominent features of the central ridge is the presence of steep-sided embayments in the escarpment known as dry valleys or coombes. Throughout its outcrop, the Chalk Group produces very thin calcareous soils usually with abundant flints. Although many parts of the central ridge are cultivated, there are still large areas of typical downland remaining, supporting species-rich grassland.

While the central ridge is dominated by the Chalk, a relatively restricted belt of Gault Clay and Upper Greensand occurs immediately south of the Chalk ridge, although their outcrop is largely masked by downwash from the Chalk above. However, in a few places, the Upper Greensand forms a narrow secondary ridge, well-defined between Shorwell and Brighstone.

The northern half of the Island is underlain by Palaeogene clays and silts, with an extensive cover of Quaternary gravels. This has produced a low plateau, which has been dissected by recent river and stream erosion to produce a region of generally low hills with gentle slopes down to the Solent shore, although in the west, cliffs occur in places. The Palaeogene sediments are everywhere prone to landslips and mudflows.

The Palaeogene sequence produces very heavy clay soils, which are prone to waterlogging. This makes cultivation difficult without extensive under-draining. Consequently, most of the Palaeogene outcrop is either grass meadow or woodland. This is the part of the Island where most of the ancient woodland has survived. While sands and silts do occur in the Eocene succession, in general they have only limited outcrops. Where they are present, they have produced acid sandy or silty soils, capable of supporting a heathy vegetation. There are also some thin limestones within the Eocene sequence, but again their outcrop is small. The exception to this is the area immediately south of Wellow, where a relatively extensive tract of the Bembridge Limestone occurs. This gives rise to thin light calcareous soils. Much of the area is under cultivation, with only tiny pockets of limestone grassland surviving.

While the solid geology has a major influence on the character of the soils over much of the Island, in some areas the drift or superficial deposits formed in the Quaternary are more important. In terms of their character, they can be divided into two groups: gravels and alluvium.

The gravels include the Angular Flint Gravels of the Downs and Older River Gravels These all produce light pebbly and sandy soils. Significantly, these soils are acidic and thus where gravels overlie the Chalk, it is possible for acid heathland to develop adjacent to typical calcareous downland vegetation. In some places, the gravels have been cemented into hard iron pans, as found on Arreton Down, which result in impeded drainage.

The alluvial silts and clays laid down by the modern rivers form low-lying and waterlogged areas, historically used for grazing and referred to as 'moors'. Here, the soils vary from neutral to slightly calcareous. However, in a few areas, peats have developed, yielding acid soils and specialised bog communities.

Most of the Island is drained by two principal rivers, the Medina and the Eastern Yar. Both rivers rise in the southern downs and flow northwards into the Solent. The Western Yar also originally rose in the southern downs but its headwaters and tributaries have been lost by coastal erosion. All of the northward draining streams and rivers have extensive estuaries at their lower ends, with well-developed saltmarshes. In contrast to the northern shoreline, the southern coast is punctuated here and there by narrow ravines, locally known as chines. These are produced by a combination of rapid coastal erosion and continuous downcutting by relatively small streams. While the process of chine formation is the same in each case, the chines are not uniform in character but vary according to the nature of the rocks in which they are formed and also the local climate. The chines eroded in the Lower Greensand Group tend to be relatively narrow and steep-sided, while those in the Wealden Group and Palaeogene are broader with gentle slopes. The chines are interesting biologically because they provide a sheltered haven for wildlife.

REFERENCES

Insole, A., Daley, B., & Gale, A. (1998) *The Isle of Wight*, Geologists' Association Guide No 60.

The Island Climate

DENIS J. SIMMONS

The diverse underlying geology determines the varied topography of the Island, which in turn influences the climate by modifying the weather systems. In particular, south-facing slopes receive more insolation, but the downlands intercept the prevailing south-westerly winds, to increase rainfall amounts in the vicinity. However, for the Island, a mere 147 square miles in area, the influence of the sea is paramount, but it is not so maritime as the Isles of Scilly, the Hebrides or Shetland Islands, which are all more remote from the mainland, and perhaps more importantly, from the Continent.

Winds off the English Channel naturally keep the day temperatures lower in summer on all windward coasts, and a little higher in winter, when compared with inland Britain at a similar altitude. Mean maxima in the warmest month, usually July or August, range from 19° C at St Catherine's Point to 21.5° C in Newport. Mean night temperatures are higher at the coast at all seasons, and there are fewer air frosts in coastal locations during the winter half year (October–March). Typically there are 22 air frosts in Ryde and 32 in Newport. Over the year, the mean minimum at St Catherine's Point is 8.1° C, while Newport averages 7.3° C. The latter is in an 'inland valley' situation, giving our capital town a slightly more 'continental' climate, with warmer days and colder nights.

Rainfall varies from around 710mm on the south-west and west coasts to about 900mm in the Shanklin and Newport areas. This is the result of orographic enhancement caused by the downs to the south-west. A rain shadow effect becomes apparent to the north-east, with the mean rainfall in Ryde being 760mm. The proportion of the annual fall in the summer half year (April-September) is roughly 40% of the total. In unsettled weather, we often escape showers in the summer, provided the wind blows off the Channel; oft-times, the sky is clear to the south and east, while heavy clouds producing showers and thunderstorms occur over the mainland. In the winter half year, the reverse applies, as the sea, rather than the sun, is the warming agent. With north-westerly winds we expect a clear sunny day in winter, while north and west facing coasts of mainland Britain have showers, probably of a wintry nature with hail or snow.

The Island has long boasted its sunshine records. Next to Jersey, which has the highest recorded figure for the British Isles (1941 hours annually, 5.3 hours daily over the year), the coast from Sandown to Ventnor in particular enjoys at least as much sun as the Sussex resorts. Daily averages range from 2 hours in December to as much as 8.7 hours at Shanklin in June (1947/1970). The average for the three resorts (Sandown, Shanklin, Ventnor) is 1860 hours annually (5.1 hours daily). Inland Hampshire has an average of approximately 1500-1550 hours annually, or 4.1 – 4.2 hours daily.

Whilst not experiencing on a regular basis the fierce Atlantic storms of north-west Ireland and Scotland, we can, on occasions get a 'good blow' mainly from the south-west or west during the winter half year. The 'Back of the Wight' between St Catherine's Point and Freshwater gets the full force of such winds. Rarely, the Needles and St Catherine's have recorded gusts of 100 mph or slightly above – for example on October 16th 1987, when a 'hurricane' struck south-east England.

Large amounts of snow have occurred from time to time, most notably in January 1881 when 12 ft drifts were formed in a violent easterly gale, but such extremes are a rare occurrence. Of recent years, heavy snowfalls have been absent. Snow or sleet, however slight, occurred on average only 6 days annually during the 1990/1999 period. During these years, on average, it was lying on only 1 day at 0900 hours at Newport; a 32 year average (1969/2000 inclusive), works out at 8 days of snowfall and 3 days with snow lying at 0900 hours respectively, still not notably wintry.

Fog is rare in Newport, averaging 19 days annually at some hour of the day, and a mere 7 days at 0900 hours. Sea fog tends to affect the coast to a greater extent than inland in the early spring and summer months, when a warm moist airstream passes over the still-chilly sea; they are usually not frequent, but seen from time to time most years between March and June.

We have relatively few thunderstorms: those that occur are generally imported from the Continent, and usually arrive in the late evening. At other times, particularly in the late autumn, the comparative warmth of the sun sets off storms when the upper air is cold. These storms, usually short-lived, affect the Island and other south coast areas, while places inland tend to escape. From records taken over 60 years at Totland Bay, the observer, Mr John Dover, stated that his area rarely had overhead thunderstorms. He attributed this to the downs coming to a point at the Needles to the south-west diverting the storms from the South and West, sending them up the Solent or down the English Channel.

The Undercliff is a remarkable area in respect of its micro-climate due to the sea's influence, southerly inclination of the ground and shelter from north and north-east winds provided by St Boniface Down. The area of the landslip lying between the sea and the foot of the escarpment averages just 12 days of frost (1985/2001), and has up to 17% less rainfall (1985/2001) than on the nearby downs. The Holm Oak, a Mediterranean species, has colonised the escarpment above Ventnor; and the town's Botanic Garden, formerly the site of a hospital for the treatment of consumption, has a particularly sheltered terrace where a wide variety of Mediterranean plants thrive outdoors throughout the year without protection.

The Palaeoecological Background

ROB SCAIFE

Students of geology will be aware of the palaeobotanical importance of the Isle of Wight that results from the range of strata spanning the Lower Cretaceous to the mid-Tertiary Oligocene periods. This Island, with its geological microcosm, includes sediments laid down in depositional habitats that were highly favourable to the preservation of fossil plant material. The lower Cretaceous Wealden series, including the 'pine raft' and fossil cycad cones at Hanover Point in Brook Bay is well known, as is the rich flora of the Eocene and Oligocene deposits found in Whitecliff and Alum Bays (Chandler 1960).

Pollen analysis (palynology) has often been regarded as a technique for dating sediments and whilst this is more true of geological sediments, as it provides stratigraphical bio-marker horizons, the use and development of absolute dating techniques has largely superceded this role for recent deposits. The science of palynology is now seen as providing evidence of past vegetation communities and changing environments; the basic premise is that pollen which becomes incorporated in sediment represents plants and plant communities which produced pollen at the time of the deposition. This is especially the case when dealing with the most recent geological era, the Quaternary, a period of markedly fluctuating warm (interglacial) and cold (stadial) periods.

With the development of radiocarbon dating, there has also been a much greater awareness of variations in vegetation and environment in different geographical areas which have taken place since the final close of the last cold stage, the Devensian period, some 10,000 years ago. To obtain pollen data, and thus information on past environments, requires sites where peat or mineral sediments have accumulated through time and where a number of other factors favouring preservation are met. These include undisturbed anaerobic sediments caused by waterlogging, acidity from local bedrock and lack of oxidation caused by draining.

Until recent years, Southern England had been an area with a paucity of detailed pollen records. The reasons for this are exemplified in the Isle of Wight, where there exists a wide combination of natural environmental (largely lithological) and anthropogenic factors, which negated detailed studies such as those carried out in the wetter northern and western regions of the country where domed (ombrogenous) mires, blanket peats and lakes are common. In contrast, those peat mires (bogs and fens) which exist in the Isle of Wight are groundwater-controlled (soligenous) and are thus subject to the base status of the bedrock lithology in the river catchment. Alkaline ground water is highly detrimental to pollen preservation, rendering many of the peat accumulations on or near the chalk unsuitable for pollen analysis. This is the case at Newchurch and Newbridge, where deep valley fen peat is largely devoid of sub-fossil pollen.

It is also now recognised that peat cutting for fuel was practised in the region throughout the historical period, and this has truncated the stratigraphical record. This is clearly evidenced at the important site of Munsley Bog, where rectangular 'peat cuts and baulks' can still be seen. This removed some 8,000 years of accumulation, leaving only sediment dating from the end of the last cold stage (the Devensian) and early part of the current interglacial period (the Holocene).

More recent human disturbances have also caused the demise of peat mire communities and changed the status of others. For example, The Wilderness region of the upper Medina valley has a well-documented acid flora, surrounded by the heathland of Bleak Down. Extension of agriculture on the adjacent valley sides apparently changed the status of the mire to neutral fen, with drained grassland and willow carr scrub woodland.

Potential exists for pollen studies on the Island where suitable preservational parameters occur - on the Lower Greensand, some areas of the Tertiary clays, in the coastal zone and from buried soil profiles underlying archaeological field monuments such as Bronze Age burial mounds.

It is interesting that the work of Clifford (1936) on the sediments of Brook Bay was a pioneer study in the field of palaeoecology, having examined both the sub-fossil pollen and plant macrofossils. This was undoubtedly due to the very visible peats exposed in the cliffs here (and also at Compton Bay), the frequent hazel nuts referred to as 'Noah's nuts' (H. Osborne-White 1921) and the associated Mesolithic artefacts.

Subsequently, no work was carried out on the Island until the early 1970s, when Devoy (1972 & 1987) examined the peats of the Western Yar at Yarmouth in relation to changing sea levels, and the author undertook a detailed survey of the Island's peat forming communities. There are also a few, but very important records of earlier Quaternary vegetation recorded in peats and sediments from the Foreland at Bembridge (Reid and Chandler 1924; Holyoak and Preece 1983; Preece *et al.* 1990). These data relate to earlier climatic phases of the middle Quaternary and the last interglacial period, the Ipswichian period.

Extensive field survey (Scaife 1980, 1980) produced a number of sites that enabled the late Devensian (11,000 to 10,000 BP)★ and Holocene (10,000 BP to present) vegetation history of the Island vegetation to be examined. The most favourable sites were found on acid, sandy Cretaceous deposits (largely Lower Greensand) to the south of the Island's central chalk downland. It was realised that some areas of remaining Alder carr woodland, such as the important site of Gatcombe Withy Bed on the River Medina, had remained little disturbed other than from traditional management by coppicing. The underlying peats were intact, whereas adjoining areas of grass-sedge fen had been largely cut or drained.

Other areas yielding information included sites on the Eastern Yar and its tributaries at Borthwood Farm, Munsley Bog, Ninham Farm and Bohemia Bog. The peats of Brook Bay were also re-examined. From these and more recent studies, there are now pollen data from a very substantial number of sites, coming from the valley peat mires of the River Medina, Eastern and Western Yar; the inter-tidal and submerged peats of Sandown Bay, Wootton-Quarr, Newtown and Bouldnor Cliff and from a number of ancient buried soil profiles and archaeological contexts. The geographical distribution of these sites is shown on the map on page 26. When sites are first referred to in the text, the number in parentheses refers to this map.

★ Where dates are given in year BP (Before Present), this refers to uncalibrated radiocarbon dates (to a base year of 1950). Calibrated dates (calendar years) based on dendrochronology are given in some cases. Because pollen morphology does not always allow the separation of taxa to species level, some trees are presented in lower case in this chapter.

THE LAST 12,000 YEARS OF VEGETATION DEVELOPMENT

To understand the way in which the Isle of Wight vegetation communities have developed into their current status, recourse must be made to studying the environmental changes from the end of the Devensian at *c.*11,000-10,000 years ago to the present. Sediments referable to this period have produced pollen and plant seed data that display the very marked vegetation and environmental changes which have occurred from the Tundra regime of the late glacial some 11,000 to 10,000 years ago through successive woodland changes and human interference, to give the present-day pattern of vegetation communities.

The end of the last cold stage (The late Devensian: 11,000-10,000 BP)

The two sites of Munsley Bog (7) and Gatcombe Withy Bed (6) (Scaife 1980, 1982, 1987) have basal sediments that date to the last phase of the Devensian (the Younger Dryas, Loch Lomond re-advance). This period witnessed a sharp return to a harsh Arctic Tundra, periglacial regime between 11,000 to 10,000 BP, after a period of amelioration between *c.*12,000 and 11,000 BP (the late glacial interstadial). Whilst the latter warmer phase certainly saw the expansion of birch and possibly pine woodland, which is just evidenced at the bottom of the Munsley Bog sequence, the reversion to cold, harsh conditions saw a return to unstable permafrost ground. The pollen evidence at these sites reflects the diversity of micro-climates, and the edaphic (soil-related) and geological controls that created a high degree of plant competition, resulting in a rich and diverse herb flora.

Within this flora is represented a range of community types typical of both Arctic and Alpine environments. The characteristic of this period is the admixture of these elements for which there is no contemporary analogue. However, by comparison with aspects of both Alpine and relict communities in northern England and Scotland, it is possible to distinguish a number of these plant communities.

Dwarf Shrub, heathland: Localised communities most probably growing in situations where they exist today on acid, sandy substrates. Taxa recorded typically include: Juniper, possibly dwarf (*Juniperus communis*), Heather (*Calluna vulgaris*), Heath types (*Erica* spp.), Crowberry (*Empetrum nigrum*), Dwarf Birch (*Betula nana*), Sea-buckthorn (*Hippophae rhamnoides*), dwarf Willows (*Salix* spp.) and Alpine Clubmoss (*Diphasiastrum alpinum*).

Short turf grassland: Analogous in many ways to the short turf grassland of the chalk downland today. Grasses (Poaceae), Rock-roses (*Helianthemum* spp.), Mountain Avens (*Dryas octopetala*), Thrift (*Armeria maritima*), Ribwort, Hoary and Sea Plantains (*Plantago* spp., *P. lanceolata*, cf. *P. media*, *P. maritima*), Daisy types (Asteraceae) and the fern, Moonwort (*Botrychium lunaria*).

Tall herb: Communities such as are found today in Alpine meadows and limited areas of northern England and in Scotland. Grasses (Poaceae), Sedges (Cyperaceae), Meadow-rues (*Thalictrum* spp.), Great Burnet (*Sanguisorba officinalis*), Meadowsweet (*Filipendula ulmaria*), Common Bistort (*Persicaria bistorta*), Alpine Bistort (*P. vivipara*), Jacob's-Ladder (*Polemonium caeruleum*), Scabious (*Scabiosa* sp.), Devil's-bit Scabious (*Succisa pratensis*), Comfrey (*Symphytum* sp.), Carrot family (Apiaceae spp.), Alpine Saw-wort (cf. *Saussurea alpina*), Common Valerian (*Valeriana officinalis*), Narrow-fruited Cornsalad (*V. dentata*), Knapweeds (*Centaurea nigra/nemoralis*) and Globeflower (*Trollius europaeus*).

Disturbed ground: Frost heaving due to periglacial phenomena, permafrost and unstable valley floor, river gravel bars. Plants typically included Goosefoots (Chenopodiaceae), Mugworts (*Artemisia* spp.), Greater Plantain (*Plantago major*) and Spurreys (*Spergula* spp.).

Fen and bog: Wet valley floor and spring flush communities supported *Sphagnum* bog and grass-sedge fen. Grasses (Poaceae), Sedges (Cyperaceae), Lesser Bulrush (*Typha angustifolia*), Bur-reed (*Sparganium* sp.), Yellow Iris (*Iris pseudacorus*), Bogbean (*Menyanthes trifoliata*), Flowering-rush (*Butomus umbellatus*), Marsh-marigold (*Caltha palustris*) and Royal Fern (*Osmunda regalis*) were common.

Aquatic: There is little evidence from these sites, but White Water-lily (*Nymphaea alba*) has been recorded and it is likely that such areas existed in other valley bottom areas.

Within the above groups are plants that are attributable to a number of phytogeographical regions. Of interest are halophytes (salt-tolerant plants such as Thrift, Sea Plantain and Sea Buckthorn) today found on mountain tops and in coastal communities, which had a broader habitat distribution during the cold stage and tall herbs showing the presence of typical Alpine type pasture (Globeflower, Common Bistort, Alpine Bistort, possibly Alpine Saw-wort, Meadowsweet and many other taxa).

The growth and survival of the former group is attributed to the high incidence of surface salts caused by the preponderance of mineral soils, and the climatic conditions (hot summers and cold winters). These conditions produced localised salt pans, allowing the growth of such salt-loving or salt-tolerant plants in much the same way that salt used on modern roads has allowed the spread of halophytes to inland areas. This very open, but otherwise rich and diverse vegetation set the scene for the rapid environmental changes which occurred with final rapid temperature amelioration at *c.* 10,000 BP, marking the end of the last cold stage and the onset of the present interglacial conditions.

The early warm period
(The pre-Boreal and Boreal: 10,000 to 7,000 BP)
Rapid temperature amelioration at *c.* 10,000 BP marked the final close of the Devensian cold stage after the fluctuations of the late glacial. This set in motion rapid successional vegetation changes and associated soil formation, which culminated in mature, climax woodland of various types.

The transitional phase at this time was of short duration and is manifested in the pollen record for the Isle of Wight (and elsewhere) by a strong expansion of Juniper seen at Munsley Bog and Gatcombe Withy Bed. This species had been locally present, but suppressed, as part of the dwarf shrub communities and was thus able to respond rapidly to temperature amelioration along with other plants such as willow species and Meadowsweet.

From *c.* 10,000 to 9,500 BP (Flandrian Ia), tree birches (*Betula* spp.) colonised the Island, rapidly outcompeting and suppressing Juniper and the more light demanding elements of the herb flora. This change is evidenced at Gatcombe Withy Bed, The Wilderness (42) and Munsley Bog (Scaife 1980, 1987). Scots Pine (*Pinus sylvestris*) closely followed, to become dominant over large areas of the Island and southern England on a wide range of soil types from *c.* 9,500 BP and continued to extend its range progressively northwards.

During the period of pine dominance, Hazel (*Corylus avellana*) similarly arrived from its glacial refugia and available evidence suggests that as a tree, it formed an understorey to open-aspect pine forest. This early importance of Hazel has been observed from pollen and macrofossils at Gatcombe Withy Bed, Munsley Bog and from a palaeochannel derived from a river, now submerged in Sandown Bay (17). The latter, at a depth of 19 to 21m below Ordnance Datum, reflects the low sea level at the close of the late glacial period, which allowed migration of the flora from continental Europe. Much of the region surrounding the Isle of Wight was lowland marsh, utilised by the Mesolithic hunting, fishing and foraging communities (the Maglemosian) as evidenced by discovery of *in-situ* artefacts underlying submerged peats at Bouldnor Cliff (23) (Momber 2001; Scaife 2001).

With this expansion of woodland, further suppression of existing dry-land light demanding herb communities occurred. The slower migration rates of oak (*Quercus* spp.) and elm (*Ulmus* sp.) is reflected by their arrival into the region after Scots Pine. Indications of their increasing proximity is shown by sporadic pollen occurring at Gatcombe and Munsley Bog from the period of pine dominance. Subsequently, they arrived and became dominant, suppressing pine after *c.* 8,500 BP (Flandrian Ic). However, Scots Pine remained and it is not clear whether it was a minor element of mixed deciduous woodland such as occurs today in the Black Forest, Germany, or remained locally dominant on poorer sandy soils.

The middle Holocene : 7,000–5,000 BP.

The period from 7,000 BP to 5,000 BP has been regarded as the climatic optimum for this interglacial but whilst there was an increase in oceanicity and humidity, the use of the term 'optimum' might be misleading, because evidence from throughout Europe now suggests that the thermal maximum occurred early in the Holocene under a more continental regime. Unlike the preceding early Holocene instability and successional vegetational changes, this period was, in contrast, one of greater stability. The pine domination of the early Holocene was superceded by dominance of deciduous woodland.

From *c.* 7,500-7,000 BP, the expansion of Alder (*Alnus glutinosa*) and the first occurrence of thermophiles, including Small-leaved Lime (*Tilia cordata*), Holly (*Ilex aquifolium*) and Mistletoe (*Viscum album*), seems a more likely response to ecological and migration factors. In the case of Alder, increased oceanicity, higher relative (to land) sea levels, the separation of the Isle of Wight from the mainland, and waterlogging of river valleys played a more important role.

It is now recognised that differing plant migration rates caused arrival at different times in different regions of the country. This is also the case within the small geographical area of the Isle of Wight. Submerged coastal peats at Bouldnor Cliff and under Norton Spit (20) show clearly the early importance of Alder at *c.* 8,000-7,500 BP and Small-leaved Lime at 7,500 BP at Tanners Hard, Lymington. In contrast, at the inland sites of Gatcombe Withy Bed and Borthwood Farm (4), radiocarbon dates show expansions at *c.* 7,000 BP for Alder and later for lime.

There was, however, a complex of interacting factors that resulted in the 'climax natural woodland'. Earlier work has described this as 'mixed oak woodland'. Pollen investigations now show that Small-leaved Lime woodland was dominant, or at least co-dominant, with oak over most of southern and eastern England. This is especially apparent from the Isle of Wight, where its dominance on the chalk and greensand soils has been demonstrated (Scaife 1980, 1987, Scaife in Tomalin *et al.* forthcoming). Sites which have produced evidence of this include Gatcombe Withy Bed, Borthwood Farm, Ninham Farm (1), Brook Bay (8), Bouldnor Cliff, Norton Spit, Freshwater Gate (24), Sandown Water Treatment Works (18), Newnham Farm (10) and Ranelagh Spit (15).

Small-leaved Lime continued to form a major element of the pollen assemblages lasting into the late-prehistoric Neolithic and early-middle Bronze Age (the Sub-Boreal) period. Records of Large-leaved Lime (*Tilia platyphyllos*) are rare, but known from southern England and in the Isle of Wight at Gatcombe Withy Bed (Scaife, 1980).

Past views anticipated that the heavier Tertiary clays to the north of the Island would have supported largely closed oak, elm and hazel woodland. More recent pollen analysis of sites at Newnham Farm, Firestone Copse (19) and the coastal submerged and inter-tidal peats at Wootton-Quarr (14), Newtown (25), Bouldnor Cliff and Yarmouth (21, 22) (Scaife in Tomalin *et al.* forthcoming; Scaife 2001) have shown that this was not necessarily the case.

These sites indicate the period *c.*7,000 to 5,000 BP was one of deciduous woodland of Pedunculate Oak (*Quercus robur*), Wych Elm (*Ulmus glabra*), Alder and Hazel, but importantly, with Small-leaved Lime also present in quantity. It is now clear that lime grew in a far wider range of habitats than previously thought. There is also evidence of other woodland trees and understorey shrubs growing in suitable habitats. These include Ash (*Fraxinus excelsior*), Holly, Buckthorn (*Rhamnus catharticus*), Dogwood (*Cornus sanguinea*) and Spindle (*Euonymus europaeus*). There is some evidence that remnants of Scots Pine possibly existed on more suitable acidic soils.

Sea level change and its effects on the past vegetation of the coastal zone:

Detailed stratigraphical, micro-palaeontological and radiocarbon dating studies have been carried out on Solent sediments during the last decade. Age/altitude curves for Holocene changing (relative) sea levels have been established for the last 7,000 years, providing information on the palaeogeography, including the separation of the Island from the mainland. These data are based on studies from the eastern Solent at Wootton Quarr, Sandown Bay (Long and Scaife in Tomalin et al. forthcoming), the western Solent at Bouldnor (Scaife and Long 2001) and from Southampton Water (Long and Tooley 1995; Long and Scaife 1996; Long et al 2000; Scaife and Long 2001). Their relevance to the Island flora lies in the creation of the broad geographical outline of the Island and the establishment of the salt marsh and valley communities.

The presence of early Holocene Hazel nuts in submerged palaeochannels (19 to 21m below Ordnance Datum) in Sandown Bay has already been noted as evidence of the early migration of Hazel into the region when Britain remained connected to France. The start of the middle Holocene at c. 7,000 BP (Atlantic period) saw the arrival of lime, Alder and Ash woodland immediately prior to inundation of the English Channel and separation of the Island from the mainland by rising relative (to land) sea level.

Marine inundation of the deeper river channels gave rise to an extension of sedge fen/reed swamp and especially Alder carr woodland upstream and salt marsh communities fringing the Solent coastline and the drowned valleys of the Western Yar at Norton Spit, Yarmouth and Yarmouth Marsh (Devoy 1972, 1987; Scaife 2001; Scaife in Tomalin et al. forthcoming;) and Newtown Creek (Scaife unpublished). Environmental stabilisation occurred in the later Atlantic period from 6,000-5,000 BP. with peat accumulating in a freshwater sedge-fen.

At Bouldnor Cliff, near Yarmouth, a sequence of underwater peats laid down within salt marsh mineral sediments exists to a depth of 11.7 to 12m below Ordnance Datum. Whilst providing valuable data on changing sea level, this site has yielded information on the development of vegetation communities prior to the submergence of the Solent.

Three peat strata are present and have been radiocarbon dated at 5,580+/-60 BP (4,452-4,359 Cal BC) for the upper, 5,870+/-80 BP (4,838-4,821 Cal BC) for the middle peat and the lowest at 7,640+/-70 BC (6,564-6,550 Cal BC). These correspond with a similar sequence recorded from under Norton Spit, with a freshwater organic sediment layer and overlying peat sequence radiocarbon dated to 7,230+/-110 BP (6,380-5,840 Cal BC). Newtown Creek has similar evidence of a freshwater habitat at 7,570+/-80 BP (6,570-6,180 Cal BC).

The lowest of the Bouldnor peats at the early/middle Holocene transition rests on an old land surface, which contains Mesolithic artefacts (Momber 2001). The development of the basal peat accumulation reflects increasing wetness as sea level rose to fill the Solent and is shown by the increasing occurrence of Alder and sedge fen with drier areas of oak, elm and Hazel woodland. Substantial oak trunks found in the basal peat attest to the woodland prior to marine inundation at c.7,000 BP. The middle peats at 4.76 to 5.36m below Ordnance Datum represent a minor standstill event when salt marsh developed. This was again inundated and mud flat remained until further reduction in relative (to land) sea level resulted in the formation of the upper peat. This is notable in that a *Sphagnum* and Royal Fern bog community developed. This was clearly an acid to mesotrophic community, similar to the remaining *Sphagnum* mires at Bohemia and Munsley Bogs, possibly due to drainage from the sandy Tertiary deposits. Now at a depth of 3.72m below Ordnance Datum, this was inundated in the final late prehistoric marine incursion.

Between Wootton and Ryde, intertidal peat accumulations and an associated submerged forest have also been studied, and are of Neolithic age. The specific ecological character of this woodland has, in the past, been poorly addressed and there is no contemporary analogue. Common Reed (*Phragmites australis*) macrofossils and pollen of Lesser Bulrush and Bur-reed indicate colonisation by reed swamp in areas of previously fresh and brackish water. This resulted from a negative sea level tendency (i.e. relative lowering of sea level in relation to land), which created new habitats.

Subsequently, woodland was able to colonise from the land over these areas. This succession culminated in mature oak woodland with Hazel of a damp poor fen character. This is evidenced along the foreshore by numerous oak root boles and tree trunks associated with peat beds radiocarbon dated at between c. 5,300 to 4,700 BP (4,000-3,000 Cal BC). Although a fen woodland, typical elements such as Alder and willow were largely subordinate to oak and other tree and understorey shrubs which included lime, Ash, Yew (*Taxus baccata*), Hazel, Bird Cherry (*Prunus padus*), Dogwood, Field Maple (*Acer campestre*), Buckthorn, Blackberry (*Rubus fruticosus*), and possibly Crab Apple (*Malus sylvestris*). A herb layer of sedges and other fen taxa including Marsh Marigold existed.

The presence of Small-leaved Lime macrofossils and pollen from these peats is enigmatic, since lime favours well-drained and rich soils. However, there are an increasing number of peat sites where its seeds and pollen have been found. These may derive from nearby dry land, but, however, the physical settings and topography do not always make this likely. Given the dominance of lime in the landscape, it now seems plausible to suggest that its ecological range may have been greater in the past than previously thought.

Dendrochronology (Hillam in Tomalin *et al.* forthcoming) has dated numerous trees from this now submerged forest to between *c.*3,200 and 2,400 BC. The dendrochronology also shows thin growth rings, indicating a stressed environment. This may have been caused by climatic adversity but more probably due to the physical stresses such as variable salinity and anaerobic soils within this ecotone. Overall, the botanical evidence obtained from sediments associated with the 'submerged forest', show the peat to be an *in situ* deposit, formed under conditions which changed from marine, salt marsh and brackish water conditions, through reed swamp with aquatics to a rather dry poor fen woodland. This was subsequently finally inundated by further marine ingress asynchronously between the late Neolithic and Iron Age (980-790 Cal BC at Ranelagh) according to local topography. Saltmarsh and mud flat were broadly established in their present locations.

The first substantial effects of human interference: 5,500-5,000 BP

From *c.* 5,500 to 5,000 BP in Britain, the arrival of the Neolithic culture marked the first clearances of the 'natural' woodland for agriculture. In pollen diagrams from the Island, this is evidenced by the first occurrence of cereal pollen and associated weeds of arable ground at the same time as there is a reduction of tree pollen, indicating small-scale deforestation.

Such clearances of the forest have often been seen as of ephemeral character, with subsequent abandonment and then regeneration of these areas by secondary woodland. It was originally thought that these Neolithic 'Landnam' clearances of slash and burn character were of 25-30 years duration (Iversen 1949). Detailed work has now demonstrated that they were not truly ephemeral, but activity may have lasted for some hundreds of years. This is the case in the Isle of Wight where Neolithic woodland clearance for occupation and agriculture has been studied at Gatcombe Withy Bed, Newnham Farm, Borthwood Farm and Bohemia Bog (3) (Scaife 1980, 1987; Tomalin and Scaife 1979).

Frequently associated with, and just prior to these clearances and cereal cultivation, is the much discussed 'elm decline', a distinctive pollen event dated more or less synchronously at *c.* 5,500-5,000 BP. This event is well represented in the Island's pollen record at Gatcombe Withy Bed, where it has been radiocarbon dated to 4,850 +/-45 years BP and at Borthwood. Many explanations have been given in the past for this significant event, including climatic change, use of elm leaves for fodder and tree clearance through ring-barking (summary by Smith 1970, 1981; Turner 1970; Scaife 1987 for discussion). In light of the spread of Dutch elm disease during the 1960-1970's and the finding of fossil records, it is now held to be the result of elm bark beetles (*Scolytus scolytus*) and fungal disease (*Ceratocystis ulmi*) carried by these beetles. The late Dr. M. Girling suggested that the arrival of Neolithic economy and opening-up of woodland promoted the geographical expansion of these beetles and fungal disease across most of the forests of Britain.

Whilst it was once thought that the Neolithic people were responsible for large-scale woodland clearances, especially in the chalklands (now downland), the overall effect of these early agriculturists was to create a mosaic landscape of agricultural clearances, set in remaining climax woodland and secondary woodland. The latter woodland elements which especially include Ash, Beech (*Fagus sylvatica*), Holly, and secondary regeneration of elm, became important particularly during a phase of late Neolithic woodland regeneration seen in pollen data from Gatcombe and Bohemia Bog at around 4,000 BP. The strong regional evidence for this could be indicative of a changing agricultural economy to one based on woodland pastoralism.

Woodland Management

Extensive survey of the inter-tidal zone of north-east Wight centred on Wootton, by The Isle of Wight Council Archaeological Unit, identified many wooden archaeological structures preserved in peat and mineral sediments. These range in age from the Neolithic to present and comprise such features as prehistoric (Neolithic) trackways across the marshes, fish traps, a Romano-British building structure and medieval post alignments. Identification of the timbers by Gale (Gale in Tomalin *et al.* forthcoming) has provided an almost unique history of the utilisation and management of woodland during the late prehistoric and historic periods, which mirrors the pollen evidence for progressive woodland clearance.

Neolithic trackways dated at 4,860 +/- 70 BP, 4,800+/- 60 BP and 4,770+/-50 BP show clearly the specific selection of Hazel wands (produced by coppicing?) with occasional Field Maple, elm, Dogwood and Spindle. A number of Bronze Age timber stakes dated to between 3,600 to 3,100 BP used oak, Field Maple, Alder and Hazel. A Romano-British timber structure dated to 1,860+/-50 BP and brushwood platform was constructed of oak and Blackthorn (*Prunus spinosa*) or Hawthorn (*Crataegus mongyna*).

The post Roman/Saxon period is characterised by longshore post alignments dated to between 1,450 BP and 1,320 BP and were composed of Ash, Field Maple, oak, Alder, willow/poplar (*Salix/Populus* spp.) and fruit trees (*Pomoideae*). From the medieval period, fish traps (1,040+/-50 BP) were constructed from Field Maple and Beech. Other later medieval structures utilised oak, elm and Alder and occasionally Hazel, all being indicative of areas of remaining and managed woodland taken from a wider geographical area.

Lime: the Island's natural woodland ?
The importance of Small-leaved Lime as a dominant or co-dominant constituent (with oak) of woodland in the Island from *c*.7,000 BP has already been noted. This woodland type remained and was, perhaps, more important during the Neolithic when clearance of other woodland for agriculture was taking place. It is by no means clear why lime or linden remained so untouched, especially when it occupied light fertile soils ideally suited to agriculture. It is highly probable that the bast fibres of the tree and its nutritious leaves were considered to be an important natural resource by the late prehistoric communities until the middle to late Bronze Age at least.

In a substantial number of pollen sites on the Island, and from other southern English pollen profiles (Scaife 1980, 2000; Waller 1994), the lime pollen declines sharply to low levels or absence during the late prehistoric Bronze Age period. This event is associated with a corresponding increase of herbs of agriculture and evidence of cereal cultivation. At some sites, (Gatcombe Withy Bed, Ninham Farm and Newnham Farm), there are also significant changes in the valley mire vegetation communities brought about by the resulting changes in the local hydrological and ground water table.

Originally, this decline was attributed to climate change, marking the division between the sub-Boreal and sub-Atlantic climatic worsening at *c*. 500 BC. However, from work previously carried out, it is clear that the reduction in lime was due to human activity (agriculture) and/or localised water-logging reducing its areas of growth. Although this phenomenon is asynchronous, the majority of radiocarbon dates fall in the late Bronze Age at *c*.3,300-3,000 BP. This is the case on the Isle of Wight as evidenced at the sites of Bohemia Bog, radiocarbon dated at 2,190+/-130 BP, Borthwood with an initial, late Neolithic decline at 4,010 +/-110 BP and a later final Bronze Age reduction at 3,280+/-80 BP at Sandown Water Treatment Works, Ninham Farm and Gatcombe Withy Bed. These sites relate to the vegetation of the Lower Greensand areas south of the downlands.

At Newnham Farm, a small, localised, spring-fed peat mire has produced the first evidence of late prehistoric and historic vegetation in the terrestrial, non-marine zone of north-east Wight. Earlier views that the vegetation of Tertiary lithologies remained as dominant oak woodland, little affected by prehistoric activity, are now confounded. Lime woodland was dominant or co-dominant with oak, elm and Hazel. It was largely felled during the Neolithic, with some woodland remaining into the Bronze Age before its final demise at the expense of increasing agriculture. Oak and Hazel remained in the region to become managed woodland.

The Heathland and Downland
Heathland, and the downland especially, are major components of the Island vegetation. These are often viewed as natural communities but they have prehistoric origins, being created by woodland clearance and subsequently maintained by human activities. The two communities are obviously contrasting, with the relatively floristically poor acid heathland on sandy substrates and the species-rich, short turf communities of the chalk. However, the origins and development of these plagioclimax (human maintained) communities are very similar, and are thus discussed together. Whilst there were undoubtedly localised areas where there were suitable micro-habitats of both heath and calcareous communities during the late-glacial period and subsequently, it was the large scale effects of Neolithic and especially Bronze Age woodland clearance for occupation and agriculture which was responsible for their expansion over a significant area of the Island.

Although pollen is not well preserved in alkaline soils and sediments, data have been obtained from downland as well as the better preserving heathland soils. Acid podzolic soils under Bronze Age barrows on St Boniface Down (32) (Cox 1986), Headon Warren (28) (Scaife 1980), Puck House (34) (Scaife in Tomalin *et al*. forthcoming) and a field boundary bank at Standen Heath (33) (Scaife 1991) demonstrate that by the early-middle Bronze Age, wet and dry heath dominated by *Erica* spp. and Heather dominated heathland was in existence. Analysis of the peat sequence at Bohemia Bog (Scaife 1980), a valley side spring flush, confirms that the area of Bleak Down was also subject to woodland clearance, soil deterioration and heathland expansion, initially during the Neolithic, but more extensively from the Bronze Age. This pattern of events is similarly evidenced from other southern English lowland heathlands. Once established, these areas have been maintained by rough grazing, fire and in the last millennium, by rabbits.

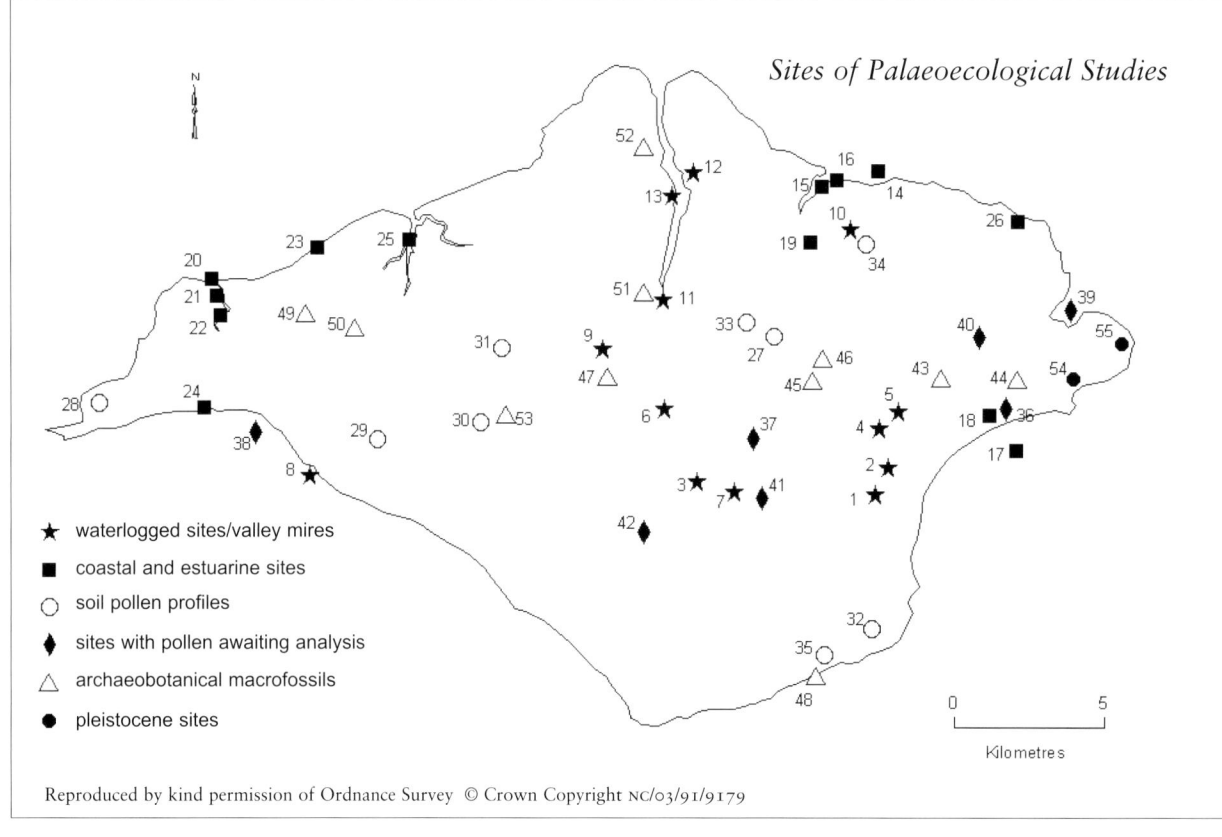

SITE GAZETTEER

1 Waterlogged sites/Valley Mires
1 Ninham Farm
2 Scotchells Brook (James Doe undergraduate dissertation Exeter University)
3 Bohemia Bog
4 Borthwood Farm
5 Alverstone Marsh (Andrea Smith undergraduate dissertation Southampton University)
6 Gatcombe Withy Bed
7 Munsley Bog
8 Brook Bay
9 Clatterford Roman Villa
10 Newnham Farm
11 Coppins Bridge, Newport
12 Whippingham Church
13 Werrar, Medina

2 Coastal Sites and Estuarine Sites
14 Wootton-Quarr foreshore
15 Ranelagh Spit
16 Ranelagh North
17 Sandown Bay
18 Sandown Sewage Treatment Works
19 Firestone Copse/Blackbridge Brook
20 Norton Spit
21 Yarmouth Marsh
22 Yarmouth Marsh (Devoy 1977, 1987)
23 Bouldnor Cliff
24 Freshwater Gate
25 Newtown Creek
26 The Duver, Seaview

3 Soil Pollen Profiles
27 Combley Villa
28 Headon Warren Bronze Age Barrow
29 The Longstone (Neolithic)
30 Newbarn/Gallibury Down (Bronze Age Barrow)
31 Apes Down valley (Bronze Age ring ditch)
32 St Boniface Down (Bronze Age barrows) Cox 1986
33 Standen Heath (? Bronze Age land boundary)
34 Puck House Bronze Age Barrows
35 Ventnor Landslide

4 Sites with pollen awaiting analysis
36 Yaverland
37 Redway Farm
38 Compton Bay
39 Bembridge Harbour
40 Eastern Yar
41 Moor Farm, Godshill
42 The Wilderness

5 Archaeobotanical-macrofossils
43 Brading Roman Villa
44 Yaverland Roman Villa
45 Newchurch Roman corn drier
46 Mersley Farm. Late Iron Age ditch fills
47 Clatterford Roman Villa (De Moulin 2001, De Roufingnac 2001)
48 Flowers Brook, Ventnor
49 Thorley medieval settlement
50 Wellow. Bronze Age
51 Newport High Street
52 Somerton, Cowes (Caruthers 2003)
53 Newbarn/Gallibury Down

6 Pleistocene sites
54 Bembridge School (Preece 1983; Holyoak and Preece 1983)
55 Bembridge Foreland (Preece et al 1990)

Except where stated otherwise, sites have been investigated by the author.

The old species-rich chalk turf downlands, once thought to be 'natural' grassland, have limited pollen evidence. Environmental information gained from sub-fossil snails is more important, showing that the first localised woodland interference occurred in the Neolithic but with extensive clearance and creation of downland pasture during the early Bronze Age from c. 3,500 BP. Pollen data from the high downs at Gallibury (30) and the Apes Down valley (31) (Scaife, 1980 and in Tomalin et al, forthcoming) show that by the early Bronze Age, short turf grassland existed on previously wooded areas.

Analysis of a deep peat sequence at Borthwood (Scaife, 1980), which was developed on Lower Greensand, indicated that it had a pollen catchment that included the downs to the north. This similarly shows that widespread woodland clearance occurred at this time and once established, these grasslands became floristically rich and were maintained through grazing. Whereas pollen is usually poorly preserved in this habitat, palaeo-molluscan evidence has provided more detailed evidence of environmental change (Preece 1980). From dry valleys at Duxmore Coombe and Newbarn Coombe, work by Allen (1993) attests to this general sequence of events, with ephemeral phases of possible arable cultivation during the Iron Age and Romano-British period. The Roman villa sites of Combley (27) and Clatterford (47) on the chalk have, however, produced valuable pollen and plant macrofossil sequences (Busby 2001) Both demonstrate an open agricultural landscape, with the latter showing some evidence for localised stands of trees.

Therefore, heathland and downland have a long and complex vegetation history. Subject to periglacial processes during the late-glacial period, the downland became colonised by woodland after the start of the present interglacial period (Holocene). The downland as we know it today was both initiated and subsequently intricately controlled by human activity since the Neolithic and early Bronze Age. Questions of exposure to wind may today play an important role in the maintenance of some short turf grassland on some south and south-west facing zones, such as Tennyson Down (Lousley, 1950), but it should be considered that the prevailing wind direction during these earlier periods (the sub-Boreal) may have been different - possibly from the east.

Bohemia Bog

It has become apparent in recent years that, although of only very small extent, this remaining bog is an extremely important element of the Island's plant communities. It is a small, spring-line mire, which may have formed in response to prehistoric (Neolithic) woodland clearance that raised the local water table. Pollen analysis provides a means of studying the development of such communities through time. Whilst it is now an acid community, this was initially not the case. From initiation of the peat until c.3,000 BP, this was a more base-rich habitat with some surrounding Alder, Willow, Bog Myrtle (*Myrica gale*) and grass-sedge fen with Royal Fern.

More extensive woodland clearance and agriculture resulted in the extension of heathland and gorse scrub, which was accompanied by increased acidity and the formation of the Bog Moss (*Sphagnum*) community and associated taxa. The latter shows interesting pollen records of Bog Asphodel (*Narthecium ossifragum*), Bog Pimpernel (*Anagallis tenella*), Marsh Clubmoss (*Lycopodiella inundata*), Round-leaved Sundew (*Drosera rotundifolia*) and Oblong-leaved or Great Sundew (*Drosera intermedia* type). Some of these taxa remain important constituents to the present day. Useful records of Meadow Saxifrage (*Saxifraga granulata*) and Adder's-tongue fern (*Ophioglossum vulgatum*) come from nearby grassland whilst Cornflower (*Centaurea cyanus*) pollen, associated with cereals and other weeds, was found in quantity in peats of around 300 years old.

The vegetation of the Romano-British period

Pollen data come from the upper peat levels at a number of Lower Greensand sites previously noted (Gatcombe Withy Bed, Borthwood Farm, Bohemia Bog and Ninham), from the Tertiary lithology north of the downs (Newnham Farm, Firestone Copse) and the downs at Combley Roman Villa (Scaife 1980) and Clatterford (9). At the latter, a peat deposit of Roman age has provided useful pollen and plant macrofossil data of some rarity (Scaife 2001 in Busby 2001), since the site is located in the middle of the downland environment and is in very close proximity to a Romano-British villa. The depositional habitat here was a herb rich fen dominated by grasses, sedges, and rushes (*Juncus* spp.). Other fen elements present include Marsh-marigold, Yellow iris, Bulrush / Bur-reed, Water-plantain (*Alisma plantago-aquatica*), and Purple-loosestrife (*Lythrum salicaria*).

Willow was also possibly growing on the site in the earlier period of organic accumulation. Interestingly, there are a small number of Bog Moss spores, which are more typical of acidic boggy habitats; these spores are likely to have come from those taxa that are more tolerant of base rich conditions (e.g. *Sphagnum palustre*). The small numbers of Heath and Heather also attest to some areas of slightly more acid conditions. Tree and shrub pollen, although more abundant in the lowest levels of the peat, are generally few, suggesting the absence of woodland locally during the period spanned by this organic accumulation (i.e. 1st to 3rd century AD) although some stands may have been present in localised valley situations. Such openness is also evidenced from Combley Villa. Pollen of cereals and weeds and seed at both sites may indicate arable cultivation on drier valley sides. In other areas of the Island, remaining lime woodland had been cleared and oak and Hazel woodland remained most important.

Vegetation during the Historic Period

With the widespread removal of lime during the Bronze Age, oak and Hazel became the principal woodland elements with localised Beech, Ash and hedgerow elements. It appears that remaining woodland became progressively diminished, ultimately remaining largely in park enclosures (for example Parkhurst Forest) and in river valleys to the north of the downs. Managed woodland on drier ground comprised typical oak standards with Hazel coppice. This is evidenced from analysis of sediments in the headwaters of Blackbridge Brook at Firestone Copse and Newnham Farm (Scaife in Tomalin *et al.* press). Identification of inter-tidal wooden structures at Wootton-Quarr also provides some evidence for available wood resources. Alder and willow remained in valley fens (carr woodland) where peat was not cut for fuel. Willow was certainly managed as osier beds for basketry and hurdles, and a radiocarbon date of 1460 AD marks the expansion of such activities at Gatcombe Withy Bed.

Further pollen evidence of woodland management comes with the planting of pine from *c*.1750 in private estates, subsequent to the writings of John Evelyn. The large pollen production and airborne pollination of pine causes widespread dispersion of its pollen and consequently, an expansion in numbers is found in the upper sections of peat and sediment sequences. This increase in pollen, also frequently accompanied by hedgerow elm, has been noted at Gatcombe, Bohemia Bog, Ninham Farm, Firestone Copse and Newnham Farm and in a substantial number of sites in Hampshire and Dorset (Barber 1975; Haskins 1978; Scaife 1980; Waton 1982; Long *et al.* 1999, 2000).

Apart from cereal and associated weeds that become increasingly important in the pollen record, there are occasional other crops including Hemp (*Cannabis sativa*), grown for fibre, and medieval Buckwheat (*Fagopyrum esculentum*) pollen found at Yarmouth. The Walnut (*Juglans regia*) is an interesting pollen record from a number of Island sites, including Gatcombe Withy Bed, Bohemia Bog (*c*. AD 1700), Clatterford valley fen (Roman) and upper salt marsh sediments at Yarmouth Marsh (*c*. AD 1600-1700), and Firestone Copse (Medieval and later). Although there are occasional records in pre-Roman sediments, walnut is generally regarded as a Roman introduction into Western Europe as a whole. Thus, most pollen records are from Roman and post-Roman sequences and in only small or individual occurrences such as found on the Island. It is not clear whether this record represents long distance pollen transport but it seems more likely that pollen came from local trees introduced for nuts or as garden ornament. More detailed examination of cultivated plants comes from seed remains found on archaeological sites.

The Archaeobotanical Record

Pollen data from the peats at Gatcombe Withy Bed, Bohemia Bog and Borthwood Farm show that the first evidence of cultivation occurred during the Neolithic period at *c*. 5,000 years BP and a progressive, increasing impact of agriculture on the natural vegetation of the Island. Unfortunately, however, there are as yet, few archaeological seed records for the Mesolithic, Neolithic or Bronze Age periods. This reflects the paucity of excavation of sites of this date undertaken in recent decades. Plant macrofossil records from the Mesolithic period purported to relate to the Mesolithic hearth at Brook Bay (Clifford 1936), but in fact, relate to the peat forming vegetation of the site. Even the presence of Hazel nuts ('Noah's nuts') is not evidence that these were used by these hunting and foraging human communities. This similarly applies to the seed data derived from the inter-tidal peats at Wootton-Quarr and submerged at Bouldnor Cliff.

Although there is pollen evidence for the first cereal cultivation from the Neolithic period at *c*. 5,000 BP, at present it can only be assumed that the pattern of cultivation within woodland clearances followed lines demonstrated for southern England as a whole (Helbaek 1952; Murphy 1977), that is, reliance on Emmer and Bread Wheat (*Triticum dicoccum* and *T. aestivum* respectively) and Barley (*Hordeum vulgare*).

The Celtic, or Horse Bean (*Vicia faba* var. *minor*), is a precursor of the Broad Bean of today. The Isle of Wight has (to date) the earliest record of Celtic Bean in England (Scaife 1982), dating to the early Bronze Age. This was identified from a pottery impression obtained from a food vessel urn excavated from Gallibury Down (53) (Tomalin 1979). A substantial quantity of well preserved, charred seeds of *Vicia faba* was also recovered during the excavation of Brading Roman

Villa in the last century. These were originally identified as Cherry pits (*Prunus domestica*) and were, for many years, displayed as such. Further examination of these (Scaife unpublished 1997), showed that these are in fact *Vicia faba* var. *minor* adding to the increasing evidence that these early Beans became progressively important though the late Bronze Age to Romano-British period.

Other Island records include a single Celtic Bean from Flowers Brook, Ventnor (48) (Scaife unpublished) found adjacent to an Iron Age skull (inhumation), and from Clatterford Roman Villa (De Moulin 2001). The classical writings of Pliny suggest that Celtic Bean was cultivated for flour used in making a type of bread and/or porridge. Its importance as a major food crop from the later prehistoric period is now generally recognised and it is concluded that, along with well-known arable cultigens such as Emmer Wheat and particularly Spelt Wheat (*Triticum spelta* type), these pulses constituted an important cultivated crop, especially during the Iron Age and Roman period.

There are more archaeobotanical records from the Iron Age and Romano-British phases than from the late prehistoric period. This reflects more rigorous recovery methods during more recent excavations including Mersley Farm (46) and Thorley (49). The latter, a middle Iron Age settlement, has produced grain of Barley, Emmer and Spelt Wheat and seeds of Bird Cherry (*Prunus*), Dock (*Rumex* sp. unidentified), Vetches and Tares (*Vicia* spp.), Wild Peas (*Lathyrus* sp.) and Fat-hen (*Chenopodium album*) (Scaife unpublished).

The most abundant Iron Age remains found to date are charred cereals and chaff debris disposed of in a late Iron Age ditch at Mersley Farm (Scaife in Trott forthcoming). This comprised largely grain and chaff of Spelt Wheat, the predominance of which is highly characteristic of the Iron Age and Romano-British periods. Jessen and Helbaek (1944) and Helbaek (1952), initially demonstrated this predilection for Spelt Wheat from the Iron Age, and its importance was maintained into the Romano-British period.

Data relating to crops from the Roman villa sites at Brading (43), Yaverland (44) (Scaife unpublished), Clatterford (De Moulin 2001), from excavations at Somerton (52) (Caruthers 2003) and from a corn drier at Newchurch (45) (Scaife in Tomalin 1988) are all comparable, comprising largely grain and chaff of Barley and Spelt. The Clatterford Villa (47) has produced an assemblage of cereals and also a range of weeds associated with the wet meadow (grasses, sedges and rushes), trees and shrubs used for fuel, Hawthorn, Hazel, oak and Traveller's-joy, (*Clematis vitalba*); and weeds of disturbed arable ground. Of special interest from this site is a cone of Stone Pine (*Pinus pinea*), which was undoubtedly imported and may have been a votive offering. Oat (*Avena* sp.) and Rye Brome or Chess (*Bromus secalinus*) from Somerton may have been cultivated or weeds of cultivation.

Quantities of clay briquettage used in salt-making have been found on the coast at Quarr and Wootton. These sites are thought to be of Romano-British date (Scaife, in Tomalin *et al.* forthcoming). The majority of the fired clay fragments examined proved to be extremely rich in plant impressions and voids, indicating that substantial quantities of plant material had been added to the clay as tempering prior to firing. They comprised solely cereal chaff remains including straw, glume bases, spikelet forks and lemmas/paleas, but with only a small number of grain impressions. The glume bases and spikelet forks, rather than the grain itself, are especially diagnostic and identifiable to specific cereal type.

About 150 of the best preserved, and thus identifiable, surface impressions were examined and these comprised roughly even proportions of Emmer Wheat and Spelt Wheat. These taxa are glume wheats and unlike free-threshing Bread Wheats (*Triticum aestivum* and *T. compactum*), require parching (heating) in ovens or on hot floors before the grain is released from the husks by threshing, and sorted by sieving and winnowing. Whilst much of the regional evidence for these wheat types comes from accidental charring during this process, here the process was clearly successful, by virtue of the almost complete absence of grain itself. The threshed grain had been removed and only the remaining chaff debris was gathered and used to temper the clay.

There are very few data of medieval and later date from the Island. However, of note are the excavations of the first piped water in Newport High Street (51), which comprised hollowed Elm logs placed end to end with metal collars (Tomalin and Scaife 1987). They form the earliest recorded piped water system in the United Kingdom. Associated medieval sediments also contained small numbers of weed seeds and grains of barley. The excavations for the pipelines of a major sewage treatment scheme, installed in 2000, revealed remains of a medieval house at Wellow (50), which produced some charred grain including Bread Wheat (*Triticum aestivo-compactum*) and Rye (*Secale cereale*). This is typical of such medieval contexts, but has some significance as the first record of this crop from this period on the Island.

REFERENCES

Allen, M.J. 1993 The land-use history of the southern English chalklands with an evaluation of the Beaker period using environmental data: colluvial deposits as environmental and cultural indicators. Unpubl. Ph.D. thesis, University of Southampton.

Barber, K.E. 1975 Vegetational history of the New Forest: A preliminary note. *Proc. Hampshire Field Club & Archaeol. Soc.* 30, 5-8.

Busby, P. 2001 Clatterford Roman Villa. An assessment report for the excavations at Clatterford Roman Villa, Isle of Wight (CAS Project 550). *Proc. Hampshire Field Club Archaeol. Soc.* 56, 95-128

Caruthers, W. 2003 Network Archaeology report to Transco: Somerton to Briddlesford Cross main.

Chandler, M.E.J. 1960 *The Lower Tertiary Floras of Southern England*. British Museum (Natural History) Press.

Clifford, M.H. 1936 A Mesolithic flora in the Isle of Wight. *Proc. Isle Wight nat. Hist. Archaeol. Soc.* 2, 582-594.

Cox, J. 1986 Pollen analysis of a Bronze Age barrow on the Isle of Wight. Unpubl B.Sc. dissertation of University of London, King's College (Botany).

De Moulin, D. 2001 Plant macrofossil remains *in* Busby, P., De Moulin, D., Lyne, M., McPhillips, S. and Scaife, R.G. Excavations at Clatterford Roman Villa, Isle of Wight, *Proc. Hampshire Field Club Archaeol. Soc.* 56, 95-128.

Devoy, R.J. 1972 Environmental changes in the Solent area during the Flandrian. Unpubl. B.A. dissertation of the University of Durham.

Devoy, R.J. 1987 The Estuary of the Western Yar, Isle of Wight: Sea Level Changes in the Solent Region *in* Barber, K.E. (ed.). *Wessex and the Isle of Wight Field Guide*: 115-122. Quaternary Research Association, Cambridge.

Gale, R. (forthcoming) The utilisation of prehistoric woodland. Section 4.6 in Tomalin, D.J., Loader, R.D. and Scaife R.G. *Coastal Archaeology in a Dynamic Environment: A Solent Case Study*. English Heritage monograph.

Haskins, L. E. 1978 The Vegetational History of South-East Dorset. Unpubl. Ph.D. University of Southampton, Department of Geography.

Helbaek, H. 1952 Early crops in southern England *Proc. Prehist. Soc.* 18 (2) 194-233.

Hillam, J. (forthcoming) Tree ring analysis of oak timbers from the Wootton-Quarr survey, in Tomalin, D.J., Loader, R.D. and Scaife, R.G. (eds) *Coastal archaeology in a dynamic setting: a Solent case study*. English Heritage monograph.

Holyoak, D. T. and Preece, R. C. 1983 Evidence of a high Middle Pleistocene sea-level from estuarinedeposits at Bembridge, Isle of Wight England. *Proc. Geol . Soc.* 94 (3) 231-244.

Iversen, J. 1949 The influence of prehistoric man on vegetation. *Danm. geol. Unders.* Ser. IV 3 (6) 1-25.

Jessen, K. and Helbaek, H. 1944 *Cereals in Great Britain and Ireland in prehistoric and early historic times*. Det. Kongelige Danske Videnskabernes Selskab Biologistie Skrifler 3.(2) Copenhagen.

Long, A. and Scaife, R.G. 1996 *Pleistocene and Holocene Evolution of Southampton Water and its tributaries*. Report of Environmental Research Centre, Department of Geography, University of Durham.

Long A.J., and Scaife, R.G. (forthcoming) Solent sea-level record in Tomalin, D.J., Loader, R.D. and Scaife, R.G. (eds) *Coastal archaeology in a dynamic setting: A Solent case study*. English Heritage monograph.

Long, A.J. and Tooley, M.J. 1995 Holocene sea-level and crustal movements in Hampshire and Southeast England, United Kingdom. *Journal of Coastal Research Special Issue* No.17, 299-310.

Long, A.J., Scaife, R.G. and Edwards, R.J. 1999 Pine pollen in intertidal sediments from Poole Harbour, U.K.: implications for late-Holocene sediment accretion rates and sea-level rise. *Quaternary International* 55, 3-16.

Long. A.J., Scaife, R.G. and Edwards, R.J. 2000 Stratigraphical architecture, relative sea level and models of estuarine development in southern England: new data from Southampton Water. in Pye, K. and Allen J.R.L. (eds.) *Coastal and estuarine environments: sedimentology, geomorphology and geoarchaeology*. Geol. Society Special Publications Vol. 175, 253-279.

Lousley, J.E. 1950 *Wild flowers of Chalk and Limestone*. New Naturalist Series. No.16. London.

Momber, G. 2001 Recent investigation of deeply submerged human occupation on the floor of the Western Solent at Bouldnor Cliff *in* McInnes, R.G. Tomalin, D.J. & Jakeways, J. (eds.) *Coastal Change, Climate and Instability: Final Technical Report*. European Commission LIFE report no. 97 ENV/UK/000510. Isle of Wight Centre for the Coastal Environment, Ventnor.

Murphy, P.L. 1977 Early Agriculture and environment on the Hampshire chalklands: circa. 800 B.C. - 400 A.D. M.Phil. thesis of University of Southampton.

Osborne White, H.J. 1921 *A Short Account of the Geology of the Isle of Wight*. HMSO London.

Poole, H.F. 1936 An outline of the Mesolithic flint cultures of the Isle of Wight. *Proc. Isle Wight nat. Hist. archaeol. Soc.* 2, 551-581.

Preece, R.C. 1980 Biostratigraphy and dating of a post-glacial slope deposit at Gore Cliff, near Blackgang, Isle of Wight. *Journal of Archaeological Science* 7,256-265.

Preece, R.C. 1987 'Pleistocene sea-level history in the Bembridge area of the Isle of Wight'. pp. 99-114 in Barber, K.E. (ed.), *Wessex and the Isle of Wight, Field Guide*. Cambridge: Quaternary Research Association.

Preece, R.C., Scourse, J.D. Houghton, S.D., Knudsen, K.L. and Penney, D.N. 1990 The Pleistocene sea-level and neotectonic history of the eastern Solent, southern England. *Philosophical Transactions of the Royal Society of London* (B) 328,425-477.

Reid, E.M. and Chandler, M.E.J. 1924 On the occurrence of *Ranunculus hyperboreus* Rottb. in Pleistocene beds at Bembridge, Isle of Wight'. *Proc. Isle Wight nat. Hist. Soc.* 1, 292-295.

Scaife, R.G. 1980 Late Devensian and Flandrian palaeoecological studies in the Isle of Wight. Unpubl. Ph.D. thesis. Univ. London, King's College.

Scaife, R.G. 1980 Pollen Analysis *in* Basford, H.V. *The Vectis Report: A Survey of Isle of Wight Archaeology*, 56-59. Isle of Wight County Council. Newport, I.W.

Scaife, R.G. 1982 Late-Devensian and early Flandrian vegetational changes in Southern England. *in* Limbrey, S. and Bell, M. (eds). *Archaeological Aspects of Woodland Ecology*. Symposia of the Association for Environmental Archaeology. No.2 B.A.R. (Internat. Ser.) pp. 57-74

Scaife, R.G. 1982 An early Bronze Age record of *Vicia faba* L. (Horse Bean) from Newbarn (Gallibury) Down, Isle of Wight. *Ancient Monuments Laboratory Report* No. 3501.

Scaife, R.G. 1984 Gallibury Down, Isle of Wight: pollen analysis of a Bronze Age downland palaeosol. *Ancient Monuments Laboratory Report* No. 4240

Scaife, R.G. 1987 The Late Devensian and Flandrian Vegetation of the Isle of Wight. pp.156-180 *in* Barber, K.E. (ed.), *Wessex and the Isle of Wight Field Guide*. Quaternary Research Association. Cambridge.

Scaife, R.G. 1987 The elm decline in the pollen record of South East England and its relationship to early agriculture. *in* Jones, M. (ed.) *Archaeology and the flora of the British Isles* Oxford Univ. Coom. Archaeol., 21-33

Scaife, R.G. 1991 Standen Heath, Isle of Wight: palynological assessment. Unpublished archival report for Wessex Archaeology, Salisbury.

Scaife, R.G. 1997 Brading Roman Villa; The charred seeds. Unpublished document.

Scaife, R.G. 2000 Chapter 8. Holocene vegetational development in London. *in* Sidell, J., Wilkinson, K., Scaife, R.G. and Cameron, N. *The Holocene Evolution of the London Thames*. Museum of London Archaeology Service Monograph 5, 111-117.

Scaife, R.G. 2001 Pollen Analysis of Valley fen peats of Romano-British date. *in* Busby, P. Clatterford Roman Villa. An assessment report for the excavations at Clatterford Roman Villa, Isle of Wight (CAS Project 550). *Proc Hampshire Field Club Archaeol Soc.* 56, 119-124

Scaife, R.G. 2001 Palaeo-environmental investigations of the submerged sediment archives in the Western Solent study area at Bouldnor and Yarmouth. pp. 13-26 *in* McInnes, R.G., Tomalin, D.J. & Jakeways, J. (eds.) *Coastal Change, Climate and Instability: Final Technical Report*. European Commission LIFE report no. 97 ENV/UK/000510. Isle of Wight Centre for the Coastal Environment. Ventnor.

Scaife, R.G. 2001 Pollen analysis of valley fen peats of Romano-British date in Excavations at Clatterford Roman Villa, Isle of Wight. *Proc. Hampshire Field Club Archaeol. Soc.* 56, 95-128.

Scaife, R.G. (forthcoming) The changing vegetation and environment. in Tomalin, D.J., Loader, R.D. and Scaife R.G. *Coastal Archaeology in a Dynamic Environment: A Solent Case Study*. English Heritage monograph.

Scaife, R.G. in Trott, K. (forthcoming) The Mersley Farm Archaeological Survey

Scaife, R.G. & Long, A.J. 2001 The Western Solent: a summary discussion on changes in the Holocene coastal environment pp. 26-28 and Biostratigraphical markers and their application to coastal sedimentary dynamics pp.28-32 *in* McInnes, R.G. Tomalin, D.J. & Jakeways, J. (eds.) *Coastal Change, Climate and Instability: Final Technical Report*. European Commission LIFE report no. 97 ENV/UK/000510. Isle of Wight Centre for the Coastal Environment. Ventnor.

Smith, A.G. 1970 The influence of Mesolithic and Neolithic man on British vegetation: A discussion *in* Walker, D. and West, R.G (eds.). *Studies in the Vegetational History of the British Isles*, 81-96. Cambridge University Press.

Smith, A.G 1981 The Neolithic *in* Simmons, I.G. and Tooley, *The Environment in British Prehistory*, 125-209 London: Duckworth.

Tomalin, D.J. 1979 Barrow excavation in the Isle of Wight. *Current Archaeol.* 68, 273-276.

Tomalin, D.J. and Scaife 1979 A Neolithic flint assemblage and associated palynological sequence at Gatcombe, Isle of Wight. *Proc Hampshire Field Club & Archaeol. Soc.* 36,25-33.

Tomalin, D.J. 1988 A mid-fourth century corn-drier at Packway, Newchurch, IW. *Proc. Isle Wight nat. Hist. Archaeol. Soc.* VIII, 43-55.

Tomalin, D.J. and Scaife, R.G 1987 The excavation of the first piped-water system at Newport, I.W. and its associated urban palynology. *Proc. Isle Wight nat. Hist. Archaeol. Soc.* VIII, 68-81.

Tomalin, D.J., Loader, R.D. and Scaife R.G. (forthcoming) *Coastal Archaeology in a Dynamic Environment: A Solent Case Study*. English Heritage monograph.

Turner, J. 1970 Post-Neolithic disturbance of British vegetation. pp.97-116 *in* Walker, D. and West, R.G. (eds.) *Studies in the vegetational history of the British Isles*. Cambridge University Press.

Waller, M. 1994 Paludification and pollen representation: the influence of wetland size on *Tilia* representation in pollen diagrams. *The Holocene* 4, 430-434.

Waton, P.V. 1982 Man's impact on the chalklands: some new pollen evidence. *in* Bell, M. and Limbrey, S. (eds.) *Archaeological Aspects of Woodland Ecology*. B.A.R. (Internat. Ser.), 46, 75-79.

Island Distinctiveness
A comparison with adjacent areas

FRANCIS ROSE

The Isle of Wight is separated from the botanically-rich county of Hampshire by The Solent, a narrow stretch of water which is nowhere wider than 8 km. Physical separation from the mainland occurred sometime around 7,000-8,000 B.P, but this period has been too short for any significant degree of endemism to have developed in the Island's flora; the bramble, *Rubus salteri*, is a possible exception. Nevertheless, the Island has a flora that is distinct from that of its neighbouring south coast counties; for example, Wood Calamint (*Clinopodium menthifolium*) is only found on the Isle of Wight today. On the negative side, many species do not occur. The absence of some species is unexpected whilst other absentees can be explained in part by the relatively small land area and scarcity of certain habitats such as wet heath, valley mires and chalk rivers.

The most notable feature of the Isle of Wight, as compared with all other south-eastern and eastern coastal counties of England, is the geomorphological complex along its southern coastline which has produced a splendid series of unstable cliffs, landslides and undercliff features. The open areas, the scrub areas, and also the older woodland, especially to the east of Ventnor, provide some excellent habitats for bryophytes and are, in effect, an extreme eastern outlier on the south coast of England of a more 'western' type of habitat.

The coastal cliffs, undercliffs and landslips have much in common with southern Dorset as a habitat for plants, although the rocks are all of Cretaceous rather than Jurassic age. Maritime communities are well developed and include species such as Wild Madder (*Rubia peregrina*), Rock Sea-spurrey (*Spergularia rupicola*), Oxtongue Broomrape (*Orobanche artemisiae-campestris*), Sea Spleenwort (*Asplenium marinum*) and Sea Campion (*Silene uniflora*) (as a cliff species). Rock Sea-spurrey does not occur to the east but is found on the mainland coast of Hampshire, west of Hurst Castle. Wild Madder extends into West Sussex, where it is rare and local, and it reappears in east Kent along the Dover to Folkestone cliffs. Sea Spleenwort is not found east of Dorset on the mainland. Shaggy Mouse-ear (*Pilosella peleteriana*) is a notable western species, which is also in Dorset but absent from Hampshire and Sussex, although there are old records from east Kent. Italian Lords-and-ladies (*Arum italicum* ssp. *neglectum*) and Ivy Broomrape (*Orobanche hederae*) are predominantly western species, common along the Undercliff. They are also both common locally in Hampshire on the chalk scarp.

The western influence is also apparent in the Bramble flora, which differs markedly from that in Hampshire. Six species with a strongly western distribution occur on the Island but not in Hampshire (*Rubus aequalidens, R. angusticuspis, R. corniubiensis, R. dumnoniensis, R. orobus* and *R. effranatus*) and other western species are commoner in the Island than in Hampshire (Allen, 1990).

The cliffs at St Catherine's Point have the maritime lichen *Anaptychia runciata*, known on the mainland no further east than Dorset. The lack of suitable hard coastal rocks may be the main reason for this, and may also account for the survival of *Lichina pygmaea* on a single large rock in Freshwater Bay.

The presence (at least until recently) of such liverworts as *Marchesinia mackaii, Cololejeunea rosettiana, Lophocolea fragrans*, Southbya nigrella, Cephaloziella baumgartneri*, C. stellulifera** and *Porella obtusata* is interesting. These are unknown on the coast of mainland Hampshire but, with the exception of *Porella*, they are all known on the coastal cliffs in Dorset. Those species marked with an asterisk are also known from the cliffs of East Sussex in the Fairlight area.

Among the mosses, *Philonotis marchica* is known nowhere else in Britain, and *Acaulon triquetrum, Microbryum davallianum* (previously *Pottia commutata*) and *Tortula atrovirens* (*Desmatodon convolutus*) also occur on the East Sussex coast. *Conardia compacta* is a moss that, on the south coast, is only known in Devon and south Wight. East Sussex has, in addition, *Dumortiera hirsuta* and *Tortula freibergii*, unknown on the mainland of Hampshire and also unknown in Wight. *Dumortiera* is a strongly oceanic – tropical species absent east of Devon, excepting at the rather special coastal ravine at Fairlight. *Bryum canariensis* is a southern species of dry,

The impressive jumble of greensand boulders which have given rise to Bonchurch Landslip provide a rich habitat for mosses and liverworts. H. Englefield, 1816. (AB)

sunny limestone found in Dorset and westwards, and in single sites in Wight, Sussex and east Kent. Another southern, largely Mediterranean species, *Scorpiurium circinatum*, is less restricted and occurs in rather sunny sites all along the south coast east to the 'Ragstone' cliffs (formerly sea cliffs), which lie to the west of Hythe in Kent. This species occurs in mainland Hampshire and Sussex on ancient limestone buildings (eg. St Cross Winchester, Pevensey Castle, Winchelsea Friary Chapel) in the absence of suitable natural hard limestone, as well as on the East Sussex cliffs.

Many other local, oceanic bryophytes however occur in Sussex, particularly on the sandrocks of the High Weald, but are unknown in Wight, where there are no massive outcrops of porous acidic sandstone. These include *Dicranum scottianum, Orthodontium gracile, Marsupella emarginata, M. sprucei, Kurzia sylvatica, Blepharostoma trichophyllum, Scapania gracilis* (also in the New Forest on wet logs), *Scapania umbrosa,* *Brachythecium plumosum* and *Dicranodontium denudatum*. The same applies to the Tunbridge Filmy-fern (*Hymenophyllum tunbrigense*), and it is strange that the Hay-scented Buckler-fern (*Dryopteris aemula*) is not known in damp ravines in Wight. It is frequent in the Sussex High Weald and in recent years has been found in several places in the New Forest and even in south Wiltshire, at Langley Woods.

The rest of the bryophyte flora of the southern part of Wight is rather unexceptional. This is due to the rarity of good heaths and bogs today, whatever may have been the case in past times. However, remarkable sites as rich as ever, like Bohemia Bog, do still survive. Many heath and bog species, such as *Dicranum spurium*, *Hypnum imponens* and *Ptilidium ciliare*, still in the New Forest and in several places on the Sussex heaths, are now unknown in Wight.

Heathland was formerly much more extensive in the Island, but it is mostly destroyed now. From old records and the small fragments left (e.g. on Headon Warren and Bohemia Bog) it is clear that there was once a respectable wet heath flora including White Beak-sedge (*Rhynchospora alba*), Broad-leaved Cottongrass (*Eriophorum latifolium*) and Pillwort (*Pilularia globulifera*). Pale Butterwort (*Pinguicula lusitanica*) and Ivy-leaved Bellflower (*Wahlenbergia hederacea*) are western species recorded from Wight but both are common locally in the New Forest and the latter species is also in the Sussex Weald. Wight is now unique among southern counties (excepting for Kent) in the small areas of heathland that remain.

The flora of the northern parts of Wight, on Tertiary strata, is essentially comparable with that of the Hampshire Basin and, at least historically, some parts would have shown similarities with Hampshire's jewel, the New Forest. Although the northern parts of the Island have, on the whole, a rather poor bryophyte flora, Parkhurst Forest in olden times when it was ancient pasture-woodland may well have had a flora close to that of the New Forest. However, today at least, such New Forest survivors as *Pallavicinia lyellii**, *Breutelia chrysocoma, Bazzania trilobata*, Saccogyna viticulosa** and *Plagiochila killarniensis* do not occur in Wight at all. (However, those marked * also occur in the High Weald of Sussex).

The lichen flora of the Island has few very distinctive features, whatever may have been the case in past times. Certainly, several woodland species have been lost, but *Lobaria pulmonaria* survives in a few sites where old woodland remains. This species is still occasional to frequent in the New Forest and hangs on in three old Sussex former deer parks. The extreme southern lichen, *Cryptolechia carneolutea*, was once found on elm when mature trees of that species were characteristic landscape features. It remains today on a few Ash and Field Maples, and it also survives on a few old Beeches in the New Forest.

Ancient woodland is reasonably well represented in the north of the Island, and yet the woodland vascular plant flora is somewhat impoverished. Small-leaved Lime (*Tilia cordata*), although once common, is restricted today to a single tiny copse in the west of the Island. It is also very rare in the New Forest but this, presumably, is due to the management regime of the Forest. Small-leaved Lime is very palatable to animals, and has probably been prevented from regenerating by grazing. However, many woodland species are completely absent from Wight. They include Lily-of-the-valley (*Convallaria majalis*), Wood Forget-me-not (*Myosotis sylvatica*), Herb Paris (*Paris quadrifolia*), Bastard Balm (*Melittis melissophyllum*) and Solomon's Seal (*Polygonatum multiflorum*). Most of Hampshire's earlier Holocene flora would have been able to spread easily to the Island but, by the time of the separation from the mainland, the full woodland flora might not have colonised the south Hampshire mainland. This may help to account for the absence of certain woodland species from the Island.

Narrow-leaved Lungwort (*Pulmonaria longifolia*) was however, able to colonise both banks of the Solent River before the sea-level rose. It is confined as a native in Britain to the basin of the prehistoric Solent River, on tertiary formations in east Dorset, south-west Hampshire and the Island, although it is frequent in north-west France. Historically, Wild Gladiolus (*Gladiolus illyricus*) had a rather similar distribution and a few Bramble micro-species may have shown a distribution confined to the Solent River basin (Allen, 1990).

The Rowridge valley is a botanically interesting area and has long been known as a good site for otherwise quite local plants. This dry chalk valley is the only British site for Wood Calamint today and it is possible that this could be another Solent River valley species. Interestingly another Calamint, the distinctive and rare form of Common Calamint (*Clinopodium ascendens* var. *baetica*), is frequent today on the chalk ridge at Lulworth in Dorset which was at one time continuous with the chalk ridge of the Island.

Downland is well represented and yet the chalk grassland is singularly lacking in some species that are widespread in Hampshire, Sussex and Dorset. Chalk Milkwort (*Polygala calcarea*) and Tor-grass (*Brachypodium pinnatum*) are quite absent and Dropwort (*Filipendula vulgaris*) is exceedingly rare. Round-headed Rampion (*Phyteuma orbiculare*) is a notable absentee. Most chalk orchids, with the exception of Pyramidal Orchid (*Anacamptis pyramidalis*) and Autumn Lady's-tresses (*Spiranthes spiralis*), are infrequent or scarce. Like Hampshire and Sussex, the Island's chalk downland has had (although very rarely) Man Orchid (*Aceras anthropophorum*) and Burnt Orchid (*Orchis ustulata*). In earlier times, Early Spider Orchid (*Ophrys sphegodes*) was apparently locally common on chalk near Bonchurch, as it is today on the coastline near Folkestone and Dover, and in East Sussex. In Dorset, it is confined to the Purbeck limestone.

By contrast, Early Gentian (*Gentianella anglica*), Dwarf Mouse-ear (*Cerastium pumilum*), Small-flowered Buttercup (*Ranunculus parviflorus*) and Bastard-toadflax (*Thesium humifusum*) are all remarkably frequent and widespread. Some of these species may well be favoured by the maritime climate, which exerts an influence right across the Island. Bastard-toadflax is unusual amongst chalk grassland species in its rather oceanic European distribution. In addition, elements of the flora exhibit a greater affinity with that of Purbeck than Hampshire or Sussex. Chalk grasslands towards the western end of the Island for instance, frequently have Saw-wort (*Serratula tinctoria*) and Betony (*Stachys officinalis*) as conspicuous members of the community. Their presence may be indicative of surface leaching of the soils.

To summarize, two major geographical features make the flora of Wight distinctive. These are the irregular coastal terraces of the south coast, with their western communities of vascular and cryptogamic plants, and the influences of the old Solent River and affinities with the New Forest seen on the Tertiary soils in the north of the Island. Floras never remain static, and recent modifications in weather patterns have resulted in changes to coastal habitats and the arrival and spread of some species. In the future, new species may colonise the shores of southern England. Cottonweed (*Otanthus maritimus*) is still abundant on the north coast of Normandy, east of Cherbourg. It formerly occurred in southern England and, with current climatic changes, it may occur again, as has Sea Knotgrass (*Polygonum maritimum*).

REFERENCES

Allen, D.E. (1990) The *Rubus* flora of the Isle of Wight. *Watsonia* 18 (1):21-31.

Brewis, A., Bowman, P. & Rose, F. (1996) *The Flora of Hampshire*. Harley Books. Colchester, Essex.

TABLE 1
Bryophytes recorded from the Isle of Wight but not recorded from mainland Hampshire.

Cephaloziella baumgartneri
Cephaloziella stellulifera
Cephaloziella turneri
Jungermannia caespiticia
Lophocolea fragrans
Lophocolea semiteres
Southbya nigrella
Porella obtusata
Marchesinia mackaii
Cololejeunea rossettiana
Fissidens celticus
Ephemerum serratum var. *minutissimum*
Leptodontium gemmascens
Gymnostomum viridulum
Pterygoneurum ovatum
Chenia leptophylla (alien)
Acaulon triquetrum
Hennediella macrophylla (alien)
Microbryum davallianum (*Pottia commutata*)
Grimmia ovalis
Ephemerum recurvifolium
Bryum canariensis
Philonotis marchica
Conardia compacta

TABLE 2A
Liverworts not recorded from Isle of Wight which occur on Hampshire mainland or in West Sussex and would perhaps be expected in Wight.

H = Hampshire mainland; NF = New Forest; S = Sussex

Bazzania trilobata
Calypogeia integristipula NF
Calypogeia neesiana
Calypogeia sphagnicola
Cephaloziella elachista
Cephaloziella turneri
Cladopodiella francisci
Cryptothallus mirabilis H/S
Diplophyllum obtusifolium H/S
Fossombronia foveolata
Fossombronia incurva
Frullania fragilifolia
Harpalejeunea molleri
Jungermannia atrovirens H/S
Jungermannia pumila H/S
Kurzia sylvatica S/NF
Leiocolea badensis H/S
Lophozia capitata H/S
Lophozia incisa H/S NF
Nardia geoscyphus H/S
Odontoschisma denudatum NF/S
Pellia neesiana H/S
Plagiochila killarniensis NF
Porella cordaeana NF
Preissia quadrata NF
Ptilidium ciliare
Riccardia incurvata
Riccardia palmata NF
Riccia cavernosa
Ricciocarpos natans
Saccogyna viticulosa NF/S
Scapania curta H/S
Scapania gracilis NF
Scapania umbrosa S
Targionia hypophylla H/S
Trichocolea tomentella NF etc

TABLE 2B
Mosses not recorded from Isle of Wight which occur on Hampshire mainland or in West Sussex and would perhaps be expected in Wight.

Aloina ambigua
Aloina brevirostris
Aloina rigida
Aphanorhegma patens
Atrichum tenellum
Brachydontium trichodes
Brachythecium plumosum
Brachythecium populeum
Breutelia chrysocoma
Bryum intermedium
Camplylostelium saxicola
Campyliadelphus elodes
Campylium stellatum
Campylium stellatum
Campylopus subulatus
Climacium dendroides
Dicranella rufescens
Dicranodontium denudatum
Dicranum flagellare
Dicranum fuscescens
Dicranum polysetum
Dicranum spurium
Dicranum tauricum
Didymodon ferrugineus
Didymodon nicholsonii
Drepanocladus cossonii
Drepanocladus polygamus
Drepanocladus revolvens
Ephemerum sessile
Ephemerum stellatum
Fissidens limbatus
Fissidens osmundoides
Funaria muhlenbergii
Hedwigia ciliata
Heterocladium heteropterum var. *flaccidum*
Hylocomium brevirostre
Hyocomium armoricum
Hypnum imponens
Leptodontium flexifolium
Leucobryum juniperoideum
Orthotrichum cupulatum
Orthotrichum sprucei
Palustriella commutata var. *falcata*
Philonotis arnellii
Philonotis calcarea
Pohlia drummondii
Polytrichum strictum
Pottiopsis caespitosa
Racomitrium elongatum
Racomitrium ericoides
Racomitrium fasciculare
Racomitrium heterostichum
Racomitrium lanuginosum
Rhytidiadelphus loreus
Sanionia uncinata
Schistostega pennata
Scorpidium scorpioides
Seligeria recurvata
Sphagnum contortum
Sphagnum flexuosum
Sphagnum magellanicum
Sphagnum molle
Sphagnum subsecundum
Sphagnum teres
Splachnum ampullaceum
Tetraplodon mnioides
Thuidium abietinum
Thuidium abietinum ssp. *hystricosum*
Thuidium delicatulum
Tortella tortuosa
Tortula cuneifolia
Tortula wilsonii
Ulota coarctata
Warnstorfia fluitans
Zygodon forsteri

A History of Botanical Recording in the Isle of Wight

DAVID E. ALLEN

CROSSING the Solent is a speedy and trivial affair these days, but for the pioneers of field botany in Britain it was a sufficiently lengthy and potentially complicated undertaking to ensure that they made the journey rarely, if at all. John Goodyer, who lived all his life in Hampshire and diligently described and recorded the plants of wide areas of that county from 1617 to 1656, has left no evidence among his numerous surviving notes of ever having visited the Island. We do know that WILLIAM TURNER did visit at least once, for he says as much in his *New Herbal* of 1562. That may have been in the course of the botanising tour he made along the Dorset coast twelve years earlier. However, it is odd that he chose to mention only one plant as having been seen by him in the Island, namely Wild Madder (*Rubia peregrina*). It is, moreover, suspicious that he recorded also seeing it in quantity in a part of Hampshire where it is much more likely to have been the cultivated *Rubia tinctorum*, which was once much grown as a crop thereabouts.

The few who did make the crossing seem invariably to have decided that the flora of the Island was unlikely to include sufficient that was novel to justify reserving more than two days at most for its investigation. That, at any rate, was all that THOMAS JOHNSON and his eleven-strong party of fellow apothecaries and their servants allowed themselves in July 1634, when they interrupted their return to London from a wide 'herbarizing' sweep of the West Country to cross from Southampton to Cowes in two boats. They devoted the first day to riding over to Newport, and engaging in some sightseeing; and the second to inspecting the south coast before hurrying on to Ryde. From there, they sailed to Portsmouth – a voyage that took 'a few hours', it is salutary to be reminded. The next year Johnson, the foremost English botanist of the day, later to lose his life in the Civil War, published his *Mercurius Botanicus*, the first real attempt at a flora of the whole of Britain. It incorporated the choicest of the inevitably meagre results of that all-too-brief foray, including Samphire (*Crithmum maritimum*) on the southern cliffs and Annual Mercury (*Mercurialis annua*) by the sea at Ryde. The latter was still 'on the sea-beach near Ryde, plentifully' a few years later, and provided JOHN RAY and his friend and patron, Francis Willughby, with the sole printed evidence of a similarly hasty trip that they are presumed to have taken in 1667. It was a brief diversion on the return lap of their second West of England journey. In the case of JOHN PARKINSON, another London apothecary, we have no clue to either the date or the circumstances of the visit that yielded two further records published in his *Theatrum Botanicum* of 1640. As he was not a member of Johnson's party, it must have been on some other occasion. Apart from Sea Campion (*Silene uniflora*) 'in many places by the seaside', he found a plant near what can be deduced to have been St. Catherine's Point that Bromfield was later able to recognize from the description as Slender Club-rush (*Isolepis cernua*). Yet another Londoner, WILLIAM COLES, also recorded Samphire, this time at Freshwater in his *Adam in Eden* (1657). He was responsible for at least one of four further plants, two of them fungi, listed for Wight in Christopher Merrett's uncritical compilation of 1666, *Pinax rerum naturalium Britannicarum*.

Given the Island's small population at that period, and the few people with learned interests that it could have been expected to include, it was probably too much to hope for that a resident botanist might emerge. It is possible that the REV. EDMUND POULTER, rector of Calbourne until 1785, was the one exception, for he was later to be unmasked as one of the anonymous authors of a list published in 1798 in a short-lived periodical, the *Annual Hampshire Repository*. Overoptimistically announced as 'the commencement of a Hampshire Flora. . . hereafter to be continued, and finally to be extended to a complete *Flora Hantoniensis*', this included a fair number of Isle of Wight records. More probably, though, his co-author was at least mainly responsible for those, for he is the only one of the two with a track record as a field botanist, with a herbarium to his name and elected (in that very year) to fellowship of the Linnean Society. This was THOMAS GARNIER, much later to be Dean of Winchester but not then yet ordained and scarcely out of university.

By that time, two very important developments had taken place. The first move had been made towards turning Ryde into a fashionable watering-place to rival Weymouth and Brighthelmstone. Secondly, as part of a more general awakening to the attraction of the countryside, field botany had risen to popularity in the higher reaches of society. It was greatly helped on its way by the simplification of the study brought about by the binomial nomenclature and the sexual system of classification devised and promoted by Linnaeus. The outcome was a steadily growing stream of elegant visitors, not a few of whom were ready and able to record plants of interest that they chanced upon in the course of their outings to admire the scenery and inspect the antiquities.

Already by 1805, as the pages of Turner and Dillwyn's gazetteer, *The Botanist's Guide through England and Wales*, bear witness, the Island was receiving visits from a goodly cross-section of the country's leading botanists: Withering, Stokes, Stackhouse, the Forsters, Joseph Woods – and various lesser lights as well. One, though, arrived from a less orthodox direction. In July 1801, on the eve of his historic voyage to Australasia on *HMS Investigator*, the young ROBERT BROWN and five of his shipmates eased their boredom during a frustrating wait at anchor in Spithead by slipping across on two occasions. The first time they walked from Ryde to Newport; the second time they explored the chalky bottoms near Shorwell. The list of eleven flowering plants, four seaweeds and a lichen that Brown jotted down was one of the more substantial made for the Island up to that time, but unfortunately it was to lie buried in his diary for two whole centuries until seeing print.

With botanical visitors now arriving in some numbers, there was a clear demand for a handlist indicating where the rarer and more interesting plants were to be found. In 1823, one duly appeared, as a section in a general guide, J. Albin's *Vectiana, or a Companion to the Isle of Wight*. Perhaps it was intended to be subsequently published separately as a book, *Flora Vectiana*, or to be turned into the equivalent of one by the binding of off-prints. This first-ever publication to be devoted exclusively to the Island's botany consisted of little more than a bare list of species with localities – what was known in those days as a 'catalogue'. Considering that the commoner plants were mostly omitted, the species total of 257 was impressive. Though the book came out anonymously, it seems to have been no great secret that the author was a local schoolmaster, WILLIAM DREW SNOOKE (1787-1857). A teacher of mathematics – the first of three who were to be prominent in Isle of Wight botany – the Dorset-born Snooke lived at that date in Godshill, but the greater part of his life was to be spent at Ryde. That he continued to be active botanically till at least the 1840s is shown by many records of that period standing to his name (or in a few cases to that of his eldest son Charles) in Bromfield's *Flora Vectensis*. Astronomy, however, was long a competing interest, and a guide to the stars and a set of tables for calculating eclipses were among the published outcomes of lectures he delivered on that subject, particularly to the Island's Philosophical and Scientific Society, of which he became a prominent member.

About the same time as the *Flora Vectiana* appeared, a regular ferry service between Ryde and the mainland was introduced, following the extension of the pier to permit access to 'the handsome and commodious vessels of the United Steam Packet Co' (to quote a publicity statement of the time). Marked increases resulted in the numbers not only of visitors, but also of those who came to live, whether all the year round or for the summer months only. Over the next forty years, the population of Ryde would treble while that of Ventnor, from a later start, grew more than tenfold. An influx of invalids, attracted by the much-trumpeted benign climate, poured into Ventnor in particular, 'the English Madeira' as it was proclaimed, and that, combined with a profusion of well-to-do gentlefolk, attracted in turn an influx of physicians.

Among these last were predictably one or two fired with an enthusiasm for botany through attending the field classes laid on at that period for medical students at most of the universities. To Ventnor came GEORGE ANNE MARTIN (1806-1867), already a recruit to the Botanical Society of Edinburgh and, in due course, to be a useful source of local records (as well as to donate a herbarium to the Philosophical and Literary Society, unfortunately long since lost to sight). To Ryde, similarly, in 1839, came the even keener THOMAS BELL SALTER (1814-1858), also fresh from an Edinburgh training. He had a catalogue published of the plants of the Poole area – and a declared ambition to produce a Flora of Dorset – to his youthful credit already. Close to his maternal uncle, Thomas Bell, a surgeon and zoologist of note who later would become one of the Linnean Society's great reforming Presidents, Salter thereupon began using Bell as a secondary surname (without the hyphen that Bromfield and others habitually added in error). This was presumably with a view to heightening his social standing in the rather grand circles in which he sought to build his practice. His father and two brothers, all naturalists as well as doctors like him, were content to remain 'Salter' *tout court*. It was an ambitiousness that found an echo in his botany too, for he decided to specialize in the newly-fashionable study of brambles. Collecting examples of those around Ryde and Shanklin that he found he could differentiate, he attempted to find matches with the descriptions and plates in a pioneering German monograph on the group. It was an attempt, though, doomed to failure, for Britain, we now know, has few kinds in common with the particular part of Germany concerned and in any case both countries possess very many more than botanists of that generation were ready to believe. The classification he devised and put forward was equivalent to trying to catch a swarm of locusts with nothing larger than a pocket-handkerchief.

A third botanically-inclined physician to arrive, three years ahead of Bell Salter, was well enough off already not to need to follow the profession for which he had trained. This was WILLIAM ARTHUR BROMFIELD (1801-1851), the outstanding figure in Isle of Wight botany in the two decades that followed. Bromfield came of an old Hampshire family, which boasted a Bow-bearer to King Charles II among its ancestors. Left independently wealthy by the early death of his clergyman father, he had spent the first four years after qualifying at Glasgow travelling on the Continent. Returning chronically restless, he and his unmarried sister then lived in turn at Hastings, Bristol and Southampton, before finding Ryde sufficiently to their taste. For the botanist in Bromfield, the fact that the Island had been investigated so far very incompletely, and that its size was large enough to offer ample scope while not so large that exploring it adequately would take dauntingly long, must also have been a particular attraction. Within a year of arriving, he was already bent on producing a Flora of the combined county of Hampshire with Wight (the two being a single entity administratively) and one, what is more, of a much greater thoroughness than any other county in Britain had seen hitherto. With few other calls on his time, studious by disposition, amiable, infectiously enthusiastic and energetic, he was admirably equipped for just such a task. In addition, he could hope to secure a good deal of assistance from among the disproportionate number of *rentiers* with time hanging heavily on their hands that the Island now held, who were conveniently well scattered around.

That band of helpers gradually grew over the years to the dimensions of a miniature army, with a social composition that ranged from a groom up to the widow of an admiral. In the eventual work, some fifty in all were to have localized records against their names, and there were no doubt still more whose finds were too run-of-the-mill to lift them out of anonymity. By present-day standards, such a number hardly seems impressive. It must be remembered though, that at that time, over the country as a whole, botanists were still

Portrait of Dr. William Bromfield, from the frontispiece of *Flora Vectensis*, 1856. (AB)

relatively few, and outside a limited circle their existence was not easy to discover.

Apart from Bromfield himself, the most prolific contributor of records – to judge by the sixty-four that stand to her name in the book – was a spinster in West Cowes, GEORGIANA ELIZABETH KILDERBEE (1798-1868). She and a sister had come to live in the Island on or before 1838. They were accompanied by their elderly father, a retired Suffolk vicar who had been sufficient of an intimate of Gainsborough in his Ipswich days to have become the owner of many of his canvases. Bromfield makes references to a herbarium of Miss Kilderbee's, which has seemingly not survived. Luckily, however, we have it substantially at one remove, for a younger cousin of hers, EMMA DELMÉ-RADCLIFFE, *née* Waddington (1811?-1880), often came over from Southampton to stay and, sharing her delight in plant-hunting, was taken to collect specimens from the same favourite haunts. The ten fat folios that comprise Herb. Delmé-Radcliffe are now among the Special Collections in Southampton University Library. They have recently been found to contain 156 specimens from the Island, about one-fifth of the entire collection, many corresponding in their localities precisely with those of finds attributed to Miss Kilderbee by Bromfield. Mrs Delmé-Radcliffe's connection with Isle of Wight botany has turned out in fact to have been a double one, for by coincidence her

The house in Dover Street, Ryde, where Dr Bromfield lived with his sister. (CP)

A sketch of Bromfield's grave in Damascus, date unknown. Reproduced courtesy of The Isle of Wight County Record Office.

husband was a nephew of Thomas Garnier. She, in her turn, evidently had her cousin over to stay with her from time to time, for many Hampshire finds by Miss Kilderbee feature in Bromfield's '*Flora Hantoniensis*'. It was to the opposite side of the Solent, outside Gosport, that the latter chose to move to spend the remainder of her days, following the death of her father in 1849.

The second of Bromfield's foremost stalwarts was an affluent bachelor in Newport, GEORGE KIRKPATRICK (1793-1861), a member of a prominent local banking family. Unlike Miss Kilderbee, not only does he appear to have refrained from collecting, but he was a practical conservationist well ahead of his time. He was seemingly the first person in Britain ever to set aside a piece of land he owned expressly in order to safeguard a rarity, one of the colonies of Galingale (*Cyperus longus*) close to St Catherine's Point. He went still further by having it ringed by a protecting wall – though misguidedly, for in the long run that has probably done more harm than good.

Another to score high among Bromfield's contributors was ALBERT JOHN HAMBROUGH (1820-1861), whose affluent architect father had built the mock-Gothic Steephill Castle, outside Ventnor, as a summer residence. This was a notable local landmark, later to be occupied by the Empress of Austria, before becoming briefly a school prior to its demolition in 1964. Hambrough's forays from that base were to yield two flowering plants new to the Island, as well as many seaweed records. It was from that same address that FREDERICK TOWNSEND (1822-1905) made his first public appearance on the Isle of Wight scene, in submitting a paper to the *Phytologist* in 1846. The two were first cousins and joint inheritors of a natural history tradition that ran in their closely interlocking families (in which two brothers married two sisters). Several of Townsend's early records from the Island that were to appear in Bromfield's book were, no doubt, made in Hambrough's company, and it was naturally to Townsend that his cousin's herbarium passed following his early death; consequently it was lodged, in due course, with Townsend's in the South London Botanical Institute (SLBI).

Although residents predominated among Bromfield's helpers, significant contributions came from visiting botanists as well. Foremost among these were two of the finest ones of that generation, both of whom lived just across in Sussex. These were WILLIAM BORRER (1781-1862) and, less well-known but a supplier of more than twice as many records, the REV GERARD EDWARDS SMITH (1804-1881). He held a series of livings in and around Chichester in 1833-43. Borrer's herbarium is in Kew (K) and Smith's in the Natural History Museum (BM), but specimens of theirs are in numerous other collections besides. The University of Oxford (OXF), on the other hand, holds the herbarium of WILLIAM WILSON SAUNDERS (1809-1879), another of Bromfield's major fellow recorders, the founder of a minor Surrey dynasty of botanists, horticulturists and flower painters. None of these or any other visitors, however, made a find to compare with that of the renowned philosopher and economist, JOHN STUART MILL (1806-1873). A holiday of his, sometime in the 1820s or 1830s, yielded the only known Isle of Wight specimen of the Purple Spurge (*Euphorbia peplis*), which still reposes, as the trophy of trophies, in Bromfield's herbarium. It is somehow fitting that the most distinguished individual ever to botanise in the

Island should have had the unique experience of chancing upon its rarest plant.

Unfortunately, after some ten years had elapsed, Bromfield's lust for foreign travel returned and he was increasingly absent for long periods, including over a year on a gruelling tour in North America. Alarmed that the projected Flora might never see print, friends seem to have prevailed on him to publish at least the principal meat. Because that consisted of numerous observations of interest to botanists in general, apart from the mass of purely local data, the leading journal of the day devoted to British botany, the *Phytologist*, proved happy to bring it out serially – for such was its length. Under the title of *Flora Hantoniensis* . . . it accordingly appeared, in 27 parts spread over three years. It was the first major Flora of any part of the British Isles to be published by that method, and still the only one to have been thus exposed to a national readership, as opposed to a merely local one.

Just as the last of these parts had come out, Bromfield's death suddenly occurred. He had contracted typhus while in Asia Minor and had never recovered. Thanks to his sister, though, much that still remained unpublished of, at any rate, his Isle of Wight work was not allowed to sink into oblivion. This included a set of very detailed descriptions of every species of flowering plant and fern that he had met with in the Island, drawn up from fresh specimens. Bell Salter and Bromfield's former botanical mentor in Glasgow, Sir William Hooker, agreed that these were too valuable for the botanical public to be deprived of them. They accordingly turned them into a publishable form, combined that with an edited version of the material serialized in the *Phytologist* that pertained to Wight, and in 1856 brought out the whole between two covers. Thus, five years after his death, one of the two halves of the full *magnum opus* that Bromfield had envisaged was effectively reincarnated, with the necessarily more limited title of *Flora Vectensis*. Its publication handsomely subsidized, one must suppose, out of Bromfield's ample estate, the volume faithfully retained his enviable, carefree expansiveness. Recorders' names were nowhere abbreviated or reduced to the vulgarity of mere initials, while the sites of rarities tended to be given with what seems today hair-raisingly detailed precision (for the author always wanted others to enjoy the excitement of seeing for themselves his premier discoveries).

Bromfield's effects also included an extensive Isle of Wight herbarium, carefully kept separate from his no less extensive general British one. These were both entrusted by his sister to the Philosophical and Scientific Society, in whose rooms in Ryde the joint collection

Portrait of Alexander More from his *Life & Letters*, 1898 (AB).

thereafter remained for many years, even after the building passed into different ownership following that body's later demise. Watched over devotedly by a succession of local botanists, and after latterly criss-crossing the Solent from one custodian to another, it was finally passed in 2001 into the secure care of Hampshire's County Museum Service at its central store in Winchester (HCMS).

The eventual publication of a county Flora is commonly succeeded by a rapid falling-away of the team brought into being by that shared objective and bound together by it. Luckily, that happened only in part after the appearance of *Flora Vectensis*, for only a month or two after Bromfield's death, another enthusiast emerged who proved able to sustain activity throughout the decade that followed. This was ALEXANDER GOODMAN MORE (1830-1895). After a childhood partly in Switzerland, he had come to live in the Island in 1842, when his well-to-do parents took a house in what was then the tiny hamlet of Bembridge. A brilliant record at boarding school had promised an equally brilliant one at university and an outstanding career in a profession afterwards. However, recurrent ill health, thereafter to be lifelong, ended the first of those prematurely and precluded the second for the foreseeable future. A passion for natural history, however, which developed in his teens and extended to ornithology and entomology no less than to botany, was to prove a lasting consolation and fruitfully fill the years of enforced leisure that followed. Like Townsend and many others of that generation, he had been inspired while at Cambridge by Babington's advocacy of a more sophisticated approach to the study of local plant distribution, and by *Cybele Britannica*, H.C. Watson's ground-breaking volumes to that end. Later in life, More was to be the author of an equivalent volume for Ireland and another for birds.

More started exploring his corner of the Island in 1852, though for a combination of reasons it was not for another four years that he was able to begin in earnest. Given much encouragement and help by Bell Salter, who chanced to be the family's doctor, he was soon making many noteworthy discoveries, in time for these to be included in *Flora Vectensis*. Unusually, the successive excitements these gave him can be re-experienced by readers today, thanks to the inclusion of many extracts from his diary in a 'Life and Letters' that was published (1898) as a memorial volume after his death.

In 1859, the Philosophical and Scientific Society published in its annual report 'a catalogue of flowering plants and ferns growing wild in the Isle of Wight; to serve as an index to Herb. Bromfield . . .', which More had devoted most of the previous summer to compiling, in the process adding to the collection many specimens of his own. Early the next year, he was invited by the Rev. Edmund Venables, a Bonchurch resident (not its rector), to edit a lengthy appendix on natural history for *A Guide to the Isle of Wight* by the latter, due to appear that July. More's personal contribution to this was a very thoughtful, 36-page essay on the vascular plants, in which he followed Babington's teaching – and repaired an omission of Bromfield's – by dividing the Island into five districts, based on the geology. This was accompanied by lists of mosses, lichens and fungi that the REV ANDREW BLOXAM (1801-1878) had noted on one or more visits from his home in Leicestershire to the Ventnor area. Of these, the list of lichens was not only the first for Wight, but historically valuable for including many species that would later fall victim to air pollution and inappropriate woodland management. Appended to that were 17 further species noted around Shanklin by another lichenologist vicar, the Shropshire-based THOMAS SALWEY (1791-1877). Venables's wife, Caroline, similarly added 22 species to the list of mosses. The first-ever list of the seaweeds, by Hambrough, rounded off this stimulating collection.

Disastrously, however, just as with Bromfield, the Island's hold on More thereafter began to loosen. A friendship dating from childhood with a family that owned an estate in Co. Galway had led to numerous stays over there and an increasing fascination with the fauna and flora of Ireland and the geographical enigmas they presented. Eventually, in 1865, he decided to go and live there permanently – though it was another two years before that became reality. Despite occasional brief return visits to his family, by then living in Ryde, Ireland commanded his more or less exclusive attention thenceforward.

To his credit, More did not sever his connection with Isle of Wight botany without first putting together the mass of records additional to those in *Flora Vectensis* that he and others had accumulated. The resulting paper came out serially (like Bromfield's 'Flora Hantoniensis') in the *Phytologist's* successor, the *Journal of Botany*, in 1871. A notable ingredient was the fruit of a fortnight's visit in 1868 by JOHN GILBERT BAKER (1834-1920), who had done much highly-regarded fieldwork in the North of England before his recent recruitment to the herbarium staff at Kew. A specialist in brambles, his brief entanglement with those of Wight was far more successful than Bell Salter's had been, for his specimens at Kew show that, despite a still straitjacketing taxonomy, he had discriminated a high proportion of the species now known to be present. The co-author with Baker of a Flora of Northumberland and Durham, GEORGE RALPH TATE (1835-1974), another Edinburgh-trained doctor, had also supplied many extra records for West Wight.

Not all of More's co-workers were aware by then of his disengagement, and at least one still entertained the hope that he would go on to produce a full-scale new Flora on the lines of *The Flora of Middlesex*, which had meanwhile appeared to much acclaim. This was ROBERT TUCKER (1832-1905), a teacher of mathematics like Snooke, but in this case at a school in London. Tucker had assisted with the Middlesex Flora too, and was deeply impressed by the very extensive coverage of the earlier literature in that volume. He set out to remedy that major deficiency of Bromfield's work by combing rare books in London's libraries to that end and reporting the outcome in *The Journal of Botany*. Alas, though, it was in vain; it was too late for More to be recaptured, even supposing that his health would have allowed his energies to be poured into something so much more demanding at that time.

After the loss of More, the investigation of the Island's flora spent the rest of that century doing little more than marking time. At the start of the 1850s there had been no fewer than five capable field botanists living in Ryde, for in addition to Bromfield, Bell Salter, Snooke and the up-and-coming More Hambrough was temporarily living there too. This was a greater number than probably any other town in the British Isles of comparable size could boast at that period. By 1870, all of those except More were dead, as were all the rest of the foremost resident members of Bromfield's team of collaborators. Only one solitary figure of any prominence still remained to link that golden era with the group that would emerge early in the century following. This was FREDERICK STRATTON (1840-1916), a native of Newport who was in practice as a solicitor there for almost all his working life. A man of great charm and kindliness, his particular helpfulness to beginners no doubt came naturally from being a father of eleven children. An expert climber in his youth, a regular painter and exhibitor of water-colours as well as deeply interested in church matters and missionary work, he nevertheless found time for botany too, from at least his mid twenties. In 1865 he recorded his acquisition of the copy of *Flora Vectensis* that is now in the Natural History Museum in London, which he was to annotate copiously for the remainder of his life. Four years later, he was elected a Fellow of the Linnean Society, began sending specimens to the Botanical Exchange Club and published the first of many short papers and notes that he contributed to the *Journal of Botany* over the years. His invariably excellent specimens, which he continued to collect up to the year before his death, have found their way into many British institutions apart from the University of Oxford (OXF), to which his herbarium itself eventually went.

For all that ceaseless activity, Stratton's published output must be accounted disappointing. Around 1875, he got round to producing the semblance of a new handlist by means of a series of weekly articles in the *Hampshire Independent*, but he was able to cite Townsend's then-impending *Flora of Hampshire with the Isle of Wight* as justification that this was not more substantial. He did indeed provide Townsend with several lists and many individual records for that work. When that Flora turned out to be scarcely more than a holding operation as far as Wight was concerned, however, the need for a fuller updating appears to have weighed increasingly on his conscience. In a pamphlet produced in 1900 for the benefit of visitors (a reprint of his chapter on the botany in Deacon & Co.'s *Court Guide to Hampshire and Dorsetshire* in 1897) he expressed 'still a hope of publishing a Flora for the Island'. That hope, however, was never in the end to be fulfilled. The reason was probably in part diffidence, for he had all along fought shy of the major critical groups, coverage of which had by then come to be regarded as obligatory for any county Flora worthy of the name and had conspicuously not been dodged by the taxonomically much more confident Townsend. To make up for his deficiency in that direction, he lured over from Bournemouth for a week in 1901 the country's acknowledged expert on the brambles, the REV WILLIAM MOYLE ROGERS (1835-1920). He put him up at his home and escorted him to the most promising localities — insofar as that was necessary, for Rogers had made two botanising trips to the Island already in past years, though with a less specialized intent. Despite such efforts, however, Stratton was to be outpaced by Townsend once again and apparently gave up the idea of a Flora, disheartened.

Townsend's *Flora*, ostensibly published in 1883 but in reality a year later, had come as a great disappointment for the botanists in the Island. When it first began to take shape (in 1872, by his own account), Frederick Townsend had been living and botanising in Hampshire for almost a decade, having settled there after his marriage. His visits to Wight to stay with his Hambrough relations were by then long past, that botanical cousin of his having died in 1861 when scarcely 40. Meanwhile, Bromfield and More between them had provided accounts of the Island's flora that gave the appearance of its having been very thoroughly worked. Initially, therefore, it struck him as sensible to omit Wight and confine his *Flora* to mainland Hampshire alone. A further consideration must have been that he had just then inherited his family estate in distant Warwickshire and moved there to live, which had placed the Island very much further beyond his reach.

At some point, however, Townsend changed his mind. Unlike him, Bromfield and More had not been concerned to tackle the critical groups — Bromfield, indeed, had been a 'lumper' and a militant one at that — and in *Flora Vectensis* the Island had been crudely divided, for ease of reference, into merely two halves instead of the considerable number of 'botanical districts' based on river basins that had since become standard in county Flora work. The need to rectify those deficiencies, coupled with the fact that copies of *Flora Vectensis* were hard to come by as well as expensive, seemed to him to justify including Wight after all. He had left that very late, though, for internal evidence suggests that the fieldwork he then felt obliged to undertake himself all had to be crammed into 1879, the year he had fixed on as the final one before writing-up. His visit that summer was even so very energetic, for he was able to assure the readers of the *Flora* that he had personally explored all six of the districts he had delineated, 'though by no means exhaustively'. His records bearing that date are from many parts of the Island little explored previously. There was no getting away from the fact, all the same, that the book, when it eventually made its much-delayed appearance (due to illness on Townsend's part), was largely a reprint of Bromfield and More so far as the Isle of Wight was concerned.

The very poorness of that coverage ought to have stimulated a renewed burst of activity, for it must have been obvious to any attentive reader of the book that the records were now quite old, and there were accordingly ample opportunities for making exciting further records. Unfortunately, too few botanists were now left to do much towards rising to that challenge. For some time now, the Island had been losing the full-time leisured with learned interests and the pensioners from the tropics to newly-emergent rivals like Bournemouth, which had the edge by not being cut off by sea. Two of Britain's then ablest field botanists, Moyle Rogers and E.F. Linton, the presence of either of whom could have revitalized the study in Wight, chose that neighbour in which to settle in those years. A shift in the cultural activity had also been proceeding within the Island itself. The more or less grand who had thronged Ryde previously tended now to be located in Ventnor, and it was from there that the little evidence of remaining botanical life most conspicuously emanated. JAMES HENRY AUGUSTUS STEUART (1834-1895), a retired army officer living there then and busy forming the Isle of Wight herbarium now in Liverpool Museum (LIV), was an Old Etonian married to the daughter of a peer. A mysterious C. PARKINSON, otherwise known only as a contributor of occasional pieces to one of the national periodicals, *Hardwick's Science-gossip*, also seems to have been among the wealthy leisured, to judge from allusions in those to frequent sojourns in the Alps. In 1880, the latter together with a no less obscure CHARLOTTE O'BRIEN (misidentified in reference works with a Co. Limerick botanist of the same name) authored a book published in Ventnor, explicitly aimed at 'temporary residents', with the misleading title of *Wild Flowers of the Undercliff*. The considerable number of localities it usefully provided extended to parts of the Island as distant as The Wilderness and even to St.Helens. That, though, was the best the resident botanists with a seeming superabundance of time managed to produce.

The new century continued at first discouragingly, for the publication purportedly in 1904 (but, more probably, again one year later) of a second edition of Townsend's *Flora* came as scarcely less of a disappointment than its predecessor had been. It might have better taken the form of merely a small supplementary volume such as Rayner later produced, thus greatly reducing the amount of proof-reading entailed. Townsend had not lived in Hampshire for 30 years by then, and had meanwhile innumerable competing calls upon his time. He had a large estate to watch over and the public duties that came with his election to the House of Commons, to say nothing of new-found enthusiasms for archery, photography and, latest of all, in his seventies, cycling! Although he had continued conscientiously to enter up additions to the Flora that came to his notice, he was an octogenarian by the time he got around to bringing them together for publication. Unsurprisingly, he evidently lacked the energy on this second occasion for writing to previous correspondents and contributors to solicit anything further they might happen to have. Apart from one or two records of his own, made on a return visit in 1883, all of the very few additional ones for Wight had the appearance of having been sent to him unprompted. The majority even of these came from a paper in a Swedish journal, of which its author, a certain THORILD WULFF, who had published the fruits of a stay of several weeks at Niton in 1876, had the initiative to send Townsend a copy. Of fresh records contributed by Stratton, at any rate directly, there was no sign. Nor was any excuse or apology proffered in the preface for this repeated, even worse neglect of the Island.

The highlight of that decade, rather, was the appearance in 1909 of Morey's *Guide to the Natural History of the Isle of Wight*. FRANK MOREY (1858-1925), a timber merchant in Newport, was primarily an entomologist but his interests were very broad and his contribution to this omnibus volume, which he also edited, extended to fungi and seaweeds. Stratton was

The first published photograph of the Island speciality, Wood Calamint, taken by Percy Wadham of Newport and published in Morey's *A Guide to the Natural History of the Isle of Wight*, 1909. (AB)

the natural choice to cover the flowering plants, ferns and stoneworts, but two new figures now made their first notable appearances on the local scene. JOHN FREDERICK RAYNER (1854-1947), a florist in one of the Southampton suburbs, was mainly responsible for the lengthy list of fungi. HERBERT MANN LIVENS (1860-1946) dealt with the bryophytes and lichens (the latter with the help of a leading national expert, J.A. Wheldon, a pharmacist in Liverpool) while also supplying some of the fungi and seaweed records.

Rayner was little short of an all-rounder botanically, though fungi, and latterly aliens, were particular enthusiasms. During the 1920s he would go on to publish further lists of Island fungi as well as two lengthy papers on the adventive flora of Hampshire and Wight combined. He is best known, though, as the author in 1929 of a book-length supplement to Townsend's *Flora*, a long-running project of the Hampshire Field Club which he was brought in to revive six years earlier. An energetic populariser of the subject, he was ever-ready to assist the natural history societies in the region, in most of which he held office at one time or another. His herbarium was given to Bournemouth Natural Science Society (BMH) but little of it can now be found.

Livens, by contrast, was interested only in cryptogams at that period. He and his brother had initially worked for their father, a colonial broker in London's Mincing Lane, but on both proving unsuited to that they had gone their separate ways. The brother became a painter of note (and a friend of Van Gogh, of whom he produced the earliest-known portrait), whilst he himself became a minister in first the Congregational and then the Unitarian Church. The frequent moves such a career entailed included a period at Newport in 1905-1908, of which his lists for Morey's *Guide* were presumably the product. While there he instituted the first labour exchange in the British Isles and began a visiting chaplaincy at Parkhurst Prison that he was to hold for 40 years. After a final ministry at Southampton he returned to the Island for his retirement in 1926, settling at Totland Bay. In those last years, he was much in demand locally as a lecturer on natural history, though chiefly on birds, his collecting of lichens having petered out in 1917, to judge from the specimens of his that survive in various herbaria, especially The Ulster Museum (BEL) and Liverpool Museum (LIV) as well as in his main collection in Bolton Museum (BON).

Livens also became one of the pillars of the Isle of Wight Natural History Society following the founding of that in November 1919. He published in the first volume of its *Proceedings* what was apparently his swansong as a cryptogamist in the shape of lists of bryophyte records additional to those he had contributed to Morey's *Guide*. The Society (which added 'and Archaeological' to its title in 1926) was another of Morey's initiatives in the main, and that man was to do the Island further great services by steering it through its first years as both honorary secretary and editor. Unlike its by then long-forgotten Ryde-based predecessor, a sedentary body with a notionally far broader disciplinary reach, this new one inherited the field tradition that had taken British natural history by storm in subsequent years and so automatically placed emphasis on an annual programme of meetings out-of-doors. That was more problematic than it sounds today, for though bus services were just coming in, most members still depended on the railway for travel to meetings. Because their homes were very scattered (those of some of the most active being in West Wight), this tended to work against good attendances. Increasingly, moreover, the resident naturalists were elderly, for the era was beginning when, in common with other islands, Wight would lose to the mainland a high proportion of its younger people, and those who remained or arrived would consist more and more of the retired. It was a matter of chance whether the latter included people with some specialist expertise.

Two botanists who came into this last category were JAMES GROVES (1858-1933) and ERIC DRABBLE (1877-1933), who came to live in Yarmouth in 1918 and Freshwater in 1925 respectively (though Groves moved to Freshwater Bay in 1923). Groves had spent his entire career in the Army and Navy Stores in London, turning himself in his spare time into a world authority on stoneworts, a specialism in which he was partnered by a brother who long predeceased him. Familiar with the Island from occasional holidays since boyhood, he chose Yarmouth because the Oligocene beds nearby had yielded fossil examples of those plants and he had hopes of finding more. Arriving with a bad reputation in botanical circles apparently for a habit of making cutting remarks, he and Drabble did not get on. That could explain why, alone of the two, Groves was active in the Natural History Society, often leading its field meetings and serving a term as President. On the other hand, he seems more than Drabble to have been at a loose end, for having handed over his herbarium to the Natural History Museum (BM) on retiring, he was unable to continue more than a modicum of research on the group that had been his consuming interest. That he joined the Moss Exchange Club at that juncture suggests that bryology was intended as a substitute, but it is unclear how far, if at all, that became a reality. Luckily, however, the aquatic species among the vascular plants had tended to be overlooked, and his habitual investigation of ponds and ditches for stoneworts yielded three new county records almost as soon as he had arrived. Luckily, too, though shy and retiring, he was effective on committees and involvement in local politics further helped to make up for the loss of his former scientific life in London, for which, however, he never ceased to yearn.

Drabble, by contrast, had all along kept a keenness for British field botany in tandem with his career as a college lecturer on quite other aspects of the science (it was a study of the root anatomy in palms that had gained him a D.Sc.). Forced to retire early by chronic asthma, he had that side-interest to fall back on when his health soon recovered in the Freshwater climate. The results were a substantial enrichment of his herbarium (now also in BM) with Isle of Wight species and a lengthy list of additional records published in the Report of the Botanical Society and Exchange Club, co-authored with the Newport based JAMES WALTER LONG (1864-1948), who retired at a similar time. Long's local botanising went back to 1914 at least, but had markedly increased following his retirement, from the civil service in his case. Specializing in adventives, he found the banks of the Medina a rich hunting-ground for those, thanks to the refuse from the docks regularly being dumped there. Acquiring a car, he extended his pursuit of alien casuals to other parts of Britain, ending up with a herbarium, when donated to BM, of some 10,000 specimens and many more besides in other institutions. Shyer even than Groves, he could be enticed for no more than the one term on to the Council of the Natural History Society, of which he had been an active member since its founding. To its *Proceedings* in 1930 he contributed a further list of records for rarer plants in the Island; many others of his are to be found as well in Rayner's supplement to Townsend's *Flora*.

Two bryologists who, unlike Groves, were certainly active at this period also merit a mention. The one who was a resident, PERCY LONG (1882-1944), lived all his life in Newport, and was employed in its Royal Brewery on the clerical staff. He, too, was a founder member of the Natural History Society and served on its Council, but his long walks all over the Island in search of mosses, which added several species to the Isle of Wight list, necessarily tended to be a lone pursuit. A member of the British Bryological Society from 1928 until his death, his forays took in other parts of Britain over the years, resulting in an extensive herbarium, which, together with his notebooks, ultimately passed to the Society. HENRY HERBERT KNIGHT (1862-1944), on the other hand, while primarily a bryologist, took an interest in lichens, fungi and vascular plants as well. He was the third leading contributor to Isle of Wight botany to have been a teacher of mathematics. After graduating at Cambridge as Seventh Wrangler and being awarded a college fellowship, he joined the staff of a school in Carmarthenshire, only to retire at merely 45 and go to live with his mother in Cheltenham. There he worked intensively on the projected Flora of Gloucestershire, covering the county systematically by bicycle. For many years he was prominent nationally, serving the British Bryological Society as referee for hepatics and eventually as President. His secondary concern with lichens concurrently involved him hardly less prominently in the British Mycological Society too. It is unclear when or how his Isle of Wight connection began, but the many records standing to his name in Livens's 1926 paper suggest they were the fruit of more than one visit. He was subsequently to revise and extend the Island's lichen list in a paper of his own in the Natural History Society's *Proceedings* in 1933.

The existence of the Society had imparted a greater degree of cohesion to the Island's natural history community, and once again provided that with a regular local outlet for placing on record finds that might otherwise have escaped notice or not even have been made in the first place. It also drew in one or two individuals who would otherwise have worked in isolation, and perhaps failed to realize that what they noted could be of wider interest and relevance if seen in a wider context. One such person was GLADYS HILDA DOROTHY BULLOCK (1899-1980), who after an upbringing at Havenstreet spent her entire working life teaching botany and, latterly, biology more broadly to generations of secondary schoolchildren in Ryde. One of them was Mary Harmsworth, the future wife of Reg Kettell, so that unwittingly she played an indirect part in bringing the next Isle of Wight Flora to fruition. Joining the Society right at its start, when scarcely out of school herself, she continued as one of its most devoted members until illness and frailty finally intervened. She was repeatedly elected to the Council and was ultimately made a Vice-President. Extremely self-effacing, volunteering little on outings unless directly asked, venturing into print to the extent of no more than the occasional brief report of a new record, she nevertheless kept a modest herbarium, eventually took on the recording in her home 10km square when the Distribution Maps Scheme came along and, betraying an underlying tenaciousness, was fond of forcibly declaring that her 'Bentham and Hooker' would accompany her to the grave. And maybe it did.

Advancing age alone would have ended the active work of those who had brought about what may appropriately be called 'the Morey revival', had not the Second World War done so in any case. For part of that the Island was in the front line, extensive areas were barred to civilians throughout and a high percentage of the population was either away from home in the services or had to exchange much of its normal leisure for war work of some kind. Anyone lingering in bushes and carrying a conspicuous tin with an unknown function was as liable to be arrested as was anyone using binoculars. On the other hand there were compensating advantages for the vegetation: as JOB EDWARD LOUSLEY (1907-1976) noticed, on resuming his pre-war visits in 1947, certain orchids had multiplied enormously where downland had been closed to public access. Lousley was a bank employee in London all his career, who turned himself from schooldays into one of the most knowledgeable and productive amateur investigators of the British flora of all time. He devoted a special five-page section to the Island in his subsequent influential 'New Naturalist' book, *Wild Flowers of Chalk and Limestone*. Active on the national scene from 1927, he was to be the chief architect of the post-war transformation of the Botanical Exchange Club into the much-broadened, far larger Botanical Society of the British Isles (BSBI), of which he was eventually President. When it introduced its system of vice-county Recorders, he agreed to act for Wight in the absence of any obvious candidate resident on the Island, having acquired a good knowledge of its vascular plants over the years, and throughout the 1950s acted in that capacity. Energetic in advancing conservation, he also ran for a period the sadly short-lived Council for Nature. Rather paradoxically, in view of that, he was unable to rein in a passion for collecting he had developed before that became generally frowned upon, eventually leaving to the University of Reading (RDG) probably the last big private herbarium in Britain, comprising nearly 25,000 gatherings, among them many from Wight.

The dominant event in British field botany in the early post-war years was the BSBI's massive Distribution Maps Scheme. This produced, in 1962, the first of what have proved to be a long succession of atlases showing the distribution within the British Isles of various plant and animal groups, all thanks to the wartime division of Britain into a grid of kilometre squares and the subsequent arrival of automatic data-processing machinery. At the start of the scheme, in 1954, ALICK WILLIAM WESTRUP (1910-1964) volunteered to co-ordinate the recording for the 10km grid-squares covering south-east Hampshire and the Island. This included helping to recruit teams in both areas for that purpose – as well as taking on the larger share of the work himself. Following a career as a research chemist, Westrup had recently come to Portsmouth to teach at what was then its Municipal Technical College (now the University). With the exceptionally long vacations that came with such posts then, almost inexhaustible energy and near-freedom from domestic ties, he was nearly as favourably placed as Bromfield had been for such a task. One modern advantage he had over Bromfield, though, was the ownership of a motor-cycle – complete with a sidecar to hold his field equipment. Out of this work for the BSBI scheme grew the idea of producing a new Flora of the two counties combined (following Townsend's example) on this novel, properly systematic basis but using the much smaller mapping unit of the 2 × 2 kilometre grid square known as a 'tetrad', then just coming into favour for compiling Floras at county level. Westrup, the initiator, lost no time in making a major start on this, only for his death to occur after work had been in full swing for less than five years, throwing the continuance of the project into jeopardy. Luckily, the data he had accumulated had been bequeathed by him to the recently formed Hampshire and Isle of Wight Naturalists' (now Wildlife) Trust, which he had been personally instrumental in establishing.

The Trust was keen that the work should be carried on, in the interest of better-informed local conservation policies, and turned to one of the local BSBI Recorders, MARGARET PHOEBE YULE (1892-1981), to take Westrup's place. Elderly by then and latterly in poor health, she unfortunately found the huge clerical burden that this entailed too much, and had to step down in 1974. The consequent further hiatus, combined with the realization that attempting to cover both counties by such a demanding method had been unrealistically ambitious, led to a decision to restrict the project thenceforward to Hampshire alone. This left the Island's botanists to work toward a Flora of their own, independently and by whatever method they considered feasible. By then, though, the Island team had disbanded, demoralized by the increasingly lengthy breaks in continuity and in the absence of any assurance that publication would crown their efforts. There were two, however, who were keen to persist. These were BILL SHEPARD (b.1921) and JAMES HENRY BEVIS (1904-1976), who had first met in 1963, at a meeting held in Newport to co-ordinate work on the mapping. The two had come to Isle of Wight botany by very different routes. Shepard was Newport-born and bred, and had found solace in natural history from a succession of unappealing jobs, as a carrier, sales representative, advisor on rodent destruction and crematorium employee in turn. Long a Natural History Society member, he had been converted from bird-watching in 1961 by the sheer enthusiasm of the group then embarking on Westrup's project. Subsequently, he was appointed BSBI Recorder for Wight, and thereby was responsible in any case for keeping a note of all new records and sending in the more notable ones for publication in one of the BSBI's journals. Using those as bricks for the building of a Flora merely invested that routine with a more immediate purpose. Jim Bevis had studied botany for his London University degree prior to over 20 years in the colonial education service, and had only just come to live at Lake in his retirement (though he had several times holidayed in the Island). His more scholarly approach usefully complemented Shepard's greater familiarity with the local scene. Unexpectedly, the pair then became a trio. Shepard learned quite by chance of the existence of REGINALD EARL KETTELL (1903-2001) and of his uniquely extensive knowledge of where the rarer plants grew, quietly built up over many years. Reg Kettell's recruitment made the mix of backgrounds more colourfully variegated still, the two halves of his working life, in the fruit and flower trade and in soft drinks respectively, having been interrupted by lengthy periods at sea.

After a decade and a half of toil by these three, with much assistance from others, the intended book, *The Flora of the Isle of Wight*, appeared in 1978, its publication made possible by the Natural History Society's financial support. It was dedicated to Bevis, who, tragically, died suddenly as the manuscript was receiving the final touches. Not a county Flora on the classic scale, with little introductory matter and very limited mention of older records, its consequently handier size fitted it well for field use. All the previous accounts of the Island's vascular plants being long out of date by then and consultable only in libraries, the very full picture it provided of what was currently known to exist had been much needed, and predictably produced a marked revival in interest and activity. So numerous were the additional finds yielded by the extra fieldwork directly generated, and so many the old, overlooked specimens in mainland herbaria that were notified to Shepard, that he soon found it necessary to 'do a Rayner' and put together a supplement, which was published in 1985 in the Society's *Proceedings*.

While the appearance of 'Bevis, Kettell and Shepard' was clearly the culminating achievement of the post-war period, and one which the Island's botanical community as a whole had had a share in, there were also some contributions in other directions that ought not to go without a mention. One of these was by ERNEST HERBERT WHITE (1877-1959), a schoolmaster 'across the water' in Southsea who frequently came over to visit a sister in Shanklin and developed a wide and deep interest in the Island, building up a valuable library on its history and other aspects. A member of

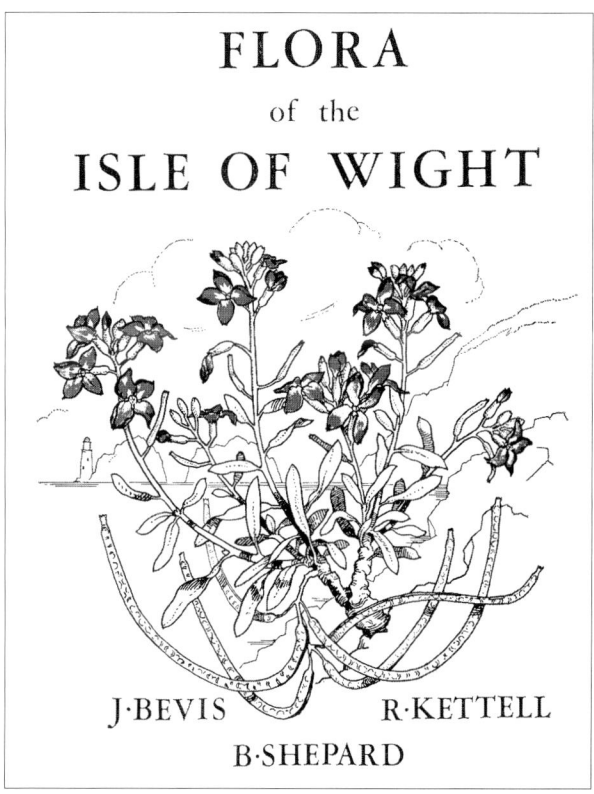

The cover of *The Flora of the Isle of Wight* features Hoary Stock against a backdrop of The Needles. The drawing was by Pamela Freeman who lived at Bembridge. She painted the first 32 colour plates of the hugely influential *Collins Pocket Guide to Wild Flowers* by Messrs. David McClintock and R.S.R. Fitter, 1956.

the Natural History Society from 1937, he became Editor of its *Proceedings* in 1949, later contributing to that a report on the current status of various local orchid species. He was serving a second term as President when his death suddenly took place in circumstances he himself would surely have considered fitting: at the end of a strenuous field meeting on a very hot day. A famously energetic leader of botanical rambles, he had a special pride in out-walking the youngest and fittest. Another Society stalwart over much the same period was THELMA WHITE (1901-1991). Elected a Vice-President in 1964 after eight years as a Council member, she also for thirteen years ran its Botany Section. A schoolteacher who had chosen to specialize in the subject from the very first, like Miss Bullock, Miss White acquired her interest during a childhood in the Weald of Kent, helped with identifications by a botanically knowledgeable local pharmacist. She came to live in the Island in 1931, having exchanged a post at Bath for one at Ventnor. A herbarium she started in her schooldays was unfortunately lost in the wartime bombing.

Looking back over the nearly four and a half centuries since the Island received its first known visit by a botanist, it is difficult not to be struck by the untypically thorough exploration it received early on, and by the equally untypical length of time that elapsed before it was thoroughly explored afresh. Alas, though, offshore islands tend to be as marginal culturally as they are physically, ever at the mercy of social and economic tides that flow in from the mainland and out again.

No less striking is how much of what has been achieved down the years has been due to twists of fate. Had not Bromfield lost his father while young and been left independently wealthy, he would have had to practise as a physician and never would have had the limitless leisure that enabled him to accomplish what he did. Had More not suffered chronic ill health from so early in life, he too would have been robbed of leisure by the demands of a profession and might well never have lived in the Island at all. Had only Townsend been persuaded by his cousin Hambrough to join him by settling in Wight instead of Hampshire on his marriage, the two editions of his Flora might have had the same beneficial lopsidedness as Bromfield's earlier venture. It was mere chance, too, that brought Livens to the area in the first place and led Groves and Drabble to choose the Island in which to recover their health in retirement. Equally, many another botanist who might have contributed much to the study of its flora instead chose not to. Truly, the Isle of Wight has had its share of luck, and it must be hoped that it will continue to do so.

Island Habitats – Past and Present

COLIN POPE

Many of our wild plants show characteristic distribution patterns resulting from the varied topography, geological formations and superficial deposits of the Island. The most distinctive features which strongly influence the distribution of plants today are the northern dissected plateau of heavy clay soils, the chalk outcrops, the southern bowl with acidic, sandy soils, and the river valleys and their flood plains.

Historically, these differences would have been far more apparent. Ancient woodlands and flower-rich meadows would have been widespread on the clays of the northern plateau. Large areas of heathland (the heaths) and acidic wetlands (the moors) were found across the southern bowl and superficial gravel deposits. On the chalk outcrops, there would have been extensive tracts of grazed downland, with older secondary woodlands in the coombes. Although this distribution of habitats is still evident today, it was more so when Bromfield was compiling his *Flora Vectensis* in the first half of the nineteenth century. Historic records have been very useful in interpreting the habitats and species present today and it is clear that the greatest habitat losses have been from the sandy soils of the southern bowl, which were ripe for agricultural improvement.

The following account describes the current, and sometimes historic, range of habitats occurring on the Island, outside the built-up areas. Many of the sites referred to can be viewed from the Island's very extensive public footpath network, and some have open public access, but this is not always the case. It should *not* be assumed that the mention of a site in this account implies that it is open to the public. Some of the richest sites are in private hands and must only be visited with the landowner's permission.

1 SOLENT COASTAL HABITATS

Archaeological research has shown that the coastline experiences alternating cycles of erosion and deposition. Currently, the Solent coastline is undergoing a period of erosion with a loss of saltmarsh and a rapid change in shoreline habitats. The numbers and distribution of plants can change dramatically from year to year as a result of these coastal processes. All of the Island's main rivers discharge into the Solent and, where they reach the sea, saltmarshes are established at the mouths of creeks and inlets. Saltmarsh dominated by Cord-grass (*Spartina*) also used to occur on the foreshore between the tidal inlets at Fishbourne and Quarr but has virtually been lost through coastal erosion. Although Common Cord-grass is the dominant species in some saltmarshes, good mixed saltmarsh survives locally in a number of places including the Western Yar estuary, Thorness Bay, a few spots in the Medina, King's Quay, St Helen's Millpond and, pre-eminently, at Newtown.

The Newtown estuary provides the greatest range of unspoilt estuarine habitats and the greatest species diversity. It is perhaps the best surviving example of an undeveloped Solent estuary, and much of it can be observed from public footpaths. Perennial Glasswort occurs in almost all of our saltmarshes and Lax-flowered Sea-lavender is more local, but not infrequent. Sea Wormwood is found locally in several spots around the Newtown Estuary but another site, at St Helen's Millpond, has recently been lost through erosion. Around the Clamerkin at Newtown, the native Small Cord-grass still survives in quantity, being particularly found around small pools in the saltmarsh. This is now its principal site on the South Coast of England. It can be easily seen on the marsh below Town Copse.

A more diverse range of species occurs in the upper saltmarshes. Marsh Mallow is a common component of this habitat whilst Golden Samphire is very local, although it can be readily found at Newtown. Divided Sedge is a particular feature and occurs in quantity in almost all of our upper marshes. Historic earth banks at the back of the marshes frequently have an interesting flora. It is here that Slender Hare's-ear grows; this can be an elusive species but the Newtown Estuary is a good place to look for it. Dry grassy banks where the vegetation is thin can also be productive places to search for clovers and annual grasses.

Shingle spits guard the mouths of all our estuaries. They are always interesting places to botanise but are particularly unstable habitats. Vegetated shingle communities also occur on the beaches at Osborne and Thorness Bays. Sea Kale has appeared with increasing

A George Brannon engraving of Ryde Dover c. 1824, looking towards Appley. This was a botanists mecca until it was destroyed in the mid 1800s with the expanding eastwards spread of the town of Ryde. (AB)

frequency, and Ray's Knotgrass appears regularly on shingle at Norton Spit, Thorness Bay, King's Quay and elsewhere. There was great excitement when several plants of Sea Knotweed were discovered at Thorness Bay in 1995, at a time when this plant was reappearing on coasts in Hampshire and Sussex. Unfortunately, hopes that it would become established have not been realised. On Hamstead Spit, where the shingle spit is migrating back onto the saltmarsh, Sea Heath can be found growing through the shingle at the interface with the saltmarsh. It appears to be particularly vulnerable to the current period of coastal instability, and it is in rapid decline.

Where shingle beaches are intermixed with sand, the flora can be diverse and St Helen's Duver, Hamstead Spit and Norton Spit are particularly rich sites. St Helen's Duver was the site of a nine-hole golf course from 1883, until it was handed over to the National Trust. It has for long been considered the richest locality for plants for its size, and it is a worthwhile site to visit. Until recently, it was the only place on the north coast to see Sea Campion, Fenugreek, Clustered Clover, Suffocated Clover, Fragrant Evening Primrose, Bur Chervil, Autumn Squill, Dune Fescue and Bearded Fescue. These plants are still found here, but several of them have also spread to other sites in recent years. At the mouth of the harbour, the spit is still growing. This is an excellent place for clovers and sand dune grasses, and it is the principal Island site for Prickly Saltwort.

Sea Holly, once common on the Duver, has gradually declined as a result of human interference and is now very rare. However, it still survives in good numbers on Norton Spit at the opposite end of the Island. This area, accessible by public footpath, is a good site for many sand dune specialists including Sea Bindweed, Dune Fescue and Bulbous Meadow-grass, and it is the only Island site for Sand Cat's-tail.

A 'Dover' or 'Duver' is an Island name given to a low-lying sandy area by the coast that is occasionally breached by the tides. Ryde Dover, the name given to the low-lying sandy ground at the eastern edge of Ryde, was made famous by Bromfield who lived nearby and visited it regularly to collect plants. The discovery of the very rare Childing Pink, together with a whole host of sand dune specialists, made Ryde Dover something of a mecca for Victorian botanists. No sooner were its riches realised, then they were lost to building development and reclamation as the popularity of Ryde as a resort grew. Remarkably, many sand dune plants can be found today on the south facing sandy bank of Ryde Canoe Lake, which is situated in the general vicinity of the original Dover. It is worth searching this bank for, amongst others,

Bulbous Meadow-grass, Clustered Clover, Suffocated Clover, Fiddle Dock, Bearded Fescue, Divided Sedge and Early Meadow-grass.

Brading Marshes is another site that underwent great change in the late nineteenth century, when a huge area of estuarine habitat was reclaimed from the sea. The mudflats and saltmarshes of the past have been replaced by one of the largest areas of coastal grazing marsh along the South Coast. Although this is a young habitat, it is of great interest and complexity and its richness is in no small part due to the surrounding semi-natural habitats. The reclaimed mudflats and sand-banks have each produced differing plant communities; the riverbed of the original River Yar meanders is also interesting. Large areas of marshy grassland have extensive tracts of Divided Sedge, and Slender Spike-rush, Narrow-leaved Bird's-foot Trefoil and Bulbous Foxtail are also frequent. A few of the wet ditches have Mare's-tail and Lesser Water-plantain, and Brackish Water-crowfoot is frequent in ponds. Water Dock, Distant Sedge and Marsh Arrowgrass occur along old river-bed meanders. Bembridge ponds, near the present day harbour, are actually brackish lagoons containing the submerged aquatics Foxtail Stonewort, in its only Island site, Fennel Pondweed and Beaked Tasselweed. The pond margins support Marsh Mallow and Brookweed. Brading Marshes is now an RSPB reserve and access is restricted.

Specimens of the rare Childing Pink collected on Ryde Dover by William Bromfield in June 1843. (AB)

2 WOODLANDS

Broadleaved woodland tends to be dominated by Pedunculate Oak and Ash in varying proportions, and Field Maple is generally frequent, particularly on woods on the heavy clay soils and chalk. Pedunculate Oak woods with birch and bracken are characteristic of the sandy soils on the south of the Island, whilst wet woods, supporting more moisture-tolerant species such as Willows and Alder, survive in the river valleys.

Ancient woodland, namely those woods which are believed to have had continuous woodland cover since at least 1600 AD, have existed for so long that they have developed a rich and diverse flora. The northern half of the Island is studded with ancient woodlands, and those on tertiary strata tend to support the greatest variety of flowering plants. Some fine old woods extend down to a natural shoreline adjoining tidal marshes, as they must have done for hundreds of years. This is well seen around sheltered inlets at Newtown and Wootton Creek. Remains of oak trunks and roots embedded in the intertidal muds between Ryde and Wootton have been radiocarbon dated back to 3000 – 4000 BC, providing evidence that at one time these woodlands extended much further out from the current coastline, and marine erosion continues to erode the wooded cliffs.

Firestone Copse, Briddlesford Copse and Combley Great Wood, situated around Wootton Creek, are our most species-rich woods. Firestone Copse and Combley Great Wood are accessible along forestry tracks, and they have Wild Service, Greater Butterfly Orchid, Broad-leaved Helleborine, Wild Daffodil, Sessile Oak, Purging Buckthorn and Narrow-leaved Lungwort. Very few vascular plant species appear to be so closely confined to ancient woodland that they can be regarded as reliable 'ancient woodland indicators' on the Island but Wild Service, Narrow-leaved Lungwort and Sessile Oak appear to be the most dependable.

Nearly all woodland on the Island was historically coppiced, a practice which has greatly declined today but is often evident by the presence of old coppice stools. The abundance of Narrow-leaved Lungwort, a native plant confined to the Hampshire basin, is a particular feature of actively coppiced woods. Wood Anemone also has a strong affinity for ancient woodland with a long history of coppice management. It occurs in carpets in Town Copse at Newtown but is rare in the adjoining Walter's Copse, which is secondary in origin. Small-leaved Lime is another good ancient woodland indicator, but it is far too rare on the Island to be of much use in this respect. It is confined to a single, small coppice wood on clay at Tapnell. Its present relict status almost certainly results from past patterns of land-use, for the sub-fossil pollen record indicates that this was once an abundant tree across the Island on a range of soil types.

Parkhurst is a fascinating woodland complex with affinities to the New Forest, and it is the largest forest on an ancient site on the Island. Even so, it represents only a fragment of the extent of the former Royal hunting forest, which in medieval times seems to have extended westwards towards Swainston. The area was gradually reduced by enclosures and its present extent dates principally from the 1812 Enclosure Act. This much-reduced Forest still encompassed ancient pasture woodland, lying largely on more fertile Tertiary clays and loams, and wet and dry heathland on the poor soils of plateau gravels. Woodland rides on acidic clays and gravels often have an interesting flora of heathland plants in addition to those characteristic of woodlands, especially if the wood is actively managed. Parkhurst Forest has by far the best examples of species-rich heathy rides, and has the advantage of being freely accessible. Dry heathland and neutral grassland survive in many of the rides in the southern part of the Forest. Heather, Bell Heather, Dwarf Gorse and Burnet Rose are widespread and Saw-wort, Devil's-bit Scabious, Meadow Thistle, Lousewort, Lesser Skullcap, Sneezewort and a range of sedges can be locally frequent. Bilberry survives in diminishing quantity beneath Beech in one or two spots. Although wet heath no longer survives, scattered plants of Cross-leaved Heath sometimes appear following plantation clearances. Rides in Fattingpark Copse are also home to scarce Island plants such as Chaffweed, Water-purslane, Creeping Willow, and Common Yellow-sedge.

In the early decades of the nineteenth century, much of Parkhurst Forest was replanted with oak, including much Sessile Oak. In the twentieth century, plantations of conifers have heavily modified the landscape. Nevertheless, a great deal of interest survives. Around Mark's Corner, at the north end of the forest, ancient pasture woodland dominated by Sessile Oak, Beech and Holly can be seen. However, the woodland has been much augmented by planting and sadly, only a handful of ancient trees survive. Scattered pockets of older (late eighteenth century) trees support ancient woodland lichens such as *Arthopyrenia ranunculospora*, *Byssoloma leucoblepharum*, *Phaeographis inusta*, *Punctelia (Parmelia) reddenda*, *Thelotrema lepadinum* and *Usnea florida*. It is for this reason that the Forest as a whole is considered to be the richest Island site for this group of lichens. The striking moss, *Hookeria lucens* grows in damp spots and the liverwort *Frullania tamarisci* is a frequent epiphyte of oak trees.

Wood-pasture was managed woodland in which farm animals or deer were allowed to graze. Although historic documents suggest that livestock were often not effectively excluded from recently-cut coppice, proper working wood pasture appears to have always been scarce on the Island, and none survives today. A distinctive and often spectacular lichen community is associated with historic wood-pasture and Victorian lichen records indicate that Appuldurcombe Park must have been such a place with trees supporting luxuriant growths of *Lobaria*, *Sticta*, *Pannaria* and *Degelia* species. There are no twentieth century records of these species from here, although the trees are believed to have survived into the mid 1900s. At Northpark Copse, a private wood at Swainston, the surviving lichen flora and the abundance of Thin-spiked Wood-sedge suggests that this wood had its origins within a deer-park. This has been substantiated by historic research. Although it lacks big, old trees, the wood is remarkable for the extremely luxuriant growth of some ancient woodland indicator lichens on ash trees, in particular *Lobaria pulmonaria*, *Dimerella lutea* and *Catillaria atropurpurea*.

There are also local concentrations of ancient woodland lichens in woods on the Briddlesford estate. In Briddlesford Copse, a privately owned wood, *Lobaria pulmonaria* survives on a single old oak tree by the Blackbridge Brook and on higher ground by the railway track, an area of oak woodland has *Thelotrema lepadinum*, *Thelopsis rubella*, *Pachyphiale corneola* and *Schismatomma niveum*. Nearby Hurst Copse has several of these species together with *Bacidia epixanthoides* and *Agonimia allobata*.

Ancient woodland lichens are particularly threatened today as they seem unable to spread either to new sites or within existing ones. The fate of the conspicuous *Lobaria pulmonaria* illustrates this well. In Northpark Copse, it has declined from 13 trees to 7 since 1980 from a combination of gale damage, tree death and overgrowth by ivy. At Mudless Copse, Swainston, it was confined to just two old ash trees at the edge of a field. The unfortunate felling of these trees in 1981 resulted in the loss of two lichens, *Agonimia octospora* and *Collema furfuracea* from the Island list, together with some of the best *Lobaria pulmonaria* and other rarities such as *Buellia griseovirens* and *Ochrolechia turneri*.

It is unlikely whether much truly ancient woodland survives on the chalk, where ancient cultivation lynchets are frequently detected within valley-side woodland. Nevertheless, some woods are clearly older and richer than others; Wood Anemone, Goldilocks Buttercup, Broad-leaved Helleborine, Woodruff and Wood Melick all appear to be good indicators of older woodland on chalk. Toothwort is characteristic of many woods on the chalk and it seems able to spread into expanding woods of less than fifty years of age with ease. Eaglehead Copse is a good place to see it in quantity from the footpath through the wood, together with Whitebeam, a rather local tree of chalk woods. Another distinctive woodland edge plant is Nettle-leaved Bellflower but this has a curiously restricted Island distribution, being confined to chalk woods in the middle of the Island. It grows in woods at Rowridge together with Wood Calamint, persisting in its only British site. The latter is really a woodland edge plant, requiring a degree of disturbance and an open, sunny aspect.

Cliff Copse at Shanklin is an interesting escarpment wood with a great deal of Wych Elm. Big old Whitebeam trees, clinging to the greensand rocks at the top of the wood, can be seen from the footpath running just above the wood. Great Wood-rush and Wall Lettuce also grow on the rocky slopes, whilst wet flushes have an abundance of Opposite-leaved Golden-saxifrage. The steep terrain has precluded this wood from being systematically managed. It is the only Island site for the Atlantic liverwort, *Lophocolea fragrans*, and a few ancient woodland indicator lichens have been recorded including *Lobaria pulmonaria*, *Dimerella lutea* and fertile *Lecanactis abietina*. However, *Lobaria pulmonaria* was lost when its host, an old ash tree, was blown over in a gale in 1988. Another old tree, supporting the only known population of *Dimerella lutea* in the wood, died in the 1990s.

Very few ancient woodlands have survived on the acidic, sandy soils of the southern bowl of the Island. The best of these are Borthwood Copse and America Wood, both accessible by public footpath. There is some evidence that these woods may have had periods of management as wood pasture and there are open grown old oak trees amongst bracken in both woods. Their woodland flora has survived less well than in woods on the north of the Island, and species such as Wood Anemone, Bitter Vetch, Betony and Goldenrod have become very scarce or have disappeared altogether. Borthwood Copse still has quantities of Climbing Corydalis, Common Cow-wheat and Great Wood-rush, the latter in particular, increasing with current management. America Wood is of historic interest as being the first (1855) recorded British site for Wild Gladiolus, 'in the midst of a wild tract of copse and heath'. Its presence here suggests that there must have been some historic management by grazing to maintain suitable open habitat.

The Undercliff is well wooded today, but most of this is recent secondary woodland containing a high

The first British specimen of Wild Gladiolus, from Bromfield's herbarium (HCMS). It was 'found at Shanklin, in Apse Woods by Mrs Phillipps, July 7 1855'. (AB)

proportion of non-native species. Remnant pockets of older woodland can be identified by the presence of Field Maple, Wych Elm and Hazel. These are damp, nutrient-rich woods, with Atlantic Ivy and Hart's-tongue fern as the dominant species. The Ivy is commonly parasitised by Ivy Broomrape; this is its Island stronghold. Italian Lords-and-Ladies, identified from Steephill in 1854 for the first time in Britain, is frequent, growing on the moist chalk debris at the base of the cliff. Sowbread is commonly naturalised. Ash, the dominant native tree, supports a rich lichen flora in the clean, moist maritime air and the area is the stronghold for two lichens that grow on tree trunks devoid of ivy, *Ramalina fraxinea* and the rare extreme southern *Cryptolechia carneolutea*, which also used to grow on mature elm trunks.

Holm Oak woodland, established on the southern slopes of Ventnor Downs, is our most unusual secondary woodland. It is the largest and longest established of its type in Britain and, although less than one hundred years old, is already beginning to acquire an interesting flora and fauna. White Helleborine, Birds-nest Orchid and Yellow Bird's-nest have become established beneath its shade, and several sclerophyllous woody exotics, including Strawberry Tree, Lauristinus and Bay, are appearing. This is the only established British site for a large, white southern agaric toadstool, *Amanita ovoidea*. Gradually, this woodland is acquiring some of the characteristics of Holm Oak woodland found in the Mediterranean.

3 CALCAREOUS GRASSLANDS

In Elizabethan times, large areas of the Island were under sheep pasture and most of the downs provided common pasture for tenants of the Island manors. At Swainston in 1630, (Jones, 2003) farm tenants were allotted grazing rights on downland stretching from Swainston Down in the west to Bowcombe Down in the east, according to the size of their tenancy. These downs fed many hundreds of sheep. At the same time, the lower slopes of Brighstone Down were being enclosed for arable farming by some enterprising farm tenants. However, sheep continued to be an important part of the mixed farming economy, and downland continued to be important for summer pasturage. The Second World War increased the demand for home-grown crops, resulting in much downland going under the plough, particularly on more gentle slopes. Further downland was lost as a result of forestry plantations, especially on Newbarn, Rowborough, Westover and Shalcombe Downs.

These changes clearly reduced the area of chalk grassland considerably, but it is interesting to note that Victorian botanical records indicate that the rarest and most localised species were found on steep, usually south-facing slopes and these were the places that have largely escaped the plough. Although many grassland species have declined through management neglect, the only one known to have become extinct as a result of land-use change is the Field Fleawort. This species was lost from Westover Down sometime around 1974, shortly after the downland slope was planted up for forestry.

Large tracts of chalk grassland still survive on the Island, and steep south facing slopes and old quarries have survived intact the best. It is difficult to single out any particular downs for comment. Very many of them are freely accessible and all are interesting. The characteristic and most widespread community of unimproved calcareous grasslands, Sheep's Fescue – Meadow Oat-grass grassland, is a very species-rich short springy grassland. It contains fine-leaved grasses and a large variety of herbs including Salad Burnet, Wild Thyme, Common Rock-rose, Rough Hawkbit and Small Scabious. Horseshoe Vetch, Stemless Thistle and Squinancywort are all typical of this grassland community. Upright Brome grassland develops in ungrazed situations on deeper, more moisture-retentive soils and this community is generally far less species-rich, although Pyramidal Orchid survives well.

The steepest drought-prone slopes, with shallow soils, have extensive areas of Carline Thistle – Sheep's Fescue grassland. This important grassland type holds nationally scarce species. Early Gentian occurs on the majority of south facing chalk slopes, in huge numbers in some years, and this is a part of its core area. Dwarf Mouse-ear, Small-flowered Buttercup and Bastard Toadflax are also remarkably widespread and frequent, in stark contrast to their distribution on the adjacent mainland.

A few species have remarkable disjunct distributions on the Island. Clustered Bellflower is confined to the downs at the western end of the Island and reappears, in smaller quantity, on Culver and Bembridge Downs. Wild Clary only grows on the chalky ramparts of Carisbrooke Castle and some of the lower chalky slopes around Ventnor and St Lawrence.

Island chalk has never been renowned for its orchids, although both Pyramidal Orchid and Autumn Lady's-tresses can be locally abundant. Frog Orchid is very rare and Fly Orchid may be extinct. Man Orchid was only ever known from a small area of Shanklin Down, where a tiny handful of plants survived up to the 1980s. Burnt Orchid, always rare, has managed to hang on in small numbers on Garstons Down but is now believed to have been lost. Early Spider Orchid occurred sparingly on the Ventnor chalk in the nineteenth century but it has only once been found on the chalk at Freshwater where its Dorset stronghold in Purbeck (on the limestone but not the chalk) is only a short distance away to the west.

The south facing slopes of Afton Down have a unique vegetation community developed over the compact, nodular chalk rock that marks the horizon between the Middle and Upper Chalk. The thin and stony soil supports grassland dominated by Saw-wort, Betony and Dwarf Thistle. *Weissia* mosses (*Weissia microstoma*) form extensive patches on bare soil and in crevices. The hard chalk pebbles have an interesting lichen flora including *Caloplaca lactea*, *Clauzadea immersa*, *C. metzleri* and *Petractis clausa*. A couple of stunted Juniper bushes cling to the slopes in their only Island site.

On Tennyson Down, close to the cliff edge, a series of whale-backed mounds supports the richest chalk grassland lichen communities recorded from this country. These include *Fulgensia fulgens*, a 'scrambled-egg' coloured lichen in its only British site on chalk, *Squamarina cartilginea*, *Megaspora verrucosa* and *Toninia lobulata*. These lichens thrive in dry, stressed, open habitats; they are much commoner in more southerly climes. On Tennyson Down, many plants grow in an extremely dwarfed state, to such a degree that a Swedish botanist, T. Wulff, visiting in 1894, incorrectly gave many of the plants varietal names.

Areas of flint gravel and clay-with-flints cap many of

our downs. In places, these have been obscured by scrub or by forestry plantations, but where open habitat survives, the combination of these more acidic soils and the chalk creates an unusual mix of plants referred to as chalk heath. Good and accessible examples of this can be seen on the brow of Tennyson Down, Brook Down and Mottistone Down.

4 NEUTRAL AND ACID GRASSLANDS

In common with the rest of the country, most fields today consist of a species-poor sward dominated by Perennial Rye-grass. Species-rich grasslands, which have largely escaped agricultural improvement, survive only as scattered, often isolated fields. Those on soils approaching a neutral pH are generally known as neutral grasslands; they are particularly associated with various clay-based soils. Their soils are deeper and less freely draining than calcareous grasslands, and are richer in plant nutrients. Provided that the grassland is managed, either by cutting or grazing, the sward is diverse. Grasses are generally more abundant than on calcareous grasslands and common species include Sweet Vernal-grass, Red Fescue, Crested Dog's-tail and Yorkshire Fog. French Oat-grass is a characteristic and frequent species of many unimproved neutral grasslands across the Island. Widespread herbs include Common Knapweed, Common Fleabane, Common Sorrel, Ox-eye Daisy, Common Bird's-foot Trefoil and Meadow Vetchling. Corky-fruited Water-dropwort has a very restricted range nationally but is characteristic of neutral grasslands on the Island. Other species are more limited to sites whose management has always been favourable and include Saw-wort, Dyer's Greenweed, Devil's-bit Scabious, Adder's-tongue Fern and Green-winged Orchid.

There are many unimproved meadows around Newtown and the most extensive, but private, surviving example is the Ministry of Defence Rifle Ranges meadow at Porchfield. It is believed that these fields were last ploughed in 1870; today the 14ha of grassland are managed as hay meadow. The sward is extremely rich and includes the largest Island population of Green-winged Orchids. Heath Dog-violet occurs locally in one of its very few Island sites and Dodder is frequent, parasitising many plants. In the absence of any fertiliser input, herbs, in particular Dyer's Greenweed, are increasing at the expense of grasses.

Neutral grasslands occur on a variety of soil types and small variations in soil chemistry and moisture can produce marked variations in vegetation. Some soils on base-rich clays can support species such as Salad Burnet, Yellow-wort, Restharrow and Dwarf Thistle, which are more characteristic of calcareous soils. Other, slightly acidic grasslands may have Sneezewort, Tormentil and Pignut. There are a group of waterlogged and heavily gleyed clay soils on the north of the Island where acid loving plants such as Heather, Bell Heather, Purple Moor-grass and Lousewort grow alongside more typically neutral grassland species and this vegetation has been termed clay heath. It is best seen in some of the rides in the southern half of Parkhurst Forest, but also occurs in Fattingpark Copse and around Cranmore.

Acidic grasslands on dry, sandy soils are overwhelmingly dominated by grasses, particularly Common Bent, Sheep's Fescue and Yorkshire Fog. Herbs are far less diverse than on other unimproved grasslands, but can include Sheep's Sorrel, Heath Bedstraw, Mouse-ear-hawkweed, Ladies Bedstraw, Pill Sedge and Buck's-horn Plantain. Interesting examples of Bristle Bent grassland, a south-western species, occur on the top of Ventnor Downs and Brighstone Down, and on Sandown golf course. However, these tend to be rather species-poor.

Thin, parched soils are able to support a diverse flora of early-flowering small plants including Early Hair-grass, Common Whitlow-grass, Bird's-foot, Common Cudweed and a variety of clovers. This type of community appears to be increasing and many of our rarer clovers, such as Fenugreek, Knotted, Rough, Subterranean and Clustered Clovers, are becoming widespread. Examples can be seen alongside the footpaths on St George's Down.

A particularly unusual acid grassland community is present in limited extent on Brading Marshes. It is dominated by Mat-grass, a scarce grass in southern England, together with a range of *Cladonia* lichens, including *C. subcervicornis* not recorded from elsewhere on the Island, in broken areas of the sward. This interesting community has become established on sandy ground resulting from the reclamation of the estuarine Brading Haven at the end of the nineteenth century.

5 HEATHLANDS

The light sandy soils of the southern half of the Island were comparatively easy for Neolithic and Bronze Age people to clear. Heathland would have been created by the deterioration of sandy and gravelly deforested soils by grazing and through leaching of nutrients by rainwater. It has been estimated that in the mid-1500s, heathland, in its widest sense, covered about one quarter of the Island. This would have included rough grassland, bracken and gorse. Economic pressures to take low-lying, flat heathland into cultivation were considerable from the late sixteenth century onwards and by 1750, the majority of heathland had been destroyed (Chatters, 1984). In 1850, the total area was estimated to be around 730ha. Further declines took place so that, by 1984, only about 133ha were left. Only one site over 40ha remains today; this is Headon Warren at the western extremity of the Island.

Typically on the Island, heathland is dominated by Heather, Bell Heather, Gorse, Dwarf Gorse and heathland grasses such as Bristle Bent. This is the typical, somewhat species-poor, lowland dry heath community of the Hampshire Basin. Heavy grazing can convert this to acid grassland. In places where the water table remains close to surface, a shallow layer of peat can develop and this supports wet heath. This community is richer in species, but there are very few remnant wet heath fragments surviving today. Characteristic species are Cross-leaved Heath, Purple Moor-grass, sphagnum mosses and, until recently, Petty Whin.

Headon Warren comprises a flat-topped hill reaching about 120 metres above sea level, with a series of landslipped undercliffs sloping down to the coast on its north and west sides. The hill is capped by plateau gravel, with base-rich horizons. Large expanses of heather-dominated heath occur on the plateau and on the broad upper part of the undercliffs. Some of this area has been damaged in recent years by severe fires, but the heathland still retains a number of interesting species. It is the only Island site for Heath Pearlwort, which grows on the grassy tracks on the plateau. Dodder, Common Yellow-sedge, Bird's-foot, Wavy Hair-grass and Upright Chickweed occur, although the latter species is particularly rare. Lost species include Small Cudweed, Smooth Cat's-ear, Chaffweed, Mat-grass and Sea Stork's-bill. The upper undercliff areas have the best *Cladonia* lichen heath on the Island with 22 taxa of *Cladonia* recorded, together with an abundance of *Cetraria aculeata*, *Parmelia saxatilis* and *Platismatia glauca* lichens. Headon Warren is accessible by public footpath but some of the less common plant species are not easy to find.

The heathland on top of Ventnor Downs is noteworthy for the abundance of Bilberry, which is not found in quantity elsewhere on the Island. It has gradually increased with sympathetic management and it fruits in most years. A heathland lichen flora is also well developed here. This used to be the site of Stag's-horn Clubmoss, found growing on peat in 1860.

Bouldnor once had extensive areas of heathland. Although most of this was obliterated by forestry plantations in the early part of the twentieth century, heathland survives as a narrow belt along the eroding cliff top, close to the coastal footpath. Interesting species here include Dodder, Pale Dog-violet, Dwarf Gorse, Burnet Rose and Flea Sedge. A remarkable survival has proved to be the aquatic fern Pillwort, recently rediscovered growing in a cleared pond within the plantation. Another surprise find was a few plants of Round-leaved Wintergreen, which appeared on the slumped cliff for a few years in the 1970s.

Lake and Blackpan Commons at Sandown comprised an extensive area of dry heath with areas of wet heath and base-poor mire. This was an historic site for Wild Gladiolus, last seen here in 1909. The area has altered a great deal and is now a golf course. Nevertheless, some wet and dry heathland does survive. There is still much Heather and small quantities of Dwarf Gorse, Mat-grass, and Cross-leaved Heath, a few plants of Sheep's-bit in one of its few Island sites, and rafts of Marsh Cinquefoil around the natural pond at the entrance to the golf course.

Both Bleak Down and St George's Down were historic heathland sites with a long history of sand and gravel extraction. The disturbed acid soils provided ideal habitats for many of the smaller exacting heathland species until recent times, but the extent of suitable habitat is very restricted today. Allseed, Dwarf Cudweed, Annual Knawel and Chaffweed thrived in these sites. Bleak Down also had wet heath, with an abundance of Round-leaved Sundew, Bog Asphodel, Ivy-leaved Bellflower, Common Cottongrass and Many-stalked Spike-rush. Many of these plants were still present into the 1960s, when the site was being used for landfill, and some struggled to survive into the early 1990s. Not all has been lost. Today, Bleak Down still has the largest extent of Cross-leaved Heath on the Island and acid pools still have Royal Fern, Alternate Water-milfoil and, until recently, Floating Club-rush. St George's Down has Lousewort, Bird's-foot, Common Cudweed and an increasing variety of clovers. Dwarf Cudweed has reappeared here in recent years and is currently locally frequent. Most of these plants can be seen from public footpaths.

Landslip heath communities occur on the Island's coast and they may represent original heathland sites. One of the best of these is on Blackgang Chine ledge where there is an expanse of Heather and Bell Heather, and Dodder and Sheep's-bit can be found growing with them. There were other landslipped coastal heaths at Luccombe and Lake but only vestigial fragments remain. At Headon Warren undercliffs, as a result of springs and seepages, the gravel deposits slip over well-lubricated clays and the mixing of basic and acidic substrata enables both calcicole and calcifuge species to grow in close proximity.

Grazed and trampled commons, such as St Helen's

St Helen's Green in Victorian times. The trampled, damp ground was home to now extinct rarities such as Small Fleabane and Pennyroyal. In 1928, the *Isle of Wight County Press* reported that the green was still deteriorating due to local residents allowing their cattle to graze. Photograph courtesy of Sylvia Taylor.

Green and Colwell Common, once supported a rich specialised flora similar to those of the New Forest Commons. The combination of dry, tightly grazed acidic grasslands, winter-flooded hollows, ponds and permanently wet flushes provided a wide range of micro-habitats for a diverse range of small, exacting species which have become very rare nationally. St Helen's Green used to be grazed by horses and geese, it had a pond and water was tapped from its wells. Today it is largely improved and heavily mown. Chamomile is characteristic of this type of habitat. It was once widespread on the Island and it continues to exist in quantity here, although prostrate and rarely flowering. Small Fleabane and Pennyroyal no longer survive but, quite remarkably, Marsh Pennywort, Bristle Club-rush and Lesser Spearwort are still to be found in a boggy area on the north-west green which once held Round-leaved Sundew, Marsh St. John's-wort, Bog Pondweed, Few-flowered Spike-rush and much else besides.

Colwell Common was a more complex site, extending to the coastal cliffs. There was neutral grassland with Chaffweed, Allseed and Field Gentian, the latter in its only Island site. Wet heath supported Petty Whin, Cross-leaved Heath and Lesser Butterfly Orchid and mire vegetation had Tawny Sedge, Marsh Helleborine, Broad-leaved Cottongrass, Common Cottongrass and Pale Butterwort, suggesting influence from base-poor and calcareous waters. Dry calcareous grassland had Bastard Toadflax. Sadly, this wonderful site has been reclaimed and developed and yet the surviving mown green still has Marsh Pennywort, Green-ribbed, Common Yellow and Pill Sedges, Bristle Club-rush and Green-winged Orchid.

The wide verges and pond at Hardingshute, near Brading, still survive but are agriculturally improved. This is a place where livestock from the Nunwell estate would be folded before sending to market and it used to hold interesting species. Horned Pondweed, Marsh Yellow-cress, Upright Goosefoot and Small Fleabane all grew in and around Hardingshute pond. The hedgerows still have the best surviving population of the local Round-leaved Dog-rose.

6 WETLANDS

Wetland habitats are principally found in the floodplains of river valleys. Traditionally, these habitats would have comprised a complex of grazing marsh, reedswamps, fen and wet carr woodlands, reflecting the management regimes and the subtle variations in water chemistry. Extensive areas of flood plain mire, often referred to locally as moors, would have provided rough grazing for livestock in the past. Poorly draining soils, along river valleys and around springs and seepages, provided ideal conditions for peat formation, and the most botanically diverse areas would have been found where water levels were close to the surface throughout the year and the vegetation was kept short by grazing. Sadly, much of this former richness has been lost, and almost all of the remaining good sites are on private land.

Rich, wet acidic habitats occurred along some 3.5 km of the Medina from Cridmore, northwards to Rookley Farm. At The Wilderness, Bog Asphodel, Round-leaved Sundew, Cranberry, Bogbean and Marsh St John's-wort were all frequent at the end of the nineteenth century. Huge quantities of Royal Fern were dug up from the ditches to satisfy the demands of Victorians. By the early twentieth century, the area was deteriorating, the result of a combination of drainage, reclamation and increasing nutrient runoff into the catchment. Today, most of The Wilderness is dominated by willow carr. Marsh Fern and Narrow Buckler-fern survive beneath the shade, and a few plants of Bog Myrtle, once abundant, struggle to survive in the only Island site where the number of plants reaches double figures. Royal Fern was last seen here in the 1970s. The hepatic *Pallavicinia lyellii*, which formerly grew in a boggy ditch here, appears to have been lost. Marsh Cinquefoil still survives, but the open areas are dominated by tall fen meadow.

At Cridmore Bog, varying depths of peat overlie less permeable clays, causing acidic water moving in from the catchment to accumulate, creating a high water table. Beneath the clays is a sand and gravel aquifer. This water is also believed to reach the surface through artesian pressure and leads to a lateral flow of base-rich waters. Cridmore Bog is the only Island site where a Bottle Sedge/Marsh Cinquefoil community is developed. It is dependent upon a water table that remains continuously at or close to the surface, and in places it develops as a floating mat. Although rather species-poor, Bogbean is constant and Common Cottongrass and Marsh Marigold are characteristic; the brown moss, *Calliergon cordifolium* forms conspicuous patches. Marsh Speedwell used to occur here in its last remaining Island site, but it is believed lost. This is a rare community in the central southern England, more characteristic of northern and western Britain. It is not found elsewhere on the Island.

Cridmore Bog also supports an interesting area of species-rich Purple Moor-grass mire growing on a moist but well-aerated, acid peat. Cross-leaved Heath is characteristic and hummocks of *Sphagnum palustre* and *S. subnitens* grow in more open areas. Common and Star Sedge, Devil's-bit Scabious, Common Cottongrass, Heath Spotted-orchid and a remarkable abundance of Dyer's Greenweed are part of the assemblage, but Petty Whin has been lost.

Floodplain mire vegetation can be very variable, as it is dependent upon water chemistry. Nutrient-poor acidic waters give rise to base-poor mires, whilst alkaline waters lead to the development of base-rich fens, and there is a continuous range of communities between the extremes. Within the past hundred years, the complex pattern of traditionally managed wetland habitats has been greatly modified as a result of changing land-use and excessive river engineering, which has hydrologically divorced wetland habitats from the river. The consequent lowering of the water table, often coupled with a reduction in water quality and a lack of appropriate management, has often resulted in a degradation of wetland habitats. Where these habitats receive more nutrient enriched waters, large areas of herb-rich marshy grassland have been modified to species-poor stands overwhelmingly dominated by tall competitive species, usually rush pasture, reed bed or Greater Pond-sedge swamp. Thankfully, some good examples of fens and mires still survive, although they tend to be located around the valley edges where they obtain much of their water supply from springs and surface run-off.

Surviving examples of species-rich base-poor mire vegetation, receiving water from springs and seepages, are found either on hillsides, such as Bohemia Bog, or at the edge of floodplains such as Munsley Bog. Bohemia Bog is a remarkable survival. It is a tiny private site of just 0.7ha and yet it is rich in specialised bryophytes and flowering plants. It is the only extant site for four bog hepatics, the mosses *Pohlia camptotrachela* and *Philonotis fontana*, and three flowering plants. A further two hepatics and two flowering plants are only known from this and one other site. The Bog Asphodel mire community is dominated by clumps of *Sphagnum* with scattered herbs in hummocks and hollows. *Sphagnum papillosum* is the dominant species but seven *Sphagna* (including *S. subnitens*, and in the wetter areas *S. auriculatum* and *S. cuspidatum*) are still present, making this the richest site for bog mosses on the Island. The bog hepatic *Cephalozia macrostachya*, a characteristic species of New Forest and Dorset valley mires, grows amongst the hummocks. Bohemia Bog is the only extant site where Round-leaved Sundew and Pale Butterwort grow together and also has Bog Asphodel, Many-stalked Spike-rush, Common Cottongrass, Bog Pimpernel, Bog Pondweed, Cross-leaved Heath and more. Lemon-scented Fern has recently been found growing at the side of an adjoining trackside ditch. The cattle-grazed site is in good condition although it will always be highly vulnerable to changes in water and to poor management. A rather similar site, Cockleton Bog at Northwood, survived until the twentieth century. It was known to support almost all of the same species as Bohemia. In addition, Brown Sedge, Broad-leaved Cottongrass and Meadow Thistle were also recorded from here, suggesting some base-enrichment of the spring-fed water supply.

Munsley Bog was another rich acidic wetland site, also private. The farmer who owned the site used to harvest sackfuls of *Sphagnum* moss from here for the florist trade in the 1950s but when cattle grazing ceased, the quality of the site declined. By the late 1960s some plants, which were formerly abundant, such as Petty Whin, were already becoming scarce. An adjoining area of species-rich Purple Moor-grass mire at Elm Court, with Meadow Thistle, Bog Pimpernel and Marsh Cinquefoil, was reclaimed for playing fields in 1988. The bog proper was the only Island site for the bog hepatic, *Mylia anomala*, but this has not been reported since 1976. By the mid 1990s, other key species were in serious decline. Purple Moor-grass mire occurs around the sloping margins of Munsley Bog (and also at The Wilderness and Bohemia Bog). Although this community tends to be species-rich when managed well, when ungrazed, Purple Moor-grass forms a thick, tussocky structure, which ultimately excludes other species. Populations of Round-leaved Sundew, Bog Asphodel, Common Cottongrass, Bog Myrtle, White Sedge and Lousewort were rapidly dwindling at Munsley in the mid 1990s. Miraculously, some of these species are still surviving, together with small quantities of Cross-leaved Heath, Dwarf Gorse, Common Sedge, Heath Spotted-orchid and Royal Fern.

At Fairfields, a tributary of the Eastern Yar has well-developed Purple Moor-grass mire, kept in check by cattle grazing, resulting in a species-rich sward. The spring-fed valley community shows a remarkable abundance of Meadow Thistle, Devil's-bit Scabious and Stalked Sedge. Within the wettest part of the mire are seasonally wet pools with low growing communities including Bog Pimpernel, Bulbous Rush, Carnation Sedge and Common Yellow-sedge.

Further examples of species-rich base-poor to neutral mire vegetation survive in the Eastern Yar valley around Alverstone. They are dependent upon water arising from small tributaries and from springs and seepages at the edge of the flood plain. It is in the vicinity of these seepages that the most interesting and diverse botanical communities survive. Localised good areas are found right along the drain below Hill Farm to the west of Alverstone, and below Borthwood Lynch and Alverstone Farm to the east of the village. Marsh Cinquefoil, Bogbean, Marsh Violet, Marsh Pennywort, Marsh Marigold, Southern Marsh-orchid, Carnation Sedge, Bottle Sedge and Brown Sedge are characteristic species in these sites.

More alkaline fen meadow is generally species-rich with a dominance of the local Blunt-flowered Rush. Calcareous fen meadow communities are found on slumped cliffs at Headon Warren and Luccombe. In addition to Blunt-flowered Rush, Marsh Helleborine, the dense-flowered subspecies of Fragrant Orchid and Southern Marsh-orchid can still be found locally in these situations.

Freshwater Marshes were by far the best example of a calcareous fen meadow. In the nineteenth century the site comprised a mosaic of tightly grazed wetlands, taller fen and reed beds, intersected by ditches containing clear, slow-flowing calcareous waters. It supported a cornucopia of local wetland species. There was Bladderwort, Opposite-leaved Pondweed, Lesser Water-plantain, Bogbean, Flowering Rush and a good range of *Potamogeton* pondweeds in the ditches. Boggy meadows at the source of the River Yar, just behind the beach in Freshwater Bay, had Bog Asphodel, Tawny Sedge, Knotted Pearlwort, Lesser Skullcap and Marsh Lousewort. The fen had Marsh Cinquefoil, Marsh Helleborine, Bottle Sedge, Blunt-flowered Rush, Greater Spearwort and perhaps Great Fen-sedge. This rich flora survived into the early part of the twentieth century but after the First World War, grazing ceased allowing the taller growth of fen vegetation and the invasion of willow carr. In the 1950s large areas of sedge swamp were burnt and drained, allowing the process of succession to proceed apace.

Today, the site is a Local Nature Reserve, accessible by footpaths, but it is very different to how it looked in the nineteenth century. The ditches contain turbid, ochre-rich water, which is not conducive to the growth of submerged aquatics, and the site is largely covered with reedbeds and carr woodland. Many wetland species have disappeared, but some still survive. The reedbeds in the South Marsh, which have grown up over fen, have scarce fen species growing amongst them, including Blunt-flowered Rush, Greater Spearwort, Lesser Water-parsnip and Water Dock. There are still stands of Lesser Pond-sedge, the food plant of a moth, Blair's Wainscot, discovered for the first time in this country breeding in the marsh by K. Blair in the 1950s but becoming extinct within a few years of its discovery. In the shade of willow carr, Marsh Fern is thriving. It grows alongside Narrow Buckler-fern and Black Currant, and there are still a few clumps of Cyperus Sedge.

Some examples of spring-fed fens occur along spring-lines at the foot of the chalk downs. Although they are small and isolated, they can support a rich flora including Southern Marsh Orchid, Marsh Marigold, Bog Pimpernel and Brown Sedge. Compton Marsh is perhaps the best of these and it still sustains a rich flora. As well as those species already mentioned, Common Sedge, Carnation Sedge, Marsh Arrowgrass and Marsh Pennywort can be found and Marsh Helleborine and Knotted Pearlwort occurred here in the nineteenth century. There is another example on the north side of Tennyson Down at Moon's Hill. At Brading Marshes, a flushed area on the marshes below Centurion's Hill is dominated by Blunt-flowered Rush with Marsh Willowherb, Common Sedge and Brown Sedge.

7 AQUATIC VEGETATION OF WATERWAYS

Island streams and rivers generally have an impoverished flora. This is as a result of impoverishment due to the 'island effect' and to historically unsympathetic management. Over-engineering has led to rivers that are often canalised, over-deepened, and carry high sediment loads. The largest rivers, the Eastern Yar and Medina, arise from the Chalk in the south of the Island but flow for most of their length through the heavily cultivated sandy soils of the Lower Greensand. They suffer from a combination of damaged structure due to drainage engineering, poor water quality and low flows. Smaller rivers, draining the gravel aquifers over the Tertiary clays in the north of the Island, have been less heavily modified by drainage engineering and some sections, where they flow through ancient woodlands, are fairly natural. These streams are highly responsive to rainfall and experience naturally low flows during summer months.

Few aquatics survive under these conditions but Spiked Water-milfoil persists through the Eastern Yar. The distribution of Wood Club-rush and Unbranched Bur-reed is also largely confined to the riverbanks of the Eastern Yar. Feeder streams and tributaries have not been kept clear and are often choked with Branched Bur-reed. The least modified stretches are often where streams flow through ancient woodlands but here they are heavily shaded and consequently have an impoverished aquatic flora.

Some streams have good water quality near their source. The Caul Bourne at Winkle Street, Calbourne is a clear chalk stream with Hybrid Monkeyflower, Blue Water-speedwell and Watercress. The Lukely Brook at Carisbrooke, another clear chalk stream, has Opposite-leaved Pondweed, Horned Pondweed and Rigid Hornwort. Stream Water-crowfoot is not native on the Island but has become well established in the River Medina as it flows through the heart of Newport below Pan Mill.

A few quiet stretches of the Eastern Yar still support a diverse marginal flora. Drainage ditches on Sandown Levels, once a very rich site, still have Brackish Water-crowfoot, Tubular Water-dropwort, Skullcap and Wild Celery. At Morton Marsh, near Brading, Water Forget-me-not, Trifid Bur-marigold, Water Chickweed, Unbranched Bur-reed and Blue Water-speedwell are still frequent. However, ditches on the marshes at Alverstone, recommended by Bevis et al (1978) for their aquatic flora, including several species of Pondweed and Water-starwort, are now overgrown and the water is turbid.

Similarly, ponds with a rich flora are scarce. Ponds recently created on the Yar floodplain at Marsh House, east of Brading, have rapidly developed a rich flora including Grey Club-rush, Spiked Water-milfoil and Small, Fennel and Lesser Pondweeds. However, one of the best collections of farm ponds survives on private land on the Elmsworth and Lambsleaze Farms on the eastern banks of the Newtown Estuary. Local rarities such as Lesser Marshwort, Lesser Water-plantain, Blunt-leaved Pondweed and the bryophyte *Riccia fluitans* can be found in some of these ponds. On the other side of the Newtown estuary, old clay pits at Cranmore and Bouldnor were known to nineteenth century botanists as sites for Pillwort, Least Bur-reed and Floating Club-rush, all plants which were believed to have become extinct. Pillwort has however recently been rediscovered from one its historic ponds at Bouldnor following clearance work.

8 SOUTH COASTAL HABITATS

The southern coastline of the Island is subject to rapid erosion and this has given rise to one of the finest series of soft cliffs, locally punctuated by steep ravines or chines, to be found along the British coastline. The steepest cliffs, where erosion is most rapid, support little in the way of vegetation. Coltsfoot, Creeping Bent and Common Reed are amongst the few plants able to find a toehold in these situations. Common Reed, which tends to grow around seepages, frequently produces long, sharply pointed rhizomes, which spread out across the surface of the cliff.

Where slumping gives rise to low terraces, the vegetation is more diverse. Common Fleabane, Wild Carrot and Great Willowherb can become established on quite narrow ledges and Rock Sea-spurrey, which is at the easternmost edge of its range here, often grows as large, rounded clumps. Gently slumping cliffs, with transient pools and wet flushes have the richest flora although these coasts are not renowned for their botanical rarities; Southern Marsh Orchid, Kidney Vetch, Sea Pink Curved Hard-grass and Sea Mouse-ear are characteristic plants. Chines are often scrubbed over with Blackthorn, but the sheltered and shaded streams can be good hunting grounds for ferns. Shanklin Chine is particularly damp and shaded and has interesting bryophytes including *Philonotis marchica*, still just surviving in its only extant British site, and the thalloid liverworts *Anthoceros punctatus*, *A. agrestis* and *Phaeoceros laevis*. At the mouth of some chines, Watercress, Brookweed and Wild Celery can be found growing at the top of the beach. At Ladder Chine, wind-blown sand has led to the formation of a dune perched at the top of the cliff and extensively colonised by Sand Sedge. Wide ledges provide the greatest interest. Whale Chine ledge has Heather and Cross-leaved Heath. Heather also grows on Blackgang Chine ledge. Also growing here are Bell Heather, Dodder, Sheep's-bit, Blunt-flowered Rush, Slender Club-rush and Royal Fern. The latter species survives in several flushed ledges along the south coast. In Victorian times, when Blackgang Chine was more extensive and surrounded by large areas of wet and dry heath, Common Cottongrass and Marsh Helleborine were also recorded.

Botanically similar ledges are found at Luccombe Chine and Marsh Helleborine has persisted here, despite on-going cliff retreat. Great Horsetail forms dense and luxuriant stands and it is a good site for Common Spotted and Southern Marsh Orchids, with hybrid swarms.

The steep cliffs of Sandown Bay between Shanklin and Sandown are particularly interesting for the bryophyte flora of flushes. Interesting species include the thalloid liverworts *Blasia pusilla*, *Anthoceros punctatus* and *Phaeoceros laevis*. These cliffs are also a rich site for naturalised plants with well-established populations of such species as Seaside Daisy, Sicilian Chamomile, Red Bistort, Hottentot-fig, Wireplant, Spanish Broom and Escallonia.

The cliff top at Redcliff, just north of Sandown has always been a good site for plants. It is here that the most accessible population of Nottingham Catchfly spreads down the slumped cliffs and it is one of the most reliable sites for Yarrow Broomrape. The sandy cliff top grassland has Clustered, Rough, Knotted, and Hare's-foot Clovers, Spring Vetch, Bird's-foot and Bulbous Meadow-grass.

There are impressive high chalk cliffs at the eastern and western extremities of the Island, and their approaches, and their flora shows many similarities. Portland Spurge, White Horehound, Yellow Horned Poppy, Knotted Hedge-parsley, Hairy-fruited Cornsalad and Bulbous Meadow-grass occur at both ends of the Island. The cliffs to the west of Freshwater Bay have, in addition, Hoary Stock, Oxtongue Broomrape, Shaggy Mouse-ear-hawkweed and Rock Sea-spurrey. Hoary Stock can form impressive shows of scented purple flowers on chalk cliffs and is easily seen on the roadside chalk cutting over Afton Down. Wild Cabbage grows on the adjoining cliff top chalk grassland. On the Needles Headland at the western extremity of the Island, a saltmarsh community is established on the flat cliff top at an elevation of around 62m above sea level. Looking out from the Needles Battery searchlight position, swards of Sea Purslane can be seen, together with smaller quantities of Sea Heath, Sea Beet and Annual Sea-blite. Common Scurvygrass has an Island outpost on these cliffs.

The Undercliff landslide complex is a wide slumped skirt of land (stretching for nearly 10 km along the coast) from Luccombe in the east to Blackgang in the west, and sheltered to the north by an inner cliff. It was activated as a consequence of aggressive coastal erosion following a rise in sea levels after the last Ice Age, about 10,000 years ago. The south facing, sheltered Undercliff has a warm, humid environment, which has given rise to its own distinctive flora. This is the headquarters for Round-leaved Crane's-bill and, at the foot of the talus, Italian Lords-and-Ladies and Ivy Broomrape. Historically, this would have been a largely open, grazed landscape interspersed with tumbled rocks and wooded pockets. It was an outpost for some plants with a more Atlantic distribution, in particular bryophytes. The coastal rocks, a scarce habitat this far east along the South Coast provide niches for Sea Spleenwort and a rocky shore lichen flora including *Anaptychia runciata*, *Ramalina siliquosa*, *Solenopsora vulturiensis* and *Xanthoria parietina* f. *ectanoides*. The principal site for *Anaptychia runciata*, a large boulder by St Catherine's Lighthouse, has been damaged by collection of rock samples by groups of geology students.

The open undercliff landscape has survived best in the area around St Catherine's Point and slumps and slippages still occur here creating open ground habitat. The soils are complex, being a jumble of calcareous chalky soils and more acidic greensand and chert. Much of the grassland is essentially similar to chalk grassland. Cowslips are frequent and Wild Liquorice survives, in greatly reduced quantity, around scrubby edges. Thin open grassland is an important site for the nationally rare bryophyte, *Acaulon triquetrum*, on south facing slopes. The rare leafy liverwort *Cephaloziella baumgartneri* grows on a calcareous greensand outcrop. Springs emerging from the inner cliff give rise to interesting streams. The liverwort *Southbya nigrella* has one of its few British sites in one of these streams. Other uncommon streamside plants just managing to survive are Galingale and Lesser Water-parsnip. Blunt-flowered Rush and Bog Pimpernel grow in flushed

Binnel Bay, near St Catherine's Point, originally photographed about 1870 when there was much more open habitat than survives today. (AB)

spots. Nit-grass has been found in small quantity by the cliff edge in Watershoot Bay and there may be larger undetected populations elsewhere. Shaded boulders in the Undercliff formerly supported an interesting western bryophyte community. The leafy liverwort *Cololejeunea rossettiana* still survives on rocks in Bonchurch Landslip but *Marchesinia mackaii* appears to have been lost from Niton undercliff.

On the inner cliff from Blackgang westwards towards St Lawrence, a maritime cliff vegetation has developed with Sea Pink, Rock Sea-spurrey, Sea Campion and the lichen *Ramalina siliquosa* creating a colourful show. This is the principal site for Sea Campion on the Island. A maritime-influenced cliff vegetation is also developed on the west facing Tolt Rocks north of Blackgang. The Atlantic liverwort, *Porella obtusata*, finds an eastern outpost here.

The mild climate of the Undercliff provides suitable conditions for many alien species to become established. Buddleja has been an invasive species for many years and is threatening some of the open ground habitats at St Catherine's Point. On the coastal cliffs at Ventnor, Shrubby Orache, Hedge Veronica, Virgin's-bower and Eastern Gladiolus are well established. Red Valerian produces a spectacular display when in full bloom west of Ventnor and attracts many butterflies. On sunny winter days, the prolific flowering of Winter Heliotrope scents the air in Wheeler's Bay. The full potential of the equable climate of the Undercliff on plant growth is best demonstrated in the Ventnor Botanic Gardens.

9 FARMLAND

Despite the regular use of herbicides in recent decades, arable plants often build up an enormous seed bank in the soil and are able to take advantage of favourable environmental changes and disturbance. Characteristic species in good sites on the chalk include Round-leaved and Sharp-leaved Fluellens, Rough and Prickly Poppies, Venus's-looking-glass and Narrow-fruited Cornsalad. Field Cow-wheat, once an arable plant of chalky fields above St Lawrence and Ventnor, is now banished to a small field edge bank, a Wildlife Trust reserve, and an open chalky cliff face both above St Lawrence. The local 'hot-spots' for arable plants correspond well with those areas historically recorded for their weed flora and some of the most interesting species are principally found on the dry, sandy soils on the south of the Island. The flora of fields around Cridmore and Appleford is particularly remarkable, being home to some of Britain's rarest arable annuals. Some of them can be seen from public footpaths. Corn Marigold is common here and Cornflower has one of its most permanent British stations. Growing with them may be found a few plants of Small-flowered Catchfly and Broad-fruited Cornsalad. Annual Knawel and Nit-grass are found in nearby fields. Small-flowered Catchfly also grows in a sandy field at Alverstone along with Loose Silky-bent. Martin's Ramping-fumitory no longer grows in fields, but it survives on allotments at Lake where it grows with Purple Ramping-fumitory. Broad-leaved Spurge is a particular speciality of arable fields on the clay soils of the north of the Island and another is Lesser Quaking-grass, found in several fields around Whippingham, Wootton and Quarr.

Some arable fields have an interesting ephemeral bryophyte flora, but this has not been studied to any degree. *Cheinia leptophylla*, a very rare moss first described as new to science (as *Tortula vectensis*) from the Isle of Wight in 1965, is still found in its type locality, an arable sandy field at Brook, together with a rich bryophyte flora.

Ancient, species-rich hedgerows are still common on the clay soils of the north side of the Island. On the greensand, hedges tend to be much poorer in species with Hawthorn and Blackthorn dominating. Around ancient settlements and along the exposed southwest coastline, Elm is the characteristic hedgerow species. It is still common as a hedge plant but mature Elms are no longer seen. Its disappearance as a mature tree from the landscape as a result of the Dutch Elm disease outbreak of the 1970s onwards has resulted in the loss and decline of many of the wayside lichens which are dependent upon its nutrient-rich bark. *Cryptolechia carneolutea*, a rare species with an extreme southern distribution in this country, had its main populations on Elm. Others, which were probably once widespread, have become extremely rare on the Island (e.g. *Anaptychia ciliaris*, *Bacidia rubella*, *Caloplaca ulcerosa*, *Candelaria concolor*, *Leptogium teretiusculum* and *Physcia clementei*) or are believed to have become extinct (*Bacidia incompta*, *Caloplaca cerina*, *Caloplaca luteoalba*, *Physcia clementii* and *P. tribacioides*). A few of these species survive in much smaller quantity on alternative tree hosts.

10 CHURCHYARDS AND CEMETERIES

The Medieval churchyards provide substitute natural rock outcrops for a rich lichen flora. The maritime influence of the climate is vividly seen in the lichen flora of old church towers and Carisbrooke Castle ramparts in the centre of the Island. Many of these ancient windswept edifices are rimmed with a shaggy growth of Sea Ivory, *Ramalina siliquosa* on the westerly faces. Godshill Church tower is also graced by bushy growths of the western maritime lichen *Roccella phycopsis*, a national rarity. Godshill churchyard is the richest site for lichens on the Island, and is the only recorded site, to date, for the moss *Grimmia trichophylla*. However, much more survey work of churchyards is required before a full picture can emerge.

Generally speaking, churchyards are not rich sites for vascular plants. Brading Churchyard is possibly the best, with Wild Clary, Yarrow Broomrape, Elecampane, Thick-leaved Stonecrop and Rustyback Fern. Shalfleet Churchyard is also good, with Wall Pennywort, Common Calamint and Grass Vetchling, and Thorley Churchyard has a good surviving area of Bembridge limestone grassland. Cemeteries tend to be less interesting for their wall flora but are frequently excellent sites of old, unimproved species-rich grassland. Mount Joy, Northwood, Fairlee, East Ashey, Lowtherville (Upper Ventnor) and East Cowes cemeteries are all very good sites in which to examine species-rich grasslands on a range of soil types and most have conservation areas that are left uncut during the main growing season.

Some plants of dry grasslands, a diminishing habitat, have found refuge on walls. The Medieval ruins of Quarr Abbey support a particularly good mural flora. This is the best Island site for Rue-leaved Saxifrage and there is also Common Calamint, Prickly Poppy and long-established populations of Wallflower and Reflexed Stonecrop. The ruins are not currently accessible to the public.

REFERENCES

Anon (2000) *Wildlife of the Isle of Wight – An audit & assessment of its biodiversity*. Isle of Wight Biodiversity Action Plan Steering Group. Isle of Wight Council.

Bevis, J., Kettell, R. & Shepard, B. (1978) *Flora of the Isle of Wight*. Isle of Wight Natural History & Archaeological Society. Newport, Isle of Wight.

Bromfield, W.A. (1856) *Flora Vectensis*. London. William Pamplin.

Chatters, C. (1984) *The Downs and Heaths of the Isle of Wight*. Countryside Heritage Study. Report for Isle of Wight County Council. Unpublished.

Jones, M.J. (2003) A survey of the manors of Swainston and Brighstone 1730 – farm buildings and farm lands. *Proc. Isle of Wight nat. Hist. Archaeol. Soc.*

Vascular Plants

COLIN POPE

The Isle of Wight, with its clearly defined geographical limits, is a convenient area for botanical survey. The records for this *Flora* stretch from a very few gleaned from sixteenth century sources up to, and including, the year 2002.

Historic records and herbaria have been researched, and although they have yielded significant data, there is considerable scope for further study in these areas. Bromfield's *Flora Vectensis* of 1856 has been a particularly valuable benchmark of the flora at that time. Many of the plant records in the first edition of Townsend's *Flora of Hampshire, including the Isle of Wight* are undated, and where presented here, they have been dated 1883, the date of publication shown on the title page of the book. It is known, however, that these records were made several years earlier, and the last records admitted were in 1879 (Brewis *et al*, 1996).

The Flora of the Isle of Wight by Bevis, Kettell and Shepard (1978) provides the most up to date account of the Island flora, but in that book, the dates of many records are not given. Where it has been possible to trace the actual dates of plant records submitted, these have been included in the present publication where relevant. Most were made several years before 1978. Bevis *et al* attempted to give an indication of the frequency of each plant species by including a count of the number of tetrads in which it was recorded. It was originally hoped that this would provide a useful indicator of change and could be repeated here. However, this has not proved possible, because the current recording has been on a 1km basis whereas Bevis *et al* used a modified tetrad (2km) recording basis. The tetrads were manipulated by combining some coastal squares, so that they arrived at exactly 100 tetrads, for economy of recording effort. Because this serves to distort the mapped distribution of species, it has not been replicated here.

For the purposes of the current flora, records made between 1987 and 2002 have been treated as 'recent'. This current phase of recording on the Island was strongly stimulated by the Botanical Society of the British Isles (BSBI) Atlas 2000 project. In May 1996, and again in May 1998, special BSBI recording meetings were held on the Island to assist with this project. Bill Shepard, a long-standing BSBI vice-county recorder and co-author of the previous county flora, was succeeded as vice-county recorder by the author in 1996. For the first time, the botanical records were computerised, using an Aditsite software database.

The most recent phase of recording has been, wherever possible, site based. However, there has been the intention of recording in every 1km square on the Island, of which there are some 430 containing at least some fragments of land. It has not proved possible to survey every square, but the overwhelming majority has been surveyed at least once, and most have been inspected more often. There has, however, been a concentration of survey effort on interesting habitats and on relocating plants not recently recorded.

PLAN OF THE SYSTEMATIC ACCOUNT

Example:
C. *sepium* (L.) R.Br.
HEDGE BINDWEED (local: Granny pop out of bed; lily!)
N. ssp. *sepium* is common in disturbed ground, hedgerows, waste places, wet woodland and fen carr. A persistent garden weed. The local name of 'Granny pop out of bed', which was in use until recently, referred to the manner in which the seeds could be squeezed out of the capsule. The pink flowered ssp. *roseata*, a western coastal subspecies in this country, was recorded in 2000 from High Grange 3784, Compton Grange 3884 and Mottistone 4084, all PS. It is apparently not uncommon along the south-west coast. It was first recorded from Yarmouth in 1929 (BM).

SEQUENCE AND NOMENCLATURE
This follows that of C.A. Stace (1997) *New Flora of the British Isles*. 2nd edition. Casual species not included in Stace are excluded from the accounts. The English names are those given in Stace. This is followed by any local names, where recorded, enclosed in rounded brackets. Many of these are taken from Bromfield (1856). The use of local plant names is rapidly diminishing today but many older Islanders still remember using colloquial names and where I have heard these still in use, albeit rarely, the name is followed by an exclamation mark (!) to indicate this.

STATUS CATEGORIES

These have been derived from Preston *et al* (2002), which attempts to distinguish native species from those that may have arrived with people a long time ago. The following categories are recognised:

N. Native. Believed to be indigenous on the Isle of Wight, although plants may have arrived in recent years by natural factors (e.g. Sea Knotgrass). There is good archaeological evidence for the native status of a number of Island plants.

Arch. Archeophyte. Believed to have been introduced intentionally or accidentally prior to A.D. 1500. These include those arable plants that are more or less confined to artificial habitats. There is very little local archaeological evidence for this group.

E. Established alien species. These are believed to have arrived since 1500, often far more recently, and are maintaining self-sustaining populations either by seed or vegetatively. Some species that are native in other parts of the country fall into this category and some species occur here as both native and alien populations. The species description should clarify this.

C. Casuals. Plants which have arisen from seed or garden outcasts and which do not currently seem able to persist in the wild. This group has been rather poorly and patchily recorded.

S. Surviving. This is a small group of mostly woody plants where individuals may have grown spontaneously from seed but are showing no signs of spreading.

P. Planted. Mostly referring to trees that have been planted in the countryside, but are showing no signs of spreading.

FREQUENCY AND HABITATS

There is a brief statement of the overall distribution and frequency of the species and the habitats in which it occurs on the Island. Reference is made to unusual habitats or situations in which the species has been recorded.

Species for which the evidence strongly suggests they are likely to be extinct have their entries enclosed in squared brackets [ ]. Also, the word Extinct appears at the start of the species account.

Species for which there is some doubt as to their identity, for instance historic records not supported by herbarium specimens, have their entries enclosed in rounded brackets ( ).

THE RECORDS

Records are usually presented in the following order: locality (generally identifiable on the current 1:25,000 scale OS map), 1km grid reference, initials of recorder, and year of record. The Isle of Wight falls completely within one 100km square of the National Grid, designated by the prefix SZ (40). Frequent and widespread species generally do not have individual records cited. Localised but widespread species may not have individual records listed, but their distribution may be indicated by a map. The map number appears at the end of the text description. For rarer species, all modern sites are given together with the most recently recorded date of the record. Wherever possible, the last recorded dates for declining and extinct species are given for specific sites. These are not, of course, necessarily the date of extinction although in some cases (for instance when the site is destroyed or for well-monitored species) they are. First records are given for some species, generally for rarities, casuals and when there is some historic interest (for instance particularly early records and noteworthy modern finds).

With the exception of some of the earliest botanists and referees, recorders are referred to by an abbreviation, which has been used throughout the species accounts for all groups. They are listed in Appendix 1. Standard abbreviations for herbaria where specimens are held have also been used, and these are listed below. A list of references cited in the text appears at the end of this chapter.

LOCAL INFORMATION

Items relating to uses, folklore and superstitions or historically interesting references have been included wherever possible.

THE MAPS

Maps have only been included where they provide additional information. Generally speaking, maps have been included where species show:
i   A particularly striking distribution.
ii A dramatic decline in distribution.
iii Rare but widely distributed species.
iv Under-recorded species, to illustrate the extent of known records.

Records have been plotted in two date classes. These are 1800 - 1986 (open circles) and 1987 - 2002 (closed circles). They are plotted on 1km squares superimposed on a 10km grid.

The maps have been produced on Aditsite software, and saved as bitmap files.

INITIALS USED FOR HERBARIA (HB)

BM British Museum (Natural History); **BEL** Belfast, Ulster Museum; **BON** Bolton Museum; **CGE** University of Cambridge; **DBN** Dublin National Museum; **E** Edinburgh Botanic Garden; **HAMU** Hancock Museum, Newcastle upon Tyne; **HCMS** Hampshire County Museum Service, Winchester; **IPS** Ipswich; **K** Kew; **LIV** Liverpool Museum; **MANCH** Manchester Museum; **NMW** National Museum of Wales, Cardiff; **OXF** Druce-Fielding herbarium Oxford; **RNG** Reading University; **SDN** Swindon Museum; **SLBI** South London Botanical Institute; **SPN** University of Southampton; **WRN** Warrington Museum.

LYCOPSIDA
CLUBMOSSES & QUILLWORTS

1. LYCOPODIACEAE
Clubmoss family

[*Lycopodium clavatum* L.
STAG'S-HORN CLUBMOSS
N. Extinct. Discovered in small quantity in dry heathland on the top of St Boniface Down in 1860 by Robert Symmens, although the specimen in the British Museum is labelled Shanklin Down, A.G. More. The herbarium material is luxuriant with cones. It was considered to have become extinct by 1900.]

2. SELAGINACEAE
Lesser Clubmoss family

Selaginella kraussiana (Kunze) A. Braun
KRAUSS'S CLUBMOSS
E. Only recorded from Shanklin Parish Church, 5881, MB 2002, where it is well established in damp, shaded grassland at the foot of the church wall.

EQUISETOPSIDA
HORSETAILS

4. EQUISETACEAE
Horsetail family

Equisetum fluviatile L.
WATER HORSETAIL
N. Very locally frequent in areas of shallow, slow moving water and swamps in many of our wet valleys. Map 1

Equisetum x *littorale* Kuhlew. ex Rupr. (*E. fluviatile* x *E. arvense*)
SHORE HORSETAIL
N. Very rare, perhaps overlooked. There is a single record from a damp roadside near a bridge at Rookley 5183, PE 1952 (K).

E. arvense L.
FIELD HORSETAIL
N. Common and widespread on waste ground, dry pastures, roadsides and ditch banks.

E. sylvaticum L.
WOOD HORSETAIL
N. Very rare, but still surviving at two long-standing stations around springs in wet acid woods and adjoining marshland at Apse Heath Withy Bed 5783, CP 1995 and Lynch Copse, Newchurch 5685, CP 1998. In some years, it grows in quantity and occasionally produces fertile shoots.

E. palustre L.
MARSH HORSETAIL
N. Locally frequent in marshes, wet meadows, pond margins, ditches and damp woodland. Map 2

E. telmateia Ehrh.
GREAT HORSETAIL
N. Common and widespread, sometimes dominant particularly on spring lines and slipping coastal cliffs. Occasional plants are found bearing small cones on lateral branches.

PTEROPSIDA
FERNS

5. OPHIOGLOSSACEAE
Adder's-tongue family

Ophioglossum vulgatum L.
ADDER'S-TONGUE
N. Infrequent and local. Although this species is undoubtedly overlooked, it has declined considerably. Found today in unimproved neutral grasslands on clay and chalky soils. Occasional in cemeteries and churchyards, e.g. East Ashey cemetery 5798, CP 1999; East Cowes cemetery 5094, CC 1984; Newtown churchyard 4290, BA 2000; Thorley churchyard 3788, PS 1999; and Wootton Common cemetery 5390, CP 2001. Many thousands of plants on an unmown bank of Seaclose playing fields, Newport 5090, CP 2001; this is the Island's largest known population. Map 3

[*Botrychium lunaria* (L.) Sw.
MOONWORT
N. Extinct. Formerly occurred sparingly in a few lightly grazed, sometimes damp, old pastures on neutral to basic soils at eight or nine different sites. More (1898) commented that 'it requires a close and careful search to find so small a plant'. Last recorded in 1913 from a meadow north of Cook's Castle near Wroxall by Fred Stratton.]

6. OSMUNDACEAE
Royal Fern family

Osmunda regalis L.
ROYAL FERN (local: snake fern)
N. Formerly, a not uncommon species of fens and carr woodland in most of our wet river valleys and on wet coastal cliffs. Referring to The Wilderness, Fred Stratton wrote (*J. Botany*, 1913), 'Though in 1870 I walked through thickets of it five or six feet high, I have not been able to find a single plant in late years. This is due to the rapacity of fern dealers and the folly of fern buyers'. Much declined everywhere since then, and rare today. Recently recorded in small quantity at the north end of Bleak Down 5182, BS 1990; in willow carr at Munsley Bog 5282, CP 2002; and

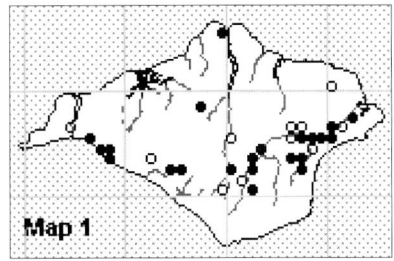

Equisetum fluviatile

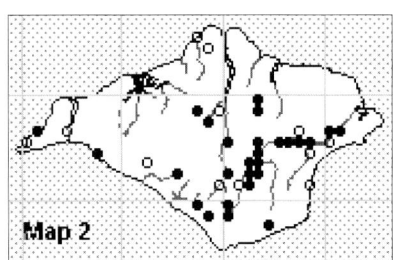

E. palustre

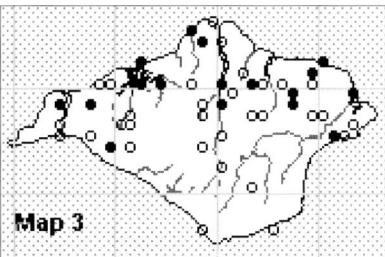

Ophioglossum vulgatum

in carr woodland alongside Bembridge ponds 6388, CP 2000. It has fared better around the coast where it occurs on flushed slopes and ledges, which are difficult to access e.g. Lake Cliffs 5983, CP 2000; Luccombe Chine 5879, JS 1988; Rocken End 4975, SC 1999; and Blackgang Chine 4877, CP 1999. Map 4

7. ADIANTACEAE
Maidenhair Fern family

Adiantum capillus-veneris L.
MAIDENHAIR FERN
E. A rare garden escape sometimes establishing itself on old stone walls within towns. Known since at least 1964 from a wall in Star Street, Ryde 5992. Also recorded from a storm drain at Atherley Road, Shanklin 5881, PS 2000; a stairwell in Cowes 4995, PS 2000; and brick walls at the base of greenhouses at Watergate Nurseries, Newport 4987, 1972 BS.

9. MARSILEACEAE
Pillwort family

Pilularia globulifera L.
PILLWORT
N. Rediscovered in 2002, growing in quantity in a pond in Bouldnor Copse 3890, GT. The overgrown pond was enlarged in 1996 and surrounding conifers were removed, and it is likely that the plant responded to this management. This is probably the site where James Groves discovered it in 1918 (BM). He found it carpeting a small heath-pool on Bouldnor Hill in the spring of that year. The following autumn he found it again, 'in considerable abundance and in fruit in two other pools, apparently the remains of old clay-pits dug out for the brickworks, some distance away but on the same hill' (*J. Botany* 1918). On 24 July 1920, he conducted members of the Isle of Wight Natural History and Archaeological Society to the spot. Ponds in this area also held Least Bur-reed (*Sparganium natans*) in the late 1800s.

11. POLYPODIACEAE
Polypody family

Polypodium L. agg.
POLYPODY
N. Locally frequent and widespread on old walls and roofs, sheltered banks and as an epiphyte on trees, especially old oaks. Very occasionally found on greensand rock outcrops. Both *Polypodium vulgare* L. and *Polypodium interjectum* Shivas, Intermediate Polypody, are widely recorded. *P. interjectum* appears to be particularly frequent along the Undercliff. *Polypodium cambricum* L., Southern Polypody, is very rare. There are records of this taxon from Mottistone Manor walls, growing with *P. interjectum*, 4083, PS 2000 and Wolverton Manor, Shorwell, 4582, MB 1999.

13. DENNSTAEDTIACEAE
Bracken family

Pteridium aquilinum (L.) Kuhn
BRACKEN (local: brake!)
N. Abundant and widespread, absent only from intensively used agricultural areas and chalk grassland. Formerly, this plant had value in the rural economy as a source of bedding, domestic fuel and fertilizer, and bracken-cutting rights were highly valued. The eastern slopes of St Catherine's Down were formerly common land within the Wydcombe Estate, and it is believed that this land was subdivided into strips for grazing and for the harvesting of bracken (Basford & Smout 2000). The 1630 survey of the parish of Swainston (Harrison, 1630) contains reference to Yatlands, near London Heath, Newtown, having four fernhouses for storing bracken.

14. THELYPTERIDACEAE
Marsh Fern family

Thelypteris palustris Schott
MARSH FERN (local: ground-fern)
N. Very rare. Formerly, it was not uncommon in fens and carr woodland but has now greatly declined. There are recent records from only two sites. At Freshwater Marshes 3486, it was once a component of a rich open fen community that has now largely been lost through succession to reedbed and willow carr. Marsh Fern has survived these changes well and still grows in quantity under the shade of the willow canopy CP 1994 and abundantly beneath reed in the south marsh, CP 2000. It is also still present in a willow carr east of Alverstone, 5785 JO & PS 1996, and may still survive in a wet copse near Langbridge Farm, Newchurch 5686, where it was last recorded in 1974, BS. Last seen in The Wilderness 5082 in 1985 and not found since, despite searching. Map 5

Oreopteris limbosperma (Bellardi ex All.) Holub
LEMON-SCENTED FERN
N. Very rare. Historically recorded in small quantity by a streamside at Apse Castle near America Woods, at intervals between 1843 and 1909, and from a damp roadside bank between Lynn and Guildford Farms, Havenstreet in 1862 and 1863 (A.G.More). No further records until discovered during a visit by the British Pteridological Society in 1990. A few plants were found by a trackside ditch at Bohemia Bog 5183, BS. The ditch banks have now become overgrown again and it could not be found in 2002.

15. ASPLENIACEAE
Spleenwort family

Phyllitis scolopendrium (L.) Newman
HART'S-TONGUE
N. Common and widespread in woods and hedge banks. It is often luxuriant in moist calcareous sites all along the

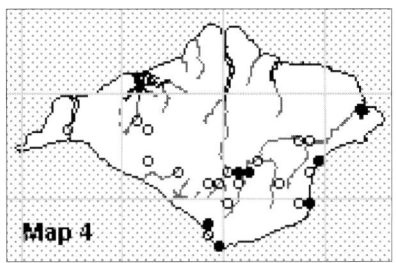

Osmunda regalis

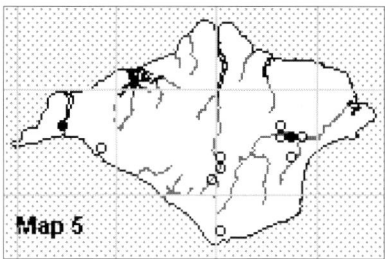

Thelypteris palustris

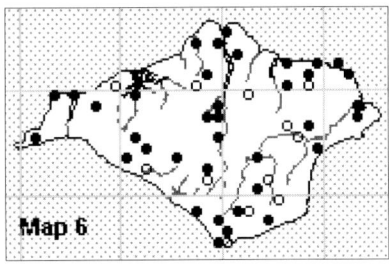
Asplenium adiantum-nigrum

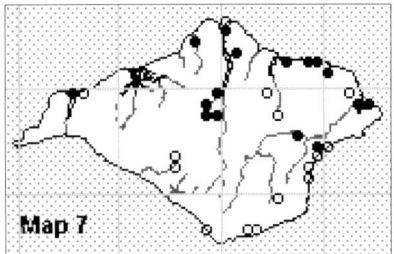
A. trichomanes

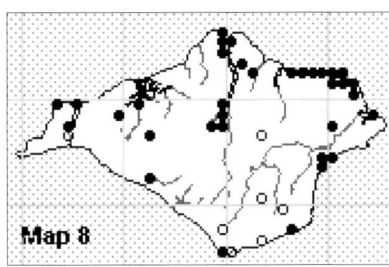
A. ruta-muraria

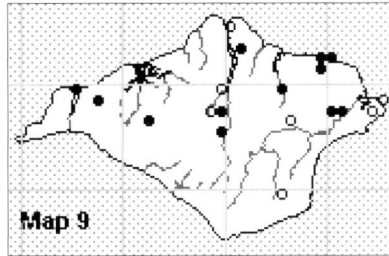
Ceterach officinarum

Undercliff. The abundance of this fern on the Island was much appreciated by the Victorians. Bromfield records that 'the fresh leaves of Hart's-tongue are applied externally in rustic practice in the island to bad legs (erysipelatous eruptions) as a cooling remedy'.

Asplenium adiantum-nigrum L.
BLACK SPLEENWORT
N. Local, and rarely in quantity, but widespread on old walls and roadside banks. Grows in crevices of greensand rock outcrops at Gatcliff 4977, CP 1996, Tolt Rocks on Gore Down 4977, CP 1995 and above St Catherine's Point 4975, CP 2002. Currently, most luxuriant on an old roadside barn wall at Basketts Farm, Rew Street 4796, JEG 1996. Map 6

A. marinum L.
SEA SPLEENWORT
N. Very rare and known today from a single large greensand boulder near the coast at Rocken End 4975, where it was discovered in 1973 BS. Numbers of plants fluctuate considerably; by 2000 the boulder had become enveloped by wind-pruned elm suckers, although the fern plants were fertile and still growing well. Sea Spleenwort has been known from this area since Victorian times. Supposedly, it was first found in 1845 when Dr. Bromfield was visiting Miss Kirkpatrick at Windcliffe, Niton. Dinner was already on the table when a young member of the party handed the doctor a crumpled piece of fern. The meal had to be taken out and kept hot in the kitchen whilst Bromfield visited the spot where the fern was growing. (W.I. 1971). The Island is the easternmost native site for this species on the south coast; the nearest mainland sites are on Purbeck. In 1969, plants appeared on a damp internal wall of Newport Roman Villa at Shide and survived for a number of years.

A. trichomanes L. ssp. *quadrivalens*
MAIDENHAIR SPLEENWORT
N. Very locally frequent on old walls, sometimes growing luxuriantly. Seems to have disappeared from some of our churches, but this may be temporary. It was considered to be rare in the nineteenth century and has certainly increased since that time, assisted no doubt by the extensive building of limestone walls during the Victorian period. On Quarr Abbey ruins (5692) it was present 'in some plenty' in Bromfield's time, but subsequently disappeared. It was refound in 1968, JH, and then lost again until 1989. In 1993, a single luxuriant plant was seen, CP and by 2002 there had been a modest increase. It is unknown today from any natural rock outcrops but More (1898) recorded it from rocks in the Undercliff. Map 7

A. ruta-muraria L.
WALL-RUE
N. Locally common on old walls. It is the most widespread of our wall ferns. Formerly recorded from natural rock outcrops but the only modern instance of this is of a single small, but fertile, plant in a greensand rock crevice above St Catherine's Point 4975, CP 2002. Map 8

Ceterach officinarum Willd.
RUSTYBACK
N. Occasional and local on old stone walls. Present on a number of churchyard walls, including Whippingham 5193, MB 2000; Brading 6087, AC 1996; Thorley 3788, MB 2000; and Gatcombe 4985, ST 1993; and at Mount Joy cemetery 4987, AC 2000. Still present in 1993 on the old seawall at Brading, near the Great Sluice, 6187. Occurs in small quantity in several locations in Newport and Ryde. On occasion, it has been lost as a result of wall renovation, only to reappear on other walls nearby. Map 9

16. WOODSIACEAE
Lady-fern family

Athyrium filix-femina (L.) Roth
LADY FERN
N. Local but widespread. Found in damp woodland, ditches and shady hedge banks. Usually grows on acidic soils but occasionally found on more basic ones. Map 10

[*Gymnocarpium robertianum* (Hoffm.) Newman
LIMESTONE FERN
E? Extinct. Two historic records from the 1860s, from a wall at Carisbrooke Castle and a wall close to greenhouses at Swainston, are recorded by A.G. More. Both are strongly suggestive of escapes from cultivation

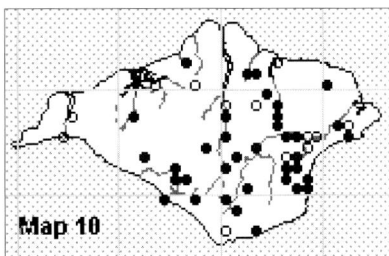
Athyrium filix-femina

at the height of Victorian pteridomania.]

[*Cystopteris fragilis* (L.) Bernh.
BRITTLE BLADDER-FERN
N. Extinct. The British Museum has a sheet from C.E. Salmon's herbarium containing four fronds of this species collected by Francis Low and labelled 'Isle of Wight 1839'. It is likely that the plant was a persisting casual, and it is still occasionally found elsewhere in south-east England in man-made habitats. However, without further information, the possibility of an escape from cultivation cannot be ruled out.]

17. DRYOPTERIDACEAE
Buckler-fern family

Polystichum setiferum (Forssk.) T. Moore ex Woyn.
SOFT SHIELD-FERN
N. Frequent and widespread in sheltered woods and hedge banks. Less often found on the chalk.

P. x bicknellii (H. Christ) Hahne
SOFT SHIELD-FERN X HARD SHIELD-FERN
N. Known only from a wooded part of Shepherds Chine 4479, a single plant, RW 1998 (confirmed Clive Jermy).

[*P. aculeatum* (L.) Roth
HARD SHIELD-FERN
N. Extinct. Despite a number of Victorian records, there is only one confirmed site, on a hedge bank opposite the Sun Inn, Calbourne 4286, FS, collected on 15 August 1871. There are herbarium specimens at BM and OXF.]

Cyrtomium falcatum (L.f.) C. Presl
HOUSE HOLLY-FERN
S. Surviving in secondary woodland amongst the ruins of a Victorian fernery in the grounds of East Dene, Bonchurch 5778, CP 2000. Long established, and slowly increasing on a north-facing bank at Ventnor Botanic Gardens, 5476 CP 2001.

Dryopteris filix-mas (L.) Schott
MALE-FERN
N. Common and widespread.

D. affinis (Lowe) Fraser-Jenk
SCALY MALE-FERN
N. Local but widespread in woods and shady hedge banks on greensand and clay soils. There are several subspecies and this complex group has been mapped together. It is believed that ssp. *borreri* is the most frequent, but ssp. *affinis* is not infrequent and ssp. *cambrensis* has also been reliably recorded. Map 11

D. carthusiana (Vill.) H.P. Fuchs
NARROW BUCKLER-FERN
N. Rare and today restricted to wet woods, principally in our main river valleys. Perhaps overlooked, but certainly greatly declined. Recently only recorded from Freshwater Marshes 3486, BS 1992; Burnt Wood 4492, BS 1996; Eastern Yar valley at Redway 5385, BS 1993; and The Wilderness 5082, TDD & CDP 1996. Map 12

D. x deweveri (J.T. Jansen)
NARROW BUCKLER-FERN X BROAD BUCKLER-FERN
N. Only recorded from The Wilderness 5082 where it grows with both parents, TDD & CDP, 1996 (conf. Clive Jermy).

D. dilatata (Hoffm.) A. Gray
BROAD BUCKLER-FERN
N. Common and widespread in wet and dry woodlands and hedge banks.

18. BLECHNACEAE
Hard-fern family

Woodwardia radicans (L.) Sm.
CHAIN FERN
S. Established on the north side of Bonchurch pond 5778 since at least 1950, and perhaps increasing slowly CP, 2000. This large fern is most probably a survival from a Victorian or Edwardian planting, but persists amongst native vegetation.

Blechnum spicant (L.) Roth
HARD-FERN (local: herringbones)
N. Local in wet acidic woodland, often on ditch banks, mostly in East Wight. Woodland sites include Combley Great Wood 5489, CP 2000; Parkhurst Forest 4789, CP 1995; Lynch Copse, Newchurch 5685, CP 1998; The Wilderness 5082, CDP & TDD 1996; and Buddle Brook, Brighstone 4283, CP 1993. Also, but rarely, on ditch sides in peaty bogs such as Munsley Bog 5282, CP 2002; Bohemia Bog 5183, CP 1993 and Pope's Farm Marsh, Newchurch 5685, CP 1994. Can show a rapid increase in population under favourable management. Referred to as "Herringbones" by E.H. White. Map 13

19. AZOLLACEAE
Water Fern family

Azolla filiculoides Lam.
WATER FERN
E. Found sporadically in ponds and ditches, widely scattered across the Island but often not persisting. First recorded in 1943 from a pond at Wootton Manor Farm 5492 where it was said to have been introduced forty years previously (*Proc. Isle Wight nat. Hist. Archaeol. Soc.* III, 1943).

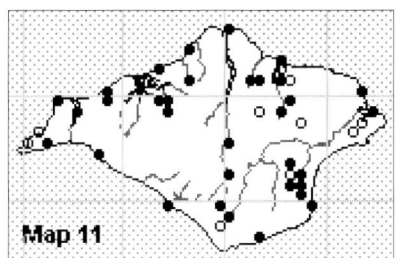

D. affinis

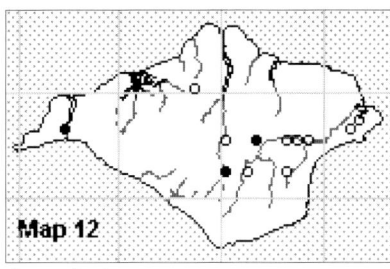

D. carthusiana

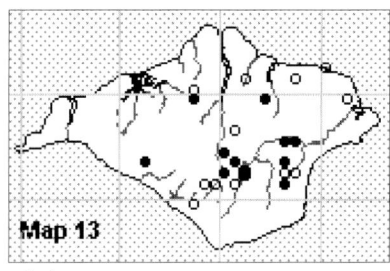

Blechnum spicant

PINOPSIDA – GYMNOSPERMS CONIFERS

A number of conifers are found in forestry plantations and some of these self-sow within the woods where they have been planted. Western Red Cedar (*Thuja plicata*) and Western Hemlock-spruce (*Tsuga heterophylla*) regenerate freely in Parkhurst Forest plantations.

20. PINACEAE
Pine family

Pinus sylvestris L.
SCOTS PINE
P. Widely planted, but very rarely naturalizing, in contrast with the situation on the New Forest heaths.

22. CUPRESSACEAE
Juniper family

Cupressus macrocarpa Hartw. ex Gordon
MONTEREY CYPRESS
E. Popularly planted as an ornamental in Victorian times and later. Very occasionally self-sown along the south coast e.g. several specimens along the cliff edge on chalk at Horseshoe Bay, Bonchurch 5777, CP 1996; a single young tree established in Shanklin Chine 5881, CP 1999.

Juniperus communis L.
COMMON JUNIPER
N. Extremely rare and only ever recorded as dwarf, solitary bushes on chalk downland. A single bush, on the down above Nunwell, was recorded by Bromfield in 1845. It occasionally bore berries and was still present in 1900, FS. A single specimen on St Boniface Down was recorded in 1893 FS. The only modern record is of two well-spaced, wind-pruned bushes on Compton Down 3685, TW 1968. They are still present but are very small specimens (the larger is just over 30cm) and do not bear fruit.

24. TAXACEAE
Yew family

Taxus baccata L.
YEW
E. Widely planted and regenerating naturally. It occurs as individual specimens in woodlands and, particularly, on the north facing slopes of many of our downs such as West High Down, Brading Down and the Ventnor Downs. Although there are no ancient specimens in churchyards, the yew in St Mary's, Carisbrooke, with a diameter at breast height of 5.15m in 2002, is possibly the largest. A yew tree illustrated in Newport's Old Ledger Book for 1567 is shown growing in Carisbrooke churchyard in the same position as the present tree and may well be the same individual.

MAGNOLIOPSIDA ANGIOSPERMS, FLOWERING PLANTS

MAGNOLIIDAE DICOTYLEDONS

25. LAURACEAE
Bay family

Laurus nobilis L.
BAY
E. Occasional but widespread. Bay was widely planted on the Island by Victorians and it is persistent in shrubberies. The abundantly produced autumn berries are attractive to birds and its present day occurrence in hedgerows, woods and scrub, is likely to be of bird-sown origin. Generally associated with woods and woodland edges but sometimes found in the open eg. a single large bush on the south slope of Arreton Down 5387, EC 2001.

27. NYMPHAEACEAE
Water-lily family

Nymphaea alba L.
WHITE WATER-LILY
E. Rather rare and considered to have been introduced at all its known sites. Sites include ponds at Dodnor 4991 & 5091, CP 1999, on the Elmsworth Farm at Porchfield 4492, CP 2000 and a roadside pond at Russell's Farm, Roud 5180, CP 1999.

Nuphar lutea (L.) Sm.
YELLOW WATER-LILY
N? Known only from the River Yar

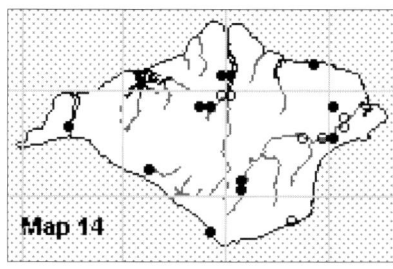

Ceratophyllum demersum

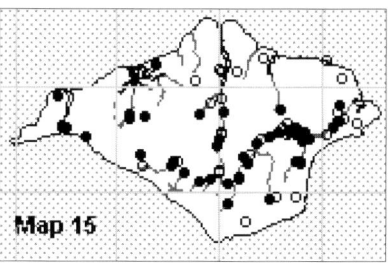

Caltha palustris

near the Great Sluice on Brading Marshes 6186, where it was first recorded in 1928 EHW. Described as 'undoubtedly introduced' in Bevis *et al* (1978) but without evidence.

29. CERATOPHYLLACEAE
Hornwort family

Ceratophyllum demersum L.
RIGID HORNWORT
N. Occasional, but sometimes abundant, in ponds, ditches and slow-flowing rivers. It is long known from the Eastern Yar at Brading 6085, BS 1991 and also in the Lukely Brook at Carisbrooke 4888, MB 1998. Possibly increasing in ponds. Map 14

30. RANUNCULACEAE
Buttercup family

Caltha palustris L.
MARSH-MARIGOLD (local: batchelor's buttons!)
N. Locally frequent. It is found occasionally in wet meadows, marshes and fens but most frequently today in wet woodland. It has declined this century, particularly from wet grasslands, as a result of drainage and agricultural improvements. Map 15

Helleborus foetidus L.
STINKING HELLEBORE
E. A rare short-lived garden escape.

More frequent in the nineteenth and early twentieth century as an established alien, principally in the Undercliff and centred around St Lawrence. Bromfield says that it used to be grown in cottage gardens as a rather violent remedy for worms.

H. viridis L. ssp. *occidentalis* (Reuter)
GREEN HELLEBORE
N. Very rare. Occurs as single isolated patches within three woods: Woodhouse Copse 5393, JC 1984, but not found subsequently; Briddlesford Copse 5589 (2 sites), CP 2000; and Sibdown Farm Copse, near Rookley 4984, CP 1995. First recorded in 1868, in great abundance in Woodhouse Copse, FS 1913; and in 1890 from a wood at Havenstreet, J.H. Steuart, (IPS). These two stations, on wet organic soil in ancient woodlands subject to flooding, have all the appearance of native sites. Possibly introduced in the third site, Sibdown Farm Copse, a small, dry secondary woodland within 150 metres of the farmhouse, where it was first recorded in the 1970s. Specimens used to be gathered by children in Havenstreet from the Briddlesford site up to the early 1960s, for wild flower competitions in the village school.

[*Eranthis hyemalis* (L.) Salisb.
WINTER ACONITE
E. Extinct. Only recorded from Thorley Manor on a roadside verge (Bevis *et al.* 1978). No recent records.]

Nigella damascena L.
LOVE-IN-A-MIST
C. Occasional on waste ground outside gardens.

Aconitum napellus L.
MONK'S-HOOD
E. Very rare. There is a single roadside clump at Westover Plantation 4184, MB 1999, from where it has been known since at least 1971. Clump by Blackbridge Road, Freshwater in verge by marshes 3486, CP 2002. Previous records (*sensu lato*): A single specimen in a wet ditch in Buckett's Copse, Swainston 4488 (Bevis *et al.* 1978) has not been seen recently. Formerly well naturalized by the Medina at Pan Mill, Newport, where it was recorded by Bromfield in 1840

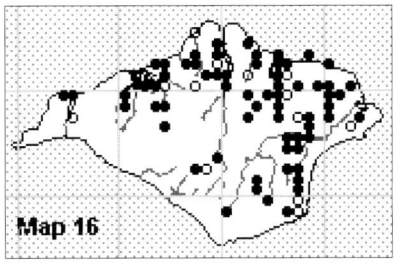

Anemone nemorosa

and persisted until 1921. Also recorded by Bromfield from the Caul Bourne near Upper Calbourne Mill in 1843, with records up to 1898.

Consolida ajacis (L.) Schur
LARKSPUR
C. Very rarely on waste ground outside gardens. In the nineteenth century, it was recorded by Bromfield as a cornfield weed in fields above the Undercliff and between Blackgang and Chale.

Anemone nemorosa L.
WOOD ANEMONE
N. Frequent in deciduous woodland, particularly ancient woodlands and old hedge banks on clay on the north side of the Island. Much less frequent in woods on the southern half of the Island, where it may be declining. Scarce in woodlands on the chalk, where it may be a good indicator of old woodland. Also recorded from ten species-rich grassland sites e.g. Harts Farm, Newtown 4290, CC 1984; Youngwoods Meadow, Alverstone 5785, IWNHAS 1998; Swiss Cottage Meadow, Osborne 5294, IWNHAS 1998. Recorded from beneath bracken in open sites at Berry Hill 4882, CP 1985 and Sibden Hill 5781, CP 1999. Map 16

A. apennina L.
BLUE ANEMONE
E. Very rare. Halletts Shute, Norton 3489, CP 2000 (it was recorded as a garden escape at Norton in 1931); well established in Bonchurch Old churchyard, with both blue and white flowered plants 5778, LS 1979, and still present; 2 or 3 small clumps long known from beneath trees at the east end of Appley Park 6092, CP 1999; Wootton St Edmund's churchyard 5492, MB 2001.

A. blanda Schott & Kotschy
BALKAN ANEMONE
E. Very rare. Occasionally found on waste ground always close to gardens.

Clematis vitalba L.
TRAVELLER'S-JOY (local: beth-twine)
N. Frequent and widespread in hedge banks, woods and scrub across the Island, but particularly on the chalk and on limy clays. Some older Island countrymen remember smoking cigar lengths of the dried stems when they were 'nippers'.

C. flammula L.
VIRGIN'S-BOWER
S. Very rare. Bromfield (1857) describes it as sparingly naturalized on shingle on Norton Spit. Recorded from a hedge on Mount Joy, Carisbrooke 4987, TF 1977. A fine plant at the top of the cliff below Salisbury Gardens, Ventnor 5677, BS 1987; still present 2000, CP.

Ranunculus acris L.
MEADOW BUTTERCUP
N. Very common and widespread.

R. repens L.
CREEPING BUTTERCUP
N. Extremely common and widespread.

R. bulbosus L.
BULBOUS BUTTERCUP
N. Common, although less so than formerly. It is abundant in chalk grassland.

R. sardous Crantz
HAIRY BUTTERCUP
N. Locally frequent in brackish marshes, coastal grassland and damp permanent pasture. It is sometimes found on waste ground and in arable fields. Map 17

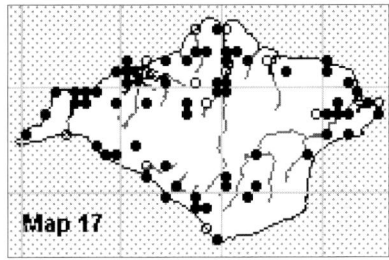

R. sardous

R. parviflorus L.
SMALL-FLOWERED BUTTERCUP
N. Occasional on dry, bare ground in grassy places, but locally in quantity. Occurs principally on south-facing downland and coastal grassland, and sometimes in arable fields. Map 18

R. arvensis L.
CORN BUTTERCUP (local: devil's claws)
Arch. Once a pernicious cornfield weed, especially on well-drained soils, and common enough to have earned a local name that refers to its distinctive fruits. It has greatly declined, and is now very rare. The most recent records are Walpen Farm, Chale 4778, TW 1959; cornfield east of Stone Farm, Blackwater 5186, BS 1974; and field west of St Helen's Church 6289, BS 1981. The only known extant site is an arable field at Carpenters Farm, St Helens 6188, RK 1973, where it appears in varying quantities in most years. It was frequent here in 1998. Map 19

R. auricomus L.
GOLDILOCKS BUTTERCUP
N. Uncommon and local in ancient woodlands; always in small quantity. Much declined, perhaps though lack of woodland management. There is only a single extant record from the southern half of the Island. Map 20

R. sceleratus L.
CELERY-LEAVED BUTTERCUP
N. Local but widespread in ditches, streamsides and ponds, especially where mud is exposed. It is particularly widespread in the Eastern Yar valley. An area of several hectares of grazing marsh, extensively disturbed in 1999/2000 by pipe-laying at Sandown Levels, 6085, was dominated by this plant the following season.

R. lingua L.
GREATER SPEARWORT
N/E Very rare, and only known as a native from Freshwater Marshes, 3486, CP 2002, where it has declined greatly and is no longer easy to see. It still occurs sparingly amongst reeds in the south marsh but has not been seen from the site where Blackbridge Road crossed the brook since 1988. Increasing as an established alien in artificial ponds and reservoirs, e.g. Branstone Farm, Winford 5583, CP 1997; farm reservoir at Scotchells Brook 5783, CP 1983; Stag Lane reservoir 4991, CP 1999.

R. flammula L.
LESSER SPEARWORT
N. Frequent and widespread in wet pastures, marshes, ponds and wet woodland rides. Absent from the chalk.

R. ficaria L.
LESSER CELANDINE
N. Common and widespread in woodlands, hedge banks, wet meadows and as a garden weed. Our common plant is ssp. *ficaria*. Ssp. *bulbilifer* is less frequent (but under-recorded) in damp, shady places. Ssp. *ficariiformis* is occasionally recorded as an established weed, Freshwater Bay 3485, TDD & CDP 1996; and Appley, Ryde 6091, CP 2001.

R. hederaceus L.
IVY-LEAVED CROWFOOT
N. Much declined and now rather scarce on wet, bare muddy ground, or in still, shallow water, principally in the Eastern Yar catchment. Sites include Roughland Cliff, Brook 3983, ST 1993; Godshill Park 5380, AC 2000; Wydcombe 5078 CP 1998; and Bohemia Bog 5183, CP 1993. Map 21

R. omiophyllus Ten.
ROUND-LEAVED CROWFOOT
N. In similar situations to the last species, and sometimes growing with it, but apparently rarer. Sites include Bohemia Bog 5183, CP 1993; Cridmore Bog 4981, CP 1998; Bridge Farm, Godshill 5181, AC 2000 and Alverstone Marsh 5885, AC 2002. Map 22

R. baudotii Godr.
BRACKISH WATER-CROWFOOT
N. Rare, but perhaps under-recorded.

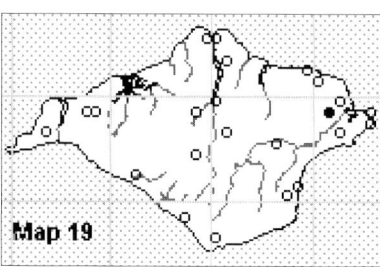

R. arvensis

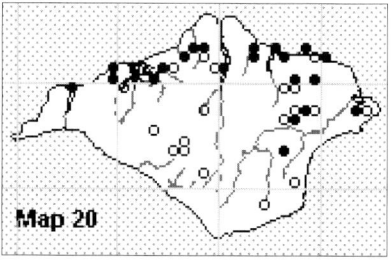

R. auricomus

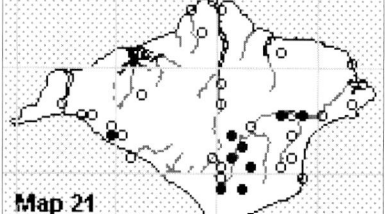

R. parviflorus

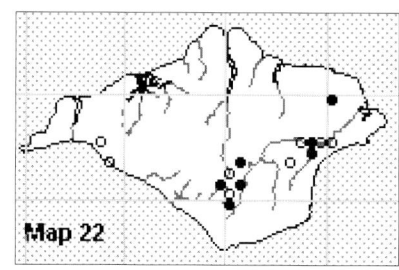

R. omiophyllus

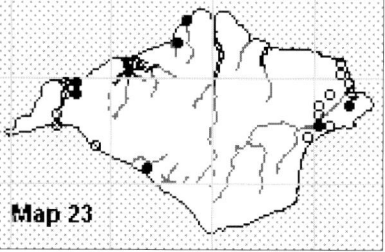

R. baudotii

R. hederaceus

Locally frequent in pools on Brading Marshes 6387, CP 1996; ditches on Sandown Levels 6085, CP 1997; Thorley Marsh 3989, CC 1989 and other pools in the vicinity; Thorness Marshes 4693, CC 1987; pond at Thorncross 4381, TDD & CDP 1996. Historically recorded from Seaview and St Helen's Duver, and other ponds along the south-west coast; it may still be present in the latter. Map 23

[R. x segretii A. Felix
BRACKISH X THREAD-LEAVED WATER-CROWFOOT
N. The origin of the report in Stace (1975) has not been traced.]

R. trichophyllus Chaix
THREAD-LEAVED WATER-CROWFOOT
N. Very local in ponds and ditches. Frequent in ponds at Elmsworth Farm, 4392, AC 2000; pond at Watershoot Bay, St Catherine's 4975, JO 1998; pond on Roughland Cliff, Brook 3983, CC 1987; Thorley and Barnsfield Brooks 3688, CC 1987; ditches on Sandown Levels 6085, CP 1997; Shalcombe Manor pond 3985, CP 2000.

R. aquatilis L.
COMMON WATER-CROWFOOT
N. Rather scarce in ponds and ditches, principally on the north side of the Island but found occasionally in pools on the undercliffs of the south-west coast. Some of the earlier records may be erroneous, through confusion with R. peltatus and R. baudotii.

R. peltatus Schrank
POND WATER-CROWFOOT
N. Rare in ponds and ditches. There are recent records from a pond at Tapnell Farm 3786, CP 1990; Sandown Golf Course 5885, MB 1996; and a reservoir at The Wilderness 5082, BG 1999.

R. penicillatus (Dumort.) Bab.
STREAM WATER-CROWFOOT
E. Not native in our streams but at one time introduced and established in the stream in Ventnor Park 5577, 1987 JB et al; and the Medina at Pan Mill 5088, 1996 TDD & CDP where it was first noticed in 1994 and is now (2002) frequent. Our form is subsp. pseudofluitans (Syme) S.D. Webster.

Recorded erroneously in Bevis et al (1978) as R. fluitans.

[Adonis annua L.
PHEASANT'S-EYE
Arch. Extinct. Formerly a scarce cornfield weed, which seems to have disappeared around the turn of the 20th century. It was widespread and particularly frequent in the upper cornfields, and on unbroken ground immediately below the cliff edge above Steephill. Last recorded in 1918 from Alverstone by Rev. D.M. Heath (BIRM).]

Myosurus minimus L.
MOUSETAIL
N. Widespread but rare, in damp corners and gateways of arable fields and trampled parts of meadows, principally on clay. Although it has probably declined, it is easily overlooked and numbers fluctuate considerably. Generally abundant at Lock's Green Farm, Porchfield 4490, 2000 BA; field behind town hall at Newtown 4290, 2000 BA; Baskett's Farm, Rew Street 4794 and elsewhere in this area, 1996 JO; Newnham Farm, Quarr 5691, CP 2001; field at Chilton Green by Military Road 4182, 2000 PS. Also recorded in the last thirty years from Sandford, Godshill 5481, BS 1970; New Fairlee Farm, Staplers 5189, BS 1981; and Havenstreet 5690, CP (until 1982) and may still survive in these places. Appeared in quantity on soil deposited by the estuary at Yarmouth 3589, 2000 JKN & VG; the soil is believed to have come from Fairlee. 2001 was a good year for this species, and a targeted search by PS & MB in gateways yielded eleven new sites, mostly in south Wight. Map 24

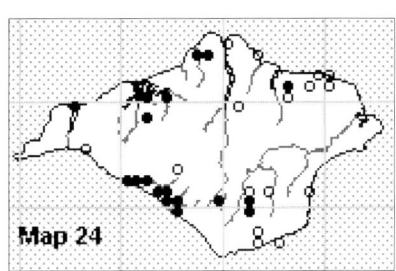

Myosurus minimus

Aquilegia vulgaris L.
COLUMBINE
N?/E Occasional in woods and thickets on chalk where it may possibly be native in some sites, eg. Chillerton Down, adjoining Tolt Copse 4884, 1998 BS; Monkham Copse, Rowridge 4586, 1978 JB *et al*. It is also a widespread garden escape, generally, but not always, near areas of habitation.

(*Thalictrum flavum* L.
COMMON MEADOW-RUE
N? Extinct. Doubt has been cast upon Bromfield's record (1856), of it growing sparingly in wet pasture at the mouth of Wootton Creek. No herbarium material has survived and it was not found subsequently by other botanists. However, the record would not be entirely unexpected, as it is still very locally frequent by Hampshire riversides, including some on the north Solent shore. A.G.More (1871) also refers to a note in Major Smith's copy of *Flora Vectensis*, 'formerly in Lee meadows but now extinct, E.M.' Lee meadows was believed to be a site in East Wight.)

31. BERBERIDACEAE
Barberry family

Berberis vulgaris L.
BARBERRY
E? Generally suspected of being introduced, although long naturalized in hedges. It has apparently always been rare, with a thin scatter of localities, although it is easily overlooked. Recent records are: a single large roadside bush at Blackwater 5086, BS 1978, still present 1992; two bushes in species-rich hedge, Whiterails Road, Wootton Common 5390, AM 1998; two bushes in the roadside hedge at Parkhurst Military cemetery 4890, MB 2001; and two bushes by Yarmouth Road at Ningwood 4089, AR 2000, which were probably originally planted and have since been grubbed up.

Mahonia aquifolium (Pursh) Nutt.
OREGON-GRAPE
S. A very occasional relic of habitation or cast-out in woods, hedges and churchyards. Not spreading.

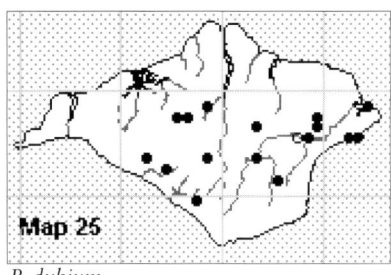
P. dubium

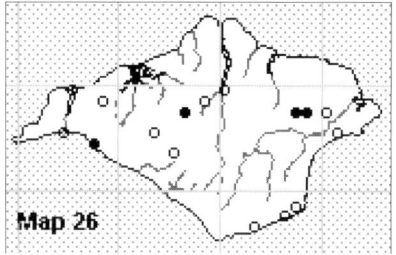
P. hybridum

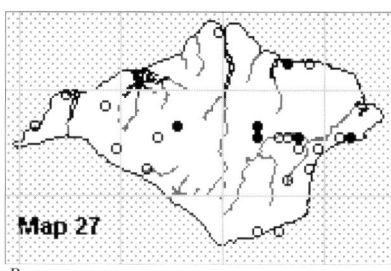
P. argemone

32. PAPAVERACEAE
Poppy family

Papaver atlanticum (Ball) Coss.
ATLANTIC POPPY
E. Very rarely established on waste ground and walls near houses. Central Ryde 5992, CP 1998.

P. somniferum L.
OPIUM POPPY
Arch. A frequent garden escape on rubbish dumps and tipped soil.

P. rhoeas L.
COMMON POPPY (local: red-weed)
Arch. Widespread and frequent on arable and fallow grounds, occasionally turning cereal fields scarlet; least frequent on clay soils. On Highdown Cliffs, 3184, CP 1986 many plants were growing down a rough track from the cliff top, well away from any nearby site; perhaps they were originally introduced from the boots of men clambering down the cliff for seagull eggs. Bromfield (1856) reported that around Godshill and elsewhere poppies used to be collected and used to feed pigs.

P. dubium L. ssp. *dubium*
LONG-HEADED POPPY
Arch. Occasional arable plant on chalky or sandy soils, also along verges and cliff edges. It is never found in quantity. Map 25

P. dubium L. ssp. *lecoqii*
YELLOW-JUICED POPPY
Arch. Rare on waste ground and dunes, but a frequent garden weed about Ventnor 5577, CP 2002. St Helen's Duver 6388, AC 1993; waste ground on St Helen's Green 6289, JO 1995.

P. hybridum L.
ROUGH POPPY
Arch. Greatly declined from arable fields in the last one hundred years, and only ever found in small quantity today. It is however, not quite as rare as suggested by Bevis *et al*. (1978). Ashey Down 5787, CP 1987 not seen since; West Nunwell Down 5887, CP 1996; Bowcombe Down 4687, SB 2000; Compton Grange 3784, BS 1990. Map 26

P. argemone L.
PRICKLY POPPY
Arch. Also greatly declined, and now extremely local in arable fields. Arreton 5386, JRM 2000; Bowcombe 4586, SB 2000; Hill Heath, Alverstone 5785, PJW 1997; Redcliff 6285, JEG 1991; Redway 5385, CP 1994. There is also a small, spring-flowering population on the old ruins of Quarr Abbey 5692, CP 2002, which is believed to be long established. Map 27

Meconopsis cambrica (L.) Vig.
WELSH POPPY
E. Very rare as a persistent weed. Occasional pavement weed on Newport estates 5089, BS 1993; numerous plants along the cliff top at Lake 5882, BS 1993.

Glaucium flavum Crantz
YELLOW HORNED-POPPY
N. Locally frequent right around our coastline, on undisturbed shingle beaches on the north coast and on chalk cliffs; very occasionally on clay undercliffs on southwest coast. Known since 1856 from the inland north-facing chalk pit on Ashey Down where it still occurs in small quantity 5887, CP 1991. It also appears occasionally on disturbed chalk elsewhere in the vicinity e.g. a chalk track in Knighton West Wood 5686, BW 1989; and by new reservoir on Ashey Down 5887, CP 1991.

Chelidonium majus L.
GREATER CELANDINE
Arch. An ancient introduction, persistent in lanes and hedge banks near villages. Formerly widespread, but much less frequent today. Still occurs at Carisbrooke Castle 4887, MB 1996; Rowborough valley 4584, MB 1998; Apes Down Farm, Rowridge 4587, DD 1998; Newchurch Shute 5685, PS 2001; Ramsdown 4882, MB 1999; and Brading 6086, GT 2000. Bromfield recorded that the expressed juice of this plant was 'in vogue with the country people of the island as a remedy for infantine jaundice'.

Eschscholzia californica Cham.
CALIFORNIAN POPPY
C. A casual from gardens on disturbed ground.

33. FUMARIACEAE
Fumitory family

[***Corydalis solida*** (L.) Clairv.
BIRD-IN-A-BUSH
C. The only record is from near Cowes, 1929, J.W. Long.]

[***C. cava*** (L.) Schweigg. & Körte
HOLLOWROOT
C. Also recorded from near Cowes, 1929, by J.W.Long. This species may have been confused with *C. solida*.]

Pseudofumaria lutea (L.) Borkh.
YELLOW CORYDALIS
E. Widespread on old walls, often in shade; sometimes a persistent garden weed. Frequent in most of our towns and villages. The first localised records date from 1929 although Bromfield recorded it as 'subspontaneous on old walls'.

P. alba (Mill.) Lidén
PALE CORYDALIS
C. Very rare garden escape. The only

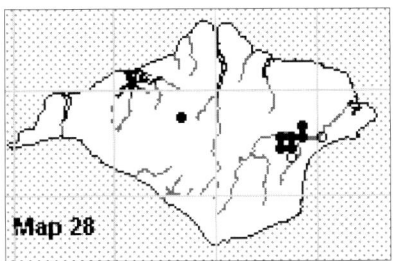
Ceratocapnos claviculata

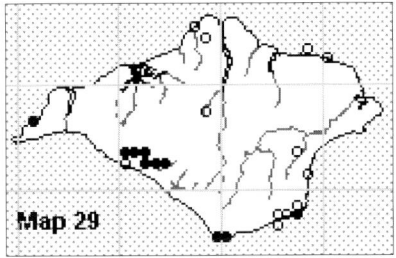
Fumaria capreolata

records are from an old wall at Weston Manor, Freshwater 3286, PS 2000; and St John's churchyard, Niton 5076, MB 2000.

Ceratocapnos claviculata (L.) Lidén
CLIMBING CORYDALIS
N. Very local in open woodland, amongst bracken and in ancient hedged field banks on greensand. This plant has a remarkably restricted distribution within an area of the Eastern Yar valley bounded by Rowdown Copse 5886, Sandown Golf Course 5885, Borthwood Copse 5784, Winford Plantation 5684 and Lynch Copse, Newchurch 5685. It is, however, abundant at suitable sites in this locality. Thirteen extant sites have been recorded. It also occurs at a single site 11.5km NE of its centre of distribution, in a sunken lane leading to Bowcombe Down 4687, CP 1998. Map 28

Fumaria capreolata L.
WHITE RAMPING-FUMITORY
N. Rare and may have declined over the last one hundred years. Generally casual in its appearance, although it can be locally frequent, It has been persistent on the slumped cliff below St Catherine's lighthouse 4975 since at least 1989, BS; and has been found quite widely around Brighstone 4282, 4283, 4382, MB 2000, and over a longer period of time. Material from Brighstone has been determined as ssp. *babingtonii*. Map 29

F. bastardii Boreau
TALL RAMPING-FUMITORY
N. Rare on waste and cultivated ground, always in small quantity. Colwell 3287, PS 1994; Norton Spit 3589, JKN, where it appeared for the first time in 1995; Springvale 6192, PS 1992; Thorley 3788, PS 1997.

Records suggest that this species may always have been uncommon on the Island.

F. reuteri Boiss.
MARTIN'S RAMPING-FUMITORY
N? Known only from Lake allotments 5883, where it was first recognized in 1963 by Alick Westrup. It remains an abundant weed and is encouraged by allotment holders. The only extant British sites for this rare fumitory are here, and a site at Pulla Cross in West Cornwall. It is a showy plant with a long flowering period. The Lake allotment site is host to several different fumitory species. It was doubtless more frequent in this area at one time, and several plants were found growing on a nearby sandy laneside bank at Blackpan 5883 in 1997, CP. In the mid 1980s, the allotment site was under threat of development and, as a safeguard, some populations are maintained in private gardens. Part of the site was lost to development but the remaining portion has been designated an SSSI for its arable weed flora. Preston *et al* (2002) consider this to be an alien species, first recorded in this country in 1904.

F. muralis Sond. ex W.D.J. Koch
COMMON RAMPING-FUMITORY
N. This is our second most frequent and widespread fumitory, possibly having increased since Bevis *et al.* (1978). It occurs on waste ground, cultivated land and hedge banks. It is more frequent on sandy soils, but by no means confined to them. The only confirmed subspecies recorded is ssp. *boraei* (Jord.) Pugsley.

F. purpurea Pugsley
PURPLE RAMPING-FUMITORY
N. Very rare. Known only from Lake allotments 5883, PJW 1997, where it

is persistent and not uncommon, although less frequent than *F. reuteri*. First recorded here in 1974, ABr, conf. P.D. Sell. A pre-1987 record for SZ47, which appears in Preston *et al* (2002), has not been verified. The Island appears to be something of an outpost, with the majority of extant sites in the west and north of Britain.

F. officinalis L.
COMMON FUMITORY
Arch. Our commonest species, widespread and not infrequent on disturbed soil, gardens and waste ground. Ssp. *wirtgenii* (Koch) Arcang has been recorded from disturbed arable ground at Northwood 4993, PS 2000 and may be overlooked elsewhere.

(*F. densiflora* DC.
DENSE-FLOWERED FUMITORY
There are two old and unconfirmed records. A single specimen found by Thomas Bell Salter in 1843 at Week's field, Ryde was later suspected of being in error. A single solitary specimen found on waste ground at Yarmouth in 1865 by Dr G. R. Tate was not preserved.)

F. parviflora Lam.
FINE-LEAVED FUMITORY
Arch. Extremely rare, and known from just one old and one modern record. A specimen was found on West High Down, Freshwater, by J. Pollard in 1929. Two plants were found growing on arable ground, which had been disturbed by trench digging, in a field opposite Wellow Farm 3988, PS 2000. This plant is largely confined to arable fields on chalk in this country, and the Wellow site is over Bembridge limestone. Its occurrence here suggests that it may have been overlooked in the past.

[*F. vaillantii* Loisel.
FEW-FLOWERED FUMITORY
Arch. Very rare, and perhaps extinct. Historically, there have been very few records of this species. The only records in the past one hundred years have been from a garden at Brighstone 4282 and an arable field between Shide and Pan Mill 5083, both reported in Bevis *et al* (1978) but without dates.]

35. ULMACEAE
Elm family

Ulmus glabra Huds.
WYCH ELM
N. Scattered across the Island in hedgerows and woodlands, but particularly associated with ancient woods, generally on basic clay soils and chalk. Characteristic of ancient woods on rock or scree of steep escarpments, e.g. Cliff Copse, Shanklin 5680 and older woodland elements along the Undercliff such as at Bonchurch Landslip 5878, Steephill 5577 and St Lawrence 5276. There were large old coppice stools in some woods, e.g. Bloodstone Copse 5887 and Tolt Copse, Gatcombe 4884. However, most larger trees died following the outbreak of Dutch elm disease in the 1970s and the majority of trees seen today were either too young at the time to contract the disease, or have grown since.

The woodland at Fort Victoria 3889 is remarkable in being largely dominated by Wych Elm, although the wood is secondary in origin. It is probably the best Island site for this species.

U. x hollandica Mill.
DUTCH ELM
P. Rare introduction but under-recorded and previously confused with *U. procera* and *U. glabra*. Believed to still survive in hedgerows Weston Manor, Freshwater 3286, JH 1966.

U. procera Salisb.
ENGLISH ELM
N? Common across the Island but with a somewhat localised distribution, centred around old settlements. No large trees survived the attack of elm disease in the 1970s. Generally a hedgerow tree, where it is frequently dominant and survives by sending up suckers: it also sometimes gives rise to dense woodlands. Bill Shepard (1978) wrote, 'my colleague Bill Wickens, who spent his childhood at Calbourne, told me it was customary to smoke elm root in preference to wild woodbine (clematis) because elm did not burn the tongue and was considered to be a better flavour by the boys.' Elmsworth, near Porchfield, derives from the Old English meaning 'the shore by which elm-trees grow' (Mills, 1996).

U. minor Mill. ssp. *minor*
SMALL-LEAVED ELM
P? Rarely planted in hedgerows but sometimes now found growing as large trees. Brook Hill 3984 1999; Toll Bar Plantation, Hulverstone 4083, CP 2000.

U. minor ssp. *angustifolia* (Weston) Stace
CORNISH ELM
P. Occasionally planted in the past. Examples include a plantation at Quarr 5692 JH 1963 and a mature tree well over 30m in height at Barfield, Ryde 5992 RL 1965. It is susceptible to Dutch elm disease and mature trees are no longer seen.

U. minor ssp. *sarniensis* (C.K. Schneid.) Stace
JERSEY ELM
P. At one time much planted in north-east Wight, probably at the time of Prince Albert, around the Osborne Estate, Whippingham railway station, Quarr Abbey and Lakeside, Wootton. These trees subsequently became a distinctive landscape feature. Some specimens have regrown from suckers to reach currently up to 10m in height e.g. in hedgerows at Woodhouse Farm, Osborne Estate 5393, BS 1991 and near Whippingham station 5291, CP 2000. There is a more recently planted line of trees near the White Lion Inn at Arreton 5386, RL 2000.

36. CANNABACEAE
Hop family

Cannabis sativa L.
HEMP
C. Very rare casual on waste ground. Sometimes found, obviously planted, in obscure locations.

Humulus lupulus L.
HOP
N. Widespread and frequent in fens, woodland edges and hedgerows, particularly on nutrient rich soils in river valleys. It is probably also a long established escape. The earliest local reference to hop cultivation is Sir John Oglander's Diary entry of 1625 where he refers to building East Nunwell and gardens, including 'hoppegardens' in 1609 (Oglander Mss.). Bromfield refers to an abandoned hop garden at Kerne, near Knighton, as the only place he knew where they were cultivated. He also refers to wild hops gathered by country people for brewing, as a substitute for cultivated hops bought in from Kent and Surrey.

37. MORACEAE
Mulberry family

Ficus carica L.
FIG
S. Known in the wild from a single bush on the southern slopes of Bembridge Down below the Fort 6285, CP 2000. It is about 3.5m high, and occasionally bears fruit. It was first recorded here, as a large bush, in 1951. It is growing on chalk rubble created when the ditches of Bembridge Fort were dug between 1862 and 1867, and its origins may date back to this time. Whitehouse (1923) in *Bembridge: An historical and general survey*, comments 'One of the outstanding characteristics of the area is the numerous fig-trees growing in the cottagers' gardens and reaching an enormous height. The green figs are gathered in August and taken to local towns to be sold.'

38. URTICACEAE
Nettle family

Urtica dioica L.
COMMON NETTLE
N. Common everywhere on nutrient-rich soil. The name of Nettlecombe, near Whitwell, is said to derive from Old English meaning 'the valley where nettles grow' (Mills, 1996).

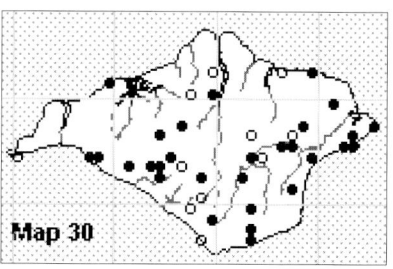

Map 30

U. urens

U. urens L.
SMALL NETTLE
Arch. Locally frequent, on light, dry arable and disturbed ground, especially on greensand. Map 30

Parietaria judaica L.
PELLITORY-OF-THE-WALL
N. Widespread but local, particularly on cliffs, shady hedge banks and walls. In churchyards and village walls, it may be a relic of herbal cultivation, but it is native on coastal cliffs. Victorian writers made reference to the abundance and luxuriance of this plant on the cliffs at Freshwater, which remains the case today. Map 31

Soleirolia soleirolii (Req.) Dandy
MIND-YOUR-OWN-BUSINESS
E. A popular Victorian garden and greenhouse plant, which thrives on shady walls and damp banks. It was first recorded as a well-established alien by J.W. Long, in 1931 from Totland and Bonchurch Old Church. It is frequent near habitations, particularly in our Victorian towns and along the Undercliff.

39. JUGLANDACEAE
Walnut family

Juglans regia L
WALNUT
P. Widely planted, and in the past, commonly near farms. A number of these still survive e.g. at the top of Plaish Lane, Bowcombe 4787, BS 2000; Burnt House Lane at entrance to Little East Standen Farm 5287, BS 2000; Idlecombe Down, well away from current habitation 4685, SB 1996; two opposite Wellow Farm 3987, CP 2000. There are several on Mottistone Green 4083, BS 2000.

Pterocarya fraxinifolia (Poir.) Spach
CAUCASIAN WINGNUT
E. Introduced, and established by suckering in woodland at The Keys, Binstead 5792, ST 1994.

40. MYRICACEAE
Bog-myrtle family

Myrica gale L.
BOG-MYRTLE
(local: sweet withy; golden osier)
N. Very rare although sufficiently familiar amongst country folk to have acquired two local names. Always confined to peaty areas in the river valleys of the Eastern Yar, and the Medina between Cridmore and Rookley and has declined as suitable habitat has been lost. A.G. More described Bog-myrtle in The Wilderness in 1860 as forming thickets 'so tall as to resemble Arbutus'. It still occurs here, but in much diminished quantity, where it struggles to survive in two patches within willow carr 5082, CP 1996. In the Eastern Yar, it grew in many sites between Sandown Waterworks and Munsley Bog, Godshill, but it seems to have been lost from most of these by the first half of the 20th century. A few plants were last recorded from a small boggy carr near Sandown Waterworks, 5885, in 1972 (RK). Four plants were struggling to survive in willow carr at Munsley Bog 5282, a former stronghold, in 1991 (CP) and three plants were still here in 2002. Material from this site currently survives in cultivation. Map 32

41. FAGACEAE
Beech family

Fagus sylvatica L.
BEECH
E. A common tree, which is widely planted. Beech woods occur in a few sites such as Brighstone Forest and Nunwell Hanger but, in direct contrast to Hampshire, beech hanger woodlands are not characteristic of the Island's landscape. There is no strong evidence to suggest that it is native on the Island, and no trees are known which pre-date the trend for planting Beech over the past three hundred years. There have been suggestions that it may be of native origin, supplemented by later plantings, in the northern half of Parkhurst Forest around Mark's Corner. The largest tree here is situated close to Mark's Corner and, with a girth of 4.8m (TR), is likely to pre-date the Forest enclosure of 1812. Parkhurst has many affinities with the New Forest, where Beech is undoubtedly native. A tree in the garden of Stonelands at Binstead 5792, RL 2000, is believed to be the largest specimen on the Island but this is a bundle-planted tree, originally of seven stems, now reduced to six. It is 42 m in height. A few of the old Beech trees in the north east corner of Parkhurst Forest have the appearance of having been bundle-planted.

Castanea sativa Mill.
SWEET CHESTNUT
Arch. A widespread and locally frequent tree, sometimes naturalising. Long planted in woodlands and parkland. It was sometimes grown as coppice as in Combley Great Wood, 5488, Parkhurst Forest 4790, and Borthwood Copse, 5784, where it is still actively managed, or in Lorden Copse, Shorwell, 4683, where some old stools survive. Sometimes old maiden trees are found in woods, e.g. America Woods 5681. Bill Shepard records an example of trees formerly planted alongside a village blacksmith's shop at the site of The Old Forge at Garretts Farm on St George's Down 5187, still growing in 1991. The largest recorded Island chestnut is in the grounds of Sydney Lodge, Shide 5088, and had a girth of 8.23m in 1993, BS.

Quercus cerris L.
TURKEY OAK
E. Locally well-established and increasing in woods, and also in open scrub.

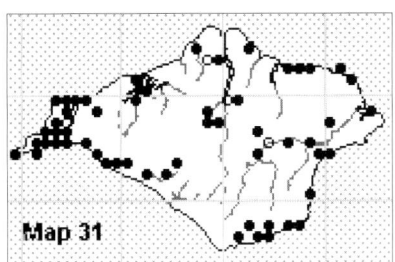

Parietaria judaica

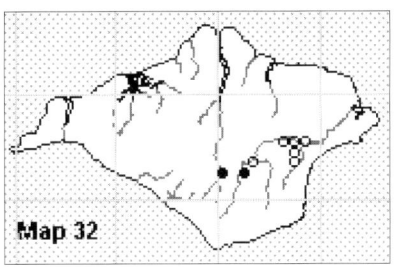

Myrica gale

Q. *ilex* L.
EVERGREEN OAK

E. A popular Victorian tree which now is widespread and increasing. It spreads invasively within established woods, on slumped cliffs and on inland cliff faces. On the Ventnor Downs, a well-established, almost pure, holm oak woodland has colonised the chalk grassland, a consequence of seed dispersal from the gardens of a nearby Victorian villa, with some deliberate planting. The first record of it establishing in the wild was in 1924, by which time it was already present in considerable numbers on the south face of St Boniface Down, but controlled by donkey grazing. During the Second World War, grazing was no longer practised and there was a substantial increase. Today, it occupies some 15 ha on the southern slopes of St Boniface Down together with over 8 ha on Bonchurch Down, making this the largest area of holm oak woodland in Britain. The National Trust has endeavoured to control its spread by a programme of cutting, flailing and poisoning. In 1993, they introduced feral goats onto the downs to control the spread and to reclaim chalk grassland, with some success.

Q. *petraea* (Matt.) Liebl.
SESSILE OAK local: white oak

N. Very local in ancient woodland on drier soils, sometimes planted. It is probably under-recorded. The best stands of lowland sessile oakwood over sparse coppice are in Briddlesford Copse 5490, but it also occurs in Firestone Copse 5591, Combley Great Wood 5489, King's Quay Woods 5493, and America Wood 5681. In the northern half of Parkhurst Forest 4792, there was much replanting of sessile oak into existing woodland following the enclosure of 1812, although some extant trees pre-date this period. The largest of these trees, near Pallance Gate, has a girth of 4.1m (TR). There are two fine trees at The Priory, Nettlestone 6390, most certainly planted; in 1985, one had a girth of 3.75m and a height of 35.7m and the second a girth of 5.18m and a height of 28.3m (RL).

The best Island example of sessile oak coppice is found in Borthwood

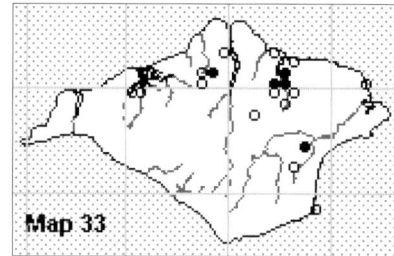

Q. petraea

Copse, 5684. Clifford Matthews recalls how the oak coppice was cut every two years in the spring, just as the sap was rising. The stripped bark was sent to the two tanneries on the Lukely stream in Newport, and any excess was shipped from Newport harbour to tanneries in Portsmouth. The tannery industry was well established in Newport as early as the seventeenth century. At that time, tannin extracted from oak bark was used in the preparation of hides for leatherworkers and shoemakers. A by-product of oak coppice was large quantities of small oak wood, which was split up for 'nicky' or kindling wood. (Jones, 1979). Map 33

Q. x *rosacea* Bechst.
SESSILE X PEDUNCULATE OAK

N. Less common than either parent but probably overlooked. Frequent in Borthwood Copse 5684, EN 1981; Alverstone 5785, TDD & CDP 1998; Bouldnor 3789, PS 1999; Whippingham 5193, PS 2000.

Q. *robur* L.
PEDUNCULATE OAK

N. Abundant and widespread almost everywhere in woods, scrub and hedgerows. It is also characteristic of open pasture on the clays of the north side of the Island. The largest trees are open grown in historic parkland situations, such as the many veteran oaks on the Nunwell Estate at Brading. About half a dozen of these have a girth in excess of 6m and the largest, in a field to the northeast of the manor, has a girth of 7.6m. Measured girths of other large surviving individuals are at Fernhill, Wootton Bridge, 5491, 6.2m; Newnham Farm, Quarr, 5691, 6.1m; Lower Knighton, 5786, 5.8m; and St Cecilia's Abbey, Appley, 6092, 5.4m. The exposed south-western quadrant of the Island is not known for large trees but there is a remarkable leaning oak by the millstream above Brighstone Mill, 4282, with a girth of 5.3m. It is known locally as the 'dragon tree'.

Oak, as a vital component of the rural economy and a venerable tree, has attracted much folklore and superstition. The local diarist, Sir John Oglander (Bamford, 1936) described a 'love feast' given in Parkhurst Forest in honour of a departing dignitary in 1596. The party was held under two oaks named 'My Lady's Oak' and 'My Lady Elizabeth's Oak', described as 'exceeding spreading and thick' trees, and 'every several arm and bodies of the trees were all stuck over with gilly flowers.' Royal Oak Day was commemorated on 29 May each year and a number of local customs are recorded. The I.W. County Press for 28 May 1910 reported, "I have vivid recollections that in the days of my youth, pinching penalties were imposed upon those luckless youths who failed to respond to the challenge to show your oak on this day". J.D. Lavers wrote in 1994, 'At Newchurch and, I believe Arreton, Oak Apple day was known as Chick Chack day. At Newchurch School, older boys hit the legs of children not wearing a sprig of oak leaves with nettle bunches. The practice ended at mid-day and by tradition anyone still with a bunch of nettles had to submit themselves to the same treatment.'

Q. *rubra* L.
RED OAK

P. Locally planted in forestry plantations. Occasionally regenerating from seed e.g. in Whitefield Woods 6089, AC 1996; and Combley Great Wood 5489, CP 2000.

42. BETULACEAE
Birch family

Betula pendula Roth
SILVER BIRCH

N. Widespread and quite frequent, generally on light acid soils in woodland clearings. Sometimes found on chalk.

(*B.* x *aurata* Borkh.

N. The hybrid between our two native

birches probably occurs where both parents grow together but has not yet been positively identified.)

B. pubescens Ehrh.
DOWNY BIRCH
N. Widespread and frequent, particularly in damp open woodland, carr and moors. Birchmoor Farm, near Rookley, derives its name from the Old English meaning 'the moor or marshy ground where birch trees grow'.

Alnus glutinosa (L.) Gaertn.
ALDER
N. Frequent, sometimes locally dominant, along river valleys and on base-poor spring lines. At Gatcombe Withybed 5085, there are old stools in the western arm of the wood where fossil pollen evidence has indicated a continued presence of alder carr woodland stretching back some eight thousand years. In recent years, alder has spread extensively here into the field at the northwest end of the withybed, RL 2000. Absent from the chalk.

Carpinus betulus L.
HORNBEAM
P. Not native. Commonly planted in woods and sometimes in hedges, mostly on the northern half of the Island. It occasionally regenerates from seed. There are shelterbelt plantings of hornbeam on the Briddlesford Estate and at Fattingpark. There are extensive coppice plantations at Noke Plantation 4891, and Stag Copse 4991, which it has been suggested are of Victorian origin, and associated with the prisons. Hornbeam used to be planted as a source of hardwood for cogwheels and as a source of charcoal. Map 34

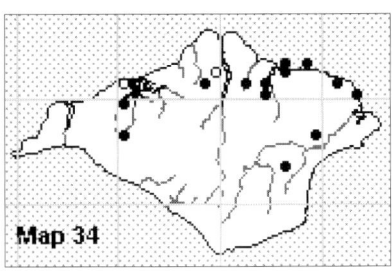

Carpinus betulus

Corylus avellana L.
HAZEL
N. Common and widespread right across the Island, excepting for some exposed parts on the south-west coast. It is an abundant understorey shrub, and the mainstay of the former coppice industry. It is also a major component of ancient hedgerows. Occasionally, it may colonise grassland near old woods invasively, as in the Rowridge valley, 4586. Hazel was important in the rural economy. Heasley, near Arreton, is Old English for 'the wood or woodland clearing where hazels grow' and the name Nettlestone, near Seaview, is derived from 'the farmstead in or near the nut-tree pasture or nut-tree wood' (Mills, 1996). The Isle of Wight County Press for 23 December 1933 recorded how, 'before the War there were four times as many men employed in the (hurdle making) trade, compared with barely a dozen in the whole Island today'. There is little coppicing today except for conservation purposes.

44. AIZOACEAE
Dewplant family

Aptenia cordifolia (L.f.) Schwantes
HEART-LEAF ICE-PLANT
C. A garden escape recently spreading from a cliff top garden onto the cliff at Forelands 6587, AC 1999.

Disphyma crassifolium (L.) L. Bolus
PURPLE DEWPLANT
E. An established garden escape around Ventnor. Known for at least twenty-five years from south-facing rock outcrops at The Cascade, Ventnor 5677, CP 2001. In recent years appearing elsewhere in the vicinity: talus below cliff, west of Bath Road, Ventnor 5577, CP 1998; large patch on earthbank at east end of Steephill Cove, present since at least 1989, 5576, CP 2001; south-facing garden wall behind Bonchurch Pond 5778, CP 2000.

Carpobrotus edulis (L.) N.E. Br.
HOTTENTOT-FIG
E. An established garden escape on cliffs on the south coast. There are large spreads on the cliff face just south of the termination of Sandown Esplanade, competing with invading holm oak and flowering well into the winter in favourable seasons 5983, CP 2000. It was introduced here in about 1936, from the Highcliffe-Bournemouth area by Ivan Hooper, in an attempt to arrest cliff erosion by planting up the slope behind the newly-made revetment. It has appeared more recently on a cliff top at Forelands 6587, AC 1999; abundantly at Steephill Cove 5577, CP 2001; and in Ventnor Bay 5677, CP 2001. These are all recent sites where the plant has yet to withstand the test of a hard winter.

Tetragonia tetragonioides (Pall.) Kuntze
NEW ZEALAND SPINACH
C. Very rare casual. The only record is of several plants on a bonfire site at St Helens Green 6289, AC 1998.

45. CHENOPODIACEAE
Goosefoot family

[*Chenopodium bonus-henricus* L.
GOOD-KING-HENRY
Arch. Extinct. Formerly uncommon on grassy verges or in farmyards, with no sites showing a continuity of records and no confirmed records since 1929.]

C. glaucum L.
OAK-LEAVED GOOSEFOOT
Arch. Rare on enriched waste ground with an historic continuity of records from the Brading Marshes area, and from Thorley stretching back to 1837. It is recorded from the following sites: several large populations on Brading Marshes 6187 and 6287, L&SS, 2002; Hill Farm, Carpenters 6189, AC 1998; Park Farm, Nettlestone 6190, AC 2000; field entrances near Thorley bridge 3689, PS 1999; footpath by shore just west of Burnt Wood 4392, AC 2000; and as a weed in a flowerbed at Medina Way, Newport 4989, PS 2000. Preston *et al* (2002) record this plant from just 36 hectads nationally (post 1987).

C. rubrum L.
RED GOOSEFOOT
N. Not infrequent on manure heaps, manured arable fields and rich mud by ponds. Not as uncommon as suggested by Bevis *et al* (1978). Map 35

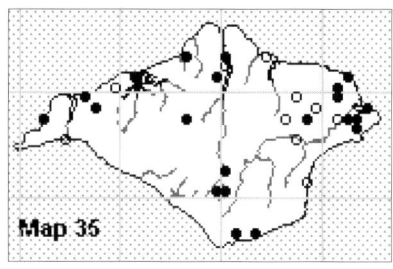
C. rubrum

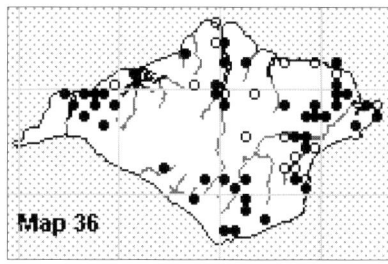
C. polyspermum

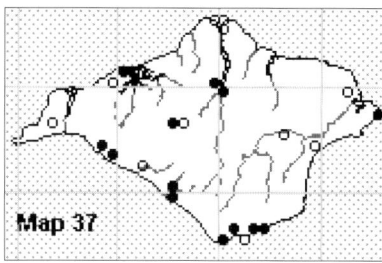
C. murale

C. polyspermum L.
MANY-SEEDED GOOSEFOOT
Arch. Locally common on manured arable fields, on mud by ponds, in farmyards and on waste ground. Map 36

[C. *vulvaria* L.
STINKING GOOSEFOOT
Arch. Extinct. Recorded by Bromfield in the nineteenth century, but only as a garden weed.]

[C. *urbicum* L.
UPRIGHT GOOSEFOOT
Arch. Possibly extinct. A casual of farmyards and waste ground, which was once not infrequent. It has not been recorded since 1964. Preston *et al* (2002) record this as a rare casual nationally since 1970.]

C. murale L.
NETTLE-LEAVED GOOSEFOOT
Arch. Scarce in farmyards, gardens and tips, particularly along the southern and south-west coasts. Although undoubtedly declined since the nineteenth century, this species is less rare than suggested by Bevis *et al* (1978). Sites include Forelands 6587, AC 2001; Bowcombe 4586, SB 2000; Dunsbury Farmyard 3884, PS 2000; coastal fields at St Lawrence 5376, CP 1999; and Atherfield 4579, PS 1999. Map 37

(*C. dessicatum* A. Nelson
SLIM-LEAF GOOSEFOOT
C. Three records belonging to this or related members of the *Chenopodium leptophyllum* group recorded by J.F. Rayner as a casual between 1924 and 1927. None since.)

C. ficifolium Sm.
FIG-LEAVED GOOSEFOOT
Arch. Scarce in manured arable fields and wasteland. Under recorded, but considered to be increasing. Map 38

C. album L.
FAT-HEN (local: lamb's quarters)
N. Cultivated and waste ground on all types of soil. It is widespread and often abundant.

Atriplex hortensis L.
GARDEN ORACHE
C. Occasionally recorded as a non-persisting garden escape on disturbed verges, waste ground and tips.

A. prostrata Boucher ex DC.
SPEAR-LEAVED ORACHE
N. Common and widespread in arable, gardens, disturbed verges, upper saltmarshes and elsewhere along the coastline.

A. x gustafssoniana Tascher.
KATTEGAT ORACHE
N. Rare but overlooked in upper saltmarshes. Recorded from Newtown estuary creek at Porchfield 4491, CP (confirmed J. Akeroyd) 2000; reedbeds behind Freshwater Church 4387; and on the eastern bank of Yar 3587, PS 2000.

A. glabriuscula Edmonston
BABINGTON'S ORACHE
N. Occasional at the strandline on pebbly beaches, often with *A. prostrata*, but probably overlooked. St Helen's Duver 6388, CP 1993; Bembridge Point 6488, AC 1996; Priory Bay 6390, AC 1999; Norton Spit 3589, JKN 1996; East Spit, Newtown 4191, CP 1997; shore near Burnt Wood 4392, AC 2000.

(*A. longipes* Drejer
LONG-STALKED ORACHE
This species may occur here; it has recently been found on the Hampshire coast and material resembling it has been seen in the Western Yar estuary. It has yet to be confirmed.)

A. littoralis L.
GRASS-LEAVED ORACHE
N. Occasional, sometimes locally common, on beaches and in estuaries, principally along the north coast of the Island. There is one inland record from the sandy beach of Leechmore Pond, Bleak Down 5080, GT 1999. Map 39

A. patula L.
COMMON ORACHE
N. Common and widespread on waste and cultivated ground, manured fields and on beaches.

A. laciniata L.
FROSTED ORACHE
N. Local and generally thinly scattered on sandy beaches, principally on the north coast. Thorness Bay 4593, CP 2000; King's Quay 5393, CP 2001; St Helen's Duver shore 6388, CP 1993; Bembridge Point 6488, AC 1996;

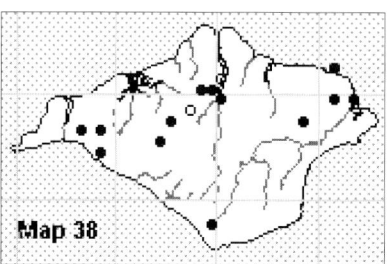
C. ficifolium

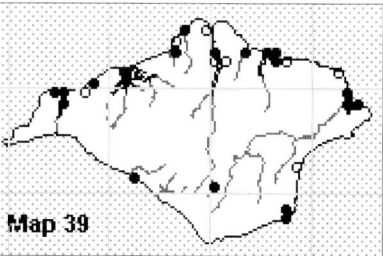
A. littoralis

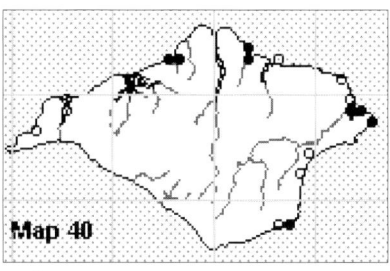
A. laciniata

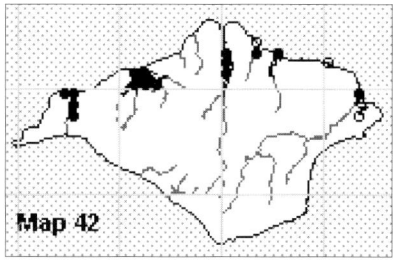
Sarcocornia perennis

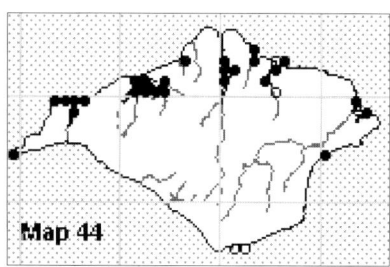
Suaeda maritima

Forelands beach 6587, AC 1996; Bonchurch shore 5777, JO 1988. Map 40

A. halimus L.
SHRUBBY ORACHE
E. Bromfield describes how this shrub was commonly grown in coastal gardens in his time and how it made 'an excellent sea fence'. Today, it is abundant and well established on the cliffs around Ventnor 5677, CP 2000. Also established on the cliffs at Orchard Bay, Steephill 5476, CP 1996. Persisting on the south facing sandy bank of Ryde Canoe Lake 6092, CP 2000.

A. portulacoides L.
SEA-PURSLANE
N. Abundant on saltmarshes in all our estuaries. It is occasionally found elsewhere behind seawalls, and rarely on clay and shingle at the base of cliffs. Abundant on chalk on the cliff top at about 60m above sea level, and north facing slopes of the Needles headland, down to sea level, 2984 CP 2000, an unusual site where it grows with other saltmarsh plants. Map 41

Beta vulgaris L. ssp. *maritima*
SEA BEET (local: wild spinach)
N. Common on upper parts of saltmarshes, shingle and sandy shores, on both chalk and clay cliffs and behind seawalls, right around our coastline. Sea beet makes a tasty vegetable; Bromfield described how, when boiled instead of greens, it was 'much relished by the poorer classes of this island'.

Sarcocornia perennis (Mill.) A.J. Scott
PERENNIAL GLASSWORT
N. This nationally scarce plant is frequent in all our estuaries, generally around creeks in upper saltmarsh. The Solent is a stronghold for this species. Map 42

Salicornia pusilla Woods
ONE-FLOWERED GLASSWORT
N. Very local in upper saltmarsh but still frequent at Norton 3589, JKN 1997; various sites in the Newtown estuary CP 1998; King's Quay 5394, CP 2002; and Medham saltmarsh, Medina, 5093 CP 1997. Perhaps overlooked elsewhere. Map 43

The annual glassworts occur in all our estuaries but, apart from *Salicornia pusilla*, they have not been critically examined. They are a taxonomically complex group and the following division into two aggregate groups follows that of Stace (1997).

S. europaea L. agg.
COMMON GLASSWORT GROUP
N. Common in all our estuaries, wherever suitable habitat exists. This aggregate includes *S. ramosissima* Woods.

S. procumbens Sm. agg.
LONG-SPIKED GLASSWORT GROUP
N. Frequent in saltmarshes in all our estuaries. This aggregate includes *S. dolichostachya* Moss.

Suaeda maritima (L.) Dumort.
ANNUAL SEA-BLITE
N. Common in all our saltmarshes. It is also occasionally found in brackish areas behind seawalls e.g. Sandown Levels 6084, CP 1997 and Fort Victoria 3389, PS 1998. Formerly occurred on clay banks above the beach at Binnel Bay 5275, CP 1986. An unusual site is on chalk at the Needles headland 2984, CP 1990, where it grows close to its altitudinal limit; it occurs both on the cliff top headland at 60m above sea level amongst fish-bones deposited by gulls, and near sea level on north-facing cliffs. Map 44

Salsola kali L. ssp. *kali*
PRICKLY SALTWORT
N. Once widespread on sandy shores about the driftline, but rare today as most suitable sites are too trampled. St Helen's Duver headland 6388, CP 2000, is by far the best site; it is plentiful here in good years. A few plants occur most years at Bembridge Point 6488, AC 1996 and at Thorness Bay 4693, AC 2000. Recorded from Norton Spit 3589, JKN 1993, and from the Newtown spits and King's Quay in the last forty years.

46. AMARANTHACEAE
Pigweed family

Amaranthus retroflexus L.
COMMON AMARANTH
C. Very rare casual of disturbed and waste land, perhaps overlooked. Known only from two early records by J.W. Long (Ryde, 1924 and Newport 1929) and, more recently

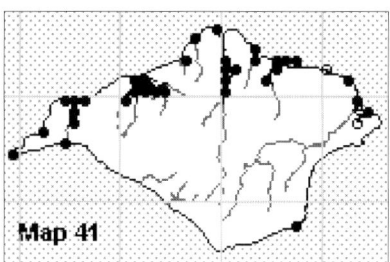
A. portulacoides

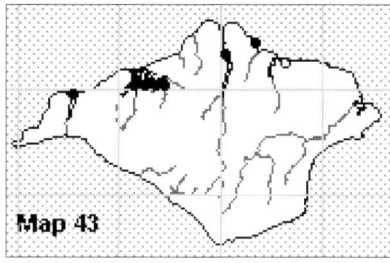
Salicornia pusilla

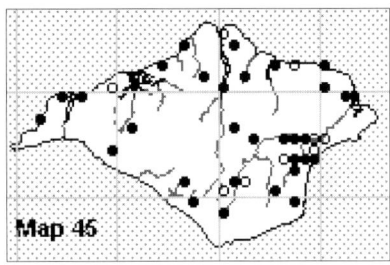
Montia fontana

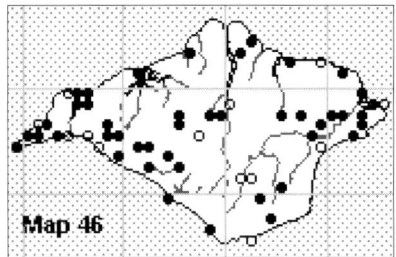
Arenaria serpyllifolia

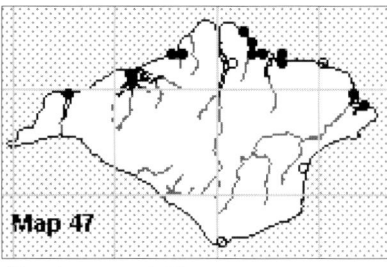
Honckenya peploides

from Horringford 5484, RJC 1978, Seaclose recreation ground, Newport 5089, PS 2001 and waste ground on St Helen's Duver 6388, PS 2001.

A. hybridus L.
GREEN AMARANTH
C. Rare casual, but less so today than *A. retroflexus*. Abandoned market garden at Barnsley Farm, near Ryde 6189, AC (conf. J.M. Mullins) 1997; weed of the margins of sweetcorn fields at Newchurch 5686, CP 1997; Stag Lane waste tip 5191, BS 1973.

A. albus L.
WHITE PIGWEED
C. The only record is from Newport, 1931, J.W. Long.

47. PORTULACACEAE
Blinks family

Claytonia perfoliata Donn ex Willd.
SPRINGBEAUTY
E. A local but increasing persistent weed of sandy soils, generally close to the coast. First recorded 1922 by J.F. Rayner from Bembridge, where it still occurs. Well established beneath sea buckthorn bushes on Bembridge Point, 6488 AC 1996, and by Bembridge Ponds, 1997; Puckpool Hill roadside 6191, CP 1986; Fernyclose Road, Seaview 6291, MW 1996. No recent records from West Wight but older records from Brook Hill, 3984 BS 1965.

C. sibirica L.
PINK PURSLANE
E. Rarely naturalised in damp woods and hedge banks; first recorded in 1936 from Ryde by Miss Bullock. Youngwoods Copse, Alverstone Garden Village, 5785, JO 1994 where it has been known since at least 1974; roadside bank at the bottom of Beacon Alley, 5181, CP 1988, where it

was first recorded 1970; large patch in a copse on Luccombe Down, 5779, ST 1994 where first recorded in 1962.

Montia fontana L.
BLINKS
N. Local and rather scarce in short turf, bare muddy places and boggy flushes, generally on acid soils. Easily overlooked. The distribution of the different subspecies is not known; ssp. *chondrosperma* (Fenzl) Walters is the only subspecies that has been reliably recorded. Map 45

48. CARYOPHYLLACEAE
Pink family

Arenaria serpyllifolia L.
THYME-LEAVED SANDWORT
N. Frequent and locally common on dry open ground: tops of walls, anthills, gardens and waste places, mostly on calcareous and sandy soils. Subspecies *serpyllifolia* is commonest; ssp. *leptoclados* is infrequent but almost certainly under recorded. Recent reports of the latter include Rowridge 4587, DD 1998; path by Bembridge Ponds and elsewhere in the vicinity 6488, AC 1996; St Helen's Duver 6389, CP 1993; Brading Down 5986, AC 1997; and Prospect Quarry 3886, CP 1996. Map 46

A. balearica L.
MOSSY SANDWORT
E. Very rare alien. Established since at least the early 1960s at the base of the north facing wall of Northcourt Manor, Shorwell 4583, CP 1994; established over 1 square metre of low brick retaining wall in a garden at Brocks Copse Road, Wootton 5392, BS 1981.

Moehringia trinervia (L.) Clairv.
THREE-NERVED SANDWORT
N. Widespread and frequent in woods

and shady hedge banks.

Honckenya peploides (L.) Ehrh.
SEA SANDWORT
N. Local on sand and shingle on the north coast of the Island. Long known from Norton Spit 3589, JKN 1996; east and west spits at Newtown 4191, AC 1997; Thorness Bay 4693, JEG 1997; Osborne Bay 5295, AM 1999; King's Quay 5394, BS 2002; Quarr beach 5693, CP 1999; St Helen's Duver 6389 CP 1993; and Bembridge Point 6488, AC 1996. It is found in small quantity at Woodside beach 5493, CP 2001. Not recorded from the Medina estuary since 1977. Map 47

[**Minuartia hybrida** (Vill.) Schischk.
FINE-LEAVED SANDWORT
N? Extinct. Known only from a specimen at the British Museum labelled 'Bembridge. Miss Bickham' and dated 1905. Possibly once a rare native or casual.]

Stellaria media (L.) Vill.
COMMON CHICKWEED
N. Very common on arable, waste and bare ground everywhere. It is always found on rich soil.

S. pallida (Dumort.) Crép.
LESSER CHICKWEED
N. Very locally frequent on bare sandy ground around the coast and inland. Overlooked and probably under recorded. Map 48

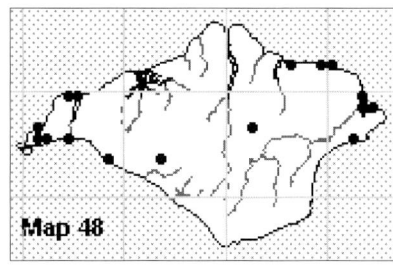
S. pallida

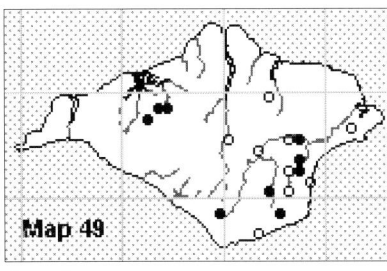

S. neglecta

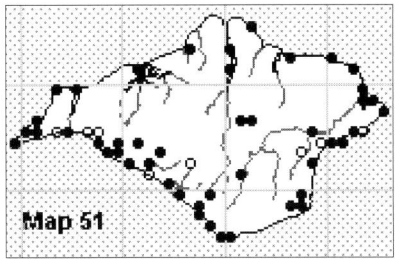

C. diffusum

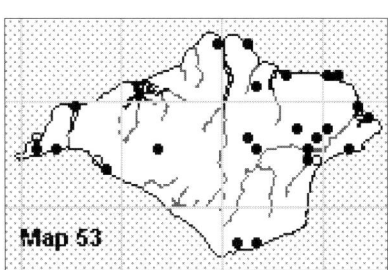

C. semidecandrum

S. neglecta Weihe
GREATER CHICKWEED
N. Very locally frequent in damp woods and shady hedge banks. Principally found in the Eastern Yar valley and around Swainston. Sites include Redhill Lane, Wroxall 5480, CP 1996; Ninham valley, Lake 5783, CP 1996; Crainges, Swainston 4287, DB 1995 and Northpark Copse, Swainston 4388, CP 2000. Map 49

S. holostea L.
GREATER STITCHWORT
(local: grandad's shirt buttons!)
N. Common in woods and hedgerows everywhere.

S. graminea L.
LESSER STITCHWORT
N. Common in rough grassland, verges and scrub in most places but infrequent on chalk.

S. uliginosa Murray
BOG STITCHWORT
N. Locally common in marshes, ditches and wet woodlands, particularly in the Eastern Yar. Map 50

[**Cerastium arvense** L.
FIELD MOUSE-EAR
N. Extinct. Possibly a rare native of sandy grassland. Only three records, all in Rayner (1924): Borthwood, Marriott 1922; Mottistone and Brighstone, Gunyon 1923.]

C. tomentosum L.
SNOW-IN-SUMMER
E. Locally well-established and persistent garden escape on walls, roadside banks, dunes and coastal rocks in parks.

C. fontanum Baumg.
COMMON MOUSE-EAR
N. Common and widespread in short grassland, sides of tracks, arable, wasteland and gardens.

C. glomeratum Thuill.
STICKY MOUSE-EAR
N. Common and widespread in dry grassland, walls, sides of tracks and arable.

C. diffusum Pers.
SEA MOUSE-EAR
N. Common around the coastline on cliff tops, undercliffs, sand dunes and beaches. Also locally frequent inland on exposed, sandy banks such as on St George's Down 5186, CP 2001. Map 51

C. pumilum Curtis
DWARF MOUSE-EAR
N. On south-facing short, open chalk grassland and around chalk pits on many of our downs, much more frequently than in Hampshire. Downs west of Freshwater Bay 3185-3385, BG 1999; Compton 3784, PS 1997; Brook Down 3985, CP 1998; Mottistone Down 4184, MB 1996;

Limerstone Down 4483, CP 1992; Arreton Down 5387, CP 1998; Brading Down 6086, AC 1996; Culver Down 6285, CP 1998. Rare on calcareous grassland around rocky outcrops at St Catherine's 4975, CP 2001. Also on limestone at Prospect Quarry 3886, CP 1996. Map 52

C. semidecandrum L.
LITTLE MOUSE-EAR
N. Locally frequent on bare, dry sandy soils such as sandy cliffs, dunes and heathy spots; occasionally in short turf on chalk. Easily overlooked and probably under-recorded. Map 53

Myosoton aquaticum (L.) Moench
WATER CHICKWEED
N. Local in marshes, streams and ditches, and in wet woods. It has declined since the nineteenth century, being found in much reduced quantity and in fewer sites. Recorded today only from the Eastern Yar and Medina river valleys. Map 54

Moenchia erecta (L.) P.Gaertn.,B.Mey. & Scherb.
UPRIGHT CHICKWEED
N. Rare on a few surviving areas of dry open sandy or gravelly turf. It is much declined since the nineteenth century and numbers fluctuate considerably from one year to the next. The only site where it can be described as abundant is St Helen's

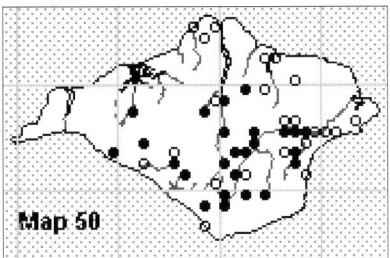

S. uliginosa

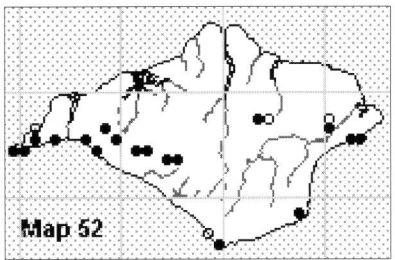

C. pumilum

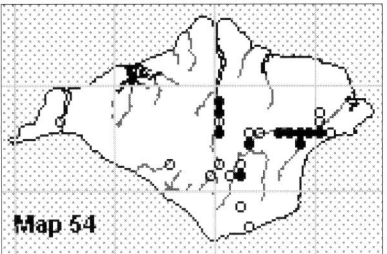

Myosoton aquaticum

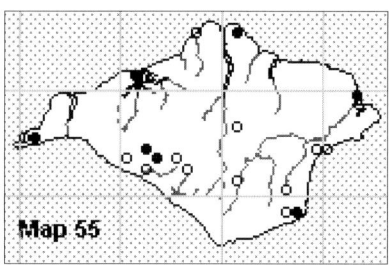
Moenchia erecta

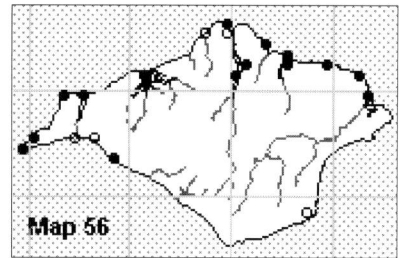
S. maritima

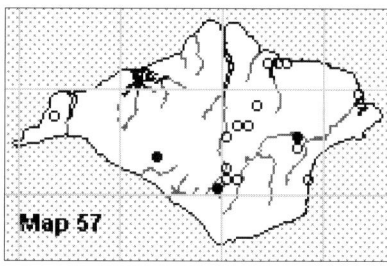
Scleranthus annuus

Duver 6389, CP 1996. Also recorded from south-facing meadow at Nodes Point 6389, CP 1996; Osborne, one plant in acid grassland 5195, NS 1996; a few in the vicinity of St Boniface Down car park 5778, GT 1996; edge of Rew Down Golf Course 5477, BS 2002; local on trackside of Limerstone Down bridleway 4383, PS 1993; one plant on Headon Warren 3185, PS 1994. Map 55

[*Sagina nodosa* (L.) Fenzl
KNOTTED PEARLWORT
N. Extinct. Formerly found on damp sandy areas, disturbed heath and basic flushes. Recorded nineteenth century sites were Compton Marsh, the sandy bog at the head of Freshwater Marsh, Wilmingham Heath, Norton Spit and St Helen's Duver. Last recorded 1931, from Norton Spit (E. Drabble & J. Long).]

S. subulata (Sw.) C.Presl
HEATH PEARLWORT
N. Very rare. Formerly on open gravelly ground in five recorded nineteenth century sites. Subsequently lost from Bleak Down, St George's Down, the gravel capping on West High Down and from a cornfield near Shanklin. It still persists at Headon Warren 3185, PS 2000, where it is locally frequent in depressions in the tracks on top of the hill.

S. procumbens L.
PROCUMBENT PEARLWORT
N. Very common on moist waste ground, lawns, paths and woodland rides throughout the Island. This is our commonest pearlwort and it is found in a wide variety of habitats.

S. apetala Ard.
ANNUAL PEARLWORT
N. Widespread and frequent on dry, bare soil, shingle beaches, walls and other dry places. Both ssp. *erecta* F. Herm. and ssp. *apetala* are recorded. The latter is considered to be the commoner, but further study is required.

S. maritima Don
SEA PEARLWORT
N. Very locally frequent in sandy and stony places around the coast, especially in our estuaries. It is sometimes found as a pavement weed on esplanades. Probably under recorded. Map 56

Scleranthus annuus L.
ANNUAL KNAWEL
N. A plant of thin, dry acid grassland and weed of dry, sandy and gravelly arable land, which has greatly declined. It was 'very common' in Bromfield's time. It had already become rare when Bevis *et al* (1978) compiled their Flora. They listed five sites, namely: St George's Down (1970); arable field on the slopes below Bleak Down (1969); Lynn gravel pit (1974); gravelly field near Fishbourne Lane (date not given); and a sandy field at Hill Heath (1968), west of Alverstone Garden village. It has further declined and today is recorded from just three sites namely, bare ground amongst thin acid grassland on Rowdown above Brighstone 4383, CP 1998; and arable fields at Hill Heath 5785, PJW 1997 and to the west of Bleak Down 5080, GT 2002. Map 57

[*Herniaria glabra* L.
SMOOTH RUPTUREWORT
C. Extinct. A very rare casual alien. First recorded in 1935 from St Lawrence undercliff. Three plants were found growing at the edge of St Helen's Green by a kerbstone in 1980, 6289 EAP. These plants survived until 1987.]

[*H. hirsuta* L.
HAIRY RUPTUREWORT
C. Extinct. A casual alien, known from only two records. St Lawrence 1910 (F. Stratton) and as a garden weed in Newport established for at least ten years 1932 (J. Long).]

Polycarpon tetraphyllum (L.) L.
FOUR-LEAVED ALLSEED
E. Long established as a weed in crevices between old paving stones in the back garden of a house in Queens Road, Ryde 5892, CP 1992.

Spergula arvensis L.
CORN SPURREY
Arch. Still not uncommon on dry arable and waste ground, mostly on sandy soil. It avoids the chalk, but otherwise is widespread. It is, however, most often encountered on the southern half of the Island.

Spergularia rupicola Lebel ex Le Jol
ROCK SEA-SPURREY
N. Frequent and often abundant, on the south-west coast from Alum Bay cliffs to Horseshoe Bay, Bonchurch, growing on chalk cliffs and sandy and clay undercliffs. It also still manages to cling onto the inner cliffs between Blackgang and Bonchurch where sufficient open cliff is exposed. It is prolific at Gore Cliff 4976, but becomes more sporadic eastwards, where it often has to compete with colonising woodland and scrub. In Hampshire, this species is only

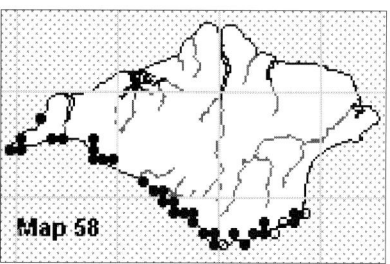
Spergularia rupicola

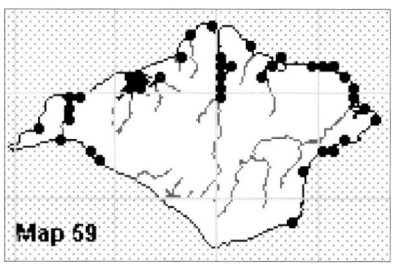

S. marina

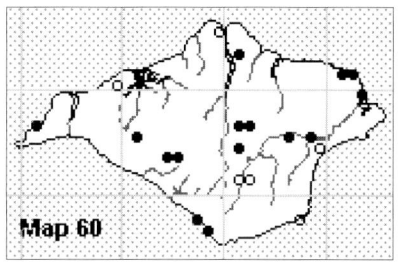

S. rubra

found around Lymington and to the west. The Island is the easternmost site along the south coast. Map 58

S. media (L.)C. Presl
GREATER SEA-SPURREY
N. Frequent in every estuary and saltmarsh on the north coast of the Island, including suitable brackish spots on Brading Marshes. Also occasionally found on undercliffs, as at Fort Victoria 3389, PS 1998 and Brook 3884, PS 2000.

S. marina (L.) Griseb.
LESSER SEA-SPURREY
N. More widespread than *S. media* occurring in all our estuaries but also locally on clay undercliffs, waste ground and esplanades right around the coast. Map 59

S. rubra (L.) J. & C. Presl
SAND SPURREY
N. Local on sandy and gravelly ground on acid soils, generally away from the coasts. Map 60

Lychnis coronaria (L.) Murray
ROSE CAMPION
E. An occasional garden escape in churchyards, verges and waste ground, sometimes persistent as at Arreton Cross 5386, AC 2000, where it has been known since 1979.

L. flos-cuculi L.
RAGGED-ROBIN
N. Widespread in damp meadows, marshes, ditches and damp woodland rides. It has undoubtedly declined in the past one hundred years and, although still frequent, is rarely seen today in such abundance as to turn fields pink.

[*Agrostemma githago* L.
CORNCOCKLE
Arch. Extinct. Bromfield (1856) described this cornfield plant as 'much too common', but Townsend (1883) says, 'Not abundant. Another instance of the change in flora since Bromfield's time'. In 1904, Ernest White wrote that he had seen Corncockle 'growing freely in the sea-sand at St Helen's Duver' but by 1907, he described the plant as being 'uncommon and rarely found' on the Island. The last record was made during a visit of the Isle of Wight Natural History and Archaeological Society to Wootton Manor Farm, 5492, on 19 June 1943. White, writing in the Society's Proceedings of 1954, described Corncockle as 'another picturesque weed of cornfields but it is becoming very scarce in the Island'. The name of Cockleton at Northwood is considered to derive from Old English meaning 'the farmstead where cockle grows', coccel being the Old English for corncockle (Mills, 1996). The occasional modern records from waste ground are casual escapes from gardens, which sometimes persist for several years.]

Silene italica (L.) Pers.
ITALIAN CATCHFLY
C. Very rare casual. Disturbed ground on St Helen's Green 6289, MB 1998.

S. nutans L.
NOTTINGHAM CATCHFLY
N. Extremely local on maritime cliffs; known from just two sites where it has persisted over the past 200 years. Locally abundant at Redcliff, Sandown 6285, AC 1996, although heavily rabbit grazed in some years; widespread over 100m of inner cliff east of High Hat, St Lawrence 5276, CP 1995.

S. vulgaris Garcke
BLADDER CAMPION (local:bull-rattles)
Ssp. *vulgaris* N. Occasional and never in quantity, in rank grassland on roadsides, waste ground and borders of arable fields. It is widespread, but principally found on the chalk and limestone. Ssp. *macrocarpa* Turrill has been long naturalised on a sandy track way near the point on St Helen's Duver 6388, CP 1993.

S. uniflora Roth.
SEA CAMPION
N. Despite being frequent on the Hampshire and Dorset coasts, this plant has always been remarkably rare and localised on the Island. Older records testify to the fact that at one time it was more widespread on the north coast: 'sparingly on Ryde Dover' 1823 W. Snooke; King's Quay 1909 F. Stratton; 'in some plenty on the shore between Cowes and Egypt (Point) 1856 W. Bromfield; shore east of Newtown 1883 F. Townsend; and Alum Bay 1909 F. Stratton. The only extant site on the north coast is on St Helen's Duver 6389, CP 1996, where it has long been known from one spot where about a dozen plants survive in sandy grassland.

On the south coast, a few plants were recorded from Monks Bay, Bonchurch by E.H. White in 1904. Otherwise, it occurs in some quantity on the exposed western end of Gore Cliff 4976, VS 1998; and on the inner cliff at St Lawrence 5176, CP 1997, where it grows towards the top of the cliff in the vicinity of Cripple path, in a narrow band below the blackthorn scrub. Perhaps at one time it was more frequent all around the Island's coasts as it was first recorded in 1640 in Parkinson's *Theatrum Botanicum* as growing in many places in the 'said Isle'.

S. noctiflora L.
NIGHT-FLOWERING CATCHFLY
Arch. Very rare arable plant on dry sandy or calcareous soil. Never particularly common but has declined considerably since 1950. Cultivated fields on the chalk ridge in the middle of the Island at Idlecombe, 4685, and Bowcombe Down, 4586, were reliable sites in the 1960s and 70s. Despite searching, the most recent records are of four plants at Bowcombe 4687, CC 1994 and one at Dunsbury Farm, Brook 3884, DF 1990.

S. latifolia Poir.
WHITE CAMPION
Arch. Common and widespread on arable, waste ground and hedgerows especially on lighter soils.

S. x hampeana Meusel & K. Werner
WHITE CAMPION X RED CAMPION
N. Occasional on roadsides, hedgebanks and waste ground, where both parents are growing together.

S. dioica (L.) Clairv.
RED CAMPION (local: tidy robins!)
N. Common, often abundant and showy, in woods and hedgerows.

S. gallica L.
SMALL-FLOWERED CATCHFLY
Arch. Very rare today, on sandy arable ground. This plant has greatly declined since the 1940s. The most recent records are of 25 plants at Brook Hill 3984, CP 1980; several at Cheverton Shute, Apse Heath 5783, JS 1993; over 50 south of Burnt House Lane, Alverstone 5885, TDD & CDP 1998; one plant at North Appleford 5081, PW 1998, and a few plants in fields north-east of Rookley 5183 and 5184, RPA 2001. In each instance, plants could not be found the following year in these locations, but they reappeared in the field at Alverstone in 2001, and in abundance in 2002, when plants with pink flowers and others with pink flowers with a deep crimson spot at the base of each petal were present. Map 61

[*S. conica* L.
SAND CATCHFLY
Extinct. The first British record for this species was made by William Sherard 'a little to the north of Sandown castle' in 1715 (Ray's Synopsis *Stirpum Britannicarum*, 1724 edition revised by Jacob Dillenius). Presumably this would have been from the sandy heathland protecting Sandown Bay. As this plant may be native on dunes in West Sussex, this could have been an historic native site. No other records.]

(*S. conoidea* L
C. A single record by J. Long from a plant gathered by the river below Newport in 1925. Eric Clement states that early records for this species were often confused with *S. conica*.)

[*S. dichotoma* Ehrh.
FORKED CATCHFLY
C. Extinct casual recorded from Rowborough, BS, in 1965.]

Saponaria officinalis L.
SOAPWORT
Arch. Small colonies of this persistent alien are occasionally established on roadside verges. At Stone Shute, Blackwater 5084, it has been established for more than 30 years.

Vaccaria hispanica (Mill.) Rauschert
COWHERB
C. Very rare casual alien. It grows on talus at the base of Lake Cliffs 5983, AC 1995. There are early records from Newport (1916 and 1925) and Brading (1965).

[*Petrorhagia nanteuilii* (Burnat) P.W.Ball & Heywood
CHILDING PINK
N. Extinct. This very rare native has always been principally confined to shingle and sand dunes in the eastern Solent and West Sussex but is long-extinct on the Island. It was the most celebrated species found growing on Ryde Dover by Bromfield in the 1840s. He wrote, 'I have seldom failed to see them on turfy parts these last ten years in some abundance', and material survives in a number of herbaria (BM, HAMU, HCMS, IPS). In 1854, the site was developed for building.]

Dianthus plumarius L.
PINK
E. Established for well over fifty years in Northcourt Manor walled garden, Shorwell 4583, CP 2002. Only a small number of plants survive on the top of the wall, but there are both pink and white flowered plants.

D. deltoides L.
MAIDEN PINK
N / E. A very rare native or established alien. A few plants in sandy grassland on St Helen's Duver 6388, CP 1995; naturalised on a sandy cutting at Sandy Way, Shorwell 4682, CP 1993; single plant by track at Little Kennerley, near Rookley 5283, BS 1993. There are two early records: 'apparently wild' on a bank near Mornhill Farm, Newport 1921 J. Groves; and 'not established' on disturbed ground at The Hermitage, St Catherine's Down 1856 E. Venables.

[*D. armeria* L.
DEPTFORD PINK
N. Extinct. Known only from a few nineteenth century records from open sandy grassland spots: a sandpit alongside Morton Road, Brading; a field near Binstead; Ventnor; and at Newtown. Last recorded in 1888 from Brading, FS]

49. POLYGONACEAE
Knotweed Family

Persicaria wallichii Greuter & Burdet
HIMALAYAN KNOTWEED
E. A rarely established garden escape on verges and waste ground. Skew Bridge at Lake 5882, JS 1988; Berrycroft Farm, Roud 5180, BS 1996; Blackgang Road, Niton 5076, JS 1984.

[*P. bistorta* (L.) Samp.
COMMON BISTORT
E? Extinct. Formerly in a few damp meadows, either as an escaped pot-herb or perhaps a rare native. Historic sites which may have been native are: Freshwater Marshes, J. Rayner 1917; wet meadow near the pond at Old Park, St Lawrence, A. Hamborough 1838; and marshy thicket near Newchurch E.H. White 1905-1928.]

P. amplexicaulis (D. Don) Ronse Decr.
RED BISTORT
E. A very rare garden escape. Recorded from the top of Lake Cliff 5882, AC 1995, where it has been known since 1931; and the Osborne Estate 5194, MB 1998.

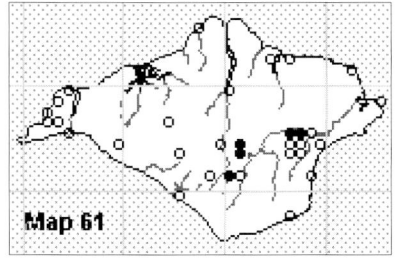

S. gallica

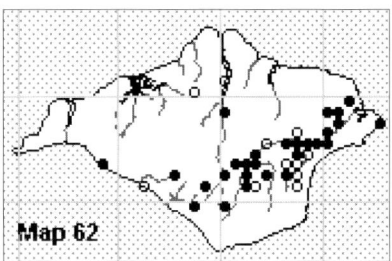

P. hydropiper

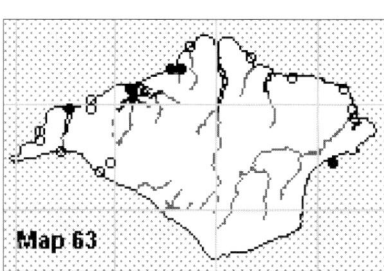

P. oxyspermum

P. amphibia (L.) Gray
AMPHIBIOUS BISTORT
N. Uncommon, but probably under recorded in its terrestrial form. The terrestrial form is recorded from Brading Marshes 6287, NS 1991; marshy ground near Quarr Abbey ruins 5692, CP 1990 (a Bromfield site); and Colwell 3287, PS 1997. The amphibious form is decidedly rare today. Two long-established stations where it still survives, but in diminished quantity, are the Blackbridge Brook at Freshwater Marshes 3486, BG 1999; and Newnham Farm pond at Quarr 5691, CP 1997. Also recorded from a pond at Thorncross 4381, MB 1999; there are historic records from nearby ponds at Kingston and Chale.

P. maculosa Gray
REDSHANK
N. A common and variable weed of arable and waste ground, particularly found on moist sandy soils and around pond margins

P. lapathifolia (L.) Gray
PALE PERSICARIA
N. In similar situations to the last species and sometimes growing with it. It is equally widespread, but less common.

P. hydropiper (L.) Spach
WATER-PEPPER
N. Locally common by ditches and streams, wet woodlands and damp fields. Particularly frequent in the Eastern Yar valley. Map 62

(*P. mitis* (Schrank) Opiz ex Assenov
TASTELESS WATER-PEPPER
N. There are five historic records, not authenticated. The most likely sites were Sandown Levels, a single specimen with *P. minor* found by A.G. More, 1858; and St Helen's Green, J. Rayner 1916.)

[*P. minor* (Huds.) Opiz
SMALL WATER-PEPPER
N. Extinct. Only recorded from Sandown Levels. Bromfield described it as occurring 'in very great profusion intersecting the meadows' and there are good herbarium specimens dated 23 September 1847 at Kew and Winchester. The habitat would have been very suitable, but it declined and was last recorded here by A.G. More in 1858.]

Fagopyrum esculentum Moench
BUCKWHEAT
C. Rare and much declined casual, still occasionally sown as pheasant food. Recorded from a field at Carisbrooke 4887, CP 1989; and Hill Farm, Carpenters 6188, CP 1998.

[*F. tataricum* (L.) Gaertn.
GREEN BUCKWHEAT
C. Extinct. A very rare casual, not seen for many years. Allotments at Sandown and Ryde, J. Long, 1931.]

Polygonum maritimum L.
SEA KNOTGRASS
N. A recent arrival discovered growing on shingle beach at Thorness Bay 4593, AC, in 1995. At this time, the species was undergoing an extension in range, and it appeared on the Dorset and Hampshire coasts. Numbers have fluctuated, but at their greatest extent, plants extended over 46m of upper shore with the largest plants 30 – 40 cm across with woody rootstocks. In 2000, only three small plants could be found, the area having been much altered by winter storms, and since then the plant has further declined. There is a single historic record from the shore at East Cowes in 1870, FS (BM).

P. oxyspermum C.A. Mey & Bunge ex Ledeb.
RAY'S KNOTGRASS
N. Very local on shingle beaches, fluctuating in numbers from year to year. Less common than in the past, although at the time of Bevis *et al* (1978), it was only known from Quarr shore. It may have increased since then, for it is now regularly recorded from Norton Spit 3589, JKN 1996; Hamstead Dover 4191, BS 1992; east spit at Newtown 4191, AC 1997; Thorness Bay 4593, AC 2000; and sandy ground in front of Sandown Zoo 6184, CP 1999. Map 63

P. arenastrum Boreau
EQUAL-LEAVED KNOTGRASS
Arch. Widespread but under-recorded in dry, bare open places, especially well compacted gateways and tracks and cracks in paving stones. It is probably common.

P. aviculare L.
KNOTGRASS (local: wire-weed)
N. Very common in cultivated ground, roadsides, waste ground and beaches throughout the Island.

P. rurivagum Jord. ex Boreau
CORNFIELD KNOTGRASS
Arch. Scarce, but easily overlooked in arable often on chalk. Arable fields at Rowborough 4484, BS 1977; Ashey Down 5787, BS 1979; Newchurch 5686, CP 1997; Roud 5080, CP 1998; Northwood 4993, where very abundant in fields, PS 2000; and Hardingshute 5988, CP 2000. On shingle beaches at Medham, River Medina 5093, CP 1996; and Thorness Bay 4693, AC 1996. On seeing this material, John Akeroyd commented that it is not unusual to see arable weeds on coastal shingle, which was probably their native habitat before farming.

Fallopia japonica (Houtt.) Ronse Decr.
JAPANESE KNOTWEED
E. Occurs as small colonies on verges, tips and waste ground scattered right across the Island. It is probably under-recorded but increasing. First recorded in 1961 from Quarr and Ryde. Var. *compacta* (Hook. F.) J.P.Bailey grows at the top of Lake Cliffs towards

Shanklin 5882, CP 1995; it was first reported here in 1968.

F. sachalinensis (F.Schmidt ex Maxim.) Ronse Decr.
GIANT KNOTWEED
E. Rarely established and in small quantity. Well established by roadside at Sandrock Road, Niton 5075, CP 1995; and in wet valley below main pond at Barton Manor 5294, CP 2000.

F. baldschuanica (Regel) Holub
RUSSIAN-VINE
E. Vigorous climber escaping from gardens and spreading rampantly over hedges and trees. Usually near habitation, but appears to be established on Lake Cliffs 5882 and cliffs near Flowers Brook, Ventnor 5576.

F. convolvulus (L.) A. Love
BLACK-BINDWEED (local: lily)
Arch. Frequent on arable and waste ground, especially on calcareous and sandy soils. Drabble & Long (1931) found that the variant with winged fruiting calyx segments (var. *subalatum* (Lej. & Courtois) D.H. Kent) is far commoner than the normal type on the Island.

[*F. dumetorum* (L.) Holub
COPSE-BINDWEED
N. Extinct? Recorded only from Youngwoods Copse, Alverstone, 5785. First recorded here in 1925 by J. Groves. There is a collection by J.W. Long made in 1929 at Hb. NMW (conf. T. Rich). Much of this wood was opened up with the development of Alverstone Garden Village but in September 1974, J.E. Lousley visited the area with Bill Shepard and refound it growing through an old hedge in front of 'The Trees' (Hb. NMW, conf. T. Rich). It was last recorded here in July 1978, BS. Searches on a number of occasions in the area since have only yielded vigorous plants of *F. convolvulus*.]

Muehlenbeckia complexa (A. Cunn.) Meisn.
WIREPLANT
E. Locally dominant as a garden escape on sandy cliffs above Shanklin Esplanade 5881, CP 1991; and the Sandown end of Lake Cliffs 5882, CP 1995.

Rumex acetosella L.
SHEEP'S SORREL
N. Frequent to common and widespread on acidic and sandy soils, and occasionally on leached soils overlying more base-rich areas.

R. acetosa L.
COMMON SORREL
N. Widespread and frequent in grasslands, particularly older meadows.

[*R. salicifolius* Weinm.
WILLOW-LEAVED DOCK
C. Apparently occurred for several years on a railway embankment and waste ground at Newport, 1929, J. Long.]

R. hydrolapathum Huds.
WATER DOCK
N. Rare and much declined. Formerly widespread and frequent in the Eastern Yar but now very local: abundant on Brading Marshes near Carpenters in a bend of the old river Yar 6187, AC 1999; a few plants in reed-choked secondary ditches on Alverstone Marshes 5785, CP 1998. Still present at Freshwater Marshes but again, much scarcer than formerly: a few plants by Blackbridge Brook and in a pond on the north marsh 3486, CP 1994; more frequent in reed in the south marsh, 2000. There are many plants at the rear of a small coastal pond at the west end of Bouldnor Copse 3790, CP 1997.

R. crispus L.
CURLED DOCK
N. Common and abundant in disturbed ground, verges and waste places. Ssp. *crispus* is the common form. Ssp. *littoreus* (J. Hardy) Akeroyd is recorded from shingle beaches on the north coast. John Akeroyd says that there is an old record for ssp. *uliginosus* (Le Gall) Akeroyd from brackish saltmarsh at Newtown estuary.

R. x schulzei Hausskn.
CURLED X CLUSTERED DOCK
N. Very rare or overlooked. Recorded in 1947 by J. Lousley near Blanket Copse, Wootton. Waste ground at St Mary's Hospital, Newport 4990, PS 2000; pasture at Knowles Farm, St Catherine's 4975, GK 2001.

R. x sagorskii Hausskn.
CURLED X WOOD DOCK
N. Very rare or overlooked. Recorded from Bembridge by A G. More (1871) and field near Bembridge Cross by J. Lousley in 1947. Path at Brook, west of Brook Church 3984, PS 2000; Brook Chine 3883, GK 2001; Forelands 6587, GK 2001; Shepherd's Chine 4479, GK 2001.

R. x pseudopulcher Hausskn.
CURLED X FIDDLE DOCK
N. Probably very rare. Field east of Brook Chine 3883, PS 1999 (confirmed E.J. Clement); dry pasture at Knowles Farm, St Catherine's 4975, GK 2001.

R. x pratensis Mert. & W.D.J. Koch.
CURLED X BROAD-LEAVED DOCK
N. The most common dock hybrid with both historic records, and over forty recent records. Probably regular on waste ground where both parents occur.

R. conglomeratus Murray
CLUSTERED DOCK
N. Frequent and widespread in open situations in disturbed grassland and verges, and in marshy places.

R. x ruhmeri Hausskn.
CLUSTERED X WOOD DOCK
N. Very rare or overlooked. Waste ground at Sudmoor, near Brook 3983, PS 2000.

R. x muretii Hausskn.
CLUSTERED X FIDDLE DOCK
N. Probably very rare. Single plants at Shorwell 4582, PS 2000; Brook 3983, PS 2000; and Knowles Farm, St Catherine's 4975, PS 2000.

R. x abortivus Ruhmer
CLUSTERED X BROAD-LEAVED DOCK
N. Rare on waste ground. There are records of single plants at Sudmoor 3983, PS 2000; St Mary's Hospital grounds, Newport 4990, PS 2000; and Brook 3883, PS 2000.

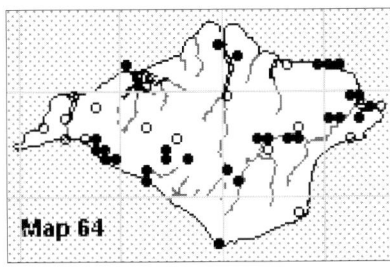

R. pulcher

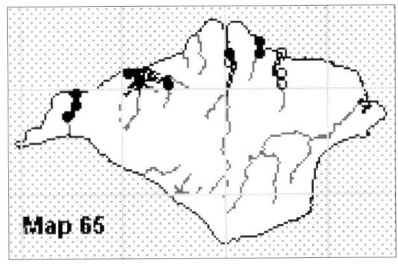

L. humile

R. sanguineus L.
WOOD DOCK
N. Common and widely distributed in hedgerows, woodland, waste ground, rough grassland and verges.

R. x mixtus Lambert
WOOD X FIDDLE DOCK
N. Very rare. The only record is of one plant in gateway at Sudmoor 3983, PS 2000.

R. x dufftii Hausskn.
WOOD X BROAD-LEAVED DOCK
N. Rare but probably under-recorded. Thirteen recent records scattered across the Island.

R. pulcher L.
FIDDLE DOCK
N. Local in old, dry trampled grassland around the coast, and on dry south-facing often sandy banks inland. This plant is less infrequent than suggested by Bevis *et al* (1978). Map 64

R. x ogulinensis Borbas
FIDDLE X BROAD-LEAVED DOCK
N. Very rare. Recorded from Brook Chine car park 3984, PS 2000; and a single plant in gateway at Sudmoor 3983, PS 2000.

R. obtusifolius L.
BROAD-LEAVED DOCK
N. Very common on disturbed ground everywhere. Our most abundant dock.

[***R. dentatus*** L.
AEGEAN DOCK
C. Recorded by J. Long as being plentiful on gravelly waste near Newport for several years until destroyed by the cold winter of 1929.]

[***R. maritimus*** L.
GOLDEN DOCK
N? Extinct. Bromfield found what he describes as this, or *R. palustris* on a roadside at Shorwell. There are also records from J. Long 1928 near Newport; and from Miss G. Bullock 1936 between Chale and Blackgang where she described it as plentiful.]

50. PLUMBAGINACEAE
Thrift family

Limonium vulgare Mill.
COMMON SEA-LAVENDER
N. Locally frequent to abundant in most of our saltmarshes on the north coast. It has, however, declined or been lost from some sites, particularly those under change. In the Eastern Yar, it was present in great abundance around Brading Haven, prior to its reclamation in the 1880s; today it is confined in this catchment to the saltmarsh of the former St Helen's millpond, 6389, where there was just one poorly flowering clump in 1997, CP. Villagers in Havenstreet remember annual pilgrimages to the marshes below Firestone Copse, 5591, to pick bunches of sea lavender. It was still frequent here in 1968, CP, but subsequently disappeared with the rapid siltation of Wootton Millpond and is currently confined to few small areas around the mouth of Wootton Creek, 5592 & 5593. It was present at the small saltmarsh at Quarr Abbey, 5692, in 1963, JH, but has since been lost from here. It has also greatly declined from the saltmarsh at Thorness, 4593, although still present in 2000 CP.

L. x neumanii C.E. Salmon
COMMON X LAX-FLOWERED SEA-LAVENDER
N. Only known from three historic records. Recorded by F. Stratton from East Medina Mill in 1913; Wootton Creek by G. Druce in 1929; and Yar estuary, Freshwater by J. Rayner in 1929.

L. humile Mill.
LAX-FLOWERED SEA-LAVENDER
N. A nationally scarce plant of muddy saltmarshes, with one of its centres of distribution being the Solent. It is scarce but has been recorded from many of our main estuaries: Western Yar, 3487, 3588, 3589 PS 1999; Newtown estuary at Clamerkin 4490, CP 1997 and Hamstead 4091, BS 1991; Medham saltmarsh on Medina 5093, CP 1999; and King's Quay 5393, CP 2002. Map 65

(***L. binervosum*** agg.
ROCK SEA-LAVENDER
N. Not confirmed from the Island but Rev. G.E. Smith, 1835, recorded plants answering this description collected on the cliffs at Scratchell's Bay, near Freshwater. There are no subsequent records and recent searches have failed to locate this plant, but it does grow on the chalk and limestone cliffs of Purbeck and Portland to the west.)

Armeria maritima Willd.
THRIFT
N. Locally frequent, sometimes abundant, in upper saltmarshes, sea cliffs, and the turf behind sandy and shingle beaches. It makes a fine display on St Helen's Duver, 6389 and around Freshwater Bay, 3485. Also grows on the inland cliff face of Gore Cliff, 4976, eastwards to Cripple path, St Lawrence, 5176 and, a few young plants, on an exposed trackside on top of Brighstone Down, 4284, CP 2002. Map 66

51. PAEONIACEA
Peony family

Paeonia officinalis
GARDEN PEONY
C. Very rare as a persistent garden outcast.

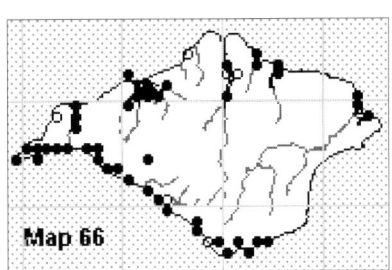

Armeria maritima

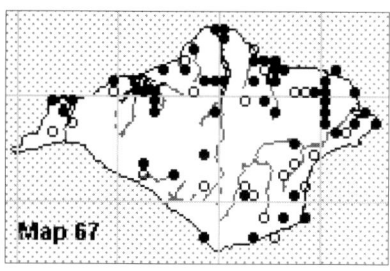
H. androsaemum

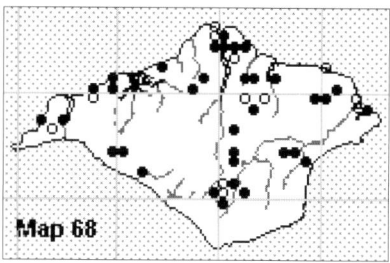
H. humifusum

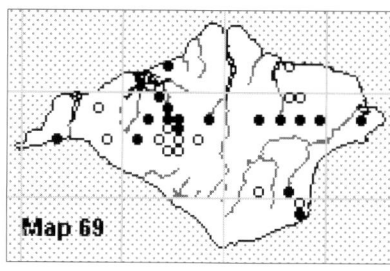
H. hirsutum

53. CLUSIACEAE
St John's-wort family

Hypericum calycinum L.
ROSE-OF-SHARON
E. Often planted in gardens, on verges and railway banks, and rather persistent. Occasionally naturalized, and recorded as such by Bromfield (1856).

H. androsaemum L.
TUTSAN
N. Widespread, but rarely in quantity, in old woods and hedge banks. It also occurs infrequently as a garden escape. Map 67

H. x inodorum Mill.
TALL TUTSAN
E. Very rarely established. Carisbrooke 4887, MB 1996.

H. hircinum L.
STINKING TUTSAN
E. Confined to the undercliffs and waste ground between Bonchurch and Luccombe. Within this area, it is abundantly naturalised and very much a feature, and clearly has been established for very many years. Very occasionally individual plants appear elsewhere and Rev. T. Salwey reports it as growing outside of a garden at Shanklin in 1871.

H. perforatum L.
PERFORATE ST JOHN'S-WORT
N. Common and widespread in grassland, scrub, woodland rides and hedge banks.

(*H. maculatum* Crantz
IMPERFORATE ST JOHN'S-WORT
There are two unconfirmed records, both from near Ninham Farm, Lake. The first by Bromfield in 1855 was subsequently discounted and the second, by A.G. More in 1871, was considered by him to be merely a form of Perforate St John's-wort.)

H. tetrapterum Fr.
SQUARE-STALKED ST JOHN'S-WORT
N. Widespread in damp places such as marshes, damp meadows and wet woodland rides across the Island, although quite often in small quantity.

H. humifusum L.
TRAILING ST JOHN'S-WORT
N. Occasional in bare sandy soil of woodland clearings and heathy places away from the chalk. Although often in rather small quantity, it is less scarce than suggested by Bevis *et al* (1987). It is sometimes rather frequent in sandy fields. Map 68

H. pulchrum L.
SLENDER ST JOHN'S-WORT
N. Widespread and fairly frequent in dry open woods, heathy places and hedge banks. Frequent in small quantities on gravelly soils capping our chalk downs.

H. hirsutum L.
HAIRY ST JOHN'S-WORT
N. Occasional in dry woodland rides, hedge banks and scrub on chalk and also, less frequently, on calcareous clay soils. Map 69

(*H. montanum* L.
PALE ST JOHN'S-WORT
N? Extinct. Recorded in Bromfield (1856) from three sites in the Undercliff, namely: Pelham Woods, St Lawrence; Bonchurch Landslip; and 'about Steephill'. E. H. White reported it to be still present 'in some plenty' in Pelham Woods in 1904 and E. Drabble & J. Long recorded it as 'still at Steephill' in 1931. There are no other records. Pale St John's-wort grows on the chalk on the mainland, where it is considered to be very rare in Hampshire and scarce and local in Dorset.)

[*H. elodes* L.
MARSH ST JOHN'S-WORT
N. Extinct. In the nineteenth century, Marsh St John's-wort was locally abundant in four or five acidic boggy sites; it seems to have become lost as an Island plant during the 1930s. The Wilderness / Cridmore bog is a Bromfield (1856) site with subsequent records up to 1931. On Lake / Blackpan Common, Sandown, it was described by Bromfield as abundant in several boggy places and recorded from here up to 1931. He also describes the plant as growing on the moors near Godshill. F. Stratton recorded the plant in 1900 from sloping boggy ground on the north-west Green at St Helens but adds a salutary note, 'unless recent drainage has changed the character of the spot'. The Natural History Society recorded it from Freshwater Marshes in September 1936, but it is surprising that there are not earlier records from here, and this may be in error.]

54. TILIACEAE
Lime family

Tilia x europaea L.
LIME (local: whitewood)
P/E. This is a frequent introduction, commonly planted in woodland and hedgerows. It is dominant in secondary woodland of the Undercliff.

T. cordata Mill.
SMALL-LEAVED LIME
N. Although the fossil pollen record shows that this was once a common and widespread species, today it is extremely rare as a native species, known only from East Afton Withy

bed, Tapnell, 3687, CP 1999. It was first recorded from here by Bromfield (1856), who described it as 'appearing to be perfectly wild there, but from having been cut as copse-wood not permitted to attain flowering size'. The site is a small wood (under 1ha) situated on the boundary of Freshwater parish and Afton Manor. There are approximately 33 old stools growing in a restricted area of the copse, with Wych Elm. The site is exposed to sea winds but flowering occurs in many years. There is also a record nearby of a few plants, in old woodland stranded within Tapnell Furze plantation 3687, CC 1985. Also occasionally planted, such as at Toll Bar Plantation, Hulverstone 4083, DB 2000.

55. MALVACEAE
Mallow family

Malva moschata L.
MUSK-MALLOW
N. Widespread but rather infrequent on light, dry soils and almost always in small quantity. Grows on verges and rough grasslands, especially on the chalk and rarely on the coast. Map 70

M. sylvestris L.
COMMON MALLOW
Arch. Common, often abundant, and widespread on hedge banks, roadsides, waste ground and on cliffs. In common with many other parts of the country, the crunchy green fruits were eaten by children and referred to as 'cheeses'.

[*M. parviflora* L.
LEAST MALLOW
C. Extinct. A short-lived casual known only from a group of records from five sites made between 1929 and 1931. Recorders were E. Drabble, J. Long and J. Rayner and there are similar records for *M. pusilla* made at the same time.]

[*M. pusilla* Sm.
SMALL MALLOW
C. Extinct. Recorded from Thorness, Apes Down, Ryde, Alverstone and Newport between 1929 and 1931, as above. There is also an earlier, 1916, record from Newport by F. Stratton and a 1960 record from Apes Down by A. Westrup.]

M. neglecta Wallr.
DWARF MALLOW
Arch. Local but widespread on dry banks, roadsides, farmyards and waste ground. Generally found in small quantity. Map 71

Lavatera arborea L.
TREE-MALLOW (local: mash mallus)
N? Well established, probably naturalised, in various places around the coast: Embankment Road, Bembridge; Forelands; Lake Cliffs; various places along the coast from Ventnor through to St Catherine's; Brook; and suitable places along the Yar valley from Freshwater Bay to Yarmouth. It has been known from some of these sites for over a hundred years but Bromfield (1856) described this plant as 'a very frequent ornament of rustic gardens' from which it escaped onto waste ground. First recorded in 1666 by Charles Merrett, 'Mr. Morgan received it from the Isle of Wight' (Merrett, 1666).

E.W. Swanton provided an interesting account of the plant in 1931. He wrote, 'Today at Brook it grows abundantly and luxuriantly in the cottage gardens on the green near the sea. Mr. William Jacobs, an old inhabitant, tells me that the tree mallow was growing wild on the coast at Brook about 130 years ago according to the testimony of his grandfather, who remembered that a French doctor used to come out from Newport to pick the leaves for making an ointment supposed to be beneficial for chapped hands, pimples, or eczema." (*Proc. Isle Wight nat. Hist. Soc.* III (1):94)

Althaea officinalis L.
MARSH-MALLOW
N. Locally frequent in upper saltmarshes of most of our principal estuaries, sometimes under the shade of fringing oaks where ancient woodland borders saltmarsh. At most sites, there is little evidence of a decline, apart from the Medina where it was described by Bromfield (1856) as occurring in many places abundantly, but today it is only recorded from the Medham saltmarsh, 5093. There are also records from the shore at Gurnard 4795 1980 and Norris Castle 5196 1960, which have not been confirmed recently. Map 72

[*A. hirsuta* L.
ROUGH MARSH-MALLOW
C. A non-persisting casual of waste ground recorded from Newport, by the Medina in 1928 by E. Drabble and J. Long (NMW)]

[*Abutilon theophrasti* Medik.
VELVETLEAF
C. A casual recorded from waste ground by the Medina at Newport in 1931 by J. Long.]

[*Hibiscus trionum* L.
BLADDER KETMIA
C. Recorded in 1931 from waste ground by a football pitch in Newport by J. Long.]

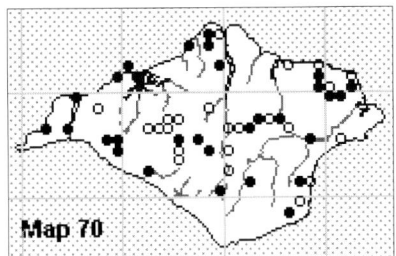

Malva moschata

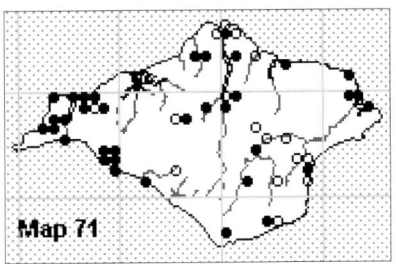

M. neglecta

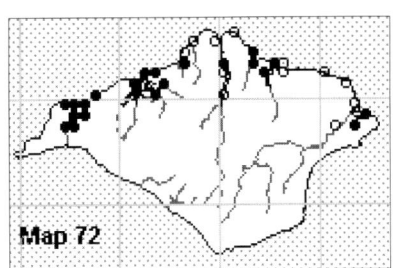

Althaea officinalis

57. DROSERACEAE
Sundew family

Drosera rotundifolia L.
ROUND-LEAVED SUNDEW
N. The decline of this species recorded by Bevis *et al* (1978) has continued, and Sundew is now very rare and seriously threatened. Known today only from Bohemia Bog 5183, CP 2001, where it is reasonably frequent over an area of 0.2 ha of cattle-grazed spring fed mire, and Munsley Bog, Godshill 5282. The latter is an unmanaged relic bog complex where it was still very locally frequent in 1988 but had declined to just 9 clumps by 1995 and 3 small patches by 2002. It was recorded from Bleak Down 5181, JB & BS, in 1969 'in wet spots, covering almost an acre'; a very few plants survived here until 1989, RPA, but this site was subsequently destroyed by tipping. It was last recorded from a small bog at Blackpan, Sandown, in 1931. Additional nineteenth century sites included the western end of St Helen's Green, The Wilderness and other moors about Rookley, Cockleton Bog at Gurnard and Freshwater Marshes. More (1898) was surprised to find it on a slipped bank at the foot of the cliff in Sandown Bay, growing with Royal Fern in 1858. It had been lost from all these sites by early in the twentieth century.

(***Drosera intermedia*** Hayne
OBLONG-LEAVED SUNDEW
N? The only record was made on 27 August 1945, found growing with *D. rotundifolia* at Bohemia Bog by Miss G. Bullock, a well-respected local botanist. This small site had not been well recorded at the time, despite its many riches. There is no herbarium material and there have been no further records, although this plant is known from the pollen record of this site. It is locally common in the New Forest and Dorset heaths.)

58. CISTACEAE
Rock-rose family

Helianthemum nummularia (L.) Mill.
COMMON ROCK-ROSE
N. Occurs on a variety of base-rich soils. It is widespread on south facing chalk grassland and scrub on most of our downs, often on anthills. Also found in coastal grassland in several spots between St Catherine's and Ventnor. It survives in remnant Bembridge limestone grassland on the verge of Quarry Lane, Newbridge 4287, CP 2000 and in East Cowes cemetery 5094, CP 1994. Formerly recorded on tertiary clays in a damp meadow near the sea at Newtown, where it was described as abundant (1931 E. Drabble & J. Long), and at Wootton Common cemetery 5391, CC 1984. It grows in base-enriched greensand grassland at Hilliards cemetery, Lake 5883, LS 1987. Occasionally, single plants with atypically coloured flowers are recorded from sites where the plants are abundant. White flowered plants are most frequent. A single pink flowered plant on Littleton Down, Ventnor 5677, AB 1997. Map 73

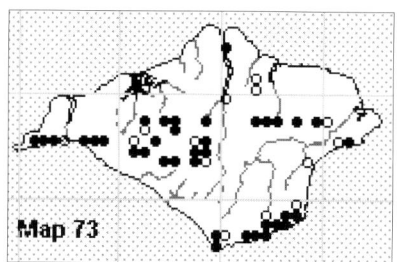

Helianthemum nummularia

59. VIOLACEAE
Violet family

Viola odorata L.
SWEET VIOLET
N / E. This is a widespread species, which occurs on chalk and rich soils in shady spots, often in hedge banks. It is often not possible to say whether the plants are wild or established but it is persistent in many sites. White flowered plants are more common than purple flowered.

V.* x *scabra F. Braun
SWEET VIOLET X HAIRY VIOLET
N. Status unknown. There are three records from the late nineteenth century by F. Stratton at Newport, Staplers and Apes Down.

V. hirta L.
HAIRY VIOLET
N. Frequent in chalk grassland and scrub; sometimes in remnant limestone grassland. Also sometimes found away from the chalk on limey clay soils.

V. riviniana Rchb.
COMMON DOG-VIOLET
N. Common and widespread in woodland, hedge banks, rough grassy slopes.

(***V.* x *bavarica*** Schrank
COMMON DOG-VIOLET X EARLY DOG-VIOLET
N. Status unknown. Said to be frequent by Bevis *et al* (1978), and likely to be so where both parents occur together in quantity, but I have not come across any authenticated records.)

V. reichenbachiana Jord. ex Boreau
EARLY DOG-VIOLET
N. Common and widespread but more closely tied to woodlands than *V. riviniana*. It is occasionally found as a garden weed.

V. canina L. ssp. *canina*
HEATH DOG-VIOLET
N. Very rare in heathy grassland. Some earlier recorders confused this species with *V. riviniana*, but nevertheless it is clear that this species has experienced a considerable decline. Known today only from a south-facing slope at Golden Hill 3787, CP 1998, where it was first noticed in 1983 and is quite frequent, and from the flower-rich meadows of Porchfield Rifle Range 4490, BA 2001, where there is a single patch together with odd specimens. The known sites from which it has been lost within the last fifty years (with their most recent records) are Headon Warren 3186, RPB 1953; area of cleared bracken near the Longstone, Mottistone 4184, BS 1973; and Smallgains Heath, Staplers 5290, BS 1970. Map 74

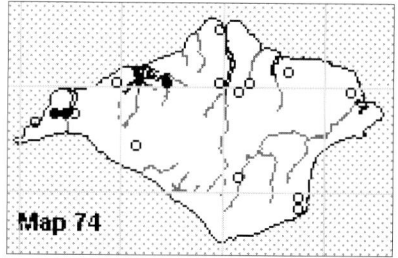

V. canina

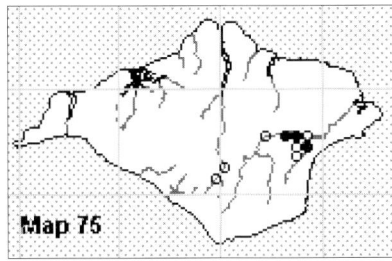
V. palustris

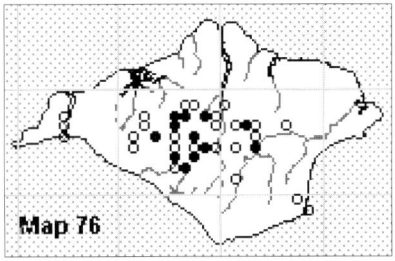
Bryonia dioica

V. lactea Sm.
PALE DOG-VIOLET
N. Very rare in dry heathy grassland. Known today only from open rides in the clay heath scrub at Cranmore 3990, CP 1990, where it was described as abundant by J. Long in 1931 and refound during a meeting of the Isle of Natural History and Archaeological Society in 1986; and nearby from heathland at the cliff edge at Bouldnor 3790, TDD & CDP 1996. Appeared abundantly in a cleared area at Golden Hill, Freshwater, in 1974 but subsequently the site was lost to scrub invasion and later clearance failed to recover it. Nineteenth century sites were at St Helen's Green, Staplers Heath, Parkhurst Forest and Backet's Copse, near Freshwater.

V. palustris L. **ssp.** ***palustris***
MARSH VIOLET
N. Rare and declining in willow carr and fen meadows in the Eastern Yar valley around Newchurch and Alverstone. Recorded from fen and willow carr below Borthwood Lynch, Alverstone 5884 & 5785, CP 1994; Pope's Farm Marsh, Newchurch 5685, CP 1994; and a boggy copse at Hill Heath, Newchurch 5785, SY 2002. Formerly, locally frequent in many spots in the Yar valley and at The Wilderness and Cridmore on the Medina. Map 75

[***V. tricolor*** L. **ssp.** ***tricolor***
WILD PANSY
N. Extinct Only known from nineteenth century records on arable land.]

V. x wittrockiana Gams ex Kappert
GARDEN PANSY
C. Very rare casual on disturbed ground but perhaps increasing. Occasionally plants are found which show characters of *V. tricolor*.

V. arvensis Murray
FIELD PANSY
Arch. Still frequent and widespread as an arable weed on a wide range of soils.

60. TAMARICACEAE
Tamarisk family

Tamarix gallica L.
TAMARISK
E. Widely planted around the coast as a windbreak and persistent old bushes are sometimes seen. Although most plants are clearly of planted origin, it occasionally naturalises on soft cliffs by suckering as at Binnel Bay, St Lawrence, and the cliff between Colwell and Totland.

61. FRANKENIACEAE
Sea-heath family

Frankenia laevis L.
SEA-HEATH
N. Now rare throughout the Solent, this plant was historically recorded from four Island sites on the north coast and is still present in three. The strongest populations are at the mouth of the Newtown estuary where it occurs along the saltmarsh/shingle interface. On the east spit 4292, it has recently been reduced by erosion CP 1999, but on the west spit 4191, it is still frequent, AC 1996. At King's Quay saltmarsh, it was formerly frequent but was reduced to three clumps on the west bank 5393, CP 1991 and had declined further by 2002. St Helen's Duver 6389, is a historic site where it was last recorded by CC in 1986. The most unusual site is on the chalk clifftop at the Needles headland 2984, where it grows on bare chalk around the searchlight position some 62m above sealevel, CDP & TDD 1996. This is the highest British site and the westernmost native British site. In 1837, it was recorded as abundant at the base of the chalk cliffs in Scratchell's Bay, just below the headland, a similar site to that at Dover in Kent where it forms a turf on eroded ledges within the splash zone. The Scratchell's Bay site is difficult to access and it is not known whether it still occurs here.

62. CUCURBITACEAE
White Bryony family

Bryonia dioica Jacq.
WHITE BRYONY
N. Very local in hedges and borders of wood on the chalk and greensand. Confined to the centre of the Island, where it is locally frequent within the core area bounded by Arreton in the east and Rowridge in the west. This is a strange distribution, and difficult to interpret for a plant which is common on the adjacent mainland. It was formerly more widespread, but the decline has not been recent. In the Western Yar valley, it was already considered to be rare in 1929. Map 76

63. SALICACEAE
Willow family

Populus alba L.
WHITE POPLAR
E. Much planted and spreading by suckers. It is widespread, often doing well around the coast, e.g. around Bembridge Harbour, Whitecliff Bay, Lake Cliffs and Norton.

P. x canescens (Aiton) Sm.
GREY POPLAR
E Much planted, especially in damp places. Large specimens can sometimes be seen in damp woods.

P. tremula L.
ASPEN (local: apse; pipple)
N. This tree has a most interesting distribution. It is quite frequent in ancient woodlands on clay on the north side of the Island, and is particularly characteristic of slumped cliffs where woodlands border the coast. Within woods, it tends to occur as discrete colonies formed by suckering, and large old trees have not been recorded. It is occasionally found in ancient species-rich hedgerows, from where it will spread into

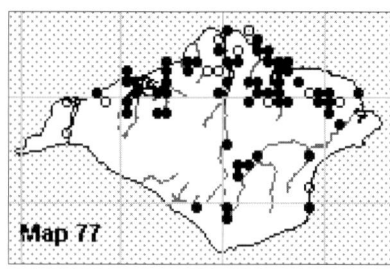

P. tremula

adjacent fields if permitted. Not found on the chalk and much rarer on the southern half of the Island, where it appears to be a relict species of damp heathland sites, occurring as very localised small clumps. Post 1987 sites recorded in south Wight are Blackpan Common, Redway and Kennerley Moor, Upper Dolcoppice, Wydcombe, Bohemia Bog, Luccombe Chine, Gatcombe Withybed and Windmill Wood at Chale. The name Apse Heath derives from Old English meaning 'The place at the aspen-tree or white poplar'. Map 77

P. nigra L. ssp. **betulifolia** (Pursh) Dippel.
BLACK-POPLAR
N? The native Black Poplar is only known from five trees. They are all in the south-east of the Island and may originate from the same, planted stock. One riverbank tree was growing alongside a tributary of the Yar between French Mill and Bobberstone Farm 5582, CC 1985 (conf. E. Milne-Redhead). It was a large tree estimated as 18.3m in height with a circumference of 3.6m, but was a casualty of the great storm of October 1987. Another old tree survives alongside a River Yar bridge at Roud 5180, CP 2002 (conf. Fiona Cooper). Two other trees were growing along the Undercliff. The largest of these was by the pond at Old Park, St Lawrence 5276, BS 1978 (conf. E. Milne-Redhead) with an estimated height of 21.3m and a circumference of 7.3m. It may well have been planted in the early 19th century as part of the landscaping alongside the newly constructed serpentine lake. It was uprooted during a storm in the late 1970s. A contorted, wind-blasted specimen still grows alongside a path at Flowers Brook, Ventnor 5577, CP 1988. Stock from both of these Undercliff trees has been kept going locally by propagation with both plants appearing identical under cultivation. Genetic analysis of the Old Park tree using RAPD markers has shown that it is most closely allied to East Anglian stock (Cottrell *et al* 1997). An old tree growing at Moorhills, Southford 5078, PSi, was identified as *betulifolia* in 2001 (conf. Fiona Cooper). All these trees were male specimens.

'Italica'
LOMBARDY-POPLAR
P. Fastigiate cultivars of Black-poplar are conspicuous planted trees in the landscape.

P. x canadensis Moench
HYBRID BLACK-POPLAR
P. Quite frequently planted in wet sites often growing to a large size. This is likely to be the Black Poplar referred to by earlier botanists.

P. x jackii Sarg.
BALM-OF-GILEAD
E. Occasionally planted and persisting by suckering. Near River Medina at Fairlee 5091, Bevis *et al* 1978; Whippingham 5193 PS, 2000; St Mary's Hospital grounds, Newport 4990, PS 2000; Luckets Farm, Bouldnor 3889, PS 2000.

P. trichocarpa Torr. & A. Gray ex Hook
WESTERN BALSAM-POPLAR
P. Believed to be a rare introduction. Growing at the foot of the cliffs at Gurnard 4695, DB 2000.

Salix fragilis L.
CRACK-WILLOW
Arch. Locally frequent in most of our river valleys and sometimes in other wet spots. Sometimes planted.

S. x rubens Schrank
HYBRID CRACK-WILLOW
P. An ancient spontaneously arising hybrid of White and Crack willows, which was probably planted for basket making. Wheelers Bay, Ventnor 5677, CP 1998. Recorded from hedgerows at Totland in 1934 by L. N. Dudley.

S. alba L.
WHITE WILLOW
Arch. Locally common in river valleys and in wet woods, where it can form large trees. Sometimes planted.

S. triandra L.
ALMOND WILLOW
P. Rare in marshland. Bromfield described this willow as 'not very unfrequent' adding that it was 'a valuable willow for the basket-maker, and making very neat fences.' There appear to have been no subsequent records until recently. Blackwater, by old railway track 5086, DB 2000; small tree by pool at Cridmore Bog 4981, DB 1999; by old railway line near Sandown Golf Course, Blackpan Common 5885, MB 1996. Surviving on the cliffs at Bembridge, a relic of cultivation for lobster pots 6588, PS 2001.

S. purpurea L.
PURPLE WILLOW
N / P. Uncommon. This species was not recorded by Bevis *et al* (1978). However it is probably under recorded in wetlands, and sometimes planted. Bouldnor 3789, PS 1999; Wilmingham 3588, PS 1999; Alverstone 5885, CDP & TDD 1998; Carisbrooke by Lukely stream 4885, SB 1995; Roud 5180, DB 1995.

S. x rubra Huds.
GREEN-LEAVED WILLOW
P. One modern record, from the foot of the cliffs at Forelands 6587, PJS 2001 (conf. R.Meikle). Recorded by L. N. Dudley from Totland in 1934. Perhaps a relic of cultivation as Dudley says that the twigs were widely used for lobster pots. The Holbrook family, who fished off Forelands, grew four different sorts of willows from slips for crab and lobster pots on the cliffs at Forelands (Anon., 2002).

S. daphnoides Vill.
EUROPEAN VIOLET-WILLOW
P. A very rare introduction. Single tree by roadside at Merstone 5384, DB 2000; single bush on west verge of Niton road near Roud 5080, BS 1980 (conf. F.H. Perring).

S. viminalis L.
OSIER
Arch. Locally frequent in wet places; formerly planted for osiers used for making baskets and lobster pots. Apparently, this was not a common plant in Bromfield's day.

[*S.* x *smithiana* Willd.
SILKY-LEAVED OSIER
N. Extinct? Only old records from Bromfield: Northwood 1843; Shanklin 1856; and Whitefield Woods 1840. Perhaps overlooked.]

S. x *calodendron* Wimm.
HOLME WILLOW
N. Status unknown. A triple hybrid recorded by Bromfield in 1843, a male tree in a wet meadow at Redhill Farm, Appuldurcombe.

S. caprea L. ssp. *caprea*
GOAT WILLOW
N. A common shrub in woodlands, hedgerows, scrub and by streams and ponds.

S. x *reichardtii* A. Kern.
GOAT WILLOW X GREY WILLOW
N. Almost certainly overlooked. Recorded from Shalfleet 4190, PS 2000; and central Newport by the Lukely Brook 4989, PS 2000.

S. cinerea L. ssp. *oleifolia* Macreight
GREY WILLOW
N. The commonest willow, widespread in woods, hedges, by water and on undercliffs. It does well in exposed places, being very tolerant of salt-laden spray.

S. x *multinervis* Döll
GREY WILLOW X EARED WILLOW
N. Status unknown. There is a single bush at Sudmoor, near Brook 3983, PS 2000. Bevis *et al* (1978) suggest that most material of *Salix aurita* is referable to this hybrid.

S. aurita L.
EARED WILLOW
N. A rare low shrub of wet heathland. Recorded from an eroding cliff at Colwell Bay 3287, BS 1993; Munsley Bog 5282, CP 1997; and dry fen in The Wilderness 5082, FR 1974.

S. repens L.
CREEPING WILLOW
N. A species of remnant damp heaths which is today on the verge of extinction, having shown a steady decline during the twentieth century with loss of suitable habitat. It was extremely abundant on Smallgains Heath, south of Fattingpark Copse, until the site was lost to agriculture in 1972, but persisted in sub-optimal habitat at the southern end of Fattingpark Copse. There were a few plants surviving here in ruts 5291, BS 1997. Formerly abundant in Parkhurst Forest, but last recorded here from a single ride behind Camphill in 1978. A few plants recorded from the heathy cliff edge at Bouldnor in 1968 but not found here since and likely to have been lost by erosion. Last recorded from Freshwater Marshes in 1970. In 1998, a single plant was unexpectedly found in a base-rich flush on the cliff slope at Headon Warren 3186, CP; there are no historic records from this site and the site is likely to be lost through erosion. Map 78

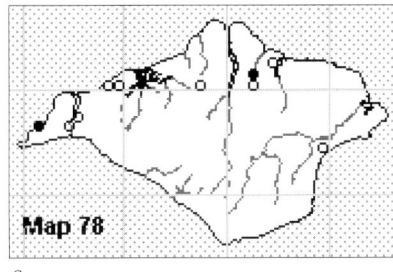

S. repens

64. BRASSICACEAE
Cabbage family

[*Sisymbrium irio* L.
LONDON-ROCKET
C. The only record is by J. Long from an old wall in Newport, destroyed in 1874.]

[*S. loeselii* L.
FALSE LONDON-ROCKET
C. A single old record by J. Long from waste ground by the Medina at Newport in 1929. Specimen at BM.]

[*S. altissimum* L.
TALL ROCKET
C. There are only two casual records, both by J. Long and E. Drabble from Newport in 1929 and Cowes in 1931.]

S. orientale L.
EASTERN ROCKET
E. Rarely established on waste ground. A recent colonist first recorded from Newchurch in 1934 by E.H. White, with no further records until 1965. Recent sites are St Helen's Duver 6388, CP 1997, where it has persisted since 1966 and occurs in various spots where the ground has been disturbed; and Bembridge Point 6488, AC 2001. There has also been a continuity of records from different sites in central Newport 4989, AC 1997.

S. officinale (L.) Scop.
HEDGE MUSTARD
Arch. Common and widespread on hedge banks and waste ground.

[*Descurainia sophia* (L.) Webb ex Prantl
FLIXWEED
C. Very rare casual of waste ground and arable. The only known records are Appley 1867, F. Stratton; Alverstone 1907, F. Stratton, and again in 1931, J. Long; Thorley 1931, J. Long; Rookley 1974, Bevis *et al*; and Brook 1974, Bevis *et al*.]

Alliaria petiolata (M.Bieb.) Cavara & Grande
GARLIC MUSTARD
N. Widespread and frequent in hedgerows and edges of damp woodland.

Arabidopsis thaliana (L.) Heynh.
THALE CRESS
N. Widespread and locally frequent right across the Island. Most often found in open, sandy and gravelly ground and on old walls.

Isatis tinctoria L.
WOAD
C. A very rare casual. In 1858, it was recorded growing in sown grass at Bembridge, by A. G. More. No further records until a plant was found in grassland at Little London, near Newport Quay 4989, CP 1997. Any

Erysimum cheiranthoides L.
TREACLE-MUSTARD

C. A very rare casual of dry waste places. There are seven records between 1856 and 1959 but Bevis *et al* (1978) were not aware of any recent records. Recorded as solitary plants from two sites since then: a garden in Newport 5088, BS 1986; and by a bridleway in the Rowborough valley 4685, SB 1997.

E. cheiri (L.) Crantz
WALLFLOWER

Arch. An established garden escape in many places. It has been remarkably persistent at some sites on old walls and ruins, and on south-facing coastal cliffs, where it always has yellow flowers. Known since 1883 from cliffs at Bonchurch and since 1931 from Freshwater cliffs. Recorded continuously since 1856 from the ruins of Quarr Abbey, and over the same period from old walls at Yarmouth. At other sites it has been lost as a result of 'restoration' and general tidying-up, for instance, many old walls in Newport and from Carisbrooke Castle.

Hesperis matronalis L.
DAME'S-VIOLET

E. A scarce garden escape, rarely established on waste ground and verges. Often white flowered. Recent sites include Moon's Hill, Freshwater; Colwell; Cranmore; Brighstone Forest; Lower Rowborough; and Kennerley Lane, Godshill.

Matthiola incana (L.) W.T. Aiton
HOARY STOCK

N? Very locally frequent. Confined to south-facing chalk cliffs between Tennyson Down and Compton Down where it was first recorded by Snooke in 1823, and calcareous cliffs between St Catherine's Point and Bonchurch where it was first recorded by Bromfield in 1841. The exquisitely scented rich purple, or sometimes white, flowers make a splendid show on our cliffs. The old woody stems of plants growing on top of the inland cliff at West cliff, Niton are encrusted with a golden growth of *Xanthoria*

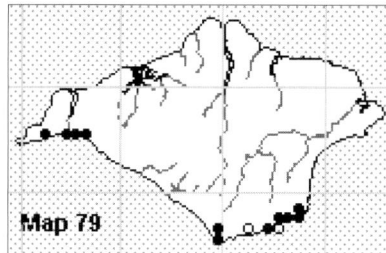

Matthiola incana

lichens. There has been much debate as to the origin of our Stocks. Bromfield says that Hoary Stock is 'probably naturalised in the above stations from gardens, of which it has been a denizen for centuries. Be that as it may, the species is now found growing abundantly in situations the least accessible and most remote from cultivation.' Because of the remote locations of some populations, the long period of known occurrence, and the constancy of characters shown by the plants, the counter argument that Hoary Stock is a native of our warm, sheltered cliffs is also sometimes put forward. Perring & Farrell (1977) considered its sites in inaccessible situations on the Isle of Wight and in Sussex as 'doubtfully native'. Bromfield used to employ Jackman, whom he describes as 'an intelligent cliffsman' to procure specimens of Hoary Stock from inaccessible ledges. Map 79

Barbarea vulgaris W.T. Aiton
WINTER-CRESS

N. Frequent in small quantity by streams, hedge banks and damp waste places across the Island. It is perhaps commoner than when recorded by Bevis *et al* (1978).

B. intermedia Boreau
MEDIUM-FLOWERED WINTER-CRESS

C. Rare casual, possibly overlooked for *B. vulgaris*. Disturbed ground at Westminster Lane, Newport 4989, GT 1999; by Atherfield Road, Yafford 4580, PS 2001; near laundry at Carpenters, St Helens 6288, MB 2001; St Helen's Green 6289, MB 2002; abundant in abandoned arable below St George's Down 5186, RK 1974; and records of 1965 from Quarr, Binstead and Fishbourne, JH.

B. verna (Mill.) Asch.
AMERICAN WINTER-CRESS
(local: land cress)

E. An escape occasionally found on verges and disturbed ground. Recent records are: Gunville 4788, TDD & CDP 1996; Wydcombe 5078, CP 1999; between Ventnor and Steephill 5577, JO 1986; Sandown 5883, GT 1997; Blacklands Farm 5289, CP 1999; Fishbourne 5592, CP 1998; Forelands 6587, AC 1996; Brading 6086, CP 1998; and St Helen's Green 6289, AC 1998. Bromfield comments on the edible qualities of this plant. He said that, 'It affords an excellent spring salad, very superior to the common Winter Cress ... The taste is much more pungent and cress-like and Mr. Loe of Newchurch tells me that it is often substituted by the people of this island for the common Water Cress, being known by the opposite cognomen of Land Cress'.

Rorippa nasturtium-aquaticum (L.) Hayek
WATER-CRESS

N. Locally frequent in marshes, streamsides and pond margins across the Island. Formerly much collected and sold by gypsies. Along the back of the Island it is still found in chines and sometimes where streams issue from the base of cliffs, localities where it used to be gathered by fishermen over generations. It also used to be collected by locals from the foot of Forelands Cliffs near Bembridge, close to the fisherman's withy beds.

R. x sterilis Airy Shaw
HYBRID WATER-CRESS

N. Only one known record, but it may well be overlooked. Recorded from marshland at Downton Farm, Brook 3983, PS 2000.

R. microphylla (Boenn.) Hyl. ex Á. & D. Löve
NARROW-FRUITED WATER-CRESS

N. Rare but probably under-recorded, in similar locations to *R. nasturtium-aquaticum*. Near St Catherine's Point 4975, GT 1999; Leechmore pond, Bleak Down 5080, BG 1999; ditches at Harbour Farm, Bembridge 6387, CP 1994; ponds at Elmsworth Farm, Porchfield 4392, AC 2000. There are early twentieth century records from near Godshill, Compton Bay and Newchurch.

R. palustris (L.) Besser
MARSH YELLOW-CRESS
N. Very rare. Known from marshes and muddy stream banks and also as a weed of cultivation. The lack of recent records suggests that this species may be under recorded. It persists in small quantity by muddy ditches at Adgestone 5985, CP 2002 and it has been recorded as a street weed in South Street, Newport 5089 SB, 2002. Also recorded from a nursery garden at Sandford 5381 1971; east of Sandown airport 5883, 1974; west of Hale Manor 5484, 1969; and a wet field south of Quarr Abbey 5692, 1975. These records are from Bevis *et al* (1978) where it is recorded as *R. islandica*. They considered the species to be in decline.

R. sylvestris (L.) Besser
CREEPING YELLOW-CRESS
N. Rare. Known only as an established weed of open ground. Found growing in rose beds at School Green, Freshwater 3387, BS 1992; Osborne walled garden 5194, CP 1995; garden in The Strand, Ryde 5992, CP 2000; garden in Newnham Road, Binstead 5792, CP 2001. There have been a number of records from the Ryde area over the years.

Armoracia rusticana P.Gaertn., B.Mey & Scherb.
HORSE-RADISH
Arch. A widespread and persistent relic of cultivation.

[***Cardamine bulbifera*** (L.) Crantz
CORALROOT
C. Extinct. There is a well-executed watercolour of the plant by Hon. Mrs. Hood dated 1862 and labelled 'Colwell Common, New fort erecting'. Otherwise, no records.]

Cardamine pratensis L.
CUCKOOFLOWER.(local: milkmaids!)
N. Frequent in damp meadows and damp woods. Although this species remains widespread, it has undergone a considerable decline and cuckooflower meadows are a rare site today.

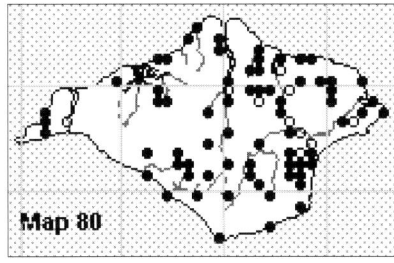

C. flexuosa

C. flexuosa With.
WAVY BITTER-CRESS
N. Locally frequent in damp woodlands, shady streams and ditches. Also occurs as an occasional weed of damp ground. Map 80

C. hirsuta L.
HAIRY BITTER-CRESS
N. An abundant and widespread weed of gardens and waste ground.

[***Arabis glabra*** (L.) Bernh.
TOWER MUSTARD
C? Extinct. There is a well-executed watercolour by Hon. Mrs. Hood dated 1871 and labelled Tapnell Farm.]

[***A. turrita*** L.
TOWER CRESS
C. Extinct. Found in a Newport garden by J. Long in 1933, specimen at Kew (conf. T. Rich).]

A. hirsuta (L.) Scop.
HAIRY ROCK-CRESS
N. Very local in short dry calcareous turf. Known only from the following sites: Carisbrooke Castle moat 4887, 1996 MB, where it is frequent; nearby Mount Joy cemetery 4987, AC 2000, where it is scarce; and Tennyson Down 3385, 1996, and West High Down 3185, 1998, both CDP & TDD, where it grows in a dwarf form. It has been known from West High Down and Carisbrooke Castle since 1823. There is also a single record from Arreton Down 5887, RDP 1987.

Aubretia deltoidea (L.) DC.
AUBRETIA
E. Very commonly grown in cottage gardens and walls and sometimes persisting and self-sowing on old walls.

Lunaria annua L.
HONESTY
C. Frequently appears in hedgerows

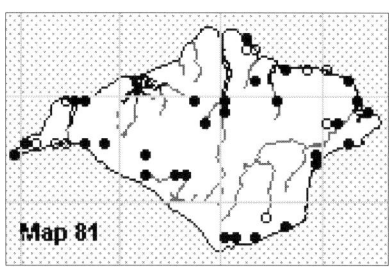

Erophila verna

and on waste ground, but rarely persisting.

Alyssum saxatile L.
GOLDEN ALISON
E. Sometimes escaping from gardens on walls and dry banks, especially around the coastline.

Lobularia maritima (L.) Desv.
SWEET ALISON
E. A common garden escape on waste ground. It is well established on the chalk cliff face at Afton Down, where it was first reported in 1927 by J. Groves. It grows here with Hoary Stock, in habitat typical of both species in France, and this has led some botanists to speculate that the populations could be native. Also well established on the cliffs at Ventnor, where it was first recorded by F. Townsend in 1909.

Erophila verna (L.) DC.
COMMON WHITLOWGRASS
N. A local but overlooked species of bare soil on wall tops, dunes, dry paths and cliff tops. It is less scarce than suggested by Bevis *et al* (1978). It is most abundant on St Helen's Duver. Other reliable sites include Norton Spit, Headon Warren, Afton Down clifftop, St Lawrence inner cliff top, Quarr Abbey ruins, Brading old sea wall and Carisbrooke Castle. Increasingly found in freely drained patches on roadsides, car parks and verges. There are old reports of var. *praecox* (Steven) Diklic, but none since 1980. Map 81

E. glabrescens Jord.
GLABROUS WHITLOWGRASS
N. Occurs in the county but, due to identification difficulties, it is under recorded. St Helen's Duver 6388, TCR 1996; gravelly ground on top

of Brighstone Down 4383, CP 1996.

Cochlearia anglica L.
ENGLISH SCURVYGRASS
N. Locally frequent in saltmarshes in the principal estuaries on the north coast. There are also a number of records from chalk cliffs. Bevis *et al* (1978) refer to these as var. *hortii* Syme (which is no longer recognised) and suggest that they are escapes from cultivation. It was apparently cultivated in the garden of The Needles Hotel garden, Alum Bay, in the nineteenth century (Rev. G. E. Smith). Records from cliff sites include chalk cliffs at the western end of Freshwater Bay 3485, CP 2000, where it has been known since at least 1931. Other sites, not recently refound, are the northern end of Colwell Bay 3287, Bevis *et al* and Monks Bay, Bonchurch 5778, Bevis *et al* 1978. It is possible that some of these plants may be referable to the hybrid of English and Common Scurvygrass, *C. x hollandica*. Map 82

C. officinalis L. ssp. *officinalis*
COMMON SCURVYGRASS
N. Common Scurvygrass has been recorded from the cliffs at Freshwater Bay since 1856 and from the cliffs around the Needles and at Scratchell's Bay since 1871. However, there has been much confusion with *C. anglica* over the years, and Common Scurvygrass has proved to be a very rare plant along the South Coast eastwards of Devon. In 1991, material was collected from the foot of the cliffs on the north side of the Needles headland 2984, and sent to P. Wyse Jackson at Kew who was able to confirm the identity as *C. officinalis*. The plant is frequent on the north facing headland cliff.

C. danica L.
DANISH SCURVYGRASS
N. Very locally frequent on cliff tops, sandy and shingle beaches, sea walls and pavements near the sea. It is very abundant along the coast from Afton to the Needles, both along the cliff edge and by the roadside over Afton Down and Freshwater Bay. Other sites include Norton Spit 3589, JKN 1996; Hamstead Spit 4191, AC 1996; Medham saltmarsh in the Medina

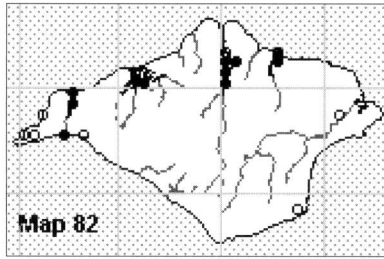
Cochlearia anglica

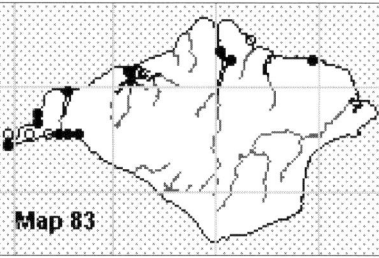
C. danica

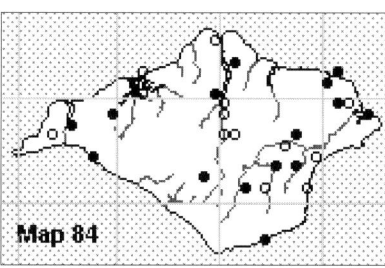
Thlaspi arvense

5093, CP 1996; King's Quay causeway 5394, CP 1986; and the track bed at Ryde Esplanade railway station 5992, CP 1996. Status little changed or possibly increasing. Map 83

[*Camelina sativa* (L.) Crantz
GOLD-OF-PLEASURE
C. Extinct. A casual of crops and on waste ground, with some ten sporadic scattered records between 1877 and 1933, but none since.]

[*Neslia paniculata* (L.) Desv.
BALL MUSTARD
C. Recorded from waste ground in Newport by J. Long in 1931. No other records.]

Capsella bursa-pastoris (L.) Medik.
SHEPHERD'S-PURSE
Arch. Abundant and widespread in a wide range of disturbed habitats.

C. rubella Reut.
PINK SHEPHERD'S-PURSE
C. Disturbed ground at edge of car park at Newport 4989, PS 2001 (conf. T. Rich). No other records.

Thlaspi arvense L.
FIELD PENNY-CRESS
Arch. A widespread but rather scarce plant of disturbed ground generally in small quantity. Map 84

Iberis umbellata L.
GARDEN CANDYTUFT
C. An occasional casual of tips and waste places. Recorded incorrectly as *I. amara* in Bevis *et al.* (1978).

Lepidium sativum L.
GARDEN CRESS
C. Apparently a very rare casual with no recent records.

L. campestre (L.) W.T. Aiton
FIELD PEPPERWORT
Arch. Very rare on dry banks and verges. Apparently quite frequent in the nineteenth century, but much declined since then. Bevis *et al* (1978) refer to six sites: Waste ground bordering the saltmarsh at Newtown, 4290; cornfield at Medham, 5093; cornfields north of West Medina mill, 5091; ditch in Mew's lane, Newport, 5190; Lynn gravel pit, 5389; and Porchfield rifle range, 4490. It was still present in the latter locality in 1981. Otherwise, only recorded from the old railway track bed around Bembridge Harbour 6388, AC where it was plentiful in 1996; and Long Lane Farm, Staplers 5288, GT 2002.

L. heterophyllum Benth.
SMITH'S PEPPERWORT
N. Found in similar situations to the previous species and like Field Pepperwort, formerly much more frequent. Bevis *et al* (1978) refer to just two sites from which it has now been lost: Lynn gravel pit 5389, 1974; and Stone Shute, Blackwater 5086, 1972. More recent records are from disturbed ground in Cowes Cemetery 4994, CP 1986; and the old railway track bed around Bembridge Harbour 6388, BS where it was plentiful in 1999.

[*L. virginicum* L.
LEAST PEPPERWORT
C. A rare extinct casual recorded by J. Long from waste land near Cowes in

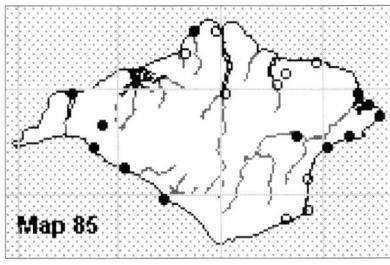
L. draba

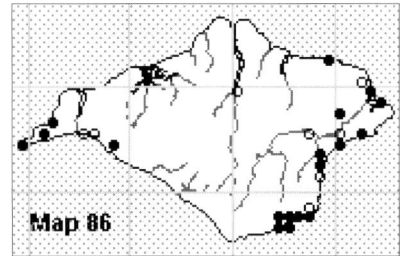
D. muralis

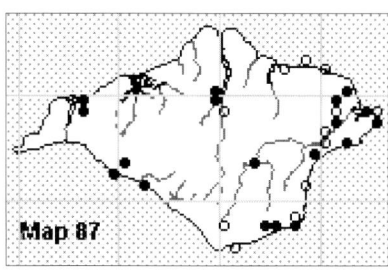
B. nigra

1927 and from waste ground between Cowes and Newport in 1928. Both specimens are at Kew.]

L. ruderale L.
NARROW-LEAVED PEPPERWORT
C? A very rare plant of waste places. Bevis *et al* (1978) reported that it was common on the actively worked Nettlestone tip, but this has now ceased operation and the plant has not been seen here recently. The only subsequent report is of a single plant growing by Newport Quay 5089, CP 2000.

[*L. perfoliatum* L.
PERFOLIATE PEPPERWORT
C. Extinct casual reported from Newport in 1924 by J. Long and from between Blackgang and Chale in 1936 by Miss G. Bullock.]

L. draba L. ssp. *draba*
HOARY CRESS
E. Locally common and increasing on waste ground, road verges and on cliffs. First recorded in 1879 at St Helens 'all around the mill and spreading along the shore' (Moyle Rogers). It was considered to have been accidentally introduced here, by barges calling at the mill, and it has persisted at this site to this day. It has also survived at Yarmouth Mill since 1909, presumably originally introduced here from a similar source. Map 85

Coronopus squamatus (Forssk.) Asch.
SWINE-CRESS
Arch. Frequent and widespread on waste ground, farmyards and well-trampled places everywhere.

C. didymus (L.) Sm.
LESSER SWINE-CRESS
E. Similar places to the previous species and almost as frequent. Bevis *et al.* (1978) suggest that it was less frequent, so the species may be increasing. First recorded at East Cowes by Bromfield (1856).

[*Conringia orientalis* (L.) Dumort.
HARE'S-EAR MUSTARD
C. Extinct. Rare casual of disturbed land with records of 1922 and 1929 near Westminster Mill, Newport; and from Havenstreet 1931 and 1947.]

[*Diplotaxis tenuifolia* (L.) DC.
PERENNIAL WALL-ROCKET
Arch. Extinct. Recorded from a high bank above Cliff End fort (Fort Albert) by A. G. More in 1865. He considered it to have been introduced, but it persisted for a few years. F. Stratton recorded this species from cliffs at Ventnor and Bonchurch in 1909. There are a few records by E. Drabble & J. Long in 1931 and three records of a few plants in 1964 at Seaclose, Newport; 1965 at St Helen's Duver; and 1966 on old railway track at Horringford (all BS).]

D. muralis (L.) DC.
ANNUAL WALL-ROCKET
E. A widespread but local plant on walls, waste ground and chalk cliffs, first recorded from the Island in 1880 at Ventnor. The coast between Steephill and Ventnor remains a stronghold for this species. Map 86

Brassica oleracea L. var. *oleracea*
WILD CABBAGE
N? On chalk cliffs and in cliff top grassland on Afton Down, 3585 to 3685, where it was first recorded in 1965 (JH) and has since increased. Also occurs in smaller quantity on chalk cliffs below Tennyson Down 3385, CP 1999. First recorded from the Island by Matthias de Lobel in 1655. In the nineteenth century, it was abundant on crumbled chalk at the foot of Culver Cliff where, according to A.G. More (1856), it was gathered by locals. It had gone from here by 1871.

B. napus L.
RAPE
C. An increasing relic of cultivation on roadsides and waste ground.

B. rapa L.
TURNIP
C. Probably a frequent relic, but under recorded. Nine post-1987 records, particularly from the south western sector of the Island.

B. juncea (L.) Czern.
CHINESE MUSTARD
C. Only recorded as an arable plant, common in unsprayed areas, at Carpenters near Bembridge, 6188, BG 1999. There is a single old record from Newport by J. Long, 1931.

[*B. elongata* Ehrh.
LONG-STALKED RAPE
C. Extinct. Known only from a single record from a chalk pit at Newport, by J. Long in 1927 (K) (conf. T. Rich).]

B. nigra (L.) W.D.J. Koch
BLACK MUSTARD (local: warlock)
N? Found occasionally on open waste ground and cliffs, most frequently along the coastal belt. It is more frequent than formerly. Map 87

Sinapis arvensis L.
CHARLOCK
N? A common and widespread plant in arable and disturbed land. Known as a pestilent weed in the past. In 1798, W. Marshall wrote, referring to the cultivation of turnips on the Island, 'Today I saw a wagon load of charlock an acre, where turnips were doubtless intended.'

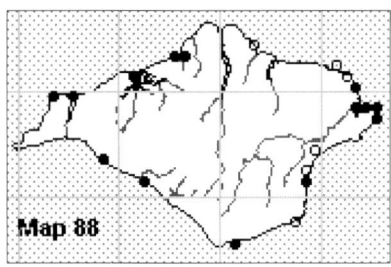

Cakile maritima

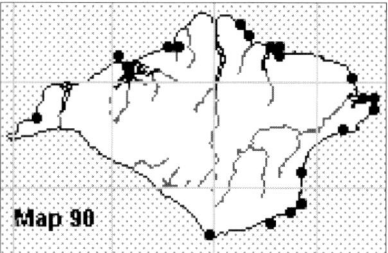

Crambe maritima

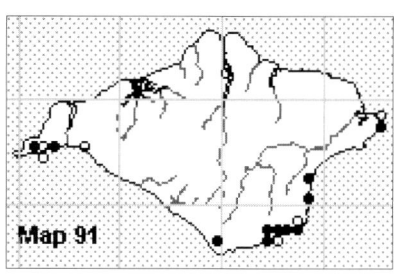

ssp. maritimus

S. alba L. **ssp.** *alba*
WHITE MUSTARD
E/C This species is a scarce relic of cultivation. According to Bevis *et al* (1978) it was 'formerly quite frequent when the practice of green manuring was widespread, but with the decline of the practice the species is now quite rare'. Few modern records, but may be under recorded. Carpenters near St Helens 6189, AC 1998; Adgestone 5986, AC 1999; and Forelands Cliff 6587, AC 1999.

[*Eruca vesicaria* (L.) Cav.
GARDEN ROCKET
C. Extinct. Known only from records by J. Long in 1925 of numerous plants at Newport and several at Cowes (conf. T. Rich). Specimens at BM.]

[*Hirschfeldia incana* (L.) Lagr.-Foss.
HOARY MUSTARD
C. Extinct. There is a single record of several plants at Quarr made by JH in 1963. They did not reappear subsequently.]

Cakile maritima Scop.
SEA ROCKET
N. Local and sporadic in small numbers on or above the strandline, on sandy or shingle beaches around the Island's coastline. Formerly common but much reduced, probably as a result of human pressure. Map 88

Rapistrum rugosum (L.) J.P. Bergeret
BASTARD CABBAGE
E. A casual or established alien on waste ground and verges. It has increased since Bevis *et al* (1978) and is continuing to spread. Current sites where it is well established include verges on Embankment Road, Bembridge 6388, AC 1998; and Three Gates Road, Cowes 4994, JEG 1998. Also seen as a casual elsewhere. Map 89

Crambe maritima L.
SEA-KALE
N. Occasional plants are found on shingle beaches right around the coast. It has increased since Bevis *et al* (1978) and large flowering plants are not uncommon. Also recorded from chalk cliffs at Horseshoe Bay, Bonchurch 5777, CP 1995, where it was first noted in 1970 (TW). Map 90

Raphanus raphanistrum L. **ssp.** *raphanistrum*
WILD RADISH
Arch. This is a locally frequent arable weed preferring sandy soils, but not confined to them. Yellow, pink and white flowered forms occur.

ssp. *maritimus* (Sm.) Thell.
SEA RADISH
N. Very locally frequent on cliffs. Largely confined to the coast between Steephill and Bonchurch, where it is frequently abundant; chalk cliffs at Tennyson Down and West High Down, where it is frequent along the cliff edge; and Forelands Cliffs. Map 91

[*R. sativus* L.
GARDEN RADISH
C. An escape from cultivation with no modern records, but perhaps overlooked.]

65. RESEDACEAE
Mignonette family

Reseda luteola L.
WELD
Arch. Frequent and widespread in disturbed ground particularly on chalk but also sometimes on gravel and waste ground elsewhere.

[*R. alba* L.
WHITE MIGNONETTE
C. Extinct Very rare casual of waste places recorded from Cowes and Ryde but not since 1931.]

R. lutea L.
WILD MIGNONETTE
N. Local on dry, disturbed ground mostly on chalk. Not infrequent in small quantity on our downs. Map 92

67. ERICACEAE
Heather family

Rhododendron ponticum L.
RHODODENDRON
E. Commonly planted on suitable soils and not infrequent in acid woods, although not overtly invasive. Found in many woods around Shanklin and Alverstone, at Osborne, Gatcombe, Brook Hill, Kingston, Appley Woods at Ryde and at the northern end of St Catherine's Down. Under recorded by Bevis *et al* (1978).

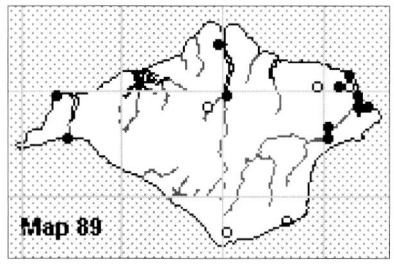

Rapistrum rugosum

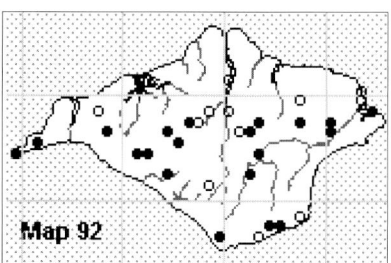

R. lutea

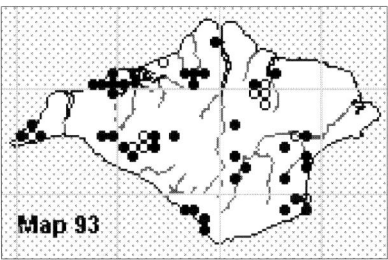

Calluna vulgaris

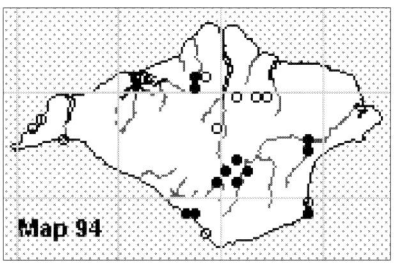

Erica tetralix

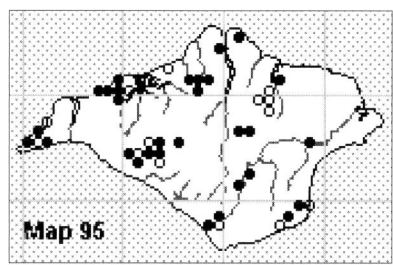

E. cinerea

Arbutus unedo L.
STRAWBERRY-TREE

E. Planted in Victorian shrubberies such as at Osborne and top of Quarr Hill. There is a naturally established tree in the Holm Oak wood on the south side of St Boniface down 5678, AB 1990.

Calluna vulgaris (L.) Hull
HEATHER

N. Almost certainly widespread in the past on heaths, moors and commons across large tracts of the Island, but poorly recorded. Now much declined and showing a relict distribution. The map gives a false impression of its frequency. At many sites, it survives as a few plants in woodland rides or scrubby areas. It is only really abundant and dominant over large areas on Headon Warren and on gravel deposits on top of Ventnor Downs. Locally frequent on roughs at Sandown Golf Course, the site of Blackpan Common. Heather occurs in a number of communities, generally with other heaths: in dry heath in woodland rides at Parkhurst, Firestone etc., in wet heath at Bohemia Bog, Blackgang Chine ledge etc; in clay heath at Cranmore and Bouldnor; and in chalk heath on top of our western downs. Both Headon Hill near Totland and Head Down near Niton derive from Old English words meaning 'the hill or down where heather grows'. Map 93

Erica tetralix L.
CROSS-LEAVED HEATH

N. Rare and declining on relic wet heaths and bogs, and wet acid coastal cliffs. The only site where it is very locally abundant is at the north end of Bleak Down 5181, CP 1996, where management is currently allowing it to increase. At all other sites, it occurs today as a small handful of plants, generally less than twenty. Wet ledges at Luccombe Chine 5879, CP 1997; temporary clearings in Parkhurst Forest 4990, CC 1996; wet ledge at Whale Chine 4678, CP 1999; Munsley Bog 5282, CP 2002; Bohemia Bog 5183, CP 2001; Cridmore Bog 4981, CP 2002; and a wet flush on north side of Headon Warren 3186, CC 1985. Map 94

E. cinerea L.
BELL HEATHER

N. This plant tends to grow with Heather and has a very similar distribution to that species. The comments under Heather apply equally here. It is now increasing on Mottistone Common 4084, under a more sympathetic management regime following the removal of conifers. Map 95

[*Vaccinium oxycoccus* L.
CRANBERRY

N. Extinct. Known in the nineteenth century from a single site described as a 'sphagnous boggy meadow by the Medina, between Cridmore and Appleford farms.' Bromfield (1856) found that the plants rarely flowered although in 1848, Bell-Salter was reported to have gathered a handful of ripe berries concealed amongst the *Sphagnum* hummocks. The site was referred to as a place to see Cranberry by succeeding botanists up until 1900, but it is unclear whether the plant was seen by any of them. Cranberry is one of the most sensitive indicators of habitat change and it is assumed that it became extinct sometime in the late 1800s or early 1900s.]

V. myrtillus L.
BILBERRY

N. A rare species of dry heaths and open, heathy ancient woods, often growing on plateau gravel deposits. It has declined in the past one hundred years and is continuing to do so. Only found in quantity locally on Luccombe Down 5779, CP 2002; management has allowed it to increase here in recent years and it produces small quantities of fruit in most years. It also occurs very locally in Parkhurst Forest, on ride sides beneath beech 4790 & 4791, CP 1995. Until recently, it persisted on Head Down, Niton 5077, CP 1996 but in 2000 the area was found to have been cleared and heavily grazed, and none could be found.

Additional nineteenth century sites were a heathy roadside at Kingston; Marvel Copse near Blackwater; St George's Down (last recorded 1943); heathy clearings in America Woods; and Blackpan Common at Sandown (last recorded 1907). It was doubtless even more widespread in the past. Hardingshute, a settlement northwest of Brading, has a name with an interesting derivation. It appears as "Hortingescet" around 1280, meaning 'the nook of land where whortleberries or bilberries grow' (Mills, 1996).

68. PYROLACEAE
Wintergreen family

[*Pyrola rotundifolia* L.
ROUND-LEAVED WINTERGREEN

N. Extinct. The discovery of three small clumps of plants on the slumped cliff between Hamstead and Bouldnor by a Mr. Baker of Yarmouth (conf. BM) in 1970 came as a complete surprise to botanists. By 1974, no sign of the plants could be found and there have been no further records. This part of the country is outside the general range of this species, but it (ssp. *maritima*) also appeared on Studland Heath in Dorset at around the same time.]

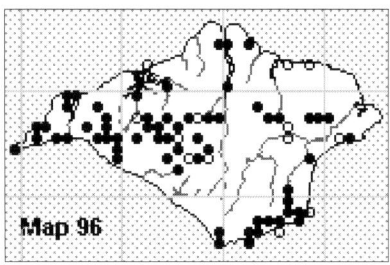
P. veris

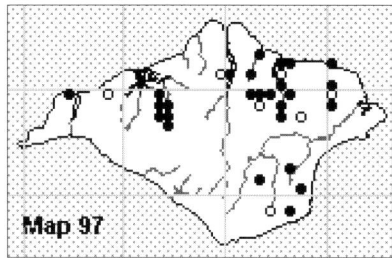
Lysimachia nemorum

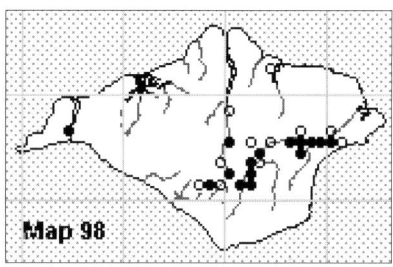
L. vulgaris

69. MONOTROPACEAE
Bird's-nest family

Monotropa hypopitys L.
YELLOW BIRD'S-NEST
N. Rare but probably overlooked, in small quantities in woodland on chalk. Known since 1985 in a beech plantation woodland at Calbourne Bottom 4284, AC 1998. The wood was planted in 1934 and around 20 plants are regularly seen. 63 plants were found beneath Holm Oaks above Ventnor industrial estate 5678, AB 1993. Our plant is believed to be ssp. *hypophegea* (Wallr.) Holmboe. Historic sites include Undercliff between Luccombe and Bonchurch; Calbourne New Barn; Carisbrooke Castle; Westover plantation; Nunwell beech wood; Players Copse, Binstead; and Firestone Copse. Bromfield (1856) said of this plant, 'The entire plant has a strong earthy smell, which has been compared to various and very dissimilar substances, as primroses, bee's-wax and vanilla. To myself the odour is far from agreeable, reminding me more of moistened rhubarb than of anything else'.

71. PRIMULACEAE
Primrose family

Primula vulgaris Huds.
PRIMROSE
N. Common in copses, hedge banks, undercliffs and churchyards across the Island. John Keats, writing on 17 April 1817 said, 'As for primroses, the Island ought to be called Primrose Island.' Although the primrose remains widespread, it has nevertheless declined dramatically and has been largely lost from field edges, verges and railway embankments and is scarce in unmanaged woods. Occasionally found as a garden escape. Forms with pink flowers with a yellow eye are not uncommon in churchyards and cemeteries.

P. x polyantha Mill.
FALSE OXLIP
N. Occasionally found where both parents grow together, usually as isolated plants.

P. veris L.
COWSLIP
N. Locally frequent in old calcareous and clay grassland and cliff tops. This is another species that has declined considerably, but remains abundant at some sites, e.g. Carisbrooke Castle moat, Mount Joy Cemetery and slopes above St Catherine's Point. Map 96

[***Hottonia palustris*** L.
WATER-VIOLET
E. Extinct. Recorded in 1956 as an introduction in a pond at Nunwell House where it survived for about thirty years.]

Cyclamen hederifolium Aiton
SOWBREAD
E. Very locally well established as a garden escape or relic. This plant naturalises particularly well in the secondary woods of the Undercliff from St Lawrence to Bonchurch, where it was first recorded in 1963 (TW).

Lysimachia nemorum L.
YELLOW PIMPERNEL
N. Local in woods, generally in ancient woodlands and largely confined to the clay woods on the north side of the Island. Map 97

L. nummularia L.
CREEPING-JENNY
N/E. Rare in wet woodland rides and damp corners of neutral meadows. All records are from the north of the chalk. Also found as a garden escape and established in cemetery grasslands. Contenders for native sites are: Cranmore 3990, VS 1998, where it has been known since at least the 1950's; boundary bank at the NW corner of Parkhurst Forest 4692, CP 1994; meadow at Whippingham 5193, CP 1993; field at Wootton 5391, BS 1990; field at Kittenocks, Havenstreet 5690, CP 1990 from where it has been known for many years but may now be lost; corner of meadow at Ryde House 5892, CP 1995; and a small roadside field at Hillway, Bembridge 6386, CP 1985.

L. vulgaris L.
YELLOW LOOSESTRIFE
N. Locally common in tall fen vegetation in river valleys. Largely confined to the Eastern Yar valley, where it occurs commonly in many sites. Other recorded sites are Freshwater Marshes in the Western Yar; Gatcombe Withy beds and The Wilderness in the Medina; and Kingston Copse. Map 98

L. punctata L.
DOTTED LOOSESTRIFE
E. An increasing garden escape, usually close to habitation but sometimes arising from throw-outs in remote locations. First recorded in 1922 from Brighstone by J. F. Rayner.

Anagallis tenella (L.) L.
BOG PIMPERNEL
N. A declining species of flushes, wet

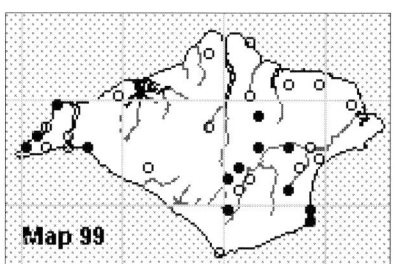
Anagallis tenella

landslipped cliffs and bogs. It was formerly widespread but has greatly declined and is now very local. There are many examples of recorded local extinctions. Map 99

A. arvensis L. ssp. *arvensis*
SCARLET PIMPERNEL
N. Abundant and widespread on arable and disturbed ground and on sandy beaches. Plants with salmon-coloured flowers are not uncommon. Plants with blue flowers are much rarer but have been recently recorded from Compton Chine cliffs 3784, CP 1999 and a field at Chillingwood Farm, Rowlands 5688, CP 2000.

A. minima (L.) E.H.L.Krause
CHAFFWEED
N. Very rare but possibly overlooked. Formerly occurred in damp woodland rides and tracks on heaths. Most recently recorded from Fattingpark Copse, where it was numerous in tractor ruts in the south side of the wood on 29 July 1973 (BS) but not found again until 2002 (CP) when it was very local. Other sites, with most recent records, are: New Copse, Fishbourne 5592, BS 1968; Smallgains Heath, Staplers 5290, BS 1967; Bleak Down 5181, Drabble & Long 1931; St George's Down 5186, FS 1913, Headon Warren 3186, FS 1900 and Freshwater Marshes, Townsend 1879.

Glaux maritima L.
SEA-MILKWORT
N. Frequent in all our estuaries on the north coast, including cattle trampled ground of former river meanders on Brading Marshes.

Samolus valerandi L.
BROOKWEED
N. Local in the upper parts of saltmarshes and, more often, in wet flushes on coastal cliffs. Sometimes very locally frequent. It would appear that, contrary to Bevis *et al* (1978), this species has shown a significant decline particularly in the last thirty years. Map 100

74. ROSSULARIACEAE
Gooseberry family

Escallonia macrantha Hook. & Arn.
ESCALLONIA
E. Very rare. Establishing on the sandy soil of Lake cliffs, south of Sandown, from gardens above 5983, AC 1997.

Ribes rubrum L.
RED CURRANT
N/E. Frequent in woods, hedges and sometimes scrub, right across the Island. Considered to occur both as a native and a bird-sown garden escape. It is particularly common as a hedgerow plant on the Swainston estate.

R. nigrum L.
BLACK CURRANT
N/E? Found locally in small quantity in alder and willow carr and other wet woods. Considered to occur both as a native (the great majority of the mapped records), and as a bird-sown garden escape. Map 101

R. sanguineum Pursh
FLOWERING CURRANT
E? A relic of cultivation, persisting usually as solitary bushes in churchyards, hedges and abandoned gardens. Apparently naturalised on the southern slope of Headon Warren until 1973 when this area was cultivated (Bevis *et al*, 1978). Occasional bushes persist on the northern side of the Warren (BS).

R. uva-crispa L.
GOOSEBERRY
N/E? Frequent in woods across the

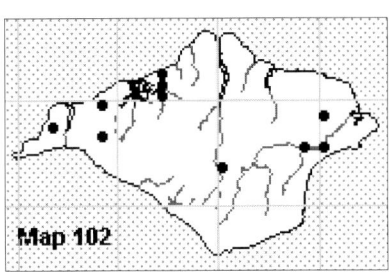

Crassula helmsii

Island and commoner than suggested by Bevis *et al* (1978). Presumably often found as a bird-sown garden escape.

75. CRASSULACEAE
Stonecrop family

Crassula helmsii (Kirk) Cockayne
NEW ZEALAND PIGMYWEED
E. A rampant colonist of mud in and around ponds. The map is likely to under-estimate its distribution. First recorded in 1965 from a pond at Marsh Farm, Newtown 4290, BS. At many sites it is a recent colonist but on Sandown Levels 6085 it has been known since 1979 and is still present in the ditches but in limited quantity, co-existing with native aquatics. Map 102

Umbilicus rupestris (Salisb.) Dandy
NAVELWORT
N. A rather rare plant of shady hedge banks on greensand and on old stonewalls. It has a highly localised distribution, which has probably remained unchanged over a long period of time. It is less rare than suggested by Bevis *et al* (1978) but this is likely to be due to it having been overlooked rather than to any increase. It can be surprisingly difficult to spot in small quantity and may occur in very localised areas. The principal extant sites are Shalfleet walls and church; stonewalls at

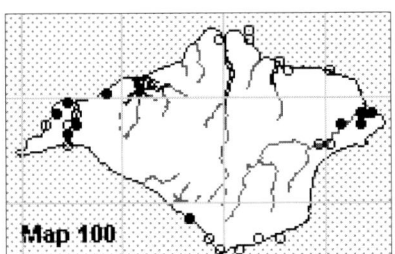

Samolus valerandi

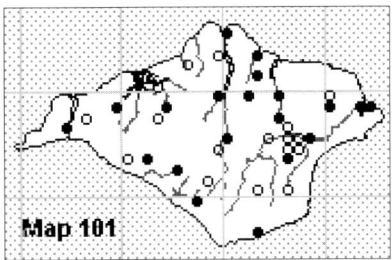

R. nigrum

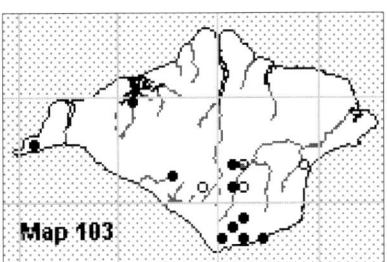

Umbilicus rupestris

Barrack Shute, Niton; walls at Old Park, St Lawrence; walls in Whitwell village; Warren Farm, Totland; and, most prolifically, in several shady sunken lanes around Godshill. Map 103

[*Sempervivum tectorum* L.
HOUSE-LEEK
E. Extinct. Formerly grown on old porches. Bevis *et al* (1978) refer to sites on old farmhouse porches at Great Whitcombe Manor outhouse 4886, Brighstone 4282, Burnt House Lane, Newport 5187, and the old Quarr Abbey 5692. Long (1886) wrote that the leaves of Houseleek or Sen-green were considered to be cooling, ' mixed with cream it cures eruptions'.]

Sedum telephium L.
ORPINE
N. Very rare in ancient woods and hedge banks. Known today from just three roadside sites where, in recent years, it is often prevented from flowering by verge cutting. On a roadside bank in Whitefield Woods 6089, BS 1999 where it was first noticed in 1963; roadside verge at Skinner's Hill, Queen's Bower 5684, JO 1996 where it was first recorded in 1904; and roadside woodland verge on Pallance Lane, Northwood 4893, PS 1999. It has been long known from Steyne Wood, Bembridge 6387, but has not been reported since 1987. It is believed that all these records refer to the wild species and not to similar escaped cultivars.

S. spurium M. Bieb.
CAUCASIAN-STONECROP
E. An occasional garden outcast. Found in churchyards, and spreading along garden walls and on verges close to houses.

S. rupestre L.
REFLEXED STONECROP
E. Frequent introduction scattered across the Island on walls, rocks and banks. First recorded by Snooke in 1823 and Bromfield said that it was particularly frequent on thatched roofs and old ruins. Well established on Quarr Abbey ruins where Bromfield (1856) recorded that ' until lately it grew rather plentifully on part of the ruins of Quarr Abbey, but is now nearly if not quite lost by the pulling down of the portion of wall on which it flourished for building purposes'.

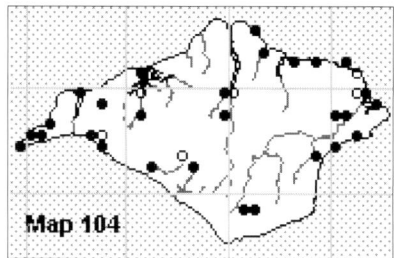

S. acre

S. forsterianum Sm.
ROCK STONECROP
E. Rare escape but some records may be confused with *S. rupestre*. Well established on the sandy verge at Arreton Cross 5386, BS 1986; Niton churchyard 5076, MB 2000.

S. acre L.
BITING STONECROP
N/E. This plant occurs as a native in dry sandy and gravelly grassland around the coast and in open, short grassland on chalk. It is also found as a persistent garden escape on old roofs and walls. At one time, it was very characteristic on roofs of old buildings in central Newport around the Quay but improvements have rendered it very rare and restricted to one or two old roofs in Holyrood and Crocker Streets. Map 104

S. sexangulare L.
TASTELESS STONECROP
E. Very rare escape first recorded in 1999 from Seaview 6291, around pavings and on lawns of a new housing estate (PS & EC). Also establishing in gravel beds at Jubilee nurseries, Branstone 5583, PS & EC 1999.

S. album L.
WHITE STONECROP
Arch. A common garden escape on walls, roofs and waste ground. It is also found on a very few chalk cliffs, where it is presumably well established but may conceivably be native. In great abundance on the south facing slopes of Culver Cliff 6385, AC 1996; chalk cliffs at Needles

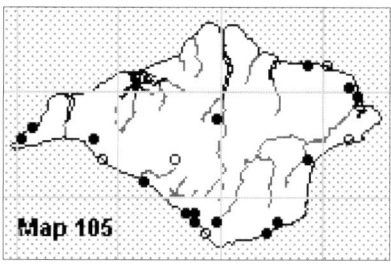

S. anglicum

Battery 2984, CDP & TDD 1996. In both situations the plants may have originated from coastguard cottage gardens above. Bromfield (1856) described it as very rare and lists just two sites on roofs.

S. anglicum Huds.
ENGLISH STONECROP
N. Very local on dry sandy places around the coast. Appears to have been lost from a few sites and largely confined to the south coast today. Frequent on St Helen's Duver, Headon Warren, and along the cliff edge on parts of the south-west coastline. The only recorded extant site away from the coast is in grassland on the exposed western face of Gore Down 4977, CP 1998. Map 105

S. dasyphyllum L.
THICK-LEAVED STONECROP
E. A rare but long-established introduction. It still grows on a roadside wall at Alverstone Mill 5785, JO 1996; and on Brading Church and churchyard walls 6087, AC 1996, both sites where it was recorded by Bromfield around 1840.

[*S. cepaea* L.
PINK STONECROP
C. Extinct Known only from a single record at Whippingham made in 1964 (JH).]

76. SAXIFRAGACEAE
Saxifrage family

Bergenia crassifolia (L.) Fritsch
ELEPHANT-EARS
E. An occasional persistent relic of old gardens.

Darmera peltata (Torr. Ex Benth.) Voss ex Post & Kuntze
INDIAN-RHUBARB
E. Well established and competing

with natural vegetation in Shanklin Chine 5881, CP 1999.

Saxifraga granulata L.
MEADOW SAXIFRAGE
E? As there are no historic records, it is presumably introduced in Mount Joy cemetery, Newport 4987, SB 2001. A few single-flowered plants near the chapel, well established in calcareous grassland and looking native. There are pollen records of this plant from peat deposits at Bohemia Bog.

S. x *arendsii* group
GARDEN MOSSY SAXIFRAGES
E. Known only as pink or white-flowered garden forms, which are occasionally persistent in churchyards and cemeteries.

S. tridactylites L.
RUE-LEAVED SAXIFRAGE
N. Very rare on tops of walls and roofs, and greatly declined in the past hundred years. There are only three known extant sites. North facing roof and porch of Yarmouth church 3589, CP 1997 where it was first noticed in 1968, and nearby on an old wall behind The Bugle PS, 1997. Known since 1856 on the ruins of Old Quarr Abbey 5692, CP 2002. Growing on an old wall in Castle Street, Carisbrooke 4888, CP 1988; the village is an historic site for this species. Other lost sites, with last recorded dates, include Wolverton Manor, Shorwell (1970); old walls in Newport (1903); Gatcombe Church (1856); and semi-natural sites on Hamstead Duver (1924) and St George's Down (1877).

Tellima grandiflora (Pursch.) Douglas ex Lindl.
FRINGECUPS
E. Well established in Shanklin Chine 5881, CP 1999.

Chrysosplenium oppositifolium L.
OPPOSITE-LEAVED GOLDEN-SAXIFRAGE
N. The most frequent of our saxifrage species, found in wet woods on spring lines and along streams, often where alder is present. Found largely in the Eastern Yar valley. There is a particularly high concentration of sites in the wet woods around Shanklin. Uncommon and very local in woods

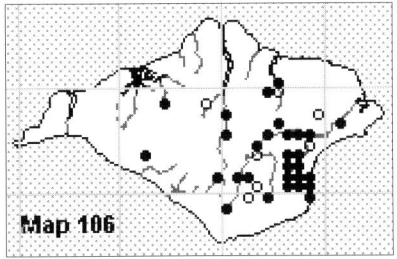

Chrysosplenium oppositifolium

north of the chalk. Since Bevis *et al* (1978), many additional sites have been located. Map 106

[*Parnassia palustris* L.
GRASS-OF-PARNASSUS
N. Extinct. Known from a single enigmatic record caught at the final stages of this plant's disappearance from the Island's flora. Bromfield (1856) says, 'Found many years ago on a piece of boggy land known as the Moor or William's Moor, at Oakfield, St John's, by Mr. John Lawrence, gardener to the late Sir R. Simeon, Bart. The meadow has since been drained, and the plant is extinct.' There is no herbarium specimen. There is also a doubtful record from 'Arreton'.]

77. ROSACEAE
Rose family

[*Spiraea salicifolia* L.
BRIDEWORT
E. Extinct. A single record from Ashey, 1930.]

S. x *pseudosalicifolia* Silverside
CONFUSED BRIDEWORT
E. Rarely persisting but under-recorded garden escape. A few suckering bushes known from at least 1991 on a roadside verge at Barton's Corner, Shalfleet 4189, BS.

Filipendula vulgaris Moench
DROPWORT
N. A species of dry chalk grassland, which has always been remarkably rare on the Island. It is currently known from two sites. It has been known since 1900 on West High Down 3285 but always in small quantity and not easy to find; only a single patch near the cliff edge, about 2 x 2 m and flowering poorly has been seen recently CP, 2002. On Brook Down 3985, SC 1994, two plants found for the first time. Extinct downland sites with last recorded dates in brackets are Bembridge Down (1952); Calbourne Down (1931); Westover Down (1900); and Rew Down (1856). Dropwort is also absent from chalk grassland in Purbeck and very rare on Portland, but much commoner inland in Dorset and Hampshire. Also rarely recorded as a garden escape.

F. ulmaria (L.) Maxim.
MEADOWSWEET
N. Common to abundant in fens, wet pastures, ditches and verges across the Island.

Rubus idaeus L.
RASPBERRY
N/E Occasional as colonies at the edges of woods and in wet woods. Sites include the woodland edge at the northern foot of Tennyson and High Downs 3185, 3285, 3385; wet woods by the Medina around Blackwater 5086, 5088; clearings in Bonchurch Landslip 5878; and Borthwood Copse 5784.

R. cockburnianus Hemsl.
WHITE-STEMMED BRAMBLE
E. Very rare escape. Two persistent stands in the moat at Puckpool Park 6192, PS & EC 1999; large patch by the Medina near St George's Way 5088, MB 1999.

Rubus fruticosus L. agg.
BRAMBLE
N/E. Very common in hedges and less shaded parts of woods, less so in plantations, heath and chalk scrub, waste places and gardens; mainly on gravel and greensand except for the ubiquitous *R. ulmifolius* and members of sect. *Corylifolii*. D. E. Allen has contributed the account that follows, updating Allen (1990); the records are his except where indicated.

This large and complex group of mostly asexually-reproducing micro-species (facultative apomicts) was studied with limited success by Bell Salter in the 1840s, earlier than in most other parts of Britain. Single, wide-ranging visits by national specialists J.G. Baker in 1868, W.

Moyle Rogers in 1901, and A. Newton in 1977, subsequently extended our knowledge very considerably. In the 1970s, DEA began the prolonged, systematic investigation on which he continues to work. During the course of these studies, complete sets of the species found have been built up by him in both BM and HCMS.

Other counties of southern England are proving to have at least 100 of those species that have distributions sufficiently 'regional' in extent to justify their being described, and thereby privileged, with a binomial. Wight is too small and too lacking in suitable habitats and terrain, at any rate today, to yield such a total. However, its position midway between two distinct *Rubus* florulas, those of south-east and south-west England respectively, has brought it goodly shares of each. But for that, it would have had many fewer than the 72 species, exclusive of garden escapes, currently accepted as reliably on record.

Unexpectedly, and in some cases inexplicably, quite a number of the common Hampshire species, including some abundant as close as Southampton, appear to be entirely absent – or patently still in the process of arriving. The Solent has evidently always constituted a more formidable barrier for blackberry-eating birds than its modest width suggests. Slightly counterbalancing that, Wight has 12 species not as yet found in either of the Watsonian counties that make up Hampshire. The majority of these are western ones, denied to that closest neighbour by its colder and drier climate.

A great asset of *Rubi* that tends to be overlooked is their manifestly subtle adaptation to slight gradations in atmospheric moisture. Some, like *R. moylei*, give out abruptly when they come up against a climatic barrier that is imperceptible on the ground, and might even be elusive after detailed meteorological study. They are indeed much more useful indicators of small-scale geographical affinities between districts than most macro-species, and accordingly ought to be treasured for their individual distributions by phytogeographers.

SECTION *Rubus*
SUBSECTION *Rubus*

(*R. divaricatus* P.J. Müller
Though this species is widespread in the New Forest, records by Bell Salter and Rogers, from near Lynn and The Wilderness respectively, have to be treated with reserve in the absence of specimens, the name having been applied too variously in the past.)

[*R. fissus* Lindley
Extinct. Hyde, near Shanklin, 5781 WMR, 1901.]

R. nessensis Hall
Rare. Still in small quantity in Apsecastle and America Woods 5681, where Bell Salter knew it in abundance. Otherwise, recently noted only in the west fragment of Youngwoods Copse 5785.

R. plicatus Weihe & Nees
Rare. Less widespread than formerly, due to the reduction of its acid habitats but still in reasonable plenty on moist alluvium in The Wilderness 5082, and Munsley Bog 5282. Otherwise now only a patch or so on Bleak Down 5181; Staplers Heath 5289; Bonchurch Down 5778; and in Parkhurst Forest 4790 and Youngwoods Copse 5785.

[*R. sulcatus* Vest
Extinct? An 1838 specimen of Bromfield's labelled 'near Alverstone *non raro in dumetis humidis*' and queried as *R. plicatus* is in E. The location was presumably 'Alverstone Copse', now Youngwoods Copse 5785, from which Hambrough sent a specimen to Babington, now in CGE. This mainly Continental species of damp woods and heath margins survives in one or two places in Hampshire, and may yet be refound.]

R. vigorosus P.J. Müller & Lef.
Formerly widespread in wet acid habitats, but now rare. Recently recorded from Golden Hill Park, Freshwater 3387; Mount Farm Copse 3989; outside Walter's Copse 4390; and Bohemia Bog 5183; also two bushes on Tennyson Down 3285, and single ones on Headon Warren 3185, and in Mill Copse, Yarmouth 3589.

This westerly pattern echoes the species' more robust survival in Dorset than in Hampshire.

SUBSECTION *Hiemales* E.H.L. Krause
SERIES *Sylvatici*, (P.J. Müller) Focke

R. albionis W.C.R. Watson
Locally plentiful in certain woods and heathy areas within a wide five-mile radius of the east side of Newport; also rare bushes in Parkhurst Forest and single ones in Cothey Bottom Copse, Ryde 6090, and Mill Copse, Yarmouth 3589.

R. boulayi (Sudre) W.C.R. Watson
Rare. In quantity in old oakwood in the north of Parkhurst Forest (4791). An outlier of the Bournemouth *Rubus* florula, this trans-channel species also occurs in west Normandy.

R. errabundus W.C.R. Watson
Very rare. A single bush on the north-facing slope of St. George's Down gravel pits, 1979, of the (genetic?) diminutive version common in the New Forest and round Bournemouth. Otherwise, this is essentially a north British species. Other records have proved to be errors for *R. riparius*.

R. leucandriformis Edees & Newton
A widespread Wessex endemic, frequent in shade across the north half of the Island, reflecting its tolerance of clay, but almost absent from the south. Particularly plentiful in Whitefield Wood 6089; Brocks Copse 5292; Thorness Wood 4593; and Newbarn Copse, Calbourne 4287. Its presence on the exposed top of Brading Down 5986 is unexpected.

R. lindleianus Lees
Frequent in hedges and woods. Particularly plentiful in The Wilderness 5082; Parkhurst Forest 4790 and adjoining squares; and Sticelett Copse 4693.

R. mollissimus Rogers
Rare on heathy ground. Much diminished on Bleak Down 5180, where it was recorded in plenty by Rogers and Stratton in 1901; elsewhere, a patch by the cottages just inside the east entrance to America Wood 5682.

R. oxyanchus Sudre
Very local and restricted to the Yarmouth area. Plentiful in Nunneys Wood 4089, and Mount Farm Copse 3989; also single patches in Cranmore Wood 3989, and Saltern Wood 3489. Like R. boulayi, an outlier here of the Bournemouth florula, this trans-Channel species also occurs in west Normandy.

R. purbeckensis W.C. Barton & Riddelsd.
Local. Doubtless it was once continuous over the entire tract of heathland that centred on Newport, and it remains plentiful on Little Lynn Common 5389, in the east part of Fattingpark Copse 5291, and in an open, once-boggy strip in Staplers Copse 5289. The odd clump or so survives also in the south-west of Parkhurst Forest 4790, at the south end of Lynbottom Copse 5389, and, in the far south, midway between St. Catherine's Down and St. Catherine's Hill 4977. Widespread in Ireland, this species is unaccountably restricted in Britain to the coastal belt of South Central England.

R. pyramidalis Kaltenb.
Rare. The typical plant only occurs in any quantity in Borthwood Copse 5684. Otherwise only a patch in Bohemia Copse 5183, and a bush in Nunneys Wood 4089, have been recorded recently. The micromorph, var. *parvifolius* K. Frid. & Gelert, the commoner variant in the New Forest, is represented by specimens in BM and SDN collected by Miss E.M. Todd in 1926 on a 'sandy common between Lake and Horringford' – presumably Apse Heath 5683.

R. riparius Newton
Local in heathy scrub, particularly round Newport: scattered along rides in Parkhurst Forest 4790 (whence came Rogers' specimen which was queried by him as R. errabundus but was presumably this); Fattingpark Copse 5291; Arreton end of St. George's Down bridleway 5286. Also plentiful in Bohemia Bog, with a patch in the nearby copse 5183, and a bush in Mill Copse, Yarmouth 3588. Doubtless it was once widespread. Until recently believed endemic to north-west Wales, this species has proved to be scattered along the coast of South Central England from westernmost Dorset to Hayling Island as well as, more surprisingly still, down the centre of Normandy.

R. salteri Bab.
First discovered in 1845 in Apsecastle and America Woods by Bell Salter, and subsequently named in his honour by Babington, this bramble with conspicuous large white flowers remains plentiful there. After a period of being confused with various similar plants elsewhere in Britain, it is still not to be found with certainty beyond the boundaries of this joint wood itself. As the only wild plant privileged with a Latin binomial that is unknown outside Wight, it has brought a long succession of *Rubus* specialists to the Island down the years usually for the sole purpose of procuring examples of it for their herbaria. That it should have received this degree of attention, however, is without justification. Throughout the central belt of Europe, it has gradually become apparent that there are innumerable more or less narrowly local entities like this, and that the only practicable course seems to be deny them taxonomic recognition.

R. silvaticus Weihe & Nees
Rare. In small quantity in both sections of Nunneys Wood 4089.

SERIES *Rhamnifolii* (Bab.) Focke

R. altiarcuatus W.C. Barton & Riddelsd.
Locally frequent. A western species much more widespread than in Hampshire, mostly scattered across the south of the Island, from Golden Hill Park, Freshwater 3387, to Burnt House Copse, Alverstone 5785, but with a further concentration in the Bleak Down area 5181 and 5182. Stray bushes have been noted in St. George's Down gravel pits 5186, and even in Quarr Wood outside Ryde 5792. Early workers included this species in R. plicatus, as shown by specimens of Baker's in K.

R. amplificatus Lees
Local. In woods and thickets mainly in the north-east, where Bell Salter knew it well, but also here and there further south: Mottistone Common 4084; Bonchurch Down 5778; Burnt House Lane 5785; and abundant round a gorse heath at Bohemia Bog 5183.

R. boudiccae A.L. Bull & Edees
Locally common on the greensand of the south-east, usually in dry oak woods or heathy scrub but in alder-willow carr on Blackpan Common 5885; trickling north to St. George's Down 5186, it reappears in strength in the Mottistone / Brighstone district. The Isle of Wight has proved to be one of the headquarters of this only recently discriminated species, as the south end of one fork of its range, which extends in a narrow belt up the middle west of England.

R. cardiophyllus Lef. & P.J. Müller
Locally common in open woodland and heathy scrub, but rare or absent in many areas.

R. cissburiensis W.C. Barton & Riddelsd.
Rare as yet. Cothey Bottom Copse 6090; Priory Bay plantation, Nettlestone 6390; plantation west of Quarr Abbey, Fishbourne 5592; Westover Down, by car park 4284. A recent arrival from Sussex, where this aggressive species is abundant. First noticed in 1998, it is likely to spread rapidly.

R. cornubiensis (Rogers ex Riddelsd.) Rilstone
Very rare, possibly extinct. A specimen from what may have been just a single, ephemeral bush was collected in 1974 at the north end of America Wood 5682. This is the only known find outside Cornwall and Devon of this English endemic.

R. curvispinosus Edees & Newton
Rare. Corfheath Common 4490 to Guyers Heath 4389; also three scattered bushes in Mill Copse, Yarmouth 3589. An acquisition from the Bournemouth florula, possibly in the recent past.

R. dumnoniensis Bab.
Rare and very scattered. St. George's Down 5086; Combley Great Wood

5488; hedge banks, Queen's Bower 5684; outside Youngwoods Copse 5784; crest of ridge midway between St. Catherine's Hill and St. Catherine's Down 4977. There are two doubtful old records. A strongly western species, absent from Hampshire.

R. elegantispinosus (A. Schum.) H.E. Weber
Very rare. A garden escape naturalised in a clay spinney at Nettlestone 6290.

R. nemoralis P.J Müller
Widely but rather thinly scattered on acid heathy ground. Rare or absent in some very suitable localities, such as the Wilderness and Mottistone Common.

R. polyanthemus Lindeb.
Locally common on the dry heathy ground of the south-east but elsewhere unaccountably scarce. Mottistone Common 4084 seems alone in the whole western half of the Island in having it in plenty.

R. prolongatus Boulay & Letendre
Locally abundant on the relics of the former heath between Yarmouth and Shalfleet, but elsewhere recorded only from St. George's Down gravel pits 5086; Staplers Copse 5289; and Bohemia Copse 5183 – and only as single bushes in the last two.

R. subinermoides Druce
Locally abundant in woods and copses, especially on the northern clay, but rare in the south and west.

SERIES *Sprengeliani* Focke

R. sprengelii Weihe
Local. Despite the shrinkage of heathland still quite widespread, especially in the south-east, including on the gravel caps of several downs. In particular plenty in Borthwood Copse 5784, and Bohemia Copse 5183. Unaccountably, unrecorded from the western third of the Island, however.

SERIES *Discolores* (P.J. Müller) Focke

R. armeniacus Focke cv. "Himalayan Giant"
A fast-increasing, already widespread escape, as elsewhere in England since the 1940s. Robust, aggressive, tolerant of most soils, it presents a growing threat to native vegetation.

R. armipotens W.C. Barton ex Newton.
Very local, probably adventive. Its anomalous and highly disjunct occurrences overall seem indicative of accidental introduction. It is abundant on Mottistone Common 4084, where conifers have been extensively planted, and in garden hedges in Shanklin and Wootton Bridge. It has a similarly suspicious look in the Bournemouth area. An abundant south-east England endemic, it could however be extending westwards, in part naturally.

R. hylophilus Rip. ex Genev.
Very rare. Among planted conifers in Firestone Copse 5591, AN 1977 and Nunneys Wood 4089 in 2002. Solitary clumps in both cases, and evidently accidental introductions.

(*R. lamburnensis* Rilstone
A very variable bramble, H234, which is widespread in the Island, and in places abundant, has been identified with this Cornubian species. However, it appears more probably to be a stabilised hybrid derivative of *R. vestitus* peculiar to this, and adjacent parts of Central South England.)

R. ulmifolius Schott
Very common in hedges and on waste ground, especially on soils avoided by most other bramble species.

R. winteri P.J. Müller ex Focke
Very rare. In open chalk scrub towards the eastern end of Tennyson Down 3385, and outside the south margin of Fattingpark Copse 5290, a solitary patch each. This species is locally common on the central Hampshire chalk, whence these possibly recent arrivals are most likely to have come.

SERIES *Vestiti* (Focke) Focke

R. adscitus Genev.
Rare and mainly in places close to the sea. Apart from an isolated colony at the north end of Barton Wood chiefly by Osborne Bay 5295, it is restricted to West Wight, where it once grew at Totland Bay 3286, and on long-vanished Colwell Heath 3367. It still survives in strength north of Colwell Chine 3288, and is patently spreading along the chalk scrub on the inland side of Tennyson Down 3385, where the high oceanicity must outweigh the unfavourable geology. Stray bushes on the eastern side of the Western Yar suggest it may be extending inland also.

R. orbus W.C.R. Watson
Very rare. One patch in a mixed oakwood clearing in the south section of Great Lynn Common 5389. This is the eastern limit of this strongly western species, endemic mainly to the Dartmoor area.

R. surrejanus W.C. Barton & Riddelsd.
Very rare. One small patch in Fattingpark Copse 5291. Another abundant species of south-east England that is clearly spreading westwards and may be a recent arrival here.

R. vestitus Weihe
Locally abundant in woods, mainly on the clay and chalk. The red-flowered variant is surprisingly rare, but abundant in Mill Copse, Yarmouth 3588.

SERIES *Mucronati* (Focke) H.E. Weber

R. mucronatiformis (Sudre) W.C.R. Watson
Very local. Abundant in three relic oak-wood fragments on opposite sides of the Blackwater valley south of Newport 4986, 5087, 5186; otherwise very scattered ranging from Cothey Bottom Copse outside Ryde 6090, to the top of St. Boniface Down 5678 and, far to the west, Headon Warren 3185. Like *R. boulayi*, *R. oxyanchus* and *R. curvispinosus*, an overspill from the Bournemouth florula.

SERIES *Micantes* Sudre ex Bouvet

R. aequalidens Newton
Rare. In great abundance on the gravel plateau capping on St. Boniface Down 5678, the probable source of

two recent-looking clumps in the centre of Parkhurst Forest 4791. Another strongly western species, here at the eastern limit of its range.

R. hantonensis D.E. Allen
Very local. Abundant in open holly woodland, Apse Heath Copse 5683; and in even greater profusion on Head Down 5077, even on exposed hedge-tops, with outlying bushes on St. Catherine's Down 4978 adjoining. Otherwise endemic to Hampshire and its Wiltshire border.

R. leightonii Lees ex Leighton
A particularly bird-dispersed species, consequently often as isolated bushes or clumps, with no marked habitat or soil preferences, and a very scattered distribution overall. Noted in quantity only in Walter's Copse, Newtown 4390; the south end of Parkhurst Forest 4789; Staplers Copse 5289; and along the crest of the ridge from Shorwell to Berry Hill 4682.

R. micans Godron
Local. In most woods and copses in the Newport-East Cowes-Ryde triangle, especially those on clay and in some in plenty. Virtually absent elsewhere, apart from adventive bushes within Newport itself. This distribution is oddly at variance with its predominantly south-western range in Britain and wide occurrence in Dorset.

R. moylei W.C. Barton & Riddelsd.
Very rare. Known only from two points on the coast between Cowes and Ryde: a colony on a plantation margin near Osborne Bay 5295 and a bush or two outside Quarr Wood 5792. As this is one of the commonest brambles over most of Hampshire, it is tempting to interpret this distribution as the product of recent dispersal by birds from across the Solent. However, a specimen in MANCH shows that it was in the second of those localities at least by 1888, and the virtual absence of the species from Dorset suggests that a slightly warmer and moister climate is inimical to it.

R. trichodes W.C.R. Watson
Rare. Scattered in chalk scrub along the inland side of Tennyson Down 3285 & 3385, achieving colony strength towards the eastern end.

SERIES *Anisacanthi* H.E. Weber

R. cinerosus Rogers
Very local. Plentiful in the south part of Nunney's Wood 4089 and throughout two of its clay-based neighbours, Atkies Copse and Cook's Copse, 4088. Also on the clay of Thorness Wood 4593 – this last doubtless being the source of the clump on the north-east margin of Parkhurst Forest 4891. The only other, more distant locality known is Little Lynn Common 5389.

R. dentatifolius (Briggs) W.C.R. Watson
Locally common in woods and heathy places. Had this variable, long-confused western species not come to be considered identical with one described earlier from Devon, it would be known as *R. vectensis*, the name Watson bestowed on it after encountering it in unusual plenty round Shanklin in 1937.

R. effrenatus Newton
Very local. Blackpan Common and Burnt House Copse 5885; and Head Down 5077, chiefly among bracken but spilling into the oak hanger, with an outlying patch by The Hermitage 4978. These are two large populations, 11km apart, and seemingly spreading. It is unrepresented in nineteenth century collections made in these localities, only being discovered in 1982. This suggests that it may be a relative newcomer. If so, however, the mode of arrival is a matter for speculation, for, remarkably, the species is not otherwise known outside north-west Wales.

R. formidabilis Lef. & P.J. Müller
Very rare. Abundant throughout Youngwoods Copse 5785, one of a handful of distant outliers in south-east England of this heath species with a distribution centred in north Surrey.

R. leyanus Rogers
Locally abundant. Virtually restricted to the Brighstone-Mottistone area. It is found in profusion on the top of Westover and Brighstone Downs and in Moortown marshes and copses alike. It was unrecorded from Moortown by Townsend when he noted other brambles there in the 1870s, and stray bushes are met with far to the east and north, at Head Down 5077, St. Boniface Down 5670, and Rowlands Wood 5689. This supports an impression that this species is gradually extending its range eastwards in England (as well as southwards across the Channel). Any colonisation of the Island, however, can hardly be all that recent, for there is a specimen in K collected by Baker on Westover Down as early as 1868.

SERIES *Radula* (Focke) Focke

R. bloxamii (Bab.) Lees
Very local. Apart from Osborne House grounds 5194 and two areas of Parkhurst Forest 4790, restricted to the western quarter of the Island, where it occurs in some quantity in all reasonably acid habitats.

R. echinatus Lindley
Local. Mostly in the northern half, as one of the few *Rubi* tolerating clay, but very patchy.

R. flexuosus P.J. Müller & Lef.
Locally abundant in woods on lighter soils and certain heathy areas in the shelter of bracken or gorse, but virtually missing from West Wight. The near-absence of nineteenth-century records for this particularly well-marked species suggests it has increased greatly, perhaps in response to extensive felling.

R. insectifolius Lef. & P.J. Müller
Surprisingly rare, in view of its profusion along the south margin of the New Forest. Except for a clump in Hill Copse, near Freshwater 3588, known only from around Newport: Bunkers Copse, Rookley 5084; St. George's Down 5187 & 5286; and two areas of Parkhurst Forest 4790 & 4891.

R. largificus W.C.R. Watson
Rare. Abundant in hedges and on wood margins at Godshill Park 5381. Elsewhere, a patch at most: Pitt Place and plantation on Strawberry Lane,

Mottistone 4183; Borthwood Copse 5784; Youngwoods Copse 5785; and Little Lynn Common 5389. Though mainly in south-east England, this species trickles west as far as Connemara, revealing a preference for oceanic conditions. This could account for its absence from Hampshire.

R. rudis Weihe
Accidentally introduced into the Rectory shrubberies in Old Shanklin 5780, from which it has spilled into nearby gardens. A widespread species of southeast and central England that could be expected to occur naturally.

(*R. rufescens* Lef. & P.J. Müller
Listed by Rogers from Pan 5088. Though a likely species for Wight, the record more probably relates to the unnamed 'H1056', which is widespread in woodland in the east of the Island. A specimen of that plant in CGE collected by Bell Salter was misdetermined by Rogers as 'my *infecundus*', the name he used for the bramble now known as *R. rufescens*.)

R. sectiramus W.C.R. Watson
Very rare. Two clumps in scrub near the crest of Brading Down 5986. Probably a recent arrival bird-brought from Hampshire or Sussex (the westernmost tip of which holds a small population).

SERIES *Hystrix* Focke

R. angusticuspis Sudre
Rare. Plentiful along the west half of the south margin of Whitefield Wood 6089, and two clumps in neighbouring Cothey Bottom Copse 6090. A strongly western species, with no other locality nearer than a solitary one in Wiltshire.

R. dasyphyllus (Rogers) E.S. Marshall
Rare and adventive, but likely to spread. A large patch on the public footpath south of Nettlestone Point, Seaview 6391, and another in cleared ash wood, Calbourne Bottom 4294. There is also an 1870 specimen of Stratton's in OXF from St. George's Down 5186.

R. phaeocarpus W.C.R. Watson
Very local but probably spreading, as a late invader from Sussex (where it is common). Noted by Rogers at Wootton Creek in 1901, it occurs in quantity today not only in the copses thereabouts, but also in part of Barton Wood 5294 to the north, and recurs eastwards as far as Nettlestone 6290. On the south coast, it has so far colonised Blackgang Chine 4876, and occurred on Ventnor Golf Course 5477.

R. rilstonei W.C. Barton & Riddelsd.
Very rare. A colony at the southern end of Beech Copse, Godshill Park 5381. A south-western species, almost absent from Dorset but with a large population at the south-east tip of the New Forest.

R. venetorum D.E. Allen
Very rare. One clump among tall bracken half-way up the east side of Great Combley Wood 5488, apparently a bird-sown waif like that other south-western species, *R. cornubiensis*.

SERIES *Glandulosi*
(Wimmer & Grab.) Focke

R. praetextus Sudre
Recently recognised as this rare species with a disjunct distribution in both southern England and France, the hitherto anonymous 'H257' is one of the Island's most abundant brambles, occurring in almost all copses and woods between Newport and Ryde and entirely carpeting several. It spills across into the south-east of Parkhurst Forest 4790, but is virtually absent everywhere else.

R. scaber Weihe
Very rare. One patch under tall pines near the west margin of Firestone Copse 5591.

SECTION *Corylifolii* Lindley

R. conjungens (Bab.)Rogers
Apparently rare and restricted to the clay and chalk. The lectotype of Babington's variety is a Bell Salter specimen in CGE from Bembridge.

R. halsteadensis W.C.R. Watson agg.
At least five different entities in Britain currently share this name. The one originally named after Halstead in Kent and widespread in south-east England occurs on the margins of Whitefield Wood 6089, and nearby Cothey Bottom Copse 6090. It has yet to be found in Hampshire.

R. hindii A.L. Bull
Rare? A recently-discriminated taxon, so far detected only along the lane bounding Staplers Copse 5288.

R. nemorosus Hayne & Willd.
Locally common in wooded districts on the clay, but also in quantity in places off it, such as about Bohemia Bog 5183 and in Moortown marshes, Brighstone 4283. Relatively much more widespread and plentiful than in Hampshire, where it is almost wholly a heath species restricted to the 'greater' New Forest.

R. pictorum Edees
Adventive? Locally common in Osborne House grounds and the west edge of Barton Wood 5194 & 5294. Probably introduced accidentally at some period with nursery stock, though *R. angusticuspis*, which is presumptively indigenous in Wight, is similarly a species of South Wales and the English counties adjacent, implying that the two have similar climatic requirements.

R. pruinosus Arrh.
Frequent in both chalk scrub and acid woodland on the greensand, but rare on the clay of the north and west.

R. transmarinus D.E. Allen
Frequent on heathy ground, especially in the west and centre.

R. tuberculatus Bab.
Common in the coastal areas, thinning out inland.

RUBI INNOMINATI
A few further brambles have distributions too restricted to have earned them names (at least as yet), but occur in sufficient quantity to merit mention:

'H107'. A member of series *Vestiti* with reddish petals and stamens, dark purple axes and undulate elliptical terminal leaflets. Thinly scattered in heathy localities, achieving greatest quantity in Fattingpark Copse 5291. Also frequent in the New Forest in Hampshire.

'H165'. A densely hairy bramble covered in red glands and with long straggly sepals. Abundant in Parkhurst Forest, but tightly confined to it.

'H252'. An apparent hybrid derivative of *R. mollissimus*, very widespread in heathy places. It is particularly plentiful in Parkhurst Forest and The Wilderness. Also widespread in southern Hampshire.

'H388'. A robust bramble with large white flowers and scattered long glands and acicles on the panicle. Conspicuous in Briddlesford Copse 5590 and its satellites, spilling into Firestone Copse in places.

'H375'. A very hairy, eglandular member of section *Corylifolii* with smallish pale pink flowers and acutely biserrate leaflets. Occurs here and there in different parts.

'H863'. A member of series *Hystrix* allied to *R. babingtonii* Salter. Its range is coterminous with the medieval bounds of Parkhurst Forest. It grows in profusion in many of its former components, both on clay and gravel soils. Also found in one copse in central Hampshire.

'H1056'. A member of series *Micantes* with small cupped flowers, all floral organs pink and obovate-cuneate terminal leaflets. In many woods and copses between Newport and Ryde. This was collected by Bell Salter and – in an outlying locality, Shanklin Chine 5881 – Baker. Also in four woods in south-east Hampshire.

SECTION *Caesii* Lej. & Courtois

R. caesius L.
DEWBERRY
N. Locally common in hedges, scrub and woodland, particularly on chalk.

Potentilla palustris (L.) Scop.
MARSH CINQUEFOIL
N. An extremely local plant of fens,

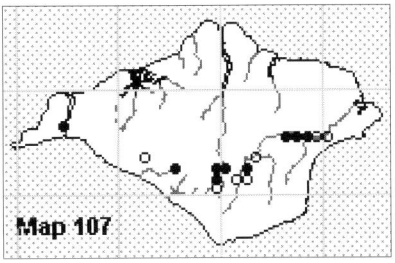
Potentilla palustris

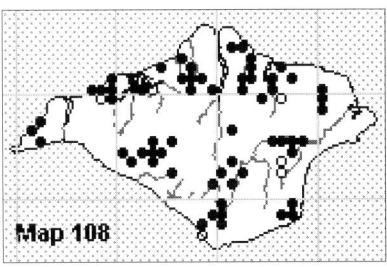
P. erecta

acid flushed bogs, and edges of pools. It has declined considerably in the last one hundred years but is less rare than Bevis *et al* (1978) suggest. Its greatest extent is at Cridmore. In the Eastern Yar, it is extant in six sites between Newchurch and Sandown Golf Course including Hill Farm drain 5785, CP 2001 and fringing the golf course pond 5885, MB 1995. In the Medina, it still occurs around The Wilderness 5082, CP 1996 and at Cridmore Bog 5082 and to the south, RK 1995. Found in small quantity in Wolverton Marsh, Shorwell 4582, CP 1990. It has been recorded from Freshwater Marshes since 1823 and it still survives in tall fen in the south marsh 3486, CP 1994. Map 107

P. anserina L.
SILVERWEED
N. Common across the Island in water-logged meadows, compacted soil, edges of ponds and coastal habitats.

P. recta L.
SULPHUR CINQUEFOIL
E. Rarely established on disturbed waste ground. Recorded from Golden Hill, Freshwater 3387, AM 1995; and Island Harbour, Binfield 5192, CP 1999.

[*P. norvegica* L.
TERNATE-LEAVED CINQUEFOIL
C. Extinct. Record from near Cowes by J. Long in 1920 and from Ryde in 1935.]

P. erecta (L.) Raeusch. ssp. *erecta*
TORMENTIL
N. Local, often in small quantity on acidic soils in woodland rides, heaths and acid grassland. Also found on chalk heath on West High Down. Although still widespread, this species

is declining as its favoured dry and wet heathland habitats deteriorate. Map 108

P. anglica Laichard.
TRAILING TORMENTIL
N. Scarce, but perhaps overlooked, in grassy rides in woods and heaths. The only recent records are: Bleak Down 5181, BS 1993; locally frequent in rides in Parkhurst Forest 4691 & 4790, CP 1997; trackside at Cranmore 3990, VS 1998; and rides in Bouldnor Copse 3890, CP 1997. There are 1960s records from Firestone and Quarr Copses.

P. x mixta Nolte ex Rchb.
HYBRID CINQUEFOIL
N. The hybrid between Trailing Cinquefoil and Creeping Cinquefoil is probably overlooked. It is frequent in Ryde Cemetery 5892, MB 2002.

P. reptans L.
CREEPING CINQUEFOIL
N. Common and widespread in disturbed grassland, verges, under-cliffs and bare ground.

P. sterilis (L.) Garcke
BARREN STRAWBERRY
N. Frequent and widespread in old woodlands, hedge banks, unimproved grasslands and sometimes old walls, e.g. Appley seawall, near Ryde.

Fragaria vesca L.
WILD STRAWBERRY
N. Frequent in open woods, scrub and in hedge banks.

F. x ananassa (Duchesne) Duchesne
GARDEN STRAWBERRY
E. An occasional escape. Historically, most frequent on railway banks and still sometimes found in these situations. This plant is likely to be the

F. ananassa referred to in Bevis *et al* (1978).

Duchesnea indica (Jacks.) Focke
YELLOW-FLOWERED STRAWBERRY
E. Occasionally naturalised in woods. First recorded from St Helen's Common hanger, 4389, in 1978 and still present. Well established by paths in Bonchurch Landslip woods 5878, CP 1998. Also recorded from around Bonchurch Pond 5778, CP 2000, and in woodland behind Steephill Castle 5177, CP 1986.

(*Geum rivale* L.
WATER AVENS
Not native on the Island. There is a single record from Blackwater in 1931 but there is no herbarium material and it is likely that the record was in error.)

G. urbanum L.
WOOD AVENS
N. Common and widespread in woods, hedge banks and shaded lanes.

Agrimonia eupatoria L.
AGRIMONY
N. Common and widespread on verges, waste places, woodland rides and rank grassland.

A. procera Wallr.
FRAGRANT AGRIMONY
N. Rare but perhaps overlooked, in woodland rides and scrub, generally on acid soils. The only known modern sites are rides in Firestone and Briddlesford Copses 5590, CP 1992, and in Combley Great Wood 5489, CP 1990; roadside verge on Wootton Common 5390, CP 1999; and Merstone old railway station 5284, DB 1999. Historic sites where it may still occur, with last recorded dates in brackets, include Lynn gravel pit (1975), Rowlands Lane (1972) and Borthwood Copse (1904).

Sanguisorba minor Scop. ssp. *minor*
SALAD BURNET
N. Common in chalk grassland and also found locally on limey clays and in neutral grassland. It continues to survive in fragments of limestone grassland. Found in compacted sandy grassland at Bembridge Point and Hamstead Duver. Map 109

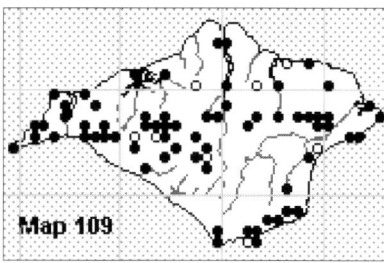

Sanguisorba minor

S. minor Scop. ssp. *muricata* (Gremli) Briq.
FODDER BURNET
C. A rare escape from cultivation, formerly sometimes grown as a fodder crop. Roadside bank at Brighstone 4282, CP 1988; fallow field south of Chillerton 4882, CP 1988.

Acaena novae-zelandiae Kirk
PIRRI-PIRRI-BUR
E. Very rare. Only known record is from a roadside verge at Brighstone 4283, MB 1998.

[***Alchemilla vulgaris*** L. agg.
LADY'S-MANTLE
E? Extinct. Not native on the Island. A single patch in the grounds of Tyne Hall, Bembridge was recorded in 1849 by Bell Salter, 'probably introduced'. No other records.]

A. mollis (Buser) Rothm.
GARDEN LADY'S-MANTLE
E. Very rarely established. Recorded from Shorwell Church 4583, MB 2000; a single plant in chalk grassland in Downend Quarry 5387, CP 1998.

Aphanes arvensis L.
PARSLEY-PIERT
N. Frequent in dry, arable fields, disturbed grassland, wall tops and dry banks generally on basic soils. Many records have not been identified to species. Parsley-piert has herbal properties and local botanist, Thelma White organised collections of the plant for the War effort. This was pointed out to members of the Natural History Society during a botany walk around Havenstreet in 1943. 'At the end of the wood, a halt was made and Miss Bullock pointed out a field in which a number of herb-gatherers had collected parsley-piert quite recently. This valuable herb is used for kidney trouble.' (*Proc. Isle Wight nat. Hist. Archaeol. Soc* III (VI) 1944)

A. australis Rydb.
SLENDER PARSLEY-PIERT
N. Locally frequent in dry, sandy turf generally on acid soils. Many records of parsley-piert have not been identified to species.

Rosa multiflora Thunb. ex Murray
MANY-FLOWERED ROSE
E. Rarely established as a bird-sown garden escape. First recorded from the old railway trackside at Dodnor in 1983, 5091 BS; the large bush still survives. Field edge of Carisbrooke High School playing fields 4889, SB 1998; old railway track near Kitbridge Road, Newport 4989, CP 1997.

R. arvensis Huds.
FIELD ROSE
N. Frequent at the edges of woods, hedges and scrub particularly on clay soils on the north half of the Island. Our second most frequent rose.

R. x pseudorusticana Crép. ex Rogers
FIELD X SHORT-STYLED FIELD ROSE
N. Occasional, sometimes with both parents and often as single bushes. Freshwater 3487, 1999; Hill Copse, Wilmingham 3588, 1999; Thorley 3788, 1997; Shalfleet 4190, 2000; Alverstone Farm, Whippingham 5292, 2000; Springvale 6191. All records PS.

R. x verticillacantha Merat
FIELD X DOG ROSE
N. Occasional. There are *arvensis* x *canina* hybrids recorded since 1998 from 3487, 3589, 3789, 3989, 4889, 4893, 5193, 5292, 5987 and 5988 (all PS) and historic records, determined by A.L. Primavesi from 3588, 5394 and 5984. There are post-1998 records of *canina* x *arvensis* from 4084, 4189, 4190, 4290, 5095 and 6191.

R. pimpinellifolia L.
BURNET ROSE
N. Very local on clay heaths and on chalk, sometimes just a few plants appearing unexpectedly following scrub clearance. Although greatly declined over the past one hundred years it is still locally frequent at the following sites: Bouldnor cliff edge 3991, JO 1998; Cranmore 3990, VS

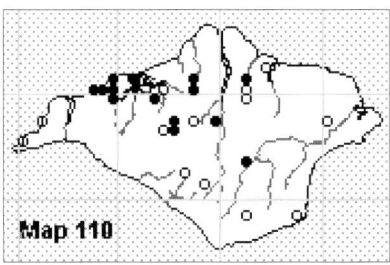

R. pimpinellifolia

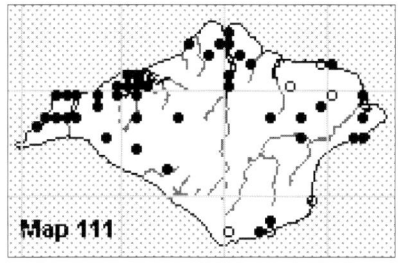

R. stylosa

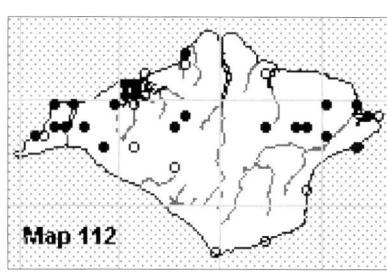

R. micrantha

1998; Parkhurst Forest rides 4790, 4791, CP 1996; and Rowridge valley on chalk 4587, DD 1998. Map 110

[**R. x coronata** Crép. ex Reut.
HARSH DOWNY X BURNET ROSE
N. Extinct. Known only from a single record from King's Quay by J. Groves in August 1931. The herbarium material (BM) has been checked by G. G. Graham.]

(**R. x sabinii** Woods
BURNET X SOFT DOWNY ROSE
N. Extinct. There is an unconfirmed record in Townsend (1883) by More. The specimen was in Watson's herbarium, which was 'consumed by fire'.)

R. rugosa Thunb. ex Murray
JAPANESE ROSE
E. Very rarely established. Spreading around the clubhouse on St Helen's Duver 6389, CP 2000; Bouldnor 3789, PS 1999.

Rosa 'Hollandica'
DUTCH ROSE
S. There is a thicket of what is considered to be this plant growing by the tidal bank of the Medina at Fairlee, 5090, for at least the last 40 years. It is regularly immersed by seawater at spring tides.

R. stylosa Desv.
SHORT-STYLED FIELD-ROSE
N. Frequent in hedges, woods and scrub. This is our third commonest rose. The map is likely to be an under-representation of its distribution. Map 111

R. x andegavensis Bastard
SHORT-STYLED FIELD X DOG ROSE
N. Probably not infrequent as single bushes in hedges and scrub. There are *stylosa* x *canina* hybrids recorded since 1996 from 3589 (MB), 3790

(TDD, CDP), 3886 (PS), 3988 (PS), 4889 (MB), 4994 (MB), 5090 (CP), 5094 (MB), 5787 (CP), 5892 (MB), 6285 (CP), 5391 (MB) and 6389 (PS). There are post 1996 records of *canina* x *stylosa* from 3389, 4190, 4284, 4482, 4885, 5183, 5193, 5293, 5387, 5988, 6389 and 6488 (all PS).

R. canina L.
DOG-ROSE
N. Our commonest rose, widespread in woods, thickets and hedges.

R. x dumetorum Thuill.
DOG X ROUND-LEAVED DOG-ROSE
N. Believed to be very rare. Known only from a single bush in a hedge at Green Lane, Hardingshute 5988, PS 2000 (conf. A.L. Primavesi).

R. x nitidula Besser
DOG-ROSE X SWEET-BRIAR
N. Believed to be rare. Three bushes found on St Helen's Duver 6389, PS 2000 (conf. A.L. Primavesi); one bush in a hedge at Whippingham 5193, PS 2000.

R. caesia Sm. ssp. *caesia*
HAIRY DOG-ROSE
E? Two plants of this northern species were found at Bleak Down 5181 in 1997, BS, SB (conf. G.G. Graham). They were growing on a bank constructed to screen the refuse tip.

R. obtusifolia Desv.
ROUND-LEAVED DOG-ROSE
N. A very rare plant in hedges and woodland edges on gravel and clay. It may be associated with historic heathland sites. The best site to date is at Little Hardingshute near Brading where it occurs along 200m of hedgerow in West Lane, 5988, PS 2001. A single pure specimen has been recorded at Corfe Camp, Shalfleet 4190, 2000. Other records are from

Cranmore 3989, 1999 and from Colwell 3287, 1997 (all PS).

R. tomentosa Sm.
HARSH DOWNY-ROSE
N. Very rare. The only modern record is of a single bush in a hedge by Whitehouse Road, Porchfield 4591, PS 2000. It may have decreased on the Island as there are historic records from nine sites between 1856 and 1904.

R. sherardii Davies
SHERARD'S DOWNY-ROSE
N. Very rare. The only modern record is of a single bush from Rowridge valley 4586, BS 1995 (conf. G.G. Graham). There are historic records of this northern species from near Ryde (1888), Rookley (18850 and Whippingham (1924). Herbarium material of these three specimens has been checked by A.L. Primavesi.

(**R. mollis** Sm.
SOFT DOWNY-ROSE
N. There is a record of this species reported by Townsend from a copse near Ryde in 1844. There is no surviving herbarium material to confirm this unlikely record.)

R. rubiginosa L.
SWEET-BRIAR
N. Rather uncommon and generally found in chalk grassland and scrub. There are recent confirmed records from 3185, 4694, 4994, 5785, 5986, 6087 and 6389.

R. micrantha Borrer ex Sm.
SMALL-FLOWERED SWEET-BRIAR
N. Occasional in chalk grassland and scrub, roadside verges, scrubby heath and coastal grassland. Also found in woods on calcareous clays. This is our fourth most frequent rose. The map is an under-representation of its range. Map 112

Prunus cerasifera Ehrh.
CHERRY PLUM
E. Introduced in hedges, sometimes bird sown but mostly originating from plantings. It is early flowering and under-recorded but considered to be widespread in small quantity.

P. spinosa L.
BLACKTHORN (local: winter kecksies)
N. Very common in hedgerows, woods and scrub everywhere. The local name refers to the fruits. For many years, 'Kixse' was adopted as the brand name for a non-alcoholic 'Wight Spice Cordial' produced by local brewers Gould, Hibberd & Randall Ltd. and popular at Christmas time.

P. domestica L. ssp. *domestica*
WILD PLUM
Arch. Occasional in hedges and thickets, often relics of cultivation or bird sown from gardens. Suckering thickets around Brickfields cottage on Newtown east spit (4292) produce succulent black fruits, and autumn pilgrimages are made to gather them.

ssp. *insitita* (L.) Bonnier & Layens
BULLACE
Arch. Occasional but locally frequent in hedges and scrub. Under recorded but believed to be widespread. Bromfield (1856) said that, 'White Bullace is very commonly brought to market for tarts and puddings'.

P. avium (L.) L.
WILD CHERRY
(local: merry-tree; merries)
N. Occurs in woodland across most of the Island on most soils with a concentration of sites in south-east Wight. Planting has obscured the native distribution. Occasional large specimens have been seen in America Wood (5682); Hungerberry Copse, Shanklin (5780); Marshcombe Copse, Yaverland (6186); and Mudless Copse, Swainston (4486). The local word 'Merries' referred to the fruits and is the origin of the name Merrie Gardens near Shanklin (it appears as Cherry Gardens on Andrew's map of 1769).

P. cerasus L.
DWARF CHERRY
E. An escape from cultivation, which is probably overlooked. There are many records in Bromfield (1856) but the only more recent records are in Bevis *et al* (1978) where two sites are listed: roadside hedge between Whippingham and Osborne, 5194, grubbed up in 1983; and Froglands Lane, 4887.

[*P. padus* L.
BIRD CHERRY
P? Extinct. Bromfield (1856) considered that this tree was quite naturalised in one or two woods, in particular the wooded slope below Cook's Castle where he described it as 'occurring in considerable plenty'. It is no longer there and the most recent record for this species is from The Hermitage at St Catherine's Down 4978, TW 1965.]

P. lusitanica L.
PORTUGAL LAUREL
P. Sometimes planted in shrubberies and plantations. Generally persistent, but not yet recorded as naturalising.

P. laurocerasus L.
CHERRY LAUREL
E. Commonly planted in woods and plantations. It sows itself and can be frequent in the understorey on acid soils.

Pyrus communis L.
PEAR
Arch. Rare in hedges, woods and scrub. It is perhaps wild, but also sometimes long established as a relic of cultivation or self-sown. It generally occurs as isolated specimens, which are sometimes large. There is a concentration of solitary trees in ancient woodlands and old hedgerows around Havenstreet, and not all of these have been critically examined. A tree still present in Rowlands Wood 5689 (1997) was known to Miss Bullock as a large tree in 1943. There are rumoured to be others in the area. Thorny bushes and trees with small, ovoid, inedible fruits which may be referable to *Pyrus pyraster* (L.) Burgsd., the Wild Pear, have been recorded from the side of the Old Millpond, Wootton 5491, BS 1999, at Corfe Camp, Porchfield 4190, PS 2000 and in a rew by Quarry Lane, Newbridge 4287, HH 2002. There are large trees surviving at Little Duxmore, Mersley 5587, CP 1999 and in Newport on the west bank of the Medina at Pan Mill 5089, BS 1987 but these are clearly relics of cultivation. Other examples are given in Bevis *et al* (1978). There is a group of five trees within Noke Plantation 4891, BS 2000 and a row of trees at the edge of Thorness Wood 4593, 2001 and these may perhaps be long-established relics of cultivation. Other individual bushes have been recorded from scrub at Thorley, Lower Hamstead, Brighstone Shute and Headon Warren, HH.

Malus sylvestris (L.) Mill.
CRAB APPLE
N. Characteristic of ancient woodlands and old hedgerows, particularly on heavy clay soils. It is generally only present in small quantity within any one wood. Probably the largest Island specimens grow in Wroxall Copse 5678. Map 113

M. domestica Borkh.
APPLE
S. Widespread as isolated trees, probably originating from discarded pips and a relic of cultivation. Found in hedges, thickets and waste ground. Hybrids with Crab Apple also occur.

Sorbus aucuparia L.
ROWAN
E. Not native on the Island but found occasionally in woods and heathy places, often as a result of planting. Sometimes occurs as a naturalised species, particularly in woods in south-east Wight. Sites include: Gatcombe Withybed 4984, CP 1988; Sibden Hill, Shanklin 5781, CP 2000; wood at Hill Heath, Alverstone 5685, PS 2000; Youngwoods Copse 5785, CP 1998; scrub and woodland on Sandown Golf Course at Blackpan

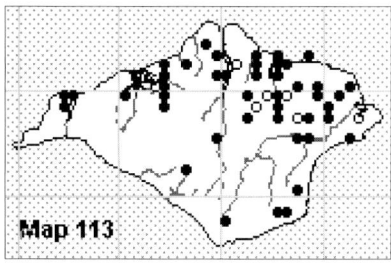

Malus sylvestris

Common 5885, MB 1996; and America Wood 5681, CP 1999, where it was recorded by Bromfield (1856).

S. intermedia (Ehrh.) Pers.
SWEDISH WHITEBEAM
E. Frequently planted but only occasionally self-sown, generally as isolated trees. Self-sown in scrubby corners of Sandown Golf Course 5885, MB 1997; a single seedling tree on Headon Warren 3186, CP 1998.

S. aria (L.) Crantz.
WHITEBEAM
(local: whipcrop, white rice)
N. Scarce and infrequent in woods on the chalk, where it is considered to be native, and sometimes found as small bird-sown bushes in chalk grassland or old chalk quarries. Also planted. Most sites are in the East Wight, such as Eaglehead Copse 5887, north side of Mersley Down 5687, Wroxall Copse 5678, Bonchurch Landslip 5878 and secondary woodland on Luccombe Down 5779. It is unusually frequent in the eastern half of Wroxall Copse 5678, where the largest specimens are to be found. Bromfield referred to twisted and contorted trees growing from the upper rock crevices of Cliff Copse, west of Shanklin 5680, and distinctive large trees are still to be found in this situation. They are growing in the fissured chert outcrops, overlain by Lower Chalk. Whitebeam grows in some woods in the central downland including Long Copse 4884 and Westridge Copse, Shorwell 4884. There are a few historic records from acid soils (see Wild Service-tree below). The local name 'Whipcrop' referred to the custom of cutting the long, straight tough shoots for whip-handles by waggoners. 'White Rice' referred to the silvery undersides of the leaves and to the low, bushy habit, rice being a local term for brushwood. However, both these terms were also applied to Wayfaring Tree, a much commoner shrub on the Island. Map 114

S. latifolia agg.
BROAD-LEAVED WHITEBEAM
E. Frequently planted. There are eighteenth century records from Carisbrooke Castle and Mount Joy cemetery and an old tree survives (2002) in a hedge in the south-west corner along the outer boundary of the Castle. Established in woodland and scrub on Sandown Golf Course at Blackpan Common 5885, BS 1983

S. torminalis (L.) Crantz
WILD SERVICE-TREE
N. Not infrequent in woods on the clay soils of the northern half of the Island, particularly around the creeks and coastline. The plant seems to propagate exclusively from suckers, which are often freely produced. This tree has a very interesting distribution; it is one of our most faithful indicators of ancient woodlands and old hedge banks.

In Walter's Copse, Newtown, a wood of secondary origin, it is unknown apart from two coppiced specimens growing on old hedge banks now well within the wood 4390, RG 1996. It has always been very rare on the greensand soils in the south of the Island although there are historic records from Borthwood Copse. It still occurs in Youngwoods Copse, an ancient woodland near Alverstone, where about five trees survive 5785, CP 1995. Surprisingly, Bromfield did not record it from here but did note a Whitebeam tree of considerable size, a species currently unknown from this wood.

Large trees are infrequent. The largest specimen known to Bromfield (1856) grew in Quarr Copse and was estimated to have a height of 40 ft (12m) and a girth of 5ft 6 in. (1.67m). A large specimen survives in Quarr Copse, within a garden, with a height of 15.76m and a girth of 2m 5692, PS 2002. However, the Island's two prime specimens are amongst the largest recorded in this country. They are growing on greensand near the bottom of an unremarkable secondary hanger woodland above Great Whitcombe Manor, 4986. In 2002, only the larger tree was still standing. The larger specimen had an estimated height of 19m and a girth of 2.04m in 1994, RL. These trees may be of planted origin as they are not clearly associated with any ancient woodland.

The only other known instance of trees that are likely to have been planted is on the east spit at Newtown estuary, where there are five trees on the now derelict Brickfields Cottage Farm boundary bank, a long way from any woodland but alongside the old farm orchard, 4292 CP 1997. However, in recent years, Wild Service has been inappropriately planted in several woods (RL).

Bromfield (1856) referring to the fruits said, 'Sold . . . in this island, in the shops and public markets, tied up in bunches, principally to children. At Ryde they go under the name of Sorbus berries, but they are not much in request'. In Sussex and Kent the berries were called Chequer-berries and Chequer Inns were often associated with old Service trees. There is a Chequers Inn south of Rookley in a part of the Island well away from the native distribution of this species. However, interestingly, there is a record in Bromfield (1856) for Whitebeam of, 'A solitary tree in the hedge on the right hand side of the road a short distance from the Star inn at Rookley, towards Bleak down.' At the time, the Chequers Inn was known as The Star Inn and Bromfield refers to the confusion of the different *Sorbus* species by countryfolk. The subsequent renaming of the public house may have its origins in the presence of this solitary tree. Map 115

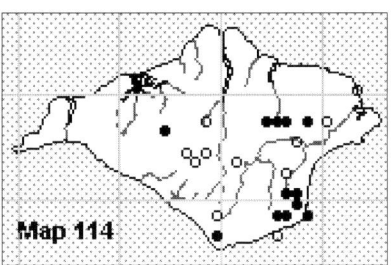

S. aria

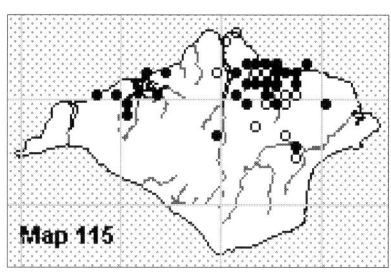

S. torminalis

Amelanchier lamarckii F.G.Schroed.
JUNEBERRY
E/S Very rare. Several specimens are self-sown on an open bracken-covered greensand hilltop at Sibden Hill, Shanklin 5781, CP 1999.

The *Cotoneaster* genus has not been critically studied on the Island and the following list is considered to be far from complete.

Cotoneaster integrifolius (Roxb.) G.Klotz
ENTIRE-LEAVED COTONEASTER
E. Long established locally in disused chalk quarries and in downland grassland right across the Island, but much confused with *C. microphyllus*. Sites include Afton Down, Pan chalk pit, Brading Down, Bembridge Down, Rew Down and St Boniface Down. There are a number of records from the early 1900s from sites where it is still present (Rayner, 1929): established on the craggy sides of Niton undercliff in 1916; thoroughly naturalised on Brading Down by 1917; and first noticed on St Boniface Down in 1921 and considered to be well established there by 1925.

C. horizontalis Decne.
WALL COTONEASTER
E. Frequently naturalised on walls, downland and chalky banks, verges and derelict gardens. First recorded in 1965, a few plants on St Boniface Down, TW.

C. simonsii Baker
HIMALAYAN COTONEASTER
E. Not infrequently naturalised on downland, old chalk quarries, scrub and woodland rides. First recorded in 1949 from Brading Down, where it still occurs.

C. bullatus Bois
HOLLYBERRY COTONEASTER
S. One plant surviving in a disused chalk quarry on Brading Down 6086, AC 1996.

C. dielsianus E. Pritz. ex Diels
DIELS' COTONEASTER
Establishing on roadside verge at Clatterford Shute, Newport 4887, GK 1994.

Mespilus germanica L.
MEDLAR
S. There is an old, thornless coppiced bush at the field edge of an old shrubbery lying to the west of the Manor House, Shanklin 5780, CP 2002, doubtless a relic of cultivation. Two specimens were seen here in 1968.

Crataegus monogyna Jacq.
HAWTHORN (local: hogiles; hogails)
N. Abundant everywhere. Occa-sional wild pink-flowered bushes occur. The newly emerging foliage used to be nibbled by children as 'bread-and-cheese'. The local word 'hogiles' referred to the haws. The ancient custom of celebrating May Day was described in the Ledger Book of Newport, Isle of Wight 1567-1799. Preserving the old custom of the town, on the Sunday after May Day the inhabitants of Newport formed a procession to Ovis Wood on the edge of the town (near Hunnyhill), there to be met by the forest keepers. It was the keeper's role to salute and to offer small green boughs of may to all the company, thereby acknow-ledging the free right of pasture granted to the Councillors by the charter of Richard de Redvers. The common people of the town then cut green boughs from the trees to decorate the town, 'placing the greenery at their doors to give a pleasant smell to their houses and comfort to passers by'. The tradition was discontinued in 1621 by decree because of the 'great damage in the King's wood of Parkhurst forest by the poor people'.

C. laevigata (Poir.) DC.
MIDLAND HAWTHORN
S. Not known as a native species but sometimes planted in hedgerows.

79. FABACEAE
Pea family

Robinia pseudoacacia L.
FALSE-ACACIA
E. Occasionally planted in woods and spreading by suckers and seeds. Recorded as established in coastal woodland at Quarr 5792, Fishbourne Copse 5692 and cliff woodland at Steephill 5577; and Barton Withybed, Apse Heath 5683.

Galega officinalis L.
GOAT'S-RUE
E. Very locally plentiful and persistent on disturbed ground, principally on the west bank of the Medina valley. Sites where this plant is abundant include Downend quarry 5387, RK 1987; Northwood camp 4992, CP 1999; Pinkmead Farm, Werrar 5092, AM 1996; and waste ground around Parkhurst Road, Newport 4990, CP 1999.

Colutea arborescens L.
BLADDER-SENNA
P? The only record is from the cliff between Bembridge Point and the lifeboat station, JB 1970, where it was presumably a garden relic.

(*Astragalus danicus* Retz.
PURPLE MILK-VETCH
There is no evidence that this plant was ever native on the Island but Snooke(1823) recorded it from Dover Spit, in plenty. It is unclear whether he was referring to St Helen's or Ryde Duver, there are no herbarium specimens and no succeeding botanists have ever recorded it. His second recorded site, Carisbrooke Castle hill, was immediately disputed by botanists)

A. glycyphyllos L.
WILD LIQUORICE
N. An Undercliff species, which has declined dramatically. It was thought to be extinct until a single plant was refound near Windy Gap car park at St Catherine's 4975, VS 1998, and four further plants have since been discovered in the vicinity growing at the edge of scrubby calcareous grass-land. For its size, it is a surprisingly inconspicuous plant and more plants probably remain to be discovered. However, in the nineteenth century it grew 'in great profusion amongst furze and brake' in several sites between Ventnor and St Catherine's Point, although particularly towards the western end. It seems to have already declined from everywhere, excepting around St Catherine's by the early twentieth century, perhaps as a result of a reduction in grazing pressure. It was still considered to be frequent at Binnel Bay 5175, 'in an almost inaccessible spot' in 1970 but the final plant was

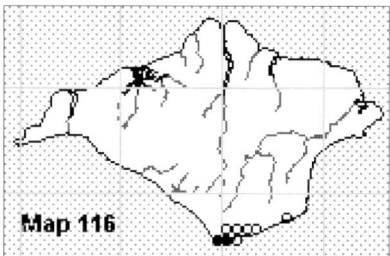

A. glycyphyllos

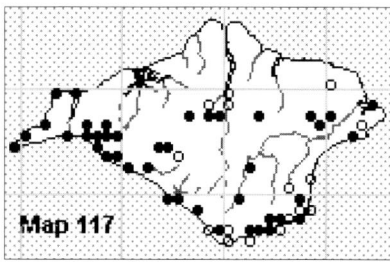

Anthyllis vulneraria

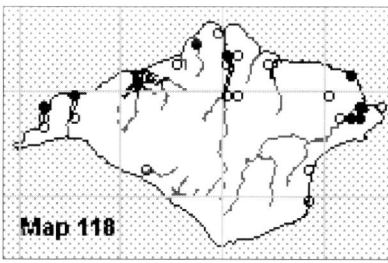

Lotus glaber

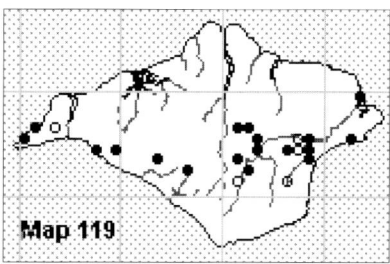

Ornithopus perpusillus

recorded here in 1979, at the cliff edge, a casualty of scrub encroachment and erosion. Map 116

Onobrychis viciifolia Scop.
SAINFOIN

N/E Very local in chalk grasslands, especially by tracksides and roadsides. Although usually a relic of cultivation or introduced with wild flower seeds, it might be native in north-facing grassland on Shanklin Down 5980, GT 1998 where it is frequent and has been recorded since 1860.

A handful of plants are recorded at High Hat, St Lawrence 5276, CP 1988 on the cliff face and the old marl pit and reservoir site on top of the cliff. It grows on a bank by the chapel at Mount Joy cemetery, Carisbrooke 4987, AM 1998. Roadside sites at Mersley Down and Merstone were deliberate introductions. Formerly, much more frequent when Sainfoin was commonly grown as a crop plant.

Anthyllis vulneraria L. ssp. *vulneraria*
KIDNEY VETCH

N. Local, forming colonies in chalk grassland, especially in disturbed sites such as chalk cliffs, old quarries and road cuttings; also frequent on undercliffs along the south coast. Grows on fixed shingle on Norton Spit and on the disused railway track at Bembridge. Map 117

Lotus glaber Mill.
NARROW-LEAVED BIRD'S-FOOT-TREFOIL

N. Scarce and local, but probably under-recorded, in brackish pastures and coastal grassland. In brackish grassland at Yarmouth, south of the mill 3589, PS 1997; and in various spots on Brading Marshes 6287, 6387, 6388, NS 1991; in open clayey grassland at Werrar 5093, CP 1999; Gurnard 4794, CP 1999; Colwell Bay cliffs 3288, JO & CC 1990; and by the Newtown estuary at Clamerkin 4391, CP 2002. Map 118

L. corniculatus L.
COMMON BIRD'S-FOOT-TREFOIL
(local: tom thumb!)

N. Frequent and widespread in dry grasslands right across the Island on all soil types.

L. pedunculatus Cav.
GREATER BIRD'S-FOOT-TREFOIL

N. Frequent in marshes, wet pastures, stream banks and wet woodland rides, generally avoiding the most calcareous soils.

(*L. angustissimus* L.
SLENDER BIRD'S-FOOT-TREFOIL

N? Two possible records are likely contenders but no herbarium material was preserved and no vouchers sent to referees. Recorded in 1963 by the harbour at Yarmouth by Mercia Seabroke, where it was also seen by Phoebe Yule. In 1968, Miss Bullock found it on St Helen's Duver and showed it to Reg Kettell who wrote in his diary, 'I had not seen it before and, being a new species, I took one for verification, and had no difficulty agreeing this to be the species she declared it to be.')

Ornithopus perpusillus L.
BIRD'S-FOOT

N. Very local but sometimes frequent in dry, disturbed sandy soil. Less scarce than suggested by Bevis *et al* (1978) but easily overlooked. Common in some years on St Helen's Duver 6389, TCR 1996; Headon Warren 3185, 3186, CP 1996; and parts of St George's Down 5286, AC 2000. Map 119

[*Coronilla scorpioides* (L.) W.D.J. Koch
ANNUAL SCORPION-VETCH

C. Extinct A rare casual which has not been seen for many years. Waste ground at Seaclose, Newport, J.W. Long 1931; and waste ground at Ryde 1962 (*Proc. IWNHAS* V(VII):318).]

Hippocrepis comosa L.
HORSESHOE VETCH

N. Characteristic of short, dry, usually south-facing chalk grassland and chalk cliffs, where it can be abundant. When in full bloom, it is a striking feature of the downs above the town of Ventnor where it has colonised grassland cleared of scrub. It occurs wherever suitable conditions are met, in long-established grassland throughout the chalk outcrop. Occasional plants with sulphur-coloured flowers are found where large populations are present. It also occurs in suitable places on the inner cliff between Gore Cliff and St Lawrence, and in limestone grassland at Prospect Quarry 3886, CP 1996 and East Cowes cemetery 5094, CP 1994.

[*Securigera varia* (L.) Lassen
CROWN VETCH

C. Known from only two old records: Newport, 1923 and Cowes, 1931, both J. W. Long.]

Vicia cracca L.
TUFTED VETCH
N. Common and widespread in rough grassland, hedges and scrub.

[*V. tenuifolia* Roth
FINE-LEAVED VETCH
C. Extinct? Very rare or extinct. The only records are Headon Warren, J. Groves 1922; Totland, J.W. Long 1931; and Bleak Down JB 1969.]

[*V. sylvatica* L.
WOOD VETCH
N. Extinct. Formerly found in two ancient woods, Luccombe Copse and Yaverland Copse. E.H. White, writing in 1904, called it 'Shanklin's pride, for it grows nowhere else in the county.' It was plentiful in Luccombe Copse and could be seen from the road, but it gradually declined. By 1950 it was considered to be still very local (EHW) and the last record was of a single patch below Nansen Hill at SZ575789 in July 1951, RPB. It is likely that it was lost as a result of lack of woodland management.]

[*V. villosa* Roth.
FODDER VETCH
E. Extinct. There are four old records of this casual: Newport and Totland, 1931, J.W. Long & E. Drabble; plentiful in a field at St Lawrence in 1947, J.E. Lousley, who mentioned that the plant was particularly frequent after the second world war as a contaminant of imported seed; and on waste ground at Freshwater with other casuals in 1979, BS (conf. E.J. Clement).]

V. hirsuta (L.) Gray
HAIRY TARE
N. Locally frequent in rough grassy places.

V. parviflora Cav.
SLENDER TARE
N. Rare in grassy places. Known from three sites: very locally frequent on Forelands slipped cliff 6486, AC 1996, where it was first recorded by Bromfield (1856); tall calcareous grassland at the edge of Ventnor football club 5477, BG 1999; and at the eastern end of Brading Down 6087, MB 1993. Historically much more frequent, but already becoming rare by the 1930s.

V. tetrasperma (L.) Schreb.
SMOOTH TARE
N. Widespread and locally frequent in long grass.

V. sepium L.
BUSH VETCH
N. Widespread and locally frequent in hedges and woodland edges.

[*V. hybrida* L.
HAIRY YELLOW-VETCH
C. Extinct. The only record was made by F. Stratton, from a field of Sainfoin in the Undercliff in 1888.]

[*V. pannonica* Crantz
HUNGARIAN VETCH
C. Extinct. The only record is from Totland, 1931 J. W. Long.]

V. sativa L.
COMMON VETCH
N./Arch. A common and widespread species in hedge banks, waste ground, woodland rides and grassland. Both subspecies, *nigra* (L.) Ehrh. N. and *segetalis* (Thuill.) Gaudin Arch. are commonly found. Subspecies *sativa* Arch. is frequent but recorded less often; it is often transient but it may be under recorded.

V. lathyroides L.
SPRING VETCH
N. Rare and very local, in dry grassy places on acidic sandy soil. Historically recorded from six stations, with four extant sites. Found in fluctuating numbers on the clifftop at Redcliff, Sandown 6285, CP 1998; a very few plants on St Helen's Duver 6389, AC 1992; many small plants on the shingle on the west spit, Newtown Harbour 4191, AC 1996; found to be not infrequent on Rowdown, Brighstone 4383, CP 1998.

[*V. lutea* L.
YELLOW-VETCH
C. Extinct. There are three records: rough ground near Sandown, H.C. Watson 1860; Newport J.W. Long 1940, specimen at BM; and on waste ground at Freshwater with other casuals BS 1979 (conf. E.J. Clement).]

[*V. bithynica* (L.) L.
BITHYNIAN VETCH
N./C. Extinct. Recorded from rough ground near Gurnard Bay, a possible native site, J.F. Rayner 1907, specimen at BM; and on waste ground at Freshwater with other casuals BS 1979 (conf. E.J. Clement). Possibly over looked.]

(*Lathyrus japonicus* Willd.
SEA PEA
N. Extinct. Bromfield (1856) referred to two sites. One of these, 'on the sands near Lord Seymour's (Norris Castle), near Cowes' was pre-1800 but quite probable, as it was not uncommon on mainland Solent shingle beaches at one time. The other site, on shingle in Sandown Bay, was later disputed by A.G. More (1871) who wrote, 'I fear some mistake about this plant. On the beach in Sandown Bay, and also on the shore at Shanklin, I have found stunted plants of *L. sylvestris*, which may have led to the error.')

L. linifolius (Reichard) Bässler
BITTER-VETCH
N. Local in ancient woodlands on clay soils in the north of the Island. It is much declined in quantity, and many woods today are far too shaded for it to survive. It remains common in a few sites, as in some rides in Parkhurst Forest. Very scarce today on sandy soils on the south of the Island where it is recorded from a damp area of Sandown Golf Course, near Scotchells Brook 5884, MB 1996; a few localised clumps in Youngwoods meadow adjoining the copse at Alverstone 5785, CP 1998; and a single plant on Bleak Down 5181, IB 1998. Not seen in Borthwood Copse for many years. Map 120

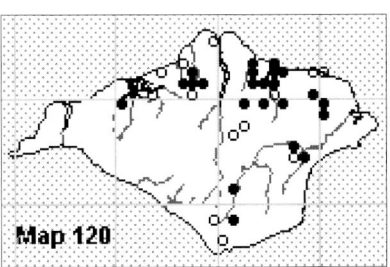

L. linifolius

L. pratensis L.
MEADOW VETCHLING
N. Common and widespread in grasslands.

(***L. palustris*** L.
MARSH PEA
N? There is a record from Golden Hill fort in 1887 by H.P. Fitzgerald but this seems a most unusual location for this plant and no herbarium material is known. The record appears in the Monks Wood database as an introduction.)

[***L. tuberosus*** L.
TUBEROUS PEA
E. Extinct. This is an introduced species, with a few historic records. Niton, F. Stratton, 1887. There is a record from Golden Common of 1838 (P.D. Radcliffe) and there are two herbarium sheets at BM of specimens labelled Totland 1931 and 1932, E. Drabble & J. Long, all of which may refer to the same site. A small colony in a hedgerow at Ashey was first recorded in 1934, but was grubbed up when the hedge was removed in 1945 (GB).]

L. sylvestris L.
NARROW-LEAVED EVERLASTING-PEA
N. Native in scrubby edges on undercliffs in south-east Wight. It is frequent at Bonchurch and Luccombe cliffs 5878 & 5879, CP 1997, and on Lake Cliffs 5880, AC 1999. It has been known from this stretch of coastline through to Sandown since 1805. Other old records away from the south-east coastline, including scattered inland records, may relate to introductions. However, it has been known for over seventy years from the woodland edge on East Cowes esplanade 5096, BS 1997. (See also the entry for Sea Pea.)

L. latifolius L.
BROAD-LEAVED EVERLASTING-PEA
E. A frequent and persistent garden outcast occurring in many places on verges, old railway banks, undercliffs and waste ground.

L. sativus L.
INDIAN PEA
C. Two records of this casual species: blue-flowered plants at Smallbrook,

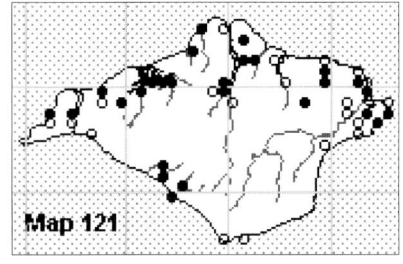
L. nissolia

Ryde 1962 and again 1963; Havenstreet 1967, TW

L. annuus L.
FODDER PEA
C. A single record from disturbed ground at Freshwater with other aliens, BS 1979 (conf. E.J. Clement).

[***L. hirsutus*** L.
HAIRY VETCHLING
C. Extinct. The only records are of a single specimen in a field near Brighstone 1856, Bromfield; waste ground at Newport 1919; and 1929 onwards at Totland, Drabble & Long, 1931.]

L. nissolia L.
GRASS VETCHLING
N. This delightful species is widespread, particularly near the coast, in tall grassland on disturbed clay soils on undercliffs, verges and grassy banks. Map 121

[***L. aphaca*** L.
YELLOW VETCHLING
C. Extinct. The only records are: a single plant at Newport, J.W. Long 1929; Gurnard, EHW 1930; and east bank of the Medina estuary, GB 1941.]

Ononis repens L.
COMMON RESTHARROW
(local: cammock)
N. Frequent and widespread, particularly in chalk grassland and coastal grasslands. Sometimes found in neutral herb-rich meadows, a habitat where it would once have been common, and it also survives in remnant limestone grassland. A spiny form, var. *horrida* Lange, is sometimes found around the coast and can be confused with Spiny Restharrow (*O. spinosa* L.) which is absent from the Island. Bromfield (1856) said that Restharrow, 'is reputed to communicate its nauseous goat-like odour to the milk and cheese of cows pastured where it abounds; cheese so tainted is said... to be cammocky.'

Melilotus altissimus Thuill.
TALL MELILOT
Arch. Widespread and frequent on disturbed ground, verges and tracks. This is our common species. It also occurs on cliffs and upper beaches where it could be native.

M. albus Medik.
WHITE MELILOT
E. Rare on waste ground and beaches. The only recent recorded site is Hamstead shingle beach 4191, AC 1996 where it was first recorded in 1931. It was recorded in a similar situation on Norton Spit between 1931 and 1957. There are old records from various scattered sites but it has always been much scarcer here than on the immediate mainland.

M. officinalis (L.) Pall.
RIBBED MELILOT
E. Very rare on the Island. Much confused with Tall Melilot in the past, but presumably introduced with crop seed from time to time. The only recent record is of a single plant on rough ground by the old railway track, Bembridge Harbour 6388, AC 1997 (conf. R.M. Burton).

M. indicus (L.) All.
SMALL MELILOT
E. Very rare. Recently only known from St Helen's Duver 6388, AC 1998 where there are very few plants. In 1968 it was frequent where the seawall had been strengthened by imported chalk, BS, but it could not be found by 1980. Also occurred on disturbed ground in Ryde 5992, RK but last seen here in 1985.

[***M. sulcatus*** Desf.
FURROWED MELILOT
C. Extinct. Apparently plentiful for a year or two near Newport, J.W. Long 1931.]

[***Trigonella corniculata*** (L.) L.
SICKLE-FRUITED FENUGREEK
C. Extinct. A single record from waste ground at Newport, J.W. Long 1925.]

[*T. foenum-graecum* L.
FENUGREEK
C. Extinct. A single record from near the corn mill at Newport, J.W. Long 1931.]

Medicago lupulina L.
BLACK MEDICK
N. Common and widespread in short grass and as a garden weed. There is an interesting note in the diaries of E.H. White (1904) where he writes, 'In the deserted backyard at Winstone, it is so abundant that the ripe black seeds give the ground an appearance of desolation, as if it had been swept by a fire.'

M. sativa L. **ssp.** *sativa*
LUCERNE
E. A relic of cultivation, which can persist for a number of years on field borders and verges. Less common today than formerly but recorded from ten post-1987 sites, mostly on roadside verges such as on Brading Down 5986, Blackwater 5086, Rew Street 4692, Thorness 4592, and Rookley 5084. Also established on the cliff top at Shanklin 5881.

M. polymorpha L.
TOOTHED MEDICK
N. Sporadic in small quantity in short coastal grassland but probably overlooked. Yarmouth Castle ramparts 3589, PS 1997; Castle Haven 5075, TDD & CDP 1996; several spots around Bembridge Harbour 6388 & 6488, AC 1999; Brading churchyard 6087, BS 1995. Probably still present in coastal grassland at Old Park, St Lawrence and at Brook, where it was recorded by Bevis *et al* (1978).

M. arabica (L.) Huds.
SPOTTED MEDICK
N. Frequent in short grassland, trackways and waste ground right across the Island, not showing any coastal preferences.

Trifolium ornithopodioides L.
BIRD'S-FOOT CLOVER
N. Locally frequent in short, parched grassland on sandy or gravelly soils. It was considered to be fairly frequent but overlooked by eighteenth century botanists but Bevis *et al* (1978) only

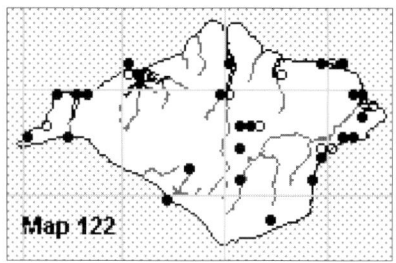

Trifolium ornithopodioides

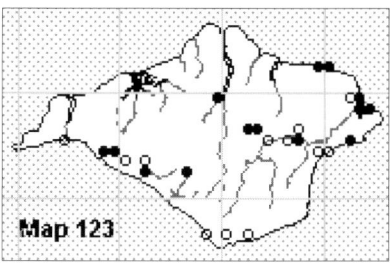

T. glomeratum

knew it from St Helen's Duver where it has always been abundant. Since the mid 1980's it has been recorded much more frequently and this is, at least in part, due to a genuine increase in abundance. The map is likely to be an under-representation of its current distribution. Map 122

T. repens L.
WHITE CLOVER
N. Abundant and widely distributed in grasslands and waste ground everywhere.

T. hybridum L.
ALSIKE CLOVER
E. Thinly scattered as a relic of cultivation in pastures, edges of fields and newly sown grass. Perhaps under-recorded.

T. glomeratum L.
CLUSTERED CLOVER
N. Very local in short, parched grassland on sandy or gravelly soils. Historically recorded from a number of scattered sites but by the time of Bevis *et al* (1978), it was one of our rarest clovers, known only from St Helen's Duver, 6389, and Redcliff 6285. It still occurs in both sites but since 1998 has been recorded more widely. This is another clover which has shown an increase in recent years but the increase seems to have started later than with the others and is continuing. It reappeared on the sandy banks of Ryde Canoe Lake 6092, PS, in 1999 after an absence from this area of nearly 150 years, and at Heath Hill, Shorwell 4682, PS 2002, from where it was last recorded by Townsend in 1883. It is spreading around Bembridge Harbour and is becoming common in suitable spots at the eastern end of St George's Down 5286 & 5386, AC 2000. It appeared on the lawn of a house in Brighstone 4282, built about twenty years previously PS 2000 and on a recent roadside verge in Newport 4989, PS 2000. At Brook Hill 3984, PS 2000, it appeared for one year in countless thousands on set-aside grassland. These plants are believed to have arisen as seed contaminants, perhaps from an Eastern European source (EC, pers. com.). Map 123

T. suffocatum L.
SUFFOCATED CLOVER
N. Rare in bare, dry sandy places. Historically recorded from a small handful of sites but Bevis *et al* (1978) only knew it from St Helen's Duver, where it can be frequent in good years. Since 1996, it has been recorded from four additional sites, suggesting that it is increasing its range. It was found on the south facing bank of Ryde Canoe Lake 6092, CP 1996 from where it has since spread; it was known from Ryde Dover in this area by eighteenth century botanists. A few plants found on Rowdown, Brighstone 4383, PS 1998; occasional, mostly on a compacted gravel path, at Brook Hill 3984, CP 1999; a few plants on earth bank in Jubilee car park, Mottistone Down 4284, PS 2002. Not yet refound on Norton Spit, where J. Groves recorded it in 1923.

T. fragiferum L.
STRAWBERRY CLOVER
N. Locally frequent in brackish pastures, coastal grassland, old pastures on alluvium and clay and in amenity grassland. Found right across the Island and by no means confined to the coast.

[*T. resupinatum* L.
REVERSED CLOVER
C. Extinct. Only two records: Newport, J.W. Long 1929; and a single plant in a field near Oaklands,

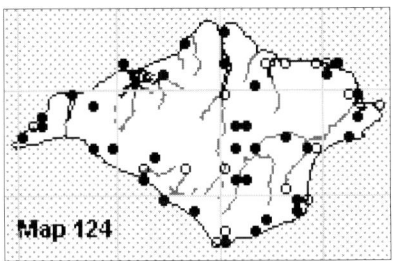

T. micranthum

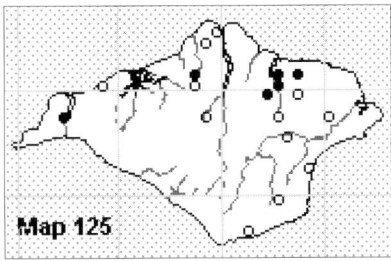

T. medium

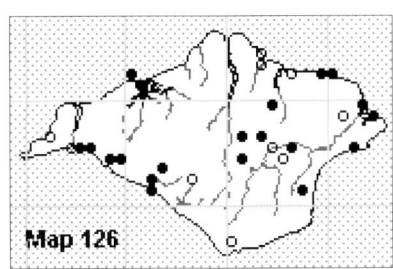

T. striatum

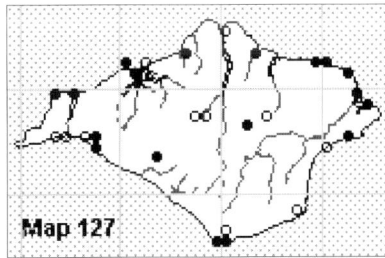

T. scabrum

Springvale, GB 1971 (conf. D. McClintock).]

[*T. tomentosum* L.
WOOLLY CLOVER
C. Extinct. A single record by J.W. Long from the riverside near Newport, 1931.]

T. campestre Schreb.
HOP TREFOIL (local: yellow clover)
N. Frequent in dry grasslands, particularly on calcareous and sandy soils.

T. dubium Sibth.
LESSER TREFOIL
N. Abundant everywhere in grassy and waste places.

T. micranthum Viv.
SLENDER TREFOIL
N. Locally frequent in lawns, playing fields, golf courses and other short turf, particularly on sandy soils. Map 124

T. pratense L.
RED CLOVER
(local: broad clover; cow-grass)
N. Common in grasslands on all types of soil but rarely in the abundance seen in the past.

T. medium L.
ZIGZAG CLOVER
N. Local in small quantity in grassy tracks, roadside verges and scrub on heathland on heavy clay soils; scarce but probably under recorded. Golden Hill, Freshwater 3487, CP 1995; several rides in Parkhurst Forest 4791, CP 1996; roadside verge in Combley Great Wood 5489, CP 2001; scrub on Dame Anthony's Common, Ryde 5791, CP 1997; roadside verge and rides in Firestone Copse 5591, CP 1998. It was recorded in Firestone Copse, at the junction of the roads to Newnham Farm and Havenstreet, by Thomas Bell Salter in 1856 and, remarkably, it still survives in this location. Map 125

[*T. ochroleucon* Huds.
SULPHUR CLOVER
C. Extinct. A single record from the roadside at Bierley, near Whitwell 5178, BS 1987.]

T. incarnatum L. ssp. *incarnatum*
CRIMSON CLOVER
C. Formerly an occasional relic of cultivation but now very rare. It was last recorded *en masse* as a crop in 1965 in a pasture south of Carisbrooke Castle. Since then it has been rarely recorded as single casual plants, most recently a few plants in a weedy field at Rowridge 4585, BS in 1990.

T. striatum L.
KNOTTED CLOVER
N. Locally frequent in dry grassland on sand or gravel. Its core sites include Hamstead Duver 4092, Redcliff 6285, Ryde Canoe Lake 6092, and sites around Bembridge Harbour 6389, 6488. Away from such sites, it has been spreading since the early 1990s. Map 126

T. scabrum L.
ROUGH CLOVER
N. Locally frequent in dry grassland, particularly around the coast. Like the previous species, Rough Clover is currently expanding its range. Map 127

T. arvense L.
HARE'S-FOOT CLOVER
N. Rare, and probably declining on sandy or gravelly soils. The only recent sites are: St Helen's Duver 6389, CP 1999, where it is common; the old railway track in Bembridge Harbour 6488, AC 1996; Bembridge Point 6488, GT 2000; Redcliff cliff top, Sandown 6285, CP 1997, where it appears in small and fluctuating numbers; a few trackside plants in Brighstone Forest 4284, MB 1999; and a few plants in an arable field margin at Heath Hill, Shorwell 4682, CP 1995.

[*T. squamosum* L.
SEA CLOVER
N. Extinct. This plant was formerly very rare in grassy places by the sea. It was recorded from 'the flat before reaching Newtown bridge from Shalfleet' (Groves 1870), grassy roadside waste between Thorley and Yarmouth (Townsend 1883), St Helen's Spit (Boys 1919) and Yarmouth (Drabble & Long 1931). There was a good stand at Wheatear Point at the western entrance to Newtown Harbour, 4191, in the 1950's, but this declined with the lack of rabbits and the invasion of coarser grasses. It was last recorded here in 1965, MMS.]

T. subterraneum L.
SUBTERRANEAN CLOVER
N. Locally frequent in short dry grassland, and sandy and gravelly places. Map 128

Lupinus arboreus Sims
TREE LUPIN
E. Local on sandy or gravelly soils. First recorded in 1947 by J.E. Lousley as well established on Lake Cliffs

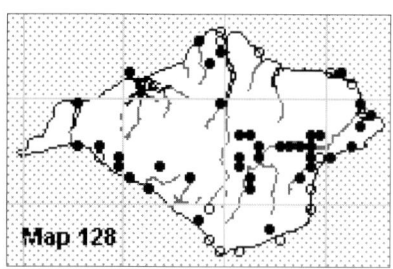
Lupinus arboreus

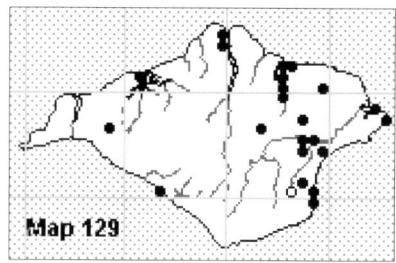
Cytisus scoparius spp *scoparius*

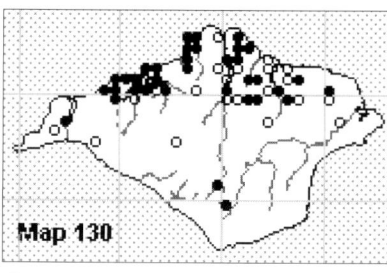
Genista tinctoria

5882, where it remains common today. First recorded on sand dunes on St Helen's Duver in 1952 6389, PB and on Bembridge Point in 1963 6488, BS. It remains common in both sites, although population growth has been severely checked by an outbreak of the large aphid *Macrosiphum albifrons*, first recorded in 1983. Occurs elsewhere in smaller quantities, such as at Ventnor and Totland.

[*L. angustifolius* L.
NARROW-LEAVED LUPIN
C. Extinct. A single record from waste land below Newport J.W. Long 1931.]

Laburnum anagyroides Medik.
LABURNUM
E? Frequently planted and occasionally regenerating on bushy waste ground. First recorded as regenerating in 1963 in Fishbourne Copse, Quarr, JH.

Cytisus scoparius (L.) Link.
ssp. *scoparius*
BROOM
N. Occasional in hedgerows, remnant heath, scrub and open ground in woods on acid soils. Interestingly, the great majority of records are from East Wight. Map 129

Spartium junceum L.
SPANISH BROOM
E. Well established on the cliffs at Lake 5882, CP 1995.

Genista monspessulana (L.) L.A.S. Johnson
MONTPELLIER BROOM
E. The only recent records are of a single small bush growing by the Downs Road at Knighton 5787, BW 2002; and a waste site near Brading Church 6087, BS where it was abundant in 1968.

Genista tinctoria L. ssp. *tinctoria*
DYER'S GREENWEED
(local: woodwax!; woad-waxen)
N. A characteristic dwarf shrub of unimproved neutral grasslands on heavy clay soils on the north of the Island. It was considered to be a weed of poor pastures. Although it has undoubtedly declined with the loss of old pastures, it frequently persists in field borders, thickets and waysides. The local name Woodwax is still in use by a few Island folk. A field at Providence Farm, Ningwood (SZ403887) is known as Woodwax; in 1958, an old farmhand remembered that a difficult to eradicate, yellow-flowered weed grew in this field. Dyer's Greenweed is also found rarely in calcareous grassland and on acid soils in the south of the Island. Records from the downs are: a single bush in scrub on the north face of Arreton Down 5487, CC 1984; a single bush on Idlecombe Down 4585, CC 1984; and on the crest of Compton Down 3785, FR 1984. Records from the south of the Island are: a lower greensand hill at Upper Dolcoppice 5079, CP 1997; and, most unusually, an alluvial mire at Cridmore Bog 4981, CP 1995. Map 130

[*G. anglica* L.
PETTY WHIN
N. Extinct. A species of rough damp moors and heaths, always local but abundant in the nineteenth century at Munsley Bog, The Wilderness, Staplers and on the north side of Parkhurst Forest. There are also a few records from other sites. At the time of Bevis *et al* (1978), it was considered to be rare and decreasing, and they list three sites. It has since been lost from the Island's flora, a combination of land reclamation and of lack of management. It was found at Lynn gravel pit (5388) until 1969, BS. It grew at Smallgains Heath between Belmont Farm and The Grange (5290) until the land was reclaimed in 1972, BS. It survived longest at Munsley Bog, Godshill (5282) but in 1982 there were only four plants left and these had gone by 1988, CP. A single large bush survived at Cridmore Bog (4981) in 1984, JC, but has subsequently been lost.]

Ulex europaeus L.
GORSE (local: vuzz!; vuzzen)
N. Common and widespread on heaths, gravel cappings on the downs, undercliffs and woodland edges. It still survives as a dominant hedgerow species along the Briddlesford Road, in the vicinity of the former Great Lynn Common. Gorse was a component of the rural economy, grown for fodder and cut for fuel. Fourteenth century records from St Cross refer to a considerable market in gorse in Parkhurst Forest (Chatters, 1991). Still referred to as Vuzz by some older Island farmers.

U. minor Roth
DWARF GORSE
N. Widespread but local on heaths and commons. Bromfield (1856) described this species as 'almost equally common' as Gorse. It is certainly not so today, but it is a characteristic component of the dry heath community. There is a paucity

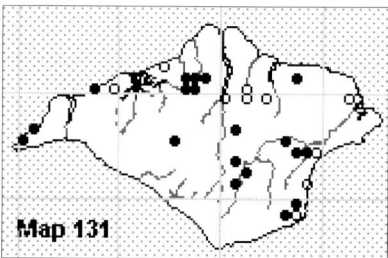
U. minor

of historic records for this species, but the map indicates the decline of this species through loss of heathland habitat. Its disappearance from St Helen's Duver is more difficult to explain. Map 131

80. ELAEAGNACEAE
Sea-buckthorn family

Hippophae rhamnoides L.
SEA-BUCKTHORN
E. Planted along the coast, where it has become established by suckering or bird-sown seeds. First introduced on St Helen's Duver in 1858 by A.G. More where it has spread and is well established. By 1886, it was extending eastwards along the coast towards Seaview, and it still grows in Priory Bay. It is also well established at Bembridge Point but there are no early records from this site. Stratton (1909) considered this shrub to be very rare, but its spread, both naturally and by planting was such that Drabble & Long described its status as 'certainly not rare now' (1931). It fruits freely (e.g. at Totland) and may be bird sown in many of its situations.' It was first recorded at Totland in 1900 and it still occurs on the cliffs here and Alum Bay; it is actively colonising the slopes at Headon Warren and the coastal slope at St Catherine's, well away from any habitation. Also recorded from the cliffs east of Ventnor (first in 1931, TW) towards Luccombe. It has been suggested that the host specific bracket fungus, *Phellinus hippophaecola*, which is found on bushes on St Helen's Duver, is only found within the native range of Sea Buckthorn.

Elaeagnus umbellata Thunb.
SPREADING OLEASTER
P. Not known to have become established. Large bushes at the bottom of Newchurch Shute 5685, DB 1998 are almost certainly a persistent relic of planting, as are bushes by the stream edge at the bottom of gardens by Freshwater Marshes 3486, BS 1997.

81. HALORAGACEAE
Water-milfoil family

Myriophyllum aquaticum (Vell.) Verdc.
PARROT'S-FEATHER
E. This plant is grown in garden ponds but it quickly establishes itself in the wild in slow moving streams, field ponds and ditches. First noticed in 1990 in the Eastern Yar at Sandown waterworks 5885, BS. Within two years it dominated several square metres of open water, and by 2002 dominated most of the wet ditches eastwards to Adgestone. In 1992, it was recorded in the River Medina at Shide 5088, BS. Since that date it has been found in an increasing number of ponds across the Island, generally as the dominant species, but it may prove to be susceptible to severe winters.

M. spicatum L.
SPIKED WATER-MILFOIL
N. One of our more frequent native aquatics, being tolerant of eutrophic conditions. Found in slow flowing rivers, ponds, drains and farm reservoirs. Most frequent in the Eastern Yar. Historic records suggest that this species has retained much the same distribution as it held one hundred years ago. Map 132

M. alterniflorum DC.
ALTERNATE WATER-MILFOIL
N. A rare aquatic of acidic, and mesotrophic ponds and ditches. In the nineteenth century it was recorded from at least seven sites. Bromfield (1856) found it 'in vast abundance' in some ditches on Sandown Levels. Within the last fifty years, the only remaining site was a flooded gravel pit at the north end of Bleak Down 5182. Bevis *et al* (1978) suggested that it was suffering from pollution from the adjoining refuse tip and unlikely to survive. However, it was still present in 2000 and flowering well. It has also colonised some recently constructed ponds: pond at Wacklands Farm, Newchurch 5585, CP 1991; and a reservoir at The Wilderness 5082, BG 1999.

82. GUNNERACEAE
Giant-rhubarb family

Gunnera tinctoria (Molina) Mirb.
GIANT-RHUBARB
S. Several plants persisting on slumped ground on Blackgang cliffs 4876, CP 1982, well away from surviving gardens.

83. LYTHRACEAE
Purple-loosestrife family

Lythrum salicaria L.
PURPLE-LOOSESTRIFE
N. Found very locally by waterside margins and in fens, but absent from many apparently suitable sites. Although always local, this species appears to have declined in the last hundred years; it is rarely found in quantity these days. Map 133

[*L. hyssopifolium* L.
GRASS-POLY
C. Extinct. Only known from two nineteenth century records: waste ground at St John's, Ryde, 1867, F. Stratton, where it was believed to have been sown by Bromfield; and on the east bank of the Medina at Newport, where it was believed to have been introduced with grass seed 1850, G. Kirkpatrick.]

L. portula (L.) D.A. Webb
WATER-PURSLANE
N. Rare on pond margins and damp tracks on acid soil; greatly declined in the last hundred years through loss of suitable habitat. Formerly abundant in The Wilderness and Cridmore, and still found around the peaty margins of a reservoir at The Wilderness 5082, BG 1999. Formerly abundant on

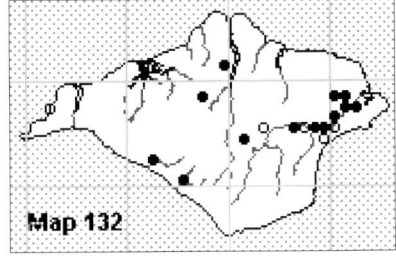

M. spicatum

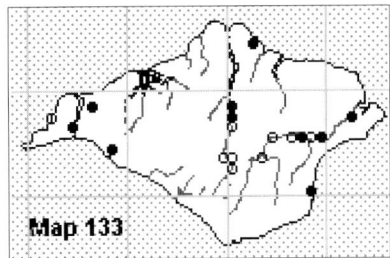

Lythrum salicaria

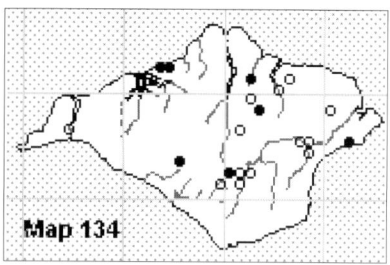
L. portula

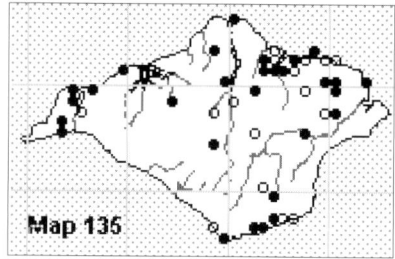
D. laureola

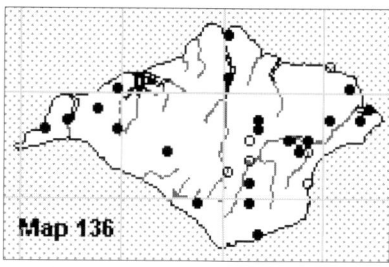
E. obscurum

Bleak Down and last recorded around the margins of old gravel pits in 1986, 5182, BS. More recent records are from a pond margin below Redcliff, Sandown 6285, BG 1999; Lynn Common 5388, CDP & TDD 1998; wet rides in Fattingpark Copse 5291, CP 2002; and pond edges at South Thorness Farm 4492, CP 1996. Map 134

84. THYMELAEACEAE
Mezereon family

[*Daphne mezereum* L.
MEZEREON
E? Extinct. Probably not native, the only records are those of Bromfield (1856) who considered that the plants were bird-sown from gardens. He sites just two records of single plants, one in a brambly thicket west of Wacklands, Newchurch and another specimen in Apse Castle wood, Shanklin, which was transplanted to Apse Farm garden in 1845.]

D. laureola L.
SPURGE-LAUREL
(local: copse or wood laurel)
N. Rather frequent in ancient and secondary woods, old hedge banks and scrub on chalk. It is particularly associated with heavy clay and chalky soils. The map is an under-representation of its distribution. Map 135

86. ONAGRACEAE
Willowherb family

Epilobium hirsutum L.
GREAT WILLOWHERB
N. Widespread and frequent along rivers, streams and ditches and around ponds. Also, sometimes found on waste ground in drier situations.

E. x *subhirsutum* Gennari
GREAT WILLOWHERB X HOARY WILLOWHERB
N. This hybrid has been recorded but the details are not known (Stace, 1975).

E. parviflorum Schreb.
HOARY WILLOWHERB
N. Less common than Great Willowherb, but still frequent and widespread in damp and wet places.

E. x *limosum* Schur
HOARY WILLOWHERB X BROAD-LEAVED WILLOWHERB
N. Recorded from St Helen's Common 6389, GK 2001, but clearly overlooked.

E. parviflorum x *E. ciliatum*
HOARY WILLOWHERB X AMERICAN WILLOWHERB
N. Recorded from Quarr in 1962, JH; and from St Helen's Common 6389, GK 2001.

E. montanum L.
BROAD-LEAVED WILLOWHERB
N. A common and widespread weed of disturbed ground everywhere.

E. montanum x *E. ciliatum*
BROAD-LEAVED WILLOWHERB X AMERICAN WILLOWHERB
N. Recorded from St Helen's Duver in 1961, BS. Probably much overlooked.

[*E. lanceolatum* Sebast. & Mauri
SPEAR-LEAVED WILLOWHERB
N. Extinct? The only record is from a roadside bank in York Avenue, East Cowes, confirmed in 1970 and still present in 1974, RK. Perhaps overlooked, although undoubtedly rare.]

E. tetragonum L.
SQUARE-STALKED WILLOWHERB
N. Fairly frequent and widespread on disturbed verges, waste ground, gardens and damp places.

E. obscurum Schreb.
SHORT-FRUITED WILLOWHERB
N. Locally frequent in hedge banks, disturbed ground and damp places. Bevis *et al* (1978) suggested that this species occurs mainly in East Wight, but this is not borne out from recent recording. Map 136

E. roseum Schreb.
PALE WILLOWHERB
N. An occasional weed recorded from a woodland ride, disturbed ground and gardens. First confirmed record from a garden in Bennett Street, Ryde 5992, RK 1980. Since recorded from Chale Green 4879, SC 1993; Steyne Wood, Bembridge 6387, AC 1997; Nunwell Down 5987, AC; Freshwater Churchyard 3487, PS 1999; Seaview 6291, PS 1999; and Ryde seafront 5992, PS 2000.

E. ciliatum Raf.
AMERICAN WILLOWHERB
E. A common and widespread weed of disturbed ground. First noticed in the early 1960's but this introduced plant had clearly been present, but overlooked, for some time before that.

E. palustre L.
MARSH WILLOWHERB
N. Local in marshes and other wet places on neutral to acid soils, particularly in the East Yar valley and the Medina valley around Cridmore and The Wilderness. It was last recorded in Freshwater Marsh in 1936 but was found on St Helen's Green 6289, another historic locality, in 1996, AC. Unlike our other willowherbs, this species is not characteristic of disturbed ground. Map 137

Chamerion angustifolium (L.) Holub
ROSEBAY WILLOWHERB
(local: tame withy)
N? Frequent, and often abundant, in woodland clearings, waste ground, verges and burnt heath across the Island, generally on rather acid soils. Rosebay Willowherb was a late arrival in lowland Britain. Bromfield (1856) said that it was one of the commonest cottage garden plants and it was known as Tame Withy, i.e. the cultivated or domesticated withy. As a wild plant, it was a great deal rarer and he recorded it from only twelve sites, first in 1840, and at most of these sites it was considered to be rare. This was actually quite a large number of sites for the plant in an English county at that time. E.H. White considered that it was still a rare plant in 1904. Its rapid spread is likely to have occurred, as elsewhere in England, during the first quarter of the twentieth century.

Oenothera L
The Evening-primroses and hybrids have not been critically investigated on the Island.

Oenothera glazioviana Micheli ex P. Mart.
LARGE-FLOWERED EVENING-PRIMROSE
E. Frequent on waste and disturbed ground and on maritime cliffs; often persisting on sandy ground. A record for this species from Niton dated 1866, J.R.E. (OXF) is the earliest wild collected specimen of this plant in this country (conf. K. Rostanski).

O. biennis L.
COMMON EVENING-PRIMROSE
E. Occasional on disturbed and sandy ground.

O. cambrica Rostanski
SMALL-FLOWERED EVENING-PRIMROSE
E? Status unknown. A specimen collected at the entrance to the refuse tip on Bleak Down in 1973 was confirmed as this species (conf. K. Rostanski).

O. stricta Ledeb. ex Link.
FRAGRANT EVENING-PRIMROSE
E. Very local. Well established in sandy ground on St Helen's Duver 6389, CP 1999 and Bembridge Point 6488, AC 1996. Also recorded behind beach huts at Sandown 6084, GT 1998. These are all sites with a long history of records. A.G. More sowed the plant on St Helen's Duver about 1858. It was recorded from Sandown Bay in 1888 (Moyle Rogers) and in 1894, plants were considered to have spread naturally from St Helen's Spit to Bembridge Point. (J.A. Preston, MANCH).

O. rubricaulis Kleb.
C. There is a record from Havenstreet in 1964, GB, which was confirmed at the time by the British Museum. The plant was considered to be a garden escape.

Fuchsia magellanica Lam.
FUCHSIA
S. A persistent outcast and survivor of derelict gardens. Recorded as surviving on slumped land on cliffs below Luccombe cliffs 5879, BS 1969, originating from gardens above. Still present here 1997, GT. Two plants which appeared to be self-sown in sandy ground on Bembridge Point 6488, AC 1997.

Circaea lutetiana L.
ENCHANTER'S-NIGHTSHADE
N. Frequent and widespread in woods and shady hedgebanks. Also occurs as a garden weed.

87. CORNACEAE
Dogwood family

Cornus sanguinea L.
DOGWOOD (local: skewerwood)
N. Common in woodland edges, scrub and old hedges. Particularly characteristic of chalky and heavy clay soils, but recorded throughout the Island. It shares the local name of skewerwood, the wood used to make skewers, with spindle.

Aucuba japonica Thunb.
SPOTTED LAUREL
S. Commonly planted in Victorian shrubberies and persisting.

Griselinia littoralis (Raoul) Raoul
NEW ZEALAND BROADLEAF
S. Commonly planted around the coast and sometimes persisting in woods as a relic of Victorian planting, e.g. Player's Wood, Ryde 5893, CP 1998 and Stroudwood Copse 5790, CP 2000.

88. SANTALACEAE
Bastard-toadflax family

Thesium humifusum DC.
BASTARD-TOADFLAX
N. Bastard-toadflax remains a common and widespread feature of short, south-facing downland turf. It has been noted colonising re-vegetating chalk where turfs have been cut (Rew Down 5577, RPB 1973) and is sometimes able to compete in fairly rank grassland. It grows prolifically along the chalk cliff edge at the two extremities of the Island. Also found along the chalky cliff edge of the inner cliff at St Lawrence. A much more unusual historic habitat for this plant was on fixed sand dunes by the coast. It was recorded in this habitat on Norton Spit by Snooke (1823) but there are no other records from this site. On St Helen's Duver there is a continuity of records from 1871 when A.G. More found it 'on the sandhills of St Helen's spit, near the millpond'. It was presumably reasonably frequent here, for a meeting of the Isle of Wight Natural History and Archaeological Society in June 1921 recorded the parasitic rust fungus, *Puciniia thesii*, growing on some of the plants. It survived here until 1987 when a single patch was found on a

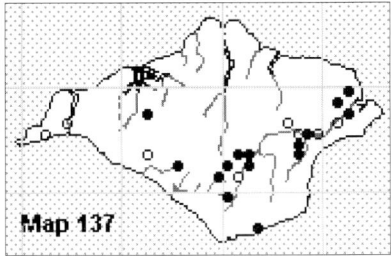

E. palustre

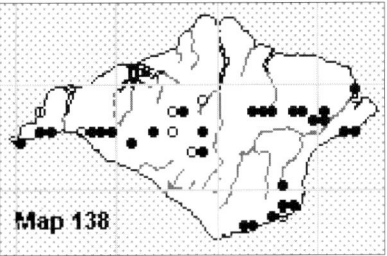

Thesium humifusum

shallow bank just north of the car park, CP. It has not been seen since. Map 138

89. VISCACEAE
Mistletoe family

Viscum album L.
MISTLETOE
N? Very local, with a distribution showing a particular concentration of sites around Carisbrooke, Northwood, Ryde westwards to Quarr, Bembridge and Arthur's Hill, Shanklin. Bromfield (1856) found this plant to be extremely rare and he quotes just four records, the first in 1841. There are few subsequent records until the mid twentieth century and many of these relate to deliberate introductions. After this, records increase but are always centred around particular sites. This pattern suggests that, on the Island, the plant originates from introductions by the Victorians and Edwardians into their orchards and that the plant has subsequently spread to neighbouring trees.

The host species recorded, with the number of post 1994 records, are apple (24), hybrid black poplar (13), lime (9), hawthorn (5), false-acacia (4), white willow (1), field maple (1), almond (1) and *Chaenomeles* (1). Mistletoe is sometimes harvested from garden apple trees and sold at Christmas. In a report in the *Isle of Wight County Press* in January 1995, the owner of mistletoe harvested at Northwood was unfortunately quoted as saying 'mistletoe kills apple trees so we always cut it down and sell it for charity'. Map 139

90. CELASTRACEAE
Spindle family

Euonymus europaeus L.
SPINDLE
(local: stinkwood!; skiverwood!)
N. Found frequently at the edges of woods, in scrub and in old hedgerows. Generally associated with calcareous or clay soils, it is scarce on the acidic sandy soils of the southern half of the Island. Together with dogwood, the wood was formerly used to make skewers (skivers) for butchers.

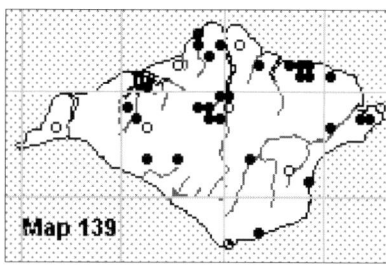

Viscum album

E. japonicus L.f.
EVERGREEN SPINDLE
S. Widely planted in hedges and shelter-belts around the coast and in churchyards. On warm slumping maritime cliffs, it spreads vegetatively and fruits freely. Ventnor east cliffs 5677, CP 1998; Binnel Bay, St Lawrence 5075, CP 2000; Lake Cliffs 5882, AC 1995.

91. AQUIFOLIACEAE
Holly family

Ilex aquifolium L.
HOLLY (local: christmas)
N. A frequent species in woods and older hedges throughout the Island. It is particularly abundant in some woods considered to have a wood pasture history e.g. the northern half of Parkhurst forest around Mark's Corner, Borthwood Copse and America Wood. The largest trees are on acid sandy soils in ancient woodlands in the south-east of the Island. Old holly trees in hedgerows away from woodland are not characteristic landscape features on the Island, but a number survive in hedgerows in and around the Chequers Inn Road at Rookley. The local name referred to by Bromfield would have applied not just to boughs cut for Christmas, but to the whole tree, in the same way that hawthorn is also called May. Holly used to be used for whip sticks within living memory.

92. BUXACEAE
Box family

Buxus sempervirens L.
BOX
S. Commonly planted in Victorian shrubberies and persisting.

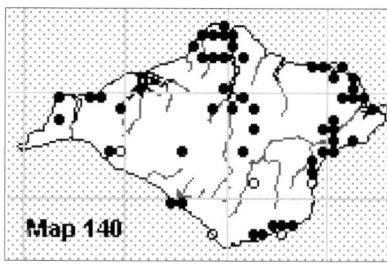

M. annua

93. EUPHORBIACEAE
Spurge family

Mercurialis perennis L.
DOG'S MERCURY
N. Widespread and often dominant in ancient and secondary woods and hedge banks, but confined to more base-rich soils. It will sometimes spread into grasslands adjoining woods and copses.

M. annua L.
ANNUAL MERCURY
Arch. A weed of disturbed ground, in arable fields, gardens and waste places. Local, but often frequent where it occurs. In 1634, Thomas Johnson, a London apothecary, published *Mercurius Botanicus*, an account of a 'herborizing journey of twelve days' around the country. During their trip they landed at Cowes, the party rode to Newport, visited Carisbrooke Castle, and went on to 'Ride' where they 'plucked annual mercury by the sea-side'. Map 140

[*Euphorbia peplis* L.
PURPLE SPURGE
N. Extinct. Recorded on sandy shores in the nineteenth century. It grew in Sandown Bay and appears to have been gathered here at least twice, in 1830 (Bromfield Hb, HCMS) and in 1842 (BEL). Also found growing on the beach at St Helen's Duver by A.B. Jackson 1872 (BM).]

(*E. oblongata* Griseb.
BALKAN SPURGE
C. According to Clement & Foster (1994), recorded as a casual in a meadow at Newport. No further details.)

E. platyphyllos L.
BROAD-LEAVED SPURGE
Arch. Arable fields on heavy soils on the north of the Island. It has declined

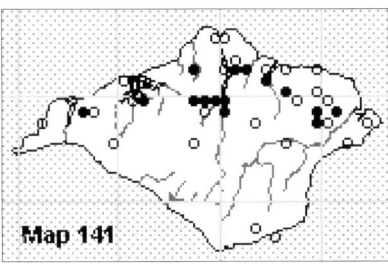

E. platyphyllos

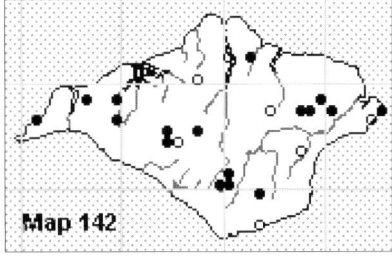

E. exigua

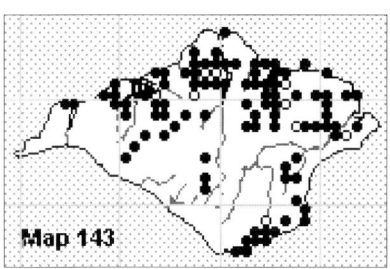

E. amygdaloides

in the last one hundred years, and is no longer recorded from the chalk and from sandy soils in the south of the Island. However, it is persistent at many of its remaining sites and it remains a widespread arable species, in contrast to its general decline nationally. It can be frequent in some fields. Frequent in a field at Carisbrooke Park 4889, CP 1995, but subsequently lost from this site following development. Recorded from cattle-grazed marshy grassland by Quarr Abbey ruins in 1986, BS. Map 141

E. helioscopia L.
SUN SPURGE
Arch. Common and widespread as a garden weed and on waste ground. Also found not infrequently in arable fields.

E. lathyris L.
CAPER SPURGE (local: caper-bush)
Arch. Occurs locally as a garden weed and on waste ground. It is recorded from most of our towns.

E. exigua L.
DWARF SPURGE
Arch. Bromfield (1856) described this arable weed as occurring 'in every part of the Island, abundantly'. Today, it is rather scarce and usually found in small quantity in fields on a variety of soil types, but more particularly on chalk. It occurred abundantly and luxuriantly in many fields following the deep digging of trenches across the Island to lay sewerage pipes in 2000. Map 142

E. peplus L.
PETTY SPURGE
Arch. A common and widespread plant of cultivated and waste ground.

E. portlandica L.
PORTLAND SPURGE
N. Confined to chalk cliffs at the two extremities of the Island, where it occurs both on the cliff face and in cliff top grassland and is currently increasing. It was first recorded from Culver in 1823, by Snooke. Not recorded from Highdown Cliffs, Freshwater until 1985 by which time it was well established. The Culver cliffs are the easternmost site along the south coast, although the plant colonised the shingle spit at West Wittering in West Sussex for a number of years up to 1953, when the spit was breached in a storm and the plant was lost.

E. paralias L.
SEA SPURGE
N. A very rare species of sandy spits. The core area is at the end of the western spit at Newtown, 4191, where numbers fluctuate. There were about 20 plants in 1996, AC. It may currently be spreading as there are recent records of a small patch at the eastern spit at Newtown, 4191, JC 2000 and two small plants at Thorness Bay 4693, CP 2000. However, there have been sporadic reports from both these sites over the last one hundred years, and they may not prove to be persistent. Sea Spurge was apparently first introduced on the Island by seed sown by Bromfield at St Helen's Duver and Norton Spit. There are many reports of the plant from Norton Spit from 1849 onwards until 1970 (JB). There is a succession of reports from St Helen's Duver from 1848 until 1951 (EHW).

[*E. x pseudovirgata* (Schur) Soó
TWIGGY SPURGE
E. Extinct. An occasional garden escape, sometimes persisting for a few years as at Newport 4988, where a clump on the verge of Watergate Road was recorded in 1963, GB, and was still present in 1970. Roadside bank at junction of Whitcombe Road and Whitepit Lane, Newport 4888 BS 1970. Also recorded in 1923 near Newport by J. Long.]

E. cyparissias L.
CYPRESS SPURGE
E. Occasionally established garden escape. No recent records but recorded from St Lawrence, AWW, in 1960 and Blackwater, near Newport, EHW, in 1944. Recorded by Bromfield (1856) as plentifully naturalised in the shrubbery at Northwood, Cowes.

E. amygdaloides L. ssp. *amygdaloides*
WOOD SPURGE
(local: cups and saucers!)
N. A common species of woodlands, hedge banks and chalk scrub. It is also characteristic of coastal cliffs in the south, between Luccombe and Binnel Bay. Frequently reappears in abundance when neglected ancient woods are managed, but this is a widespread species and by no means confined to ancient woods. Map 143

ssp. *robbiae* (Turrill) Stace
LEATHERY WOOD-SPURGE
E. A garden escape spreading into adjoining secondary woodland at Golden Hill, Freshwater 3387, CP 1998.

E. characias L.
MEDITERRANEAN SPURGE
S. Occasionally surviving as an outcast on waste ground, e.g. alongside Seaview Duver Road 6291, PS 1999.

94. RHAMNACEAE
Buckthorn family

Rhamnus cathartica L.
BUCKTHORN
N. Remarkably rare on heavy clay soils and even rarer on chalk. Bromfield (1856) recorded this shrub from nine sites, mostly on the chalk. Modern sites are: a few bushes in scrub on the north side of Tennyson Down, 3285, JC 1984, cut down during scrub clearance the following year; Tolt Copse 4884, RPA 1985; Cranmore 3990, VS 1998; Corfe Camp, Newtown 4190, PS 2000; Clamerkin Bridge, Newtown 4490, BS 1970; several bushes in scrub on Hamstead Spit 4191, BS 1996; Brickfields, Newtown 4292, RG 1997; by old railway track at Whippingham 5291, BS 1998; Wootton Common 5391, BS 1998; New Copse 5592, BS 1998. Probably the largest specimen is by the roadside at Yaverland 6186, CP 1996. Unless stated otherwise, all of these records refer to single bushes. Buckthorn is also uncommon on the immediate south Hampshire mainland and in Purbeck. Map 144

[*R. alaternus* L.
MEDITERRANEAN BUCKTHORN
S. Extinct. Known only from a single bush in a hedgerow on Pan Down, recorded by F. Stratton in 1909 and still present in 1938.]

Frangula alnus Mill.
ALDER BUCKTHORN (local: black alder)
N. A rare shrub of damp woods on acid soils. Bromfield (1856) described this plant as occurring 'abundantly in the more interior and level districts' and it was sufficiently well known to acquire a local name. Today, it is something of a rarity, and almost always occurs in small amounts. It is only found in quantity at Cranmore 3990, VS 1998, where it is a component of the mixed scrub colonising clay heath. It occurs in small numbers in rides in Parkhurst Forest, Firestone Copse, Briddlesford Copse, Combley Great Wood and Borthwood Copse. A few plants at Bohemia Bog 5183, CP 2001. The remaining sites are in the Eastern Yar valley and its tributaries, where it is scarce but found in several sites, principally around Alverstone. At some sites, such as Borthwood Copse, the population has been augmented by planting. Map 145

95. VITACEAE
Grape-vine family

Vitis vinifera L.
GRAPE-VINE
E. Rarely established in hedges and fences e.g. covering a chain link fence of industrial estate, Newport 4989, SB 1995. The records for this species in Bevis *et al* (1978) should be referred to Virginia-creeper (*see below*).

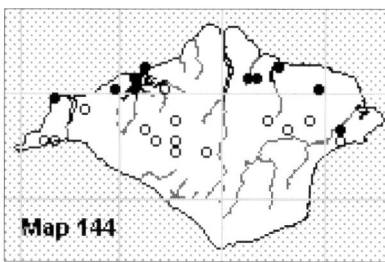

Map 144

Rhamnus cathartica

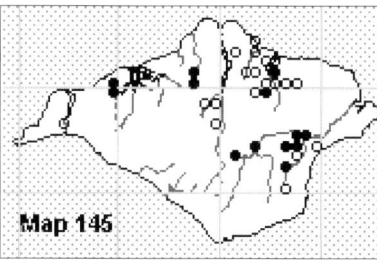

Map 145

Frangula alnus

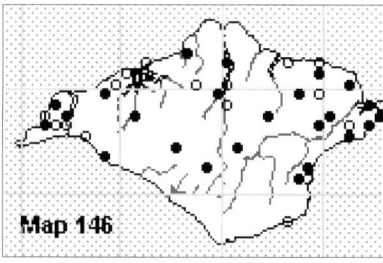

Map 146

Linum bienne

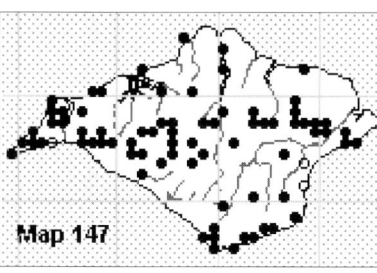

Map 147

L. catharticum

Parthenocissus quinquefolia (L.) Planch.
VIRGINIA-CREEPER
E. Occasionally established and persistent on tall hedges, railway embankments and old tips. Recorded in 1964 on rough ground adjoining the old railway track south of Arctic Road, Cowes 4994, BS, and still present here. Known since before 1989 growing over blackthorn scrub at the top of Whale Chine 4778, BS by the car park, probably originating from dumped waste. Overgrowing bushes by the River Medina at Fairlee 5090, CP 1997. Well established over scrub at a long derelict garden site by Rowborough Lane, Brading 6087, CP 1997. Overgrowing scrub to east of Yarmouth old railway station 3589, BS 1980. By footpath near The Priory, Seaview 6390, CP 1989.

P. inserta (A. Kern.) Fritsch
FALSE VIRGINIA-CREEPER
E. The only record is from waste ground at Rookley Country Park 5184, PS 2000. Records for this and the previous species may have been confused.

96. LINACEAE
Flax family

Linum bienne Mill.
PALE FLAX
N. Thinly scattered across the Island in thin, dry calcareous or sandy grassland. Local and not always persisting. Much declined in the last one hundred years with the loss of suitable habitat and general tidying up of the countryside. Map 146

L. usitatissimum L.
FLAX
C. Now a frequent casual by fields and on roadsides.

L. catharticum L.
FAIRY FLAX
N. Frequent in well drained soils with short turf. It is virtually ubiquitous in unimproved and semi-improved chalk grassland. Also found in relic limestone grassland, along heath tracks and on calcareous clays on cliff slopes. Occasionally found in dry sandy grassland and neutral herb rich meadows. This plant has declined

considerably in the last one hundred years from sites off the chalk. Map 147

[*Radiola linoides* Roth
ALLSEED
N. Probably extinct. Always very local in winter-wet, bare gravelly spots; now probably extinct as no suitable habitat remains. Known in the nineteenth century from four sites, namely: Blackpan Common, Sandown; Bohemia Bog, near Godshill; Bleak Down; and Colwell Heath. It appears to have occurred in a number of spots over a large area on Bleak Down, and this was the only site where the plant is known to have persisted into the twentieth century. Ironically, the last time it was seen here in abundance was in 1973 in disturbed gravelly ground which resulted from a ditch, dug to drain the remaining acidic wetland area. Plants were still growing here in the 1980s, RPA. It survived on some rough gravelly ground at Bleak Down until 1995, RPA, but this area was then destroyed by tipping.]

97. POLYGALACEAE
Milkwort family

Polygala vulgaris L.
COMMON MILKWORT
N. Occurs widely on chalk grassland, although usually in small quantity. Also found, less commonly, in unimproved neutral clay meadows, relic limestone grassland, coastal calcareous clays and neutral to slightly acid soils on clay heaths. Commonly found in a variety of colour forms. Map 148

P. serpyllifolia Hosé
HEATH MILKWORT
N. Local in heathy woodland rides and other open acidic sites. The

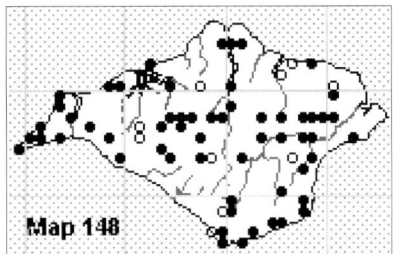
Polygala vulgaris

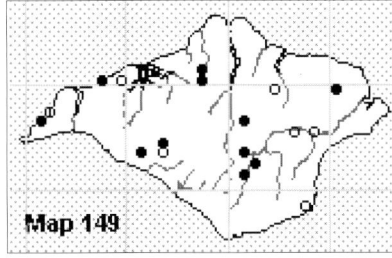
P. serpyllifolia

principal sites for this species include: St George's Down; Bleak Down; Bohemia Bog; Parkhurst Forest rides; Grammars Common, Brighstone; Headon Warren; and Bouldnor. Only deep blue flowered forms have been reliably recorded. Map 149

100. HIPPOCASTANACEAE
Horse-chestnut family

Aesculus hippocastanum L.
HORSE-CHESTNUT
E. Commonly planted in woods, parkland and around buildings. Often a relic of habitation. Self-sown saplings are frequently found. It freely regenerates in secondary woodlands along the Undercliff.

101. ACERACEAE
Maple family

Acer platanoides L.
NORWAY MAPLE
E. A frequently planted tree in gardens and woodlands, often self-seeding. First recorded as naturalised in plantations at Quarr Abbey, JH 1971. Subsequently recorded as increasing in a number of woods where originally planted.

A. cappadocicum Gled.
CAPPADOCIAN MAPLE
E. Reproducing from planted specimens by seed in Player's Copse at Binstead 5792, CP 1991.

A. campestre L.
FIELD MAPLE
N. Well distributed across the Island but least frequent in the open arable landscapes of south Wight. It is characteristic of ancient woodlands and old hedgerows. Large specimens, frequently ancient coppice stools, are sometimes found on old boundary banks and elsewhere in sheltered locations. However, they do not compare with a specimen described by Bromfield, who wrote, 'The largest specimen I am acquainted with in the island grows at Nunwell, and when measured in February 1845, girded 10 feet at 5 feet from the ground, branching into a rounded head of about 30 or 40 feet in height.' There is a row of old coppiced trees on the Brighstone/Calbourne parish boundary bank, which runs through the middle of Brighstone Forest plantation from Calbourne Bottom.

A. pseudoplatanus L.
SYCAMORE
E. A very common alien, completely naturalised in many woods and scrub, and in hedges far from houses. Very large old trees are uncommon.

102. ANACARDIACEAE
Sumach family

Rhus hirta (L.) Sudw.
STAG'S-HORN SUMACH
E. Occasionally established by suckering from gardens onto wasteland. Invading open, bracken-covered eastern slope of Sibden Hill, Shanklin 5781, CP 1985.

103. SIMAROUBACEAE
Tree-of-heaven family

Ailanthus altissima (Mill.) Swingle
TREE-OF-HEAVEN
E. Spreading by prolific suckering from a planted tree in Player's Wood west of Ryde 5893, CP 1998. May be unrecorded elsewhere in similar circumstances.

105. OXALIDACEAE
Wood-sorrel family

Oxalis valdiviensis Barnoud
CHILEAN YELLOW-SORREL
C. Several plants in waste ground at Seaclose recreation ground, Newport 5089, BS 2001 (conf. E.J. Clements). It reappeared here in 2002.

O. corniculata L.
PROCUMBENT YELLOW-SORREL
E. A frequent and increasing weed of gardens, paths, walls and waste ground. It was first recorded by Bromfield (1856), above the grounds

at Steephill and as a weed in the garden at Alverstone Mill.

O. stricta L.
UPRIGHT YELLOW-SORREL
E. Occasional but under recorded. Recently seen as a weed in gardens in Ryde, Newport and Freshwater. First recorded in 1856 as a garden weed at Thorley, near Freshwater (Bromfield, 1856).

O. articulata Savigny
PINK-SORREL
E. A frequent established garden escape across the Island in gardens, roadsides and churchyards. Well established at Lake Cliffs. First recorded in 1960 as being frequent on rubbish tips, BS.

O. acetosella L.
WOOD-SORREL
N. Very local in ancient woodlands and shady, damp hedge banks particularly on the acid, sandy soils of south-east Wight. It can be abundant in favoured locations but in other sites is restricted to small wet areas within woods. Grows in open situations in natural greensand rock crevices at Gatcliff and on the north side of the Appuldurcombe estate wall, 5380, CP 1996, and at the north end of St Catherine's Hill 4978, CP 2000. Map 150

O. debilis Kunth
LARGE-FLOWERED PINK-SORREL
E. Frequent as a persistent plant established in sandy ground in several spots around Bembridge Harbour. Probably occurs elsewhere.

O. latifolia Kunth
GARDEN PINK-SORREL
E. Scarce but probably under recorded. Growing on a bank outside of garden at Orchard Bay, Steephill 5476, DB 1997; abundant in flowerbeds on Ryde seafront 5992, PS 2000.

O. incarnata L.
PALE PINK-SORREL
E. Under recorded. There are recent records of it as an established garden weed from Yarmouth, Totland, Cowes, Ryde and Bonchurch.

106. GERANIACEAE
Crane's-bill family

Geranium endressi J. Gay
FRENCH CRANE'S-BILL
E. An occasional persistent garden outcast on waste ground. First recorded in 1961 from Shanklin, E. Newnham.

G. x oxonianum Yeo
DRUCE'S CRANE'S-BILL
E. Recorded as an outcast at Cranmore 3990, VS 1998; and Totland 3285, PS 1999.

G. versicolor L.
PENCILLED CRANE'S-BILL
E. A well established garden escape on roadside verges and waste ground. It has been recorded from various sites, but has persisted on a verge at the foot of Beacon Alley, Godshill 5181, since at least 1961, and on a bank at Rookley 5083 since at least 1974. First recorded in 1871 when it was considered to be well-established on hedge banks at Arreton and Wootton.

G. rotundifolium L.
ROUND-LEAVED CRANE'S-BILL
N. A frequent and increasing species. It has always been much commoner on the Island than in Hampshire and this was the only place where Townsend (1883) knew it. At that time, the plant was restricted to the Ventnor / St Lawrence area. It is still frequent here but, by the early 1900s, it had spread to the Alverstone / Newchurch and Freshwater areas. It continues to expand its range and has now been found in Ryde, Newport, Bembridge, Wootton, Thorley and Cranmore. In Hampshire, Round-leaved Crane's-bill is largely restricted as a native to Hayling Island. In Dorset it also has a restricted distribution, being largely concentrated around Portland.

G. pratense L.
MEADOW CRANE'S-BILL
E. A very rare garden escape. First recorded in 1871 by A.G. More, a single plant by a field border close to Sandown barracks. There have been a handful of scattered records since, most recently on a hedge bank at Combley 5488, DB 1985; and by the roadside on Mersley Down 5587, MS 1993. It is becoming established in the latter locality (2002) where it is believed to have originated from a wild flower mix.

G. himalayense Klotzsch
HIMALAYAN CRANE'S-BILL
E. A well-established flowering clump on the cliff slopes at Blackgang, on the former site of a slumped garden 4976, CP 1999.

G. sanguineum L.
BLOODY CRANE'S-BILL
E. A very rare garden escape, persisting in some churchyards. First recorded in 1913 by F. Stratton from three sites on hedge banks and waste ground.

G. columbinum L.
LONG-STALKED CRANE'S-BILL
N. Occasional, found on woodland edges, hedge banks and verges, always in small quantity. Status probably unchanged in recent years, but under recorded. Map 151

G. dissectum L.
CUT-LEAVED CRANE'S-BILL
Arch. Common and widespread on cultivated and waste ground and verges.

G. x magnificum Hyl.
PURPLE CRANE'S-BILL
E. A single flowering clump by the roadside on Mersley Down 5587, CP 2002. Meadow Crane's-bill is also

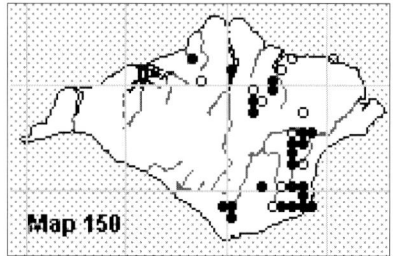

O. acetosella

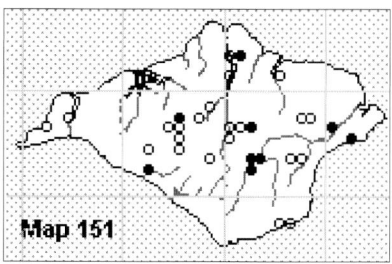

G. columbinum

established nearby on the southern verge.

G. pyrenaicum Burm. f.
HEDGEROW CRANE'S-BILL
E. An increasing species of verges and waste places, often in long grass. Bromfield (1856) found it to be rare and recorded it only from Steephill and Bonchurch, but it is now widespread.

G. pusillum L.
SMALL-FLOWERED CRANE'S-BILL
N. An occasional but widespread species of bare, dry places, often on sandy soils. Its status has probably changed little in recent years. Map 152

G. molle L.
DOVE'S-FOOT CRANE'S-BILL
N. Common and widespread on cultivated ground, waste places and verges. Var. *aequale* Bab. Status unknown but there is herbarium material from Bonchurch, May & June 1843, J.A. Hankley, OXF (conf. Serena Marner) and an unconfirmed record from St Helen's Duver (EC, pers. comm).

G. lucidum L.
SHINING CRANE'S-BILL
N. An increasing species of verges, wall tops, roadsides and gardens. Since the mid 1980s, its range has extended considerably and it is now widespread across the Island on a range of soil types, although often in small quantity. Its status as described by Bevis *et al* (1978) was little changed from that of Bromfield's time. It was very locally frequent but restricted to calcareous soils, principally on old walls in the Undercliff and on hedge banks at Cheverton Shute and Calbourne Bottom.

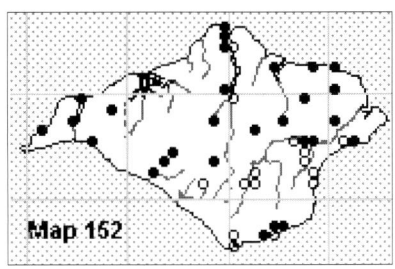

G. pusillum

G. robertianum L.
HERB-ROBERT
N. Common and widespread in woods, hedgebanks and as a garden weed. It is only occasionally found on shingle beaches today, as at Osborne 5295, MB 1999. The white flowered form is recorded sometimes.

(*G. purpureum* Vill.
LITTLE-ROBIN
N? Extinct. This species, in the form recognised by some authorities as ssp. *forsteri* (Wilmott) H.G. Baker, is endemic to shingle beaches of the Solent and can be locally plentiful in some mainland sites. However, there are only a handful of dubious old records from the Island. Bromfield (1856) recorded it from the shore near The Priory, Seaview; and there are further records from the foot of Culver Cliff (W. Newbould, 1871), Steephill Cove (J. Baker, 1871) and the downs at Totland (B. King 1901). Some of these are unlikely localities and, at the time it was suggested that there might have been confusion with the coastal form of Herb-Robert, ssp. *maritimum* (Bab.) H.G. Baker. No herbarium material is known and, to date, there are no confirmed records for this species.)

[*Erodium maritimum* (L.) L'Hér.
SEA STORK'S-BILL
N. Extinct. This species seems to be spreading in dry sandy coastal spots in Dorset but on the Island is known from principally three historic locations. It was recorded at Brook Chine by Bromfield (1856). It also grew on Headon Warren, where it was last recorded in 1931 when E. Drabble and J.W. Long described it as rare. At Alum Bay Chine, Bromfield described it as being, 'plentiful in the narrow gorge forming the descent into Alum bay, especially abundant and luxuriant at the mouth of rabbit burrows'. E.H. White last recorded it here in 1947. On 16 July 1955, an excursion of the Isle of Wight Natural History and Archaeological Society walked along the old Sandrock road at Niton. 'After crossing the very rough path where the old road had been, the members rested while flowers were identified and discussed by Mr. E.H. White, M.B.O.U., F.R.H.S. These included a specimen of the Sea Storksbill, which was thought to have become extinct in the Island'. (*Proc. Isle Wight nat. Hist. Archaeol. Soc.* IV (X): 370). No herbarium specimen of this find is known but E.H. White was very familiar with the plant at Alum Bay Chine over a period of almost fifty years.]

E. moschatum (L.) L'Hér.
MUSK STORK'S-BILL
E This plant was a very rare casual at the time of Bromfield (1856). There are no further records until 1990, when it was found on roadside verges at Marlborough Close, Ryde 6091, RK, a site where it still persists. Since then, it has started to appear in an increasing number of waste places right around the coast. It is now locally frequent in suitable spots.

E. cicutarium (L.) L'Hér.
COMMON STORK'S-BILL
N. Local but widespread on dry sandy soils around the coast and inland. It can be locally frequent but is no longer so abundant as to 'turn fields rose coloured', as noted on a Natural History Society walk in fields east of The Lynch at Newchurch in 1942. Map 153

107. LIMNANTHACEAE
Meadow-foam family

Limnanthes douglasii R.Br.
MEADOW-FOAM
C. An occasional, non-persisting garden outcast.

109. BALSAMINACEAE
Balsam family

[*Impatiens capensis* Meerb.
ORANGE BALSAM
E. Extinct. Briefly established for a few years in damp locations along the Solent shore. It was recorded as

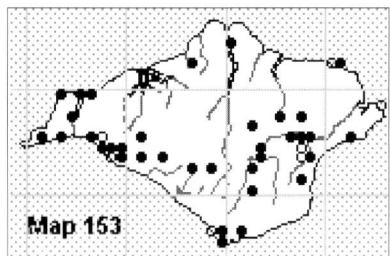

E. cicutarium

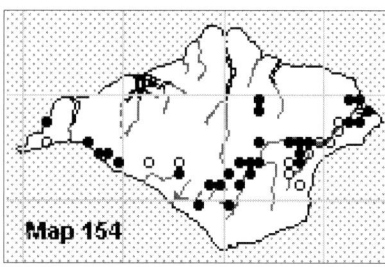
Hydrocotyle vulgaris

growing amongst saltmarsh tussocks at the mouth of a small stream on Binstead shore in 1947, GB, and it persisted here for around fifteen years. In 1964, plants were found on the edge of the shore, just above high water mark, on the beach below Quarr Abbey. There are no further records until a single plant was located at the top of the shore at King's Quay 5494, in 1986, BS & CP. It is assumed that these records originate from seed brought across the Solent; the plant is invasive and increasing in many Hampshire rivers.]

I. parviflora DC.
SMALL BALSAM
E. Very rarely established in gardens and pavements. Only ever recorded from Ryde, first in 1936, GB. It subsequently persisted on waste ground in Albert Place, off the High Street until at least the mid 1970s. Recently found off Monkton Street 5992, GT 1996.

I. glandulifera Royle
INDIAN BALSAM
E. Rare on waste ground and generally found as a short-lived casual. Not currently known as an invasive species. First recorded in 1961 from Shalfleet Manor and the following year from waste ground on the roadside between Ashey and Brading Downs, GB. Subsequently recorded from a variety of sites where it persisted for a number of years before dying out. However, it still survives (1998) in small quantity by the Yar riverbank along the railway track west of Sandown waterworks, 5885, where it was first recorded in 1971, RK. Currently established by the ditch south of Landguard Manor, Shanklin 5782, AC 2000. A public survey by Sheffield Hallam University, 2000 produced a report that in 1948, Miss Welch collected seed from Beauchief in Sheffield and took it to the Isle of Wight. Here she released it to a riverside at Newport. There are however no subsequent records from this area.

110. ARALIACEAE
Ivy family

Hedera helix L. ssp. *helix*
COMMON IVY
N. Found in woods and hedgebanks across the Island, but less frequently than Atlantic Ivy.

ssp. *hibernica*
(G.Kirchn.)D.C. McClint.
ATLANTIC IVY
N. This western subspecies is the common ivy on the Island. In the shelter and humid conditions of the Undercliff, it develops large leaves and dominates the woodland floor. It was first recognised as occurring here in 1978 by H.A. McAllister.

111. APIACEAE
Carrot family

Hydrocotyle vulgaris L.
MARSH PENNYWORT
N. Local, but sometimes frequent, in neutral to acid marshes and fens; very occasionally in wet lawns and dune slacks. Most sites are in the Eastern Yar and the Medina at Cridmore Bog and The Wilderness. Frequent in dune slacks at St Helen's Duver 6389 until 1993, when it could not be refound following saline incursion. It survives the mowing regime of historic grazed commons at St Helen's Green west 6289, CP 1994 and Colwell Common 3287, AM 2000. Map 154

H. ranunculoides L. f.
FLOATING PENNYWORT
E. Found in 1998, completely choking two contiguous roadside ponds opposite Yafford Farm 4481, MB. A determined effort was made in 2000 to exterminate the plant by dredging.

Sanicula europaea L.
SANICLE
N. Frequent in woods, particularly ancient woods, and old hedgebanks across the Island on a range of soil types.

Eryngium maritimum L.
SEA-HOLLY
N. A declining species of sandy ground generally at the mouth of estuaries, along the Solent shores of the Island. In recent years, its core areas have been Norton Spit and St Helen's Duver with occasional, short-lived colonies elsewhere. It continues to survive at Norton Spit 3489, where it has increased in recent years despite increased visitor pressure and changes in beach levels. 1726 plants were counted here, in four areas in 1997, JKN & VG. Sea-holly was formerly abundant on St Helen's Duver but by the 1980s was confined to just two spots. The plants on the dunes suffered from trampling and picking, but those south of Ferry Point fared better. However, by 1995, only 3 or 4 tiny plants were left. One flowering plant survived in 2001, MJ. Formerly on dunes at Bembridge Point 6488, but last recorded here in 1963. However, a single small plant was found in 2001, AC. Other sporadic sites, with last recorded dates, are: Quarr beach, 1986; Thorness Bay, 1981; Hamstead shingle beach, 1966; and Newtown east spit 1959.

Chaerophyllum temulum L.
ROUGH CHERVIL
N. Frequent in hedgebanks and scrub across the Island.

Anthriscus sylvestris (L.) Hoffm.
COW PARSLEY
(local: queen anne's lace!)
N. Abundant and widespread in hedgebanks, waste places and nutrient enriched woods. Long (1886) refers to the local name of 'Kecks' to describe the dried stalks of cow-parsley or hemlock.

A. caucalis M.Bieb.
BUR CHERVIL
N. Very local but increasing on dry sandy and calcareous soils around the coast. Bromfield (1856) recorded it from Freshwater Down and today it is frequent right along the cliff edge from West High Down, through to Afton Down, east of Freshwater Bay. It was first recorded on St Helen's Duver 6389, in 1964 when a few specimens were present, BS. It has since increased considerably and is

now common here. It was found on the sandy slope above Ryde Canoe Lake 6092, in 1997, CP. This occupies the former site of Ryde Dover where Bromfield and Bell Salter both recorded this plant in the nineteenth century.

[*Scandix pecten-veneris* L.
SHEPHERD'S-NEEDLE
(local: crow needles)
Arch. Extinct. In Bromfield's time this plant was described as abundant in cultivated land and sufficiently well known by farm workers to acquire a local name. By the 1930s it was becoming scarce and there are just four records post 1950. These are: cornfield above St Lawrence, RPB 1953; Ventnor, TW 1958; ploughed field at the top of Havenstreet, GB 1963; and a specimen collected from a field at East Ashey farm in 1967, BW. This specimen was collected and pressed and has only recently come to light.]

[*Coriandrum sativum* L.
CORIANDER
C. Extinct. Rare casual of waste ground, not seen for many years. Ryde, G.C. Druce 1929; Newport, Drabble & Long 1931; and from outside of a baker's shop in Shanklin, JB 1967.]

Smyrnium olusatrum L.
ALEXANDERS
Arch. Now abundant and spreading invasively on hedge banks, cliffs, permanent grassland and waste places. It has for long been frequent around the coast but in recent years has been spreading inland.

Conopodium majus (Gouan) Loret
PIGNUT
N. Local but not infrequent in deciduous woodland on clay and sandy soils and in old pastures on acidic sandy soils. Bromfield (1856) wrote, 'Though the tubers of this plant are a delicacy that boyish appetite disdains not, there is an acrimony . . . with their sweetness, better fitted to the digestion of the respectable quadrupeds whose name they share, than for Christian bipeds of tender years.'

Pimpinella saxifraga L.
BURNET-SAXIFRAGE
N. Frequent in unimproved grassland. Widespread, but commonest on the chalk, frequent on calcareous clay soils and least common on the sandy soils in the south of the Island.

Aegopodium podagraria L.
GROUND-ELDER
Arch. A frequent and persistent garden weed. Also commonly found in the countryside where waste has been dumped or as a relic of habitation.

Berula erecta (Huds.) Coville
LESSER WATER-PARSNIP
N. This plant has always been rare and restricted in its distribution to several spots in the Undercliff, the Buddle Brook at Brighstone and at Freshwater Marshes. It was found in ditches, pools and fens. It continues to survive in Afton Marsh at Freshwater 3486 where, in 2001, it was found in two well-separated populations in the south marsh. The other modern site is the stream at Windy Gap, St Catherine's 4975. This area is currently heavily overgrown with willow and the most recent record is 1984, CP.

Crithmum maritimum L.
ROCK SAMPHIRE (local: samper)
N. Locally abundant on maritime cliffs on the south coast, and much less so on shingle beaches and seawalls on the north coast. Frequent on the chalk cliffs and greens at Culver and from West High Down to Afton Down, and also growing on the roadside chalk cutting over Afton Down and by the seawall in Freshwater Bay. Frequent on the cliffs and on seawalls between St Catherine's Point and Bonchurch. It formerly grew on the inner cliff at Bonchurch. Elsewhere around the coast it is much more local. It is restricted to one or two spots on St Helen's Duver. Other sites away from the cliffs include the mouth of Wootton Creek, Thorness Bay, the east and west spits at Newtown and Norton Spit, but at all these sites it occurs in small quantity.

First recorded by Thomas Johnson in 1634 'on the white cliffes on the south side of the Isle of Wight'. Rock

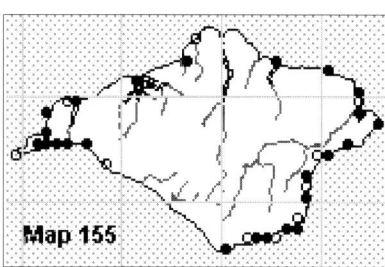

Crithmum maritimum

Samphire provided a valuable source of income for cliffsmen. Robert Turner, in 1664, wrote of the plant growing on cliffs 'where it is incredibly dangerous to gather; yet many adventure it, though they buy their sauce with the price of their lives'. Bromfield (1856) referred to the samphire harvest. He wrote, 'The warm aromatic pickle prepared with this plant is greatly esteemed and commonly seen at table in this island. The herb minced is also served up with melted butter in lieu of caper-sauce. For the purpose of pickling it is annually collected in large quantities from the cliffs at Freshwater, and sent up to some wholesale houses in London, by the cliffsmen, who make samphire-gathering a part of their summer occupation, and for which, when cleaned and sorted, they receive 4s. per bushe.' The practice appears to have died out towards the end of the nineteenth century. Map 155

Oenanthe fistulosa L.
TUBULAR WATER-DROPWORT
N. A rare plant of ditches and ponds in the Eastern Yar floodplain. In the last thirty years it has been reported from Morton Common, Brading 5985, in small quantity, BS 1971 and from the pond by the 18th green on Sandown Golf Course 5984, BS 1971 but there are no recent records from either site. It remains locally frequent in three ditches on Sandown Levels, behind Brown's Golf Course 6085, CP 2001. Bromfield (1856) recorded it from marsh ditches near Freshwater Gate.

[*O. silaifolia* M.Bieb.
NARROW-LEAVED WATER-DROPWORT
N. Extinct. A plant of unimproved damp hay meadows. Townsend (1883) described this plant as very rare in Freshwater meadows and marshes, but there is no confirmation

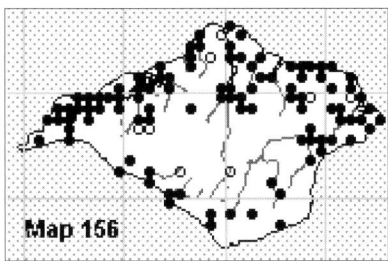
O. pimpinelloides

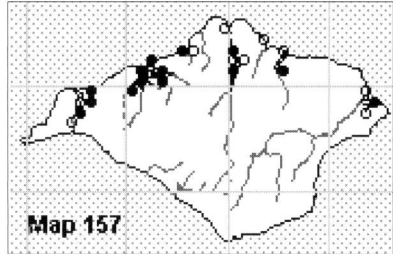
O. lachenalii

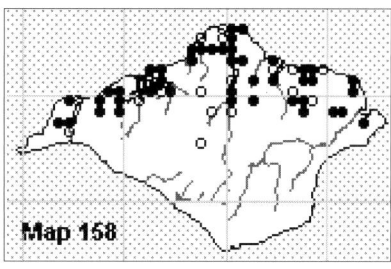
Silaum silaus

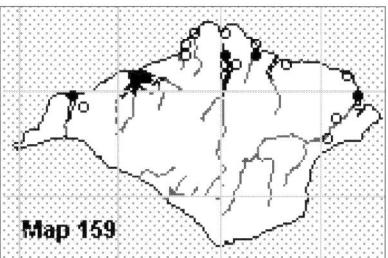
Bupleurum tenuissimum

of this. Drabble & Long (1931) recorded *O. lachenalii* as occurring in The Wilderness, but they may have been confusing it with *O. silaifolia*. On 20 June 1974, Francis Rose found the plant in small quantity in a meadow northeast of The Wilderness (SZ507826). It has never been refound here and, indeed the plant is not known from Hampshire or Somerset, and there are very few good records from Dorset.]

O. pimpinelloides L.
CORKY-FRUITED WATER-DROPWORT
N. A common species of unimproved neutral meadows and verges, particularly on the clays of the northern half of the Island. It is tolerant of brackish conditions and of some agricultural improvement. Absent from the chalk and considerably more local on the southern half of the Island. It has declined with agricultural intensification but is often persistent in field borders. The Island, along with south Hampshire, Dorset and Somerset is the stronghold of this species. Map 156

O. lachenalii C.C.Gmel.
PARSLEY WATER-DROPWORT
N. Very local in upper saltmarsh and brackish grassland around the mouths of estuaries on the north coast. Perhaps overlooked. Map 157

O. crocata L.
HEMLOCK WATER-DROPWORT
(local: belder-root)
N. Abundant, sometimes dominant, in marshes, ditches, stream banks and wet woods in suitable places right across the Island. Bromfield (1856) wrote, 'The roots resemble those of the Dahlia, and instances have been related to me of their having been sold to credulous persons for that handsome plant.'

Aethusa cynapium L.
FOOL'S PARSLEY
N. Occasional in dry arable fields on all soils, in gardens and waste ground. It is widespread across the Island and is commoner than suggested by Bevis *et al* (1978).

Foeniculum vulgare Mill.
FENNEL
E? Not uncommon in waste places and on sandy soils but largely confined to the coasts. It has increased since Bevis *et al* (1978). Sites include Bembridge Point, Forelands, Whitecliff Bay, Seaview Duver, Ventnor cliffs, Norton Spit and Freshwater Bay.

Silaum silaus (L.) Schinz & Thell.
PEPPER SAXIFRAGE
N. Locally frequent in unimproved grasslands on clay soils, including clay heaths. Confined to the northern half of the Island. Map 158

Conium maculatum L.
HEMLOCK
Arch. Widespread but local in hedgebanks, waste ground, dredged river mud and woodland edges.

Bupleurum tenuissimum L.
SLENDER HARE'S-EAR
N. Scarce on dry banks, particularly seawalls, at the mouths of estuaries. Usually present in small quantity and this, together with the fact that numbers fluctuate considerably from year to year, means that it is easily overlooked. However, this plant has declined and the extent of suitable habitat has been reduced. It is most frequent in the Newtown estuary where it occurs in a number of stations and can be locally abundant. This is a plant which is particularly at risk from rising sea-levels and increased storminess. Map 159

[***B. rotundifolium*** L.
THOROW-WAX
C. Extinct. Formerly a scarce weed of cornfields on chalk. Last recorded from Cowes in 1921 (Drabble & Long, 1931).]

[***B. subovatum*** Link ex Spreng.
FALSE THOROW-WAX
C. A rare bird seed casual which was confused with Thorow-wax. First noticed in 1964, but not recorded recently.]

Apium graveolens L.
WILD CELERY
N. Very local and usually in small quantity by saltmarsh ditches and wet seepages along undercliffs. Its distribution is predominantly coastal but it has been recorded from the following inland localities: ditch at Yafford 4481, MB 1998; St Helen's Green west 6298, AC 1998; ditch near Park Farm, Nettlestone 6189, AC 1998. It was much more frequent in the nineteenth century. Map 160

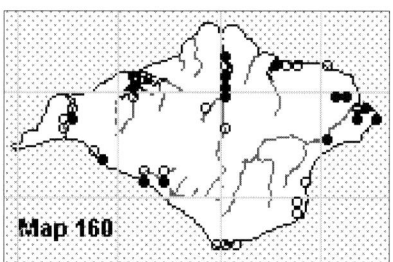
Apium graveolens

A. nodiflorum (L.) Lag.
FOOL'S-WATER-CRESS
N. Widespread and common in wet places everywhere, in both fresh and brackish water.

A. inundatum (L.) Rchb.f.
LESSER MARSHWORT
N. A very rare plant of shallow ponds in moderately acidic water. Known today only from a field pond at Shepherd's Hill, Elmsworth 4392, AC where it was refound in 2000, growing with *Baldellia* but not flowering, and a nearby field pond near Brickfields 4292, CP 2002. There are additional nineteenth century records from Leechmore pond on Bleak Down; a pond in a field at the back of Thorness Wood; pool at Golden Hill, Freshwater; coastal pond at Hamstead; and The Wilderness. A series of ponds on the north-west coastline appear to have been a stronghold for this species, together with *Pilularia globulifera* and *Sparganium natans*. There are three early twentieth century records: Sandown Levels 1931, Drabble & Long; pond on Sandown Golf Course 1943; and pond at Colwell, 1944 (possibly the Golden Hill site).

Petroselinum crispum (Mill.) Nyman ex A.W. Hill
GARDEN PARSLEY
C. Very occasionally established as a garden escape and generally not persistent.

P. segetum (L.) W.D.J. Koch
CORN PARSLEY
N. Occasional at the edges of fields and on dry banks near the coast. Bromfield (1856) described this plant as 'very frequent on chalk or clay in various parts of the Island' whilst Bevis *et al* (1978) considered it to be a very rare species. Current information would suggest that it has increased considerably and it can be locally abundant in suitable places such as in the Medina valley. It is however, an overlooked species and is likely to be under recorded. Bromfield (1856) wrote, 'The plant would probably be found well worthy of cultivation as a winter salad, as it remains green and tender throughout the severest season of the year, and the leaves are without acrimony'.

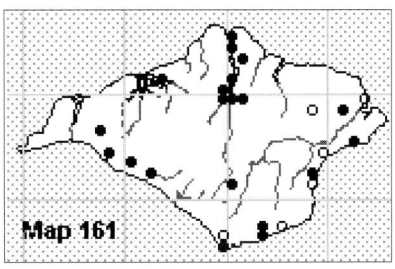

P. segetum

The first British record of Corn Parsley was made around 1620 by John Goodyer who noticed a maid of his employer gathering the plant in a field near Petersfield, Hampshire. On enquiring the reason for this, he was told that she had been taught during her upbringing at Brading on the Island that it was a useful herb for treating the cheek swellings known as 'hones' (Gunther, 1922). John Gerard referred to Honewort as an alternative name for Corn Parsley in the 1633 revised edition of his Herbal. Map 161

Sison amomum L.
STONE PARSLEY
N. Locally frequent in hedge banks, waste ground and as a garden weed, particularly on clay soils. Bevis *et al* (1978) considered that it was widespread but less common than in Bromfield's time. It is frequently abundant today and therefore may perhaps have increased in recent years.

[***Ammi majus*** L.
BULLWORT
C. Extinct. Recorded from a garden at Binstead, 1964 and from a flowerbed on Ryde Esplanade 1971, both GB.]

[***Carum carvi*** L.
CARAWAY
C. Extinct. Only recorded as a casual from Newport and Cowes, Drabble & Long, 1931.]

Angelica sylvestris L.
WILD ANGELICA
N. Local but widespread in damp woods, fens, marshy grassland and riversides. Only very occasionally found in abundance.

Pastinaca sativa var. ***sativa*** (Mill) DC.
WILD PARSNIP
N. Frequent in rank grassland, verges, scrub and waste ground on dry soil. It is particularly abundant on the chalk.

Heracleum sphondylium L.
HOGWEED
N. Common and widespread everywhere on verges, wasteland, woodland rides and rank grassland.

H. mantegazzianum Sommier & Levier
GIANT HOGWEED
E. A rare, persistent escape but not invasive on the Island. Not recorded before 1965. It was well established on the 'Donkey Bank' at Bonchurch 5778, in 1965 BS, and persisted here at least until 1992 but now appears to have been shaded out. However, it survives nearby on the north bank of Bonchurch pond. Also long established at Station Road, Freshwater 3487 (pre 1973) and alongside the bakery site in School Green, Freshwater 3387 (pre 1971) and persisting at both sites. A few plants by the roadside at Sandrock Road, Niton 5075, CP 1998.

Torilis japonica (Houtt.) DC.
UPRIGHT HEDGE-PARSLEY
N. Frequent and widely distributed on roadside verges and in scrub.

[***T. arvensis*** (Huds.) Link
SPREADING HEDGE-PARSLEY
Arch. Extinct. Bromfield (1856) considered this to be a common cornfield weed but it had become very rare by the middle of the twentieth century. The last record was in 1958, from a field at Whitwell, AWW.]

T. nodosa (L.) Gaertn.
KNOTTED HEDGE-PARSLEY
N. Very local on dry banks and grassland, particularly around the coast. Bevis *et al* (1978) knew of three sites but it is more frequent today and

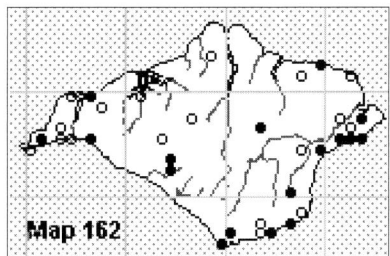

T. nodosa

may have increased. Frequent along cliff edges on the chalk from West High Down to Afton Down at the western end and Culver Down at the eastern end. Also locally frequent on dry banks in the south from St Catherine's Point to Ventnor. Occasionally found in arable fields: St Martin's Down 5680, CP 1991 and at Arreton 5386, JRM 2000. Map 162 (see overleaf).

Daucus carota L. ssp. *carota*
WILD CARROT

N. Common and widespread in grassland, roadside verges and around the coast.

ssp. *gummifer* (Syme) Hook.f.
SEA CARROT

Plants with succulent leaves growing on chalk cliffs at the western end of the Island and in exposed grassland on the south west coast are probably Sea Carrot.

[*Caucalis platycarpos* L.
SMALL BUR-PARSLEY

C. Extinct. Recorded by Townsend under the walls of St Helen's Mill on 22 June 1879, in the company of F. Stratton. No other records.]

[*Turgenia latifolia* (L.) Hoffm.
GREATER BUR-PARSLEY

C. Extinct. The only record is reported to be from Newport, Drabble & Long 1931.]

112. GENTIANACEAE
Gentian family

Centaurium erythraea Rafn
COMMON CENTAURY

N. Frequent and widespread in dry grassland on chalk, heaths, undercliffs, woodland rides and sandy and shingly beaches. Sometimes white flowered. The dwarf, congested form known as Tufted Centaury var. *capitatum* (Willd. Ex Cham.) Melderis, was first described as a British species by Townsend in 1879, from the downs at Freshwater. It was illustrated in colour as the frontispiece to his Flora (1883). Although now relegated to subspecific status, it is not infrequent at the western end of the Island on a variety of soils. It is most frequent on the chalk and is found in suitable spots from West High Down continuously eastwards to Brook Down. There is a record from Mottistone Down in 1965, DF. It is recorded from Prospect Quarry on limestone. It also grows on gravelly marls at Headon Warren undercliffs and on greensand at Compton Chine.

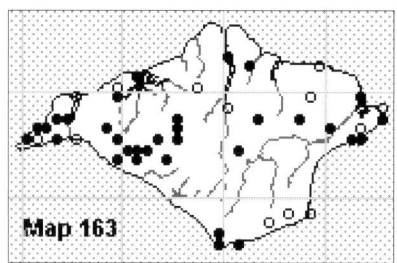

C. pulchellum

C. pulchellum (Sw.) Druce
LESSER CENTAURY

N. Found in short open turf on chalk and gravelly clay soils, sometimes in seasonally wet hollows. Bevis *et al* (1978) were not able to give a complete picture of its distribution but it is widespread and often frequent in south facing chalk grassland turf. Map 163

[*C. tenuiflorum* (Hoffmanns. & Link) Fritsch
SLENDER CENTAURY

N. Extinct. A rarity of estuaries and clay banks on the north coast of the Island, probably extinct. There is a continuity of records from two Island sites. Our plants all appear to have been pink flowered, unlike those of the extant Dorset undercliff populations.

The first British record was made by F. Stratton, from 'salt marshy ground' at King's Quay in 1870 (Townsend, 1883). Lousley saw it here in 1952, growing on a mud bank at the eastern end of the eastern spit with *Calamagrostis epigejos* (SZ 539940). After this, records from this site become confused but plants found growing on partly wooded clay landslips just east of King's Quay in 1982 were considered by Francis Rose to be referable to this species.

The second site was on a grassy bank adjoining saltmarsh on the Medina estuary, south of Medham. It was first recorded here by Townsend, 1883. Although Drabble & Long

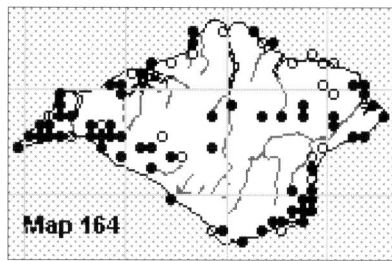

Blackstonia perfoliata

(1931) considered that it was 'probably now extinct except on private land', a specimen was illustrated from here in Butcher & Strudwick's *Further Illustrations of British Plants*, published in 1944. In 1967, Bill Shepard found plants growing on a grassy bank by a small saltmarsh near Newport (SZ 501901). The marsh was drained and infilled in 1975.

There are no records from elsewhere around the Island coasts, but John Edmondson has come across an undated herbarium sheet originating from Cheltenham College (LIV) labelled '*Erythraea linarifolia*, Freshwater, Isle of Wight'. He comments 'there can be little doubt that Freshwater, IOW was the locality and that *Centaurium tenuiflorum* the species ... although the specimen is dwarf, it matches our other material of that species collected from Dorset and reliably named by Ted Wallace.' Slender Centaury is a Mediterranean species on the edge of its range.]

Blackstonia perfoliata (L.) Huds.
YELLOW-WORT

N. Locally frequent in short chalk grassland and on calcareous clays. It is widespread on south-facing downs, but it is also common on suitable undercliffs right around the coast in base-enriched areas. Occurs in limestone grassland at Prospect Quarry, Tapnell 3866, CP 1996 and Lacey's Quarry, Totland 3286, CP 2000. Map 164

[*Gentianella campestris* (L.) Börner
FIELD GENTIAN

N. Extinct. This northern species was a former rarity only reliably recorded from dry heathy grassland on Colwell Heath. It was first referred to by Snooke (1823) who described it as plentiful. A number of specimens were

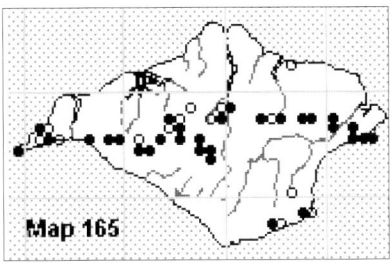
G. amarella

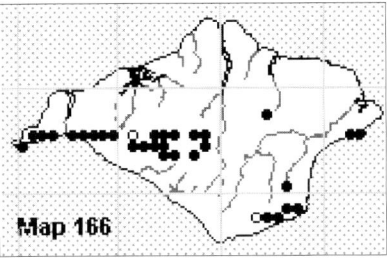
G. anglica

collected between 1860 and 1879 by H.C. Watson and A.G. More, and deposited in herbaria (BM, HCMS, OXF). They were all gathered near the Nelson Inn (now renamed Colwell Bay Inn) on Colwell Heath. F. Stratton collected a few specimens on 6 September 1879 but there are no further records. Around this time, the Common was sold to Edmund Granville Ward who let parcels of land for building. The loss of this species was undoubtedly due to habitat destruction.]

G. amarella (L.) Börner **ssp.** *amarella*
AUTUMN GENTIAN
N. Widespread in short grassland on downland across the Island, fluctuating greatly in abundance from year to year. There are a few historic records from the calcareous clays on the north of the Island but the only known extant site off the chalk is on Headon Warren undercliffs 3186, AM 1998. In favourable years, plants are regularly in flower by late May. White flowered plants were recorded on Tennyson Down 3285, BS 1999. Map 165

G. x davidiana T.C.G. Rich, *hybr. nov.*
AUTUMN GENTIAN X EARLY GENTIAN
N. Very rare on chalk downs. The flowering seasons of Early Gentian and Autumn Gentian occasionally overlap and for some time it has been suggested that hybrids could occur on the Island. Rich *et al* (1997) proposed a name for this hybrid. The holotype was collected from the Island, on St Boniface Down, 26 June 1925, R. Melville (NMW). Some plants seen recently are suspected of being this hybrid, but have not been confirmed. Recent genetic studies have shown that Early Gentian and Autumn Gentian are genetically identical, so that the naming of this hybrid may be invalid.

G. anglica (Pugsley) E.F. Warb. **ssp.** *anglica*
EARLY GENTIAN
N. Widespread and remarkably frequent in thin open ground on south facing downs, frequently sharing sites with Autumn Gentian. It is absent from the eastern downland ridge apart from Bembridge and Culver Downs. Recorded in small quantity on Arreton Down 5487, RDP in 1987 but not since. The Island, together with Dorset and Wiltshire, is the stronghold for this endemic plant, although numbers fluctuate considerably from year to year. In 1994, considered to be a good year for the plant, an approximate total of plants was calculated (Telfer, 1994). Many sites held thousands of plants; the biggest population was on West High Down 3084-3285 where an estimated 1.8 million plants were present in 1994, but only 2700 in 1995 (ST). The earliest recorded plant in flower was on 27 March on Culver Down in 1999, AC. Occasionally plants with cream coloured petals are found, e.g. Brook Down 3985, ST 1997.

The first British record of Early Gentian comes from the Island, as *Gentiana amarella* var. *praecox*, Townsend (1883). Townsend cites four localities and for one of them, Steephill, refers to specimens collected by F. Stratton on 27 May 1878. The Steephill type locality probably refers to Rew Down, a site where Early Gentian still grows. The other localities were Afton Down, Brighstone Down and Culver Down. Stratton (1878) stated that, 'I have dried a good number of specimens for distribution through the Botanical Exchange Club.' Material survives in herbaria at OXF, K, BM and CGE. Recent research has shown that this plant is not genetically distinct from Autumn Gentian and this has thrown our understanding of the taxonomic status of Early Gentian into confusion. Map 166

113. APOCYNACEAE
Periwinkle family

Vinca minor L.
LESSER PERIWINKLE (local: sengreen)
Arch. Well established in a few woods and hedge banks, often near ruined buildings. Although probably originally a garden escape, it has been long established at some sites. Bromfield (1856) recorded it from 'a remote part of Centurion's copse' and E.H. White (1904) found it to be 'abundant in the north west corner of the copse'. It still occurs here and is particularly associated with old earth banks in this part of the wood, 6286. Other sites, where it was known to be well established in the early 1970s and remains frequent today, include the following: Atkies Copse, Ningwood 4089, CP 2000; small copse by the caravan park at Whippingham 5193, PS 2000; coastal woodland on the Osborne estate, where plants are white flowered 5295, CP 1995; and the sunken lane between Standen House and St George's Down 5187, CP 1995.

V. major L.
GREATER PERIWINKLE
E. A not infrequent garden escape in hedge bottoms, refuse tips and road banks.

114. SOLANACEAE
Nightshade family

Nicandra physalodes (L.) Gaertn.
APPLE-OF-PERU
C. Occurs sporadically and in small quantity on waste or disturbed ground and in gardens. Known for a long time on the Island, it was first recorded by Bromfield (1856) as partly naturalised in waste ground at Ryde, Shanklin and other parts of the Island.

Lycium agg. *sp.*
DUKE OF ARGYLL'S TEAPLANT
E. A persistent garden hedge plant, established in waste places around the coast. The identities of the two species *L. barbaratum* and *L. chinense* are much confused.

Atropa belladona L.
DEADLY NIGHTSHADE
N. This plant has always been a rarity on the Island, in chalk scrub and by old buildings. The only modern station is in Westover Plantation, Calbourne Bottom. Plants appeared when a forestry track was created and were discovered by Miss Bullock in 1964. It was present in reasonable numbers in the 1970s but gradually declined and was thought to have been lost. However, it was refound in 1998, 4185 HH. In 2001, there were four substantial plants growing at the edge of a ride.

Hyoscyamus niger L.
HENBANE
Arch. A scarce plant of coastal sites, where it is generally regular in fluctuating numbers, and on disturbed ground on a variety of soil types, where it is sporadic and usually impermanent. Coastal localities are: Thorness Bay shingle beach 4593, which holds the largest Island population, AC 2000; Hamstead shingle spit 4092, BS 1991; and around rabbit burrows on the cliff edge at West High Down 3084, CDP & TDD 1998. A few plants also appear on the chalk on Afton Down 3685, TT 1998 and on Mottistone Down 4184, TT 1996. Henbane is no longer found as a coastal plant in Hampshire or West Sussex but does occur on coastal limestones in Purbeck. Map 167

Salpichroa origanifolia (Lam.) Thell.
COCK'S-EGGS
E. Established in a sheltered laneside by a footpath to Woody Bay, St Lawrence 5376. First recorded here in 1927 (E.H. Drabble hb. BM), and still present in 2001. This is the oldest known and longest established population in the country.

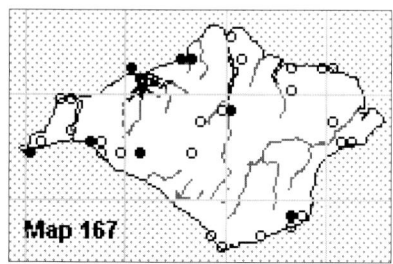

Hyoscyamus niger

Lycopersicon esculentum Mill.
TOMATO
C. A late summer casual on waste tips and sometimes on the strandline of beaches.

Solanum nigrum L. ssp. *nigrum*
BLACK NIGHTSHADE
N. A common plant of cultivated ground, waste places and rubbish tips.

S. physalifolium Rusby
GREEN NIGHTSHADE
C. First recorded as an abundant weed in a potato crop in sandy ground at Compton Grange 3884, PS & EC 2000.

S. sarachoides Sendtn.
LEAFY-FRUITED NIGHTSHADE
C. A rare casual first recorded in 1980 growing on the railway embankment at Bembridge Harbour 6483, BS (conf. E.J. Clement). On sandbank at south end of St Helen's Duver 6389, BS 1992. Dunsbury farmyard 3884, PS 2000.

S. dulcamara L.
BITTERSWEET
N. A common plant occurring in a range of habitats. It is found in hedgerows, wet woodlands and fens, waste ground, streamsides and on coastal shingle.

S. tuberosum L.
POTATO
C. A casual on waste tips and as a relic of cultivation.

S. laciniatum Aiton
KANGAROO-APPLE
C. A persistent bird sown casual from gardens around Ventnor and St Helens, surviving for several years and sometimes reaching flowering size but destroyed in cold spells. A few plants surviving for about four years on St Helen's Duver had reached a height of 2m in 1999. First recorded in 1993 as small driftline plants in the same locality.

S. rostratum Dunal
BUFFALO-BUR
C. A rare casual of waste tips and cultivated ground first recorded in 1925 by J. Long. Refuse tip at Stag Lane 5091, BS 1973 (conf. D.

McClintock); allotments at Carisbrooke 4888, BS 1989; garden at Havenstreet Station 5589, BS 1989.

Datura stramonium L.
THORN-APPLE
C. An occasional casual on waste ground and arable ground. Sometimes grows in unexpected places from long buried seed brought to the surface. An established weed in fields at Redway, 5384. The mauve flowered var. *chalybaea* W.D.J. Koch sometimes occurs. Thorn-apple was known as a not infrequent casual by Bromfield (1856).

Nicotinia rustica L. Wild
TOBACCO
C. Once grown for tobacco and, in 1922, it was considered to have become established by the river at Cowes and was shown at the annual Wild Flower Exhibition of the Natural History Society. Not recorded for many years.

N. x sanderae W. Watson
SWEET TOBACCO X RED TOBACCO
C. Occasional casual on disturbed ground, e.g. Seaclose recreation ground, Newport, 2001 EC.

115. CONVOLVULACEAE
Bindweed family

Convolvulus arvensis L.
FIELD BINDWEED
N. Common in cultivated and waste ground and a persistent garden weed.

Calystegia soldanella (L.) R.Br.
SEA BINDWEED
(local: scurvy-grass)
N. Very local on sandy dunes. The two principal extant sites, both known to Bromfield, are St Helen's Duver 6389, where it is frequent in several places on stabilised dunes and Norton Spit 3489 where it occurs in a limited area on the sandy headland. Recent records suggest a modest spread of this species at the eastern end of the Island. It reappeared in small quantity at Bembridge Point 6488, a Bromfield site, in 1997, AC and a single plant was found in Priory Bay 6390, AC 1996. Occasional and impermanent at the southern end of the Island: Watershoot Bay, St

Catherine's 4975, AB 2000; west of Monks Bay, Bonchurch 5778, JO 1988. Formerly also occurred in Sandown Bay (last recorded 1914) and the east spit at Newtown (last recorded 1965).

C. sepium (L.) R.Br.
HEDGE BINDWEED
(local: granny pop out of bed; lily!)
N. Ssp. *sepium* is common in disturbed ground, hedgerows, waste places, wet woodland and fen carr, and is a persistent garden weed. The local name of 'Granny Pop out of Bed', which was in use until recently, referred to the manner in which the seeds could be squeezed out of the capsule. The pink flowered ssp. *roseata*, a western coastal subspecies in this country, was recorded in 2000 from High Grange 3784, Compton Grange 3884 and Mottistone 4084, all PS. It is apparently not uncommon along the south-west coast. It was first recorded from Yarmouth in 1929 (BM).

C. x *lucana* (Ten.) G.Don
HEDGE BINDWEED X LARGE BINDWEED
N. Rare or overlooked in hedges, first recorded in 2001. With both parents at Whale Chine carpark 4778; laneside at St Lawrence 5376, both GK.

C. pulchra Brummitt & Heywood
HAIRY BINDWEED
E. A rare introduction in hedgerows and scrubby areas, usually near to habitation. Sudmoor hedgerow 3983, PS 2000 and waste ground at Ventnor 5577, GT 1997. Well established away from dwellings at Luccombe Chine 5879, CP 1996 and disturbed sandy ground at Knock cliff, Shanklin 5880, CP 1981. First recorded from Limerstone in 1960, DF.

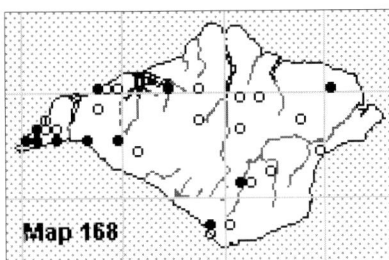

C. epithymum

C. silvatica (Kit.) Griseb.
LARGE BINDWEED
(local: hedge bells; lily!)
E. Common in hedgerows and waste places and a persistent weed in gardens.

116. CUSCUTACEAE
Dodder family

[*Cuscuta europaea* L.
GREATER DODDER
N. Extinct. A former rare parasite of nettles and hops. It is likely that some records were confused with *C. epithymum*. Bromfield's station of hedgerows between Alverstone and Kern, with records from 1840 to 1871, certainly referred to Greater Dodder. Last recorded in 1928 from near Ventnor by J. Long.]

C. epithymum (L.) L.
DODDER
N. A local, and sometimes sporadic, parasite that has declined considerably over the past one hundred years with the loss of heathy grassland habitat. It occurs today in three distinct habitats. It is found on heathland, parasitising heathers and gorses on Headon Warren 3085 & 3186, CP 1997; cliff edge at Bouldnor 3790, TDD & CDP 1996; on Bleak Down in small quantity 5181, IB 1997; on Blackgang Chine ledge 4877, CP 1993; and on gorse bushes on Westridge Golf Course roughs, south of Ryde 6090, IB 1998. It has been recorded sporadically in short, south facing chalk grassland on Tennyson Down 3385, CP 1997; Afton Down 3685, BS 1992; Brook Down Quarry edge 3985, SC 1994; and Knighton Down 5787, BS 1986. It also occurs in species-rich neutral grassland, sometimes in abundance, over some 3 ha at Porchfield Ranges meadow 4490, PJS 2001. In Parkhurst Forest, it was last recorded from a ride in 1965. Map 168

117. MENYANTHACEAE
Bogbean family

Menyanthes trifoliata L.
BOGBEAN
N. A rare plant of acidic fen meadows, which has declined over the last one hundred years and is rarely found in flower today. It is most abundant at Cridmore Bog 4981, growing in ditches within the Bottle Sedge-Marsh Cinquefoil fen community. Also recorded, in much smaller quantity, from High Grange Marsh, Compton 3784, CP 1999; and on the Eastern Yar at Alverstone in the Hill Farm drain 5785, SY, and below Borthwood Lynch 5785, CP 1994. Other sites where it has recently been recorded, including Freshwater Marshes and Sandown Golf Course pond, are considered to be introductions. In 2002, plants were flowering well at the Eastern Yar sites but not at Cridmore. Map 169

Nymphoides peltata Kuntze
FRINGED WATER-LILY
E. A rare introduction in ponds and ditches. Well-established in the ditch to the west of Sandown Waterworks 5885, where it was first noticed in 1990, BS. Abundant in a field pond at Berrycroft Farm, Roud 5180, CP 2001. First recorded in 1846 from a pool on Barrett's Common, near Ryde where it was apparently introduced by Thomas Bell Salter (Bromfield, 1856).

118. POLEMONIACEAE
Jacob's-ladder family

Polemonium caeruleum L.
JACOB'S-LADDER
C. An occasional garden outcast. First recorded in 1942 at Ashey by E.H. White.

119. HYDROPHYLLACEAE
Phacelia family

Phacelia tanacetifolia Benth.
PHACELIA
C. Occasional on disturbed verges and edges of arable fields, sometimes persisting for several years. Cultivated in recent years as an aid in controlling

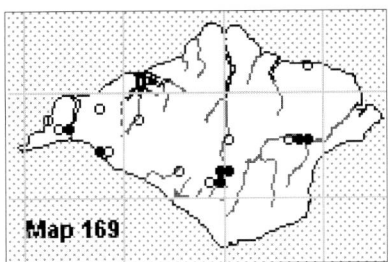

Menyanthes trifoliata

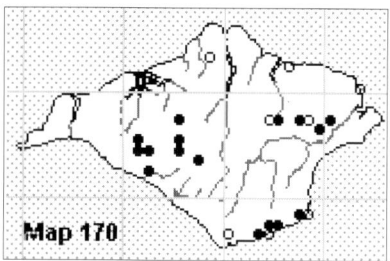
Lithospermum officinale

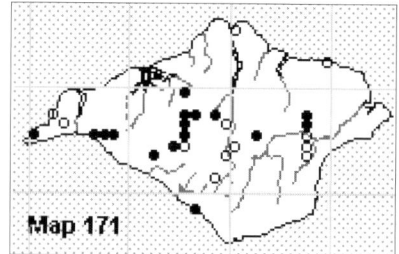
Echium vulgare

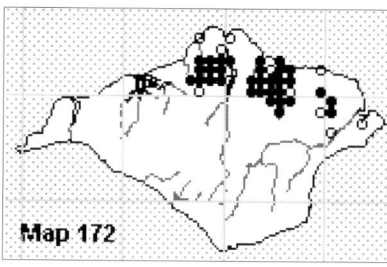
P. longifolia

120. BORAGINACEAE
Borage family

Lithospermum officinale L.
COMMON GROMWELL
N. Local in open scrub and woodland edges on the chalk. It is not infrequent on many downland sites between Mottistone Down in the west and Brading Down in the east, but neither further west nor east of these sites. Also present on the downs around Ventnor. Map 170

[*L. arvense* L.
FIELD GROMWELL
Arch. Extinct. Formerly a scarce weed of calcareous arable land. There were few listed stations and it was last referred to as occurring at St Lawrence by Drabble & Long (1931). This reference may relate to a field above Pelham Woods where E.H. White found the plant to be plentiful in 1904.]

Echium vulgare L.
VIPER'S-BUGLOSS
(local: viper-grass, snake-flower)
N. A plant with a scattered distribution found on disturbed dry soils on both chalk and sand. Behaves as a casual at some sites, but regular and known over a long period from others, such as Brook/Compton Down 3885, CP 2000; Rowridge valley 4587, DD 1998; and banks around Carisbrooke Castle 4887, MB 1996. Not recorded from sandy or shingle beaches. Map 171

E. pininana Webb & Berthel.
GIANT VIPER'S-BUGLOSS
C. Increasingly grown in gardens and sometimes self-sown and persisting for several years in sheltered spots at Ventnor and in and around Ventnor Botanic Gardens, sometimes elsewhere. Bromfield (1856) was probably referring to this species when he wrote, 'one of the most splendid species, *E. candicans*, bears the ordinary winters of this island in the open ground, and ripens seed abundantly'.

Pulmonaria officinalis L.
LUNGWORT
E. Long grown in gardens and occasionally persisting as a relic of cultivation or escaping into waste places and waysides.

P. longifolia (Bastard) Boreau
NARROW-LEAVED LUNGWORT
(local: blue cowslip; good friday flower!)
N. A faithful indicator of ancient woodlands on the base-rich clay soils of the north of the Island. This beautiful plant is confined as a native to the Hampshire Basin, formerly the Pleistocene Solent River. It survives on soil derived from Tertiary rocks here, in the New Forest and around Poole Harbour. Although it has declined in the last one hundred years, it remains abundant at many sites where coppicing or ride clearance is practised and, in the absence of grazing animals, flowering can be prolific. Lungwort was one of the spring flowers that was once picked in bunches in coppiced woodland. Large populations show great variation in blue flower colour and in leaf spotting.
Within its restricted range of Parkhurst Forest in the west, to Lower Rowborough Copse in the east, it is found in the great majority of ancient woods and sometimes in ancient hedgebanks. East of here, it was last recorded from Steyne Wood, Bembridge in 1971 (only 2 plants, JB). In the New Forest, this plant is also found in species-rich heathlands but this does not seem to be the case on the Island. In Parkhurst Forest, for instance, it is common at the northern end of the forest which is ancient woodland but absent from the species-rich heathland rides in the south (Chatters, 1991). It does however occur in Wootton Common cemetery 5390, MB 2001, which is currently woodland but would almost certainly once have been grazed open land at the edge of woodland. First recorded from the Island in 1804 by Mr Griffiths in a wood between Ryde and Newport. Map 172

Symphytum officinale L.
COMMON COMFREY
N. Widespread and frequent, particularly in the Eastern Yar valley, on riverbanks, edges of ponds, marshy places and roadside verges. Members taking part in a Natural History Society walk at Binstead in 1942 lead by Mr. E. Pollard, a Ryde chemist, had pointed out to them, 'many large clumps of Comfrey growing in a steeply sloping field below Pitts lane . . . a relic of the efforts to promote the home production of medicinal herbs during the last (1914-18) war, as again at the present time.' (Proc. for 1942 **III** (V): 311).

S. x uplandicum Nyman
RUSSIAN COMFREY
E. An increasing species on waste ground and roadside verges, once grown for forage. There are many scattered sites across the Island. First recorded in 1961 from Binstead (JH).

S. 'Hidcote Blue' hort. ex G.Thomas
HIDCOTE COMFREY
E. Well established by footpath below Buddle Inn, Niton 5075, CDP & TDD 1996.

S. grandiflorum DC.
CREEPING COMFREY
E. A garden escape occasionally naturalised by roadsides and on riverbanks. Week Farm Lane, Whitwell 5377, GT 1999; streambank at Chale Green 3784, CP 1999; field edge by Shide to Blackwater cycleway 5086, SB 1998.

S. orientale L.
WHITE COMFREY
E. An occasional but increasing garden escape on waste places and disturbed verges. It is most frequent around Bonchurch, Ventnor and St Lawrence. First recorded in 1961.

S. caucasicum M.Bieb.
CAUCASIAN COMFREY
C. The only recorded site well away from gardens is on dumped spoil by the roadside at Thorncross, near Yafford 4381, PS 2002.

Brunnera macrophylla (Adams) I.M. Johnst.
GREAT FORGET-ME-NOT
E. Very rare escape. Single plant by disused railway bank at Steephill 5577, RPB 1973; roadside at St Lawrence 5376, MB 2000.

Anchusa azurea Mill.
GARDEN ANCHUSA
C. Rare garden escape. Recorded once from the Undercliff at Niton 5176, GT 1998.

A. arvensis (L.) M.Bieb.
BUGLOSS
Arch. This plant is still locally frequent in arable fields and wasteland on sandy or gravelly soils. It is rare on the chalk but has been recorded from Bembridge Down 6285, AC 1996; and the rabbit disturbed cliff edge on West High Down 3185, CP 1999. Map 173

Pentaglottis sempervirens (L.) Tausch ex L.H. Bailey
GREEN ALKANET
E. An occasional and persistent escape on roadsides and waste places. It is frequent right along the Undercliff between St Catherine's and Bonchurch, a part of the Island where it has been known since the nineteenth century. Also occurs in scattered localities across the Island. First recorded in 1824 as naturalised at Niton by John Curtis in *British Entomology*.

Borago officinalis L.
BORAGE
E. An occasional escape on waste and disturbed ground, generally not persisting.

Trachystemon orientalis (L.) G.Don
ABRAHAM-ISAAC-JACOB
E. Rarely well established in coastal woodlands. Locally abundant but poorly flowering in Player's Wood, west of Ryde 5893, CP 1998; it was first recorded here in 1957 but plants were probably introduced as part of Victorian or Edwardian plantings. The same is likely to be true for the woodland site at Fort Victoria 3489, CP 1998 although the first record for here was in 1973. Also recorded from a roadside verge at Brook House 3984, CP 1998.

Myosotis scorpioides L.
WATER FORGET-ME-NOT
N. Frequent on streamsides and marshes in all our river valleys.

M. secunda Al. Murray
CREEPING FORGET-ME-NOT
N. Rare, but perhaps under recorded in streams and ditches in the Eastern Yar catchment. The only recent records are: marshy valley at Fairfields, north of Whitwell 5079, CP 1997; and in Scotchells Brook, Ninham 5782, BS 1995. It was also recorded from Hill Farm drain, Newchurch 5685, in 1974, BS and it may still occur here.

M. laxa Lehm.
TUFTED FORGET-ME-NOT
N. Fairly frequent in marshy streamsides and pond margins. More widespread than *M. scorpioides*. Map 174

M. sylvatica Hoffm.
WOOD FORGET-ME-NOT
C. Not native. Only known as a garden escape on waste ground.

M. arvensis (L.) Hill
FIELD FORGET-ME-NOT
Arch. Common and widespread on cultivated and waste ground.

M. ramosissima Rochel
EARLY FORGET-ME-NOT
N. Local on dry, bare ground around the coast and on sandy soils inland. Occasionally found on anthills on chalk. Abundant in some spots such as Headon Warren, Compton Chine cliffs, St George's Down and Redcliff. Map 175

M. discolor Pers.
CHANGING FORGET-ME-NOT
N. Rather frequent and widespread on a wide range of dry, light soils. Also sometimes found in marshes.

Cynoglossum officinale L.
HOUND'S-TONGUE
N. Locally frequent in disturbed chalk grassland. The two strongholds are the West Wight downs and the Undercliff, where it grows both on the downs and in calcareous grassland pockets in the slumped ground below. Map 176 (*see overleaf*)

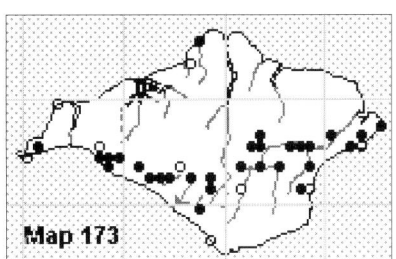

A. arvensis

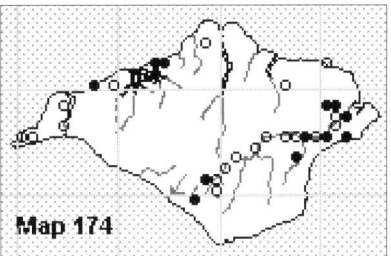

M. laxa

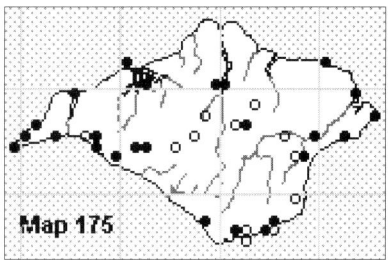

M. ramosissima

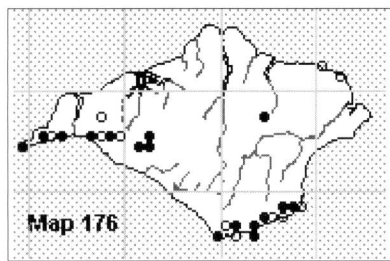
Cynoglossum officinale

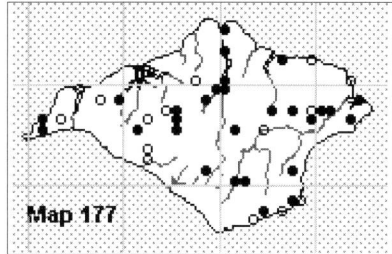
Verbena officinalis

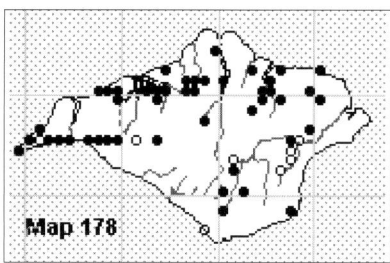
Stachys officinalis

121. VERBANACEAE
Vervain family

Verbena officinalis L.
VERVAIN
Arch. Widespread, often in small quantity, in dry soils. Most frequent on the chalk, where it occurs in grassland and in woodland rides. Also found on gravelly soils. Map 177

V. bonariensis L.
ARGENTINIAN VERVAIN
C. First recorded in 2000, on waste ground near a supermarket at Freshwater 3486, MB. Commonly grown in gardens where it can become invasive.

122. LAMIACEAE
Dead-nettle family

Stachys officinalis (L.) Trevis.
BETONY
N. A plant of rides in ancient woodland and neutral to acid unimproved grasslands. This plant is most frequent in the ancient woodlands and clay heaths north of the chalk but it does extend onto the chalk in several sites, notably on the exposed West Wight downs from Brook to the Needles headland. In places, it is a conspicuous feature of the grassland, similar to the calcareous grasslands in Purbeck and further west in Dorset and Wiltshire. It is a species of unimproved habitats with poor powers of dispersal, and is in decline, particularly so in the southern half of the Island, where suitable habitat is scarce and it is present in very small quantity in its surviving sites. Map 178

S. byzantina K. Koch
LAMB'S-EAR
E. Occasionally established close to habitation. Well established on Brading churchyard wall 6087, SB 1995. There is an old record of this garden escape from a chalk pit above Steephill by Stratton in 1909.

S. sylvatica L.
HEDGE WOUNDWORT
N. A common and widespread plant of hedge banks, woodland edges and waste ground.

S. x ambigua Sm.
HYBRID WOUNDWORT
N. Occasional in hedge banks, cultivated ground and ditches generally in the presence of one of the parents. Sudmoor 3983, PS 2000; Hardingshute 5988, CP 1999; Newport 4989, CP 1997. First recorded from Norton and Northwood in Drabble & Long 1931.

S. palustris L.
MARSH WOUNDWORT
N. Widespread but scattered on river banks and ditches, and in marshy places. Also found locally on the borders of arable fields in drier places.

[***S. annua*** (L.) L.
ANNUAL YELLOW-WOUNDWORT
C. Extinct. A single record from Cowes in Drabble & Long, 1931.]

S. arvensis (L.) L.
FIELD WOUNDWORT
Arch. Local and generally in small quantity in arable fields, generally avoiding the chalk. This plant has declined in the past one hundred years. Map 179

Ballota nigra L. **ssp. *foetida*** (Vis) Hayek
BLACK HOREHOUND
Arch. Common and widespread in hedge banks and waste ground.

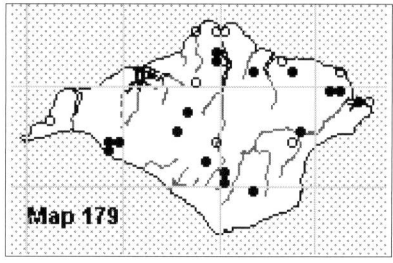
S. arvensis

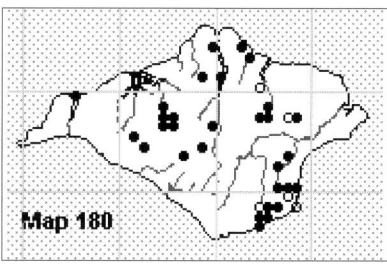
Lamiastrum galeobdolon

Lamiastrum galeobdolon (L.) Ehrend. & Polatschek
YELLOW ARCHANGEL
Ssp. *montanum* (Pers.) Ehrend. & Polatschek N. A locally frequent plant of ancient and sometimes secondary woodlands, particularly on damp and calcareous soils. Map 180 Ssp. *argentatum* (Smejkal) Stace E. Spreading invasively from gardens and from dumped garden waste. First recorded in 1996 but increasing rapidly. It has been recorded as actively colonising ancient woodland at New Copse, Fishbourne 5592, America Wood 5682, Firestone Copse 5591, Woodhouse Copse 5393 and Youngwoods Copse 5785. Map 180

Lamium album L.
WHITE DEAD-NETTLE
Arch. Frequent on verges, hedge banks, waysides and waste ground.

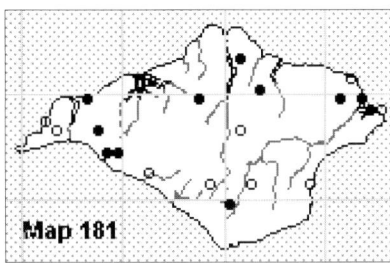
L. hybridum

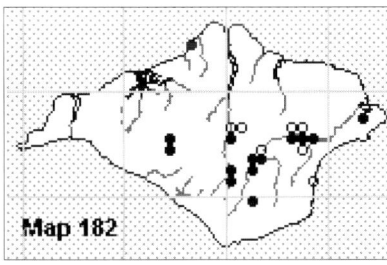
G. tetrahit

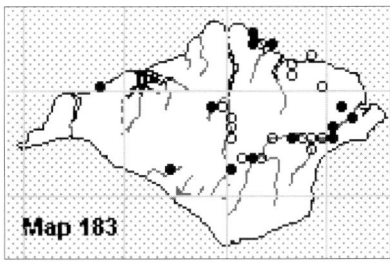
Scutellaria galericulata

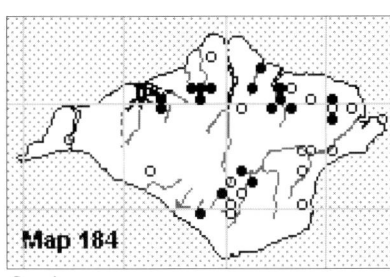
S. minor

L. maculatum (L.) L.
SPOTTED DEAD-NETTLE
E. A garden escape sometimes naturalised on verges near houses or on waste ground from dumped garden refuse. First recorded 1966 on Ventnor cliffs.

L. purpureum L.
RED DEAD-NETTLE
Arch. Widespread and abundant on waste and cultivated ground.

L. hybridum Vill.
CUT-LEAVED DEAD-NETTLE
Arch. An infrequent, but probably under-recorded, plant of cultivated ground. Two reliable areas are sandy fields in the vicinity of Brook and sandy ground around Bembridge Harbour. Several new records were made on disturbed arable ground when an extensive series of sewerage pipes were laid in 2000. Map 181

L. amplexicaule L.
HENBIT DEAD-NETTLE
Arch. A plant of waste and cultivated ground on dry soils. It has a scattered distribution and is most frequent on the sandy soils of the southern half of the Island.

[*Galeopsis angustifolia* Ehrh. ex Hoffm.
RED HEMP-NETTLE
Arch. Extinct. Historically, this was a common arable weed of cornfields on chalk or sandy soils. By the middle of the twentieth century it had become very rare, and there are only two post 1950 records. Collected from a cornfield at East Ashey Farm 5888 in 1966 (BW, hb CRP) and about one hundred plants found in a cornfield on Arreton Down 540874 in 1975 (RK, conf. J.E. Lousley).]

(*G. speciosa* Mill.
LARGE-FLOWERED HEMP-NETTLE
Arch. There is a single unconfirmed record from St George's Down in 1895 (Stratton, 1909).)

G. tetrahit L.
COMMON HEMP-NETTLE
N. Occurs in two distinct habitats, namely arable fields and wet grassland and streamsides. Widespread but local, or under recorded. The map represents the aggregate species. Map 182

G. bifida Boenn.
BIFID HEMP-NETTLE
N. In similar places to the above, but under-recorded. Scotland Farm, Godshill 5282, AC 2000. During a BSBI field meeting to the Island in 1965, it was considered to occur fairly generally in many parts of the Island.

Marrubium vulgare L.
WHITE HOREHOUND
N. Considered to be native on chalk cliffs, particularly in disturbed ground around rabbit burrows. Frequent on the cliff edge at West High Down from the Tennyson monument westwards to above Main Bench 3084, 3184, 3285. Also occurs in very small quantity in a similar situation on Culver Down 6385 CP 1997, and on a rocky knoll just south of Windy Gap car park at St Catherine's 4975, JO 1998. It is occasionally recorded from other sites where it is likely to be casual. However, it was found near a chalk pit on Chillerton Down 4782, BS 1993, and this might be a native site. Bromfield (1856) considered it to be abundant on many of the downs, 'usually along the earthen or stone fences that traverse them'.

Scutellaria galericulata L.
SKULLCAP
N. Local on banks of streams and by ponds but not common. Most sites are in the Eastern Yar valley at Newchurch Marshes, Kennerley Moor, and pools on Brading Marshes. Also found in the Medina valley, but in recent years only at The Wilderness in overgrown tall fen 5082, CP 1996. Present on the Grange Chine brook at Yafford 4482, MB 1999. Skullcap also occurs in a rather different habitat, in brackish wet ground behind shingle banks on the Solent shore of the Island. It has been recorded in this situation from the following sites: Bouldnor shore 3790, CP 2001; Osborne beach 5295, BS 1988; Woodside shore 5494, CP 2001; Quarr Abbey foreshore 5692, BS 1978; and Player's beach, west of Ryde 5892, CP 1970. Drabble & Long (1931) referred to plants from these sites as var. *littoralis*. Map 183

S. minor Huds.
LESSER SKULLCAP
N. Although undoubtedly declined, particularly from boggy ground in the south of the Island, this plant is still widespread in wet rides in ancient woodland and in damp areas in secondary woodland developed over heath. It is abundant in many rides in Parkhurst Forest. Probably the only sites where it survives in open wet ground are at Bohemia Bog 5183, CP 2001 and Cridmore Bog 4981, TH 1999. Map 184

Teucrium scorodonia L.
WOOD SAGE
N. A widespread and frequent plant of open well drained woods, plantations, scrubby areas and hedge banks.

(*T. chamaedrys* L.
WALL GERMANDER
Bromfield (1856) refers to it having grown at Carisbrooke Castle (*Hants Repos.* 1799) but he was unable to find it here, and there are no further records. The castle walls would have been a likely site for this plant in the late eighteenth century but, if present, it would have been an established escape from a physic garden, perhaps within the castle.)

Ajuga reptans L.
BUGLE
N. Frequent in damp woods, old meadows with impeded drainage and roadside verges. Pink flowered plants are not uncommon; white flowered plants less so.

(*A. chamaepitys* (L.) Schreb.
GROUND-PINE
N. Extinct. The only record is from Bromfield (1856) who said 'reported to me as growing in fields about Week Farm, near Niton, along with *Melampyrum arvense*, but, though a very likely station to produce it, this species has never occurred to my observation there or elsewhere in the island.')

Nepeta cataria L.
CAT-MINT
Arch. A rare plant of hedgerows and road banks, on chalk or gravelly soils. Formerly an occasional plant with a scattered distribution, by the time of Bevis *et al* (1978) it was considered to have become extinct. However, it has since been found in small quantity at three sites. Freshwater Bay 3485, RPB 1981; 4 to 6 plants on chalk bank at White Lane, Bowcombe 4687, an historic site, SB 1997; single plant on gravelly bank at Embankment Road, Bembridge 6388, AC 1998.

N. x *faassenii* Bergmans ex Stearn
GARDEN CAT-MINT
E. Spreading from garden into sand at St Helen's Duver 6388, BS 1996.

Glechoma hederacea L.
GROUND-IVY
N. Common and widespread in woods, hedgerows, scrub and in grasslands on calcareous and clay soils.

Prunella vulgaris L.
SELFHEAL
N. Common and widespread in grasslands, verges, woodland rides and waste places.

Melissa officinalis L.
BALM
E. An increasing garden outcast in waste places and waysides, generally near habitations. Known as a naturalised plant by Bromfield (1856).

Satureja montana L.
WINTER SAVORY
C. A casual garden escape. Single plant on wall of old radar station, Knowles Farm 4975, DB 1997.

Clinopodium menthifolium (Host) Stace
WOOD CALAMINT
N. This Island speciality continues to survive in woodland edge habitat in the Rowridge valley, 4586, its sole British station. There is some dispute as to when it was first found here. Apparently, Sir David Brewster and other eminent naturalists discovered the plant in 1841. However, credit is usually given to Dr. Bromfield, who saw it on 29 August 1843 and found it 'growing amongst the long herbage and under the shade of bushes, in vast quantity, for a great part of the way towards the head of the vale, scattered over the hillside copses wherever there is shade and shelter sufficient, but . . . always avoiding open and exposed situations, or where there is plenty of herbage and undergrowth' (Bromfield, 1843).

This description suggests a woodland edge habitat and, when More (1898) first came across it with his sister in 1853, he described it as 'in some places quite colouring the more open spaces in the copse, close under the down'. With lack of woodland management, the plant declined in the twentieth century.

Edward Lousley noticed a sharp decline between his visits of 1930 and 1947, and local botanists also recorded declines about this time. Lousley (1950) suggested that, 'the shrubs had grown up there and there is reason to believe that the Greater Calamint does not survive under such conditions'. The situation deteriorated until 1959 when work to the highway resulted in lay-bys being cut into the chalk at the edge of the wood. Plants colonised the bare chalk and the population increased. Since 1960, these lay-by populations have been maintained by annual volunteer working parties of the Isle of Wight Natural History and Archaeological Society and, in recent years, the plants have shown a modest increase. The main population remains, however, concentrated within two contiguous lay-bys.

Open ground, free of competing vegetation, appears to be the key to the recovery of this plant. In cultivation, the plant spreads rapidly both by seed and vegetatively. Bromfield (1856) remarked that 'this beautiful species grows readily from slips, and when treated as a greenhouse plant, or kept entirely within doors, becomes extremely showy'. Wood Calamint is fully protected under the Wildlife and Countryside Act (1981).

C. ascendens (Jord.) Samp.
COMMON CALAMINT
N. Very local on calcareous soils on dry bare hedge banks and old walls. Although it has declined, it continues to survive in many of the Bromfield stations. Extant sites are Thorley churchyard wall 3788, MB 2000; roadside wall at Wellow 3888, BS 1989; roadside at Plaish 4787, SB 2002; Shalfleet churchyard 4189, MB 2000; Carisbrooke Castle 4887, BS 1991; Quarr Abbey ruins 5692, CP 1989; and rocky ground at Ventnor 5777, AB 1994 and Bonchurch 5778, JS 1993. Map 185

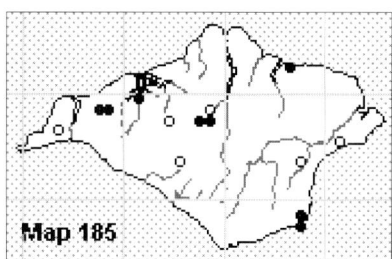

C. ascendens

Wood Calamint (*Clinopodium menthifolium*) is the only wild plant found on the Island and nowhere else in Britain. Dewberry and Nettle-leaved Bellflower, which grow with it, are shown as line drawings. Painting by Dolly Norledge.

ABOVE Shaggy Mouse-ear hawkweed (*Pilosella peleteriana*) has one of its few British sites on steep south-facing chalk cliffs near Freshwater. (KM)

TOP LEFT Red Valerian (*Centranthus ruber*), spectacularly naturalised on the cliffs at Wheeler's Bay, Ventnor, where the flowers attract large numbers of visiting insects. (AB)

TOP RIGHT Marsh Helleborines (*Epipactis palustris*) grow in fluctuating numbers on the unstable coastal ledges at Luccombe. (AB)

LEFT Oxtongue Broomrape (*Orobanche artemisiae-campestris*) has been known from the chalk cliffs west of Freshwater since Victorian times but is otherwise only known from cliffs in Kent. (SB)

ABOVE Sea Spleenwort (*Asplenium marinum*), a maritime fern with a small population on a single boulder at St Catherine's Point, is found no further east along the south coast of England. (AB)

BELOW Sea Campion (*Silene uniflora*) is a scarce Island plant but grows in quantity on Gore Cliff above St Catherine's Point. (KM)

ABOVE Royal Fern (*Osmunda regalis*) survives on several inaccessible flushed coastal cliffs, as here on Lake cliffs. (AB)

BELOW Vegetation on the inner cliff above St Catherine's Point, more typical of coastal communities further west of the Island with Sea Pink (*Armeria maritima*) and Sea Campion (*Silene uniflora*). Chalk cliffs at Freshwater in the background. (KM)

ABOVE Sea Knotgrass (*Polygonum maritimum*), a recent colonist to our shores from further south, growing above the tide line at Thorness Bay (KR)

RIGHT Small Cord-grass (*Spartina maritima*), once our common native cord-grass, still grows in quantity on saltmarshes in the Newtown Estuary, but is otherwise almost extinct along the south coast of England. (KM)

BELOW Mixed saltmarsh in the Newtown estuary at its most colourful with a fine show of flowering Common Sea-lavender *(Limonium vulgare)*. (KM)

SALTMARH AND SAND DUNES / 5

ABOVE Sea Holly (*Eryngium maritimum*), a declining coastal species, is still frequent on the tiny sandy headland at Norton Spit. (KM)

BELOW Autumn Squill (*Scilla autumnalis*) grows in abundance on the stabilised sand dune grassland at St Helen's Duver but does not occur in neighbouring coastal counties. (KM; inset AB)

ABOVE An autumnal view of species-rich chalk turf on Brook Down, looking westwards, with Autumn Lady's-tresses (*Spiranthes spiralis*), Burnet Saxifrage (*Pimpinella saxifraga*) and Small Scabious (*Scabiosa columbaria*). (KM)

LEFT Early Gentian (*Gentianella anglica*) grows in abundance in good years on the downs of the Island, one of the main headquarters of this tiny British endemic. (KM)

BELOW Bluebells (*Hyacinthoides non-scripta*) provide a spectacular spring show on the leached soils of the north slopes of Ventnor Downs before disappearing beneath unfolding Bracken fronds. (AB)

ABOVE A fine display of Horseshoe Vetch (*Hippocrepis comosa*) has colonised disturbed ground on St Boniface Down, Ventnor, an area now kept free of colonising Holm Oak (*Quercus ilex*) by grazing feral goats. (AB)

BELOW Thin broken turf near the cliff edge on Tennyson Down provides ideal conditions for the richest maritime chalk grassland lichen community in the country, dominated by the lemon-yellow *Fulgensia fulgens* and greyish-cream *Squamarina cartlaginea*, here growing on cushions of the moss *Trichostomum crispulum*. (KM)

ABOVE *Microbryum curvicolle* is an uncommon moss of open chalky grassland, here at the Needles headland. (JDS)

ABOVE A fine display of Horseshoe Vetch (*Hippocrepis comosa*) has colonised disturbed ground on St Boniface Down, Ventnor, an area now kept free of colonising Holm Oak (*Quercus ilex*) by grazing feral goats. (AB)

BELOW Thin broken turf near the cliff edge on Tennyson Down provides ideal conditions for the richest maritime chalk grassland lichen community in the country, dominated by the lemon-yellow *Fulgensia fulgens* and greyish-cream *Squamarina cartlaginea*, here growing on cushions of the moss *Trichostomum crispulum*. (KM)

ABOVE *Microbryum curvicolle* is an uncommon moss of open chalky grassland, here at the Needles headland. (JDS)

ABOVE Characteristic neutral grassland flora with Dyer's Greenweed (*Genista tinctoria*) and Corky-fruited Water-dropwort (*Oenanthe pimpinelloides*) growing with Oxeye Daisy (*Leucanthemum vulgare*) and Common Spotted-orchid (*Dactylorhiza fuchsii*). Meadow by the Medina at Dodnor. (KM)

ABOVE RIGHT Adder's-tongue (*Ophioglossum vulgatum*) is a small, distinctive fern of unimproved grasslands. (SB)

RIGHT Yarrow Broomrape (*Orobanche purpurea*) is a colourful rare plant of dry grasslands, here growing in cliff-top grassland at Redcliff, near Sandown. (AB)

ABOVE Field Cow-wheat (*Melampyrum arvense*), a colourful plant confined to two calcareous grassland sites above the Undercliff. (KM)

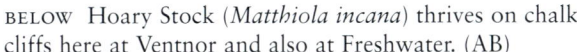

BELOW Hoary Stock (*Matthiola incana*) thrives on chalk cliffs here at Ventnor and also at Freshwater. (AB)

ABOVE Italian Lords-and-Ladies (*Arum neglectum* ssp. *neglectum*), a rare plant and a shy flowerer, which thrives in woodland along the Undercliff. (KM)

BELOW South-facing chalky slopes above St Catherine's Point are home to the tiny, very rare moss, *Acaulon triquetrum*, with distinctively coloured shoots in early spring. (JDS; inset by RDP)

ABOVE Bohemia Bog is a tiny, but incredibly rich, acidic wetland site, here being visited by members of the British Bryological Society in March 2002. (JDS)

BELOW All of our sphagnum mosses have experienced dramatic declines; *Sphagnum cuspidatum* is one of the rarest today, photographed here at its principal site of Bohemia Bog. (JDS)

BELOW Round-leaved Crowfoot (*Ranunculus omiophyllus*) is a much-declined species of acidic ditches, photographed here at Bohemia Bog. (CNP)

ABOVE *Hookeria lucens* is a distinctive moss of wet woodlands, here growing on a stream bank of the Buddle Brook at Brighstone. (JDS)

RIGHT Tufa-forming mosses have encouraged the build-up of deposits around the lime-rich waters of the water-spout in Fountain Cottage at Bonchurch, creating a feature admired since Victorian times. (DC)

BELOW A water-colour painting of the now extinct Bladderwort (*Utricularia australis*) flowering in the ditches at Freshwater Marshes in July 1872. Painted by Elizabeth Hood, who lived at Norton, in the margin of her flower book. (AB)

BELOW Field Gentian (*Gentianella campestris*) was only ever known from Colwell Common in the nineteenth century. This painting, dated 8 October 1861 in the margin of Bentham's *Flora*, was by Elizabeth Hood who painted very accurate pictures of West Wight flowers. (AB)

ABOVE Wild Daffodils (*Narcissus pseudonarcissus*) are an early spring feature of many coppiced woodlands on the Island. They are growing here at Centurion's Copse. (AB)

LEFT Wood Horsetail (*Equisetum sylvaticum*) is a delicate and rare northern plant of boggy, acidic woodland, photographed here at Apse Heath Withy Bed. (KM)

BELOW Wild Service-tree (*Sorbus torminalis*), a tree of ancient woodlands fringing the Solent coast, with colourful autumn foliage and berries which used to be gathered and eaten. Photographed here in Firestone Copse. (KM)

ABOVE Early-purple Orchids (*Orchis mascula*), known locally as 'kettle-cases', are commonly found in ancient woodlands in the spring. (AB)

ABOVE The tree lungwort lichen (*Lobaria pulmonaria*), an ancient pasture-woodland species, is a rarity today but grows spectacularly in Northpark Copse where some plants produce orange fruiting bodies. (KM)

BELOW The moss *Zygodon rupestris*, growing here in Briddlesford Copse on oak, is an ancient woodland indicator species. (JDS)

BELOW Green Hellebore (*Helleborus viridis*) is locally a very rare early spring flowering plant of base-rich wet woodland, here in Briddlesford Copse. (KM)

ABOVE A wheat crop at Cridmore full of Cornflower (*Centaurea cyanus*), one of a tiny handful of sites in this country where it persists in arable fields. (SB; KM inset)

BELOW Martin's Ramping-fumitory (*Fumaria reuteri*) is tolerated as a weed in allotment plots at Lake; it is known from only one other British site. (KM)

ABOVE Early Meadow-grass (*Poa infirma*) may well have been over-looked growing early in the year in warm, disturbed coastal sites; once a national rarity confined to Cornwall, it appears to be spreading. (KM)

ABOVE Rue-leaved Saxifrage (*Saxifraga tridactylites*) has its main Island site on the medieval walls of Quarr Abbey. (DDa)

RIGHT Cushions of the south-western maritime lichen, *Roccella phycopsis*, adorn the medieval walls and tower of Godshill Church, our richest churchyard for lichens. (KM)

BELOW *Chenia leptophylla* is a tiny moss which grows in an arable field near Brook, where it was discovered as new to this country in 1964. (JDS)

Narrow-leaved Lungwort (*Pulmonaria longifolia*) is a beautiful spring flower of ancient coppice woodlands in the north of the Island. Painting by Dolly Norledge.

C. vulgare L.
WILD BASIL
N. Widespread on hedge banks, rough grassland, scrub and dry woodland rides. Particularly frequent on calcareous soils but not confined to these.

C. acinos (L.) Kuntze
BASIL THYME
N. A rare plant of bare, chalky ground, much declined in the past one hundred years. The most recent record from arable ground is from a field at East Ashey Farm in 1966 (BW). The only site known to Bevis *et al* (1978) was an old chalk pit at New Barn Farm, Calbourne but it has not been seen here for many years. Recent sites are an old, north facing chalk quarry on Nunwell Down 5987, AC 1997; and Michal Morey's Hump on Arreton Down 5387, CP 1998.

Origanum vulgare L.
WILD MARJORAM
N. Frequent in dry grassland, scrub and hedge banks on calcareous soils. Also sometimes found elsewhere on old walls and on clay soils by the coast.

Thymus vulgaris L.
GARDEN THYME
E. A single, well-established plant in open slipped land at Bonchurch Landslip, well away from any habitation 5878, CP 1989. Bromfield (1856) recorded this plant as naturalised on the wall of a garden at Niton.

[*T. x citriodorus* Pers.
LEMON THYME
E. Extinct. There is a record in Bromfield (1856) from a roadside near the Sandrock Spring, Niton.]

[*T. pulegioides* L.
LARGE THYME
N. Extinct? A rare plant of thin, dry grassland. The heathland ecotype is considered to be extinct. It was last recorded from a small gravel pit at the east end of St George's Down, 5286. This is the site given in Bevis *et al* (1978) and it survived here in declining numbers until 1992, when there was a single plant overgrown by brambles. The status of the calcareous ecotype is unknown, but there is a record from Limerstone Down west 4384 in 1987, RDP.]

T. polytrichus A. Kern. ex Borbàs
WILD THYME
N. Frequent in short turf of dry grasslands. Most abundant on the chalk downs but also present more locally in heathy vegetation on gravels and in free-draining coastal grassland along the south-west coast and on St Helen's Duver.

Lycopus europaeus L.
GYPSYWORT
N. Locally frequent by streams, ponds, marshes, and in wet woodlands. Particularly frequent in the Eastern Yar valley but found in scattered localities right across the Island. It grows in dune slacks on St Helen's Duver.

Mentha arvensis L.
CORN MINT
N. Locally frequent in damp woodland rides and damp borders of arable fields and ditches on a variety of soil types, but somewhat under recorded.

M. x verticillata L.
WHORLED MINT
N. An uncommon natural hybrid occurring in ditches and disturbed grassy verges. Several confirmed records, but under-recorded and no recent records.

M. x smithiana R.A. Graham
TALL MINT
E. Occasional garden outcast in damp places and waste ground. Recent records include banks of river Yar at Roud 5180, BS 1996; marshy grassland at Whitwell 5268, CP 2000; near Sandown sewage works 6085, GT 1998.

M. aquatica L.
WATER MINT
N. Locally frequent and widespread on stream banks, ponds, marshy grassland and damp woodland rides.

M. x piperita L.
PEPPERMINT
E. An established garden escape, rare and generally growing on stream and ditch banks or as a garden outcast. Near St Mary's hospital, Newport 4990, PS 2000; Bouldnor 3789, PS 1999; Stenbury near Whitwell 5279, GT 1998; Yarmouth 3589, PS 1998.

M. spicata L.
SPEAR MINT
Arch. An increasing garden outcast, naturalised in rough and waste ground, generally near habitations. First recorded by Bromfield (1856).

M. x villosa Huds.
APPLE-MINT
E. Occasional garden escape on rough and waste ground and river banks. Brook 3883, PS 1997; Forelands 6587, AC 1996; by Afton Road, Freshwater 3485, BS 1992; by Gatcombe Mill 4985, BS & SB 2002.

M. suaveolens Ehrh.
ROUND-LEAVED MINT
(local: horse mint)
N. A native or long-established plant found widely scattered on verges and waste places and also more frequently in damp pastures in the Eastern Yar valley. Sites include pastures between Horringford and Perreton Farm 5385 to 5485, CP 1996; Great Budbridge 5382, CP 2000; and Bagwich to Lavender's Farm, Godshill 5181 to 5182 BS 1992. There are old records for *M. rotundifolia* but it is unclear whether they refer to this species or to Apple-mint.

[*M. pulegium* L.
PENNYROYAL
N. Extinct. A former rare native of grazed and trampled, seasonally inundated grassland. The best-known site was St Helen's Green. Bromfield (1856) found it to be very sparing here in 1838/39 but reported that villagers found it to be abundant in some years. Also recorded in 1883 from Apse (Stratton, 1909) and this is probably the site described as Pine Tree Hill at Ninham in 1904. E.H. White wrote of this site in his diary, 'This is no doubt pennyroyal. August 1905. Has been ascertained as such by Mr. Stratton'. There are no further records.]

M. requienii Benth.
CORSICAN MINT
E. Well established between paving

stones at Godshill model village but not cultivated there 5281, PS 2000.

Lavandula x *intermedia*
GARDEN LAVENDER
C. Found growing in a pavement crack at Carisbrooke Road, Newport 4988, SB 2000.

Rosmarinus officinalis L.
ROSEMARY
C. Occasionally self sown on walls and rough ground close to habitation. Seaview 6291, PS 1999; Freshwater church 3487, PS 1999.

[*Salvia pratensis* L.
MEADOW CLARY
C. Extinct. Only known from records in Bromfield (1856), as casual plants, which were not found subsequently. One plant found in a pasture field at Niton in 1854 and a second subsequently (this could be the same site as that of 'Mrs. Vine's grounds at Puckaster') and found in an old chalk pit in Appuldurcombe Park in 1838.]

S. verbenaca L.
WILD CLARY
N. A much declined plant of dry grassy banks and bare soil, with extant sites usually on chalk. Today it is very locally abundant. It is common in the grassland and on the old walls at Carisbrooke Castle 4887, MB 1996, a long established site. It is also locally frequent in coastal chalk grassland in and around Ventnor 5577 to 5677 but no longer grows on the downs behind Ventnor. Formerly found in suitable spots along the Undercliff from Luccombe through to Niton and still survives on the cliff at St Lawrence 5376, RG 1999. Wild Clary also shows an association with old churchyards as it does in some other southern counties. Recorded yards, with last recorded dates, are:

All Saints, Freshwater churchyard (1864); Whippingham churchyard (1883); and Brading churchyard, where it still survives. Map 186

123. HIPPURIDACEAE
Mare's-tail family

Hippuris vulgaris L.
MARE'S-TAIL
N. A rare plant of ditches and large ponds, only ever known from the Bembridge Harbour area where it was first recorded in Bromfield (1856). Frequent in a pond, now used for fishing, at Carpenters 6288, DDa 2001; also in main ditch here flowing beneath railway line, BS 1992; in a wet ditch at Harbour Farm 6387, BS 1985; and in a nearby wet area at the southern end of Harbour Farm ponds, CP 1983, now invaded by reeds. Introduced into a small pond in Firestone Copse 5591, DDa 2002.

124. CALLITRICHACEAE
Water-starwort family

Callitriche stagnalis Scop.
COMMON WATER-STARWORT
N. By far the commonest species, frequent and widespread in ponds, ditches and wet tracks in woods and meadows.

C. platycarpa Kütz.
VARIOUS-LEAVED WATER-STARWORT
N. Occurs in rivers, streams and ditches but status uncertain owing to confusion with other species. Recorded from the Medina at Blackwater 5086, and Shide, 5087/88; and the River Yar at Langbridge 5585, all PDG 1978; ditch at Carpenters, St Helens 6188, AC 1995.

C. obtusangula Le Gall
BLUNT-FRUITED WATER-STARWORT
N. Local in ponds and streams. Records include farm ponds at Thorncross 4381 and Compton Grange 3884, CDP & TDD 1996; ditch on Sandown levels 6085, CP 1997; River Medina between Shide and Blackwater 5087, CP 1999; and Shalcombe Manor pond, Chessell 3985, CP 2000.

C. hamulata sens. lat.
INTERMEDIATE WATER-STARWORT
N. Scarce in rather acid ponds and ditches. Recorded by Bevis *et al* (1978) from a ditch at Alverstone and present in nearby Scotchells Brook 5885, CP 1989; pond in slumped bowl at Redcliff 6285, AC 1996; ditch at Carpenters, St Helens 6187, AC 1999. Likely to be under-recorded, due to identification difficulties.

125. PLANTAGINACEAE
Plantain family

Plantago coronopus L.
BUCK'S-HORN PLANTAIN
N. Common right around the coast in short turf or bare ground. More local inland but often frequent on dry soils.

P. maritima L.
SEA PLANTAIN
N. Frequent on muddy shores and salt marshes in all our estuaries. There are old records from Shippards Chine and Brook Chine on the south-west coast (JB, 1969) but, despite searching, it has not been found here recently.

P. major L. ssp. *major*
GREATER PLANTAIN
N. Common along verges and tracks, in lawns and bare ground everywhere.

P. media L.
HOARY PLANTAIN
N. Common in chalk grassland and in relict limestone grassland. Also found, much more locally, in base enriched short grassland elsewhere. Often surviving in churchyards and cemeteries. Map 187

P. lanceolata L.
RIBWORT PLANTAIN
(local: soldiers!)
N. Common everywhere in grasslands on all soils and in disturbed land. As

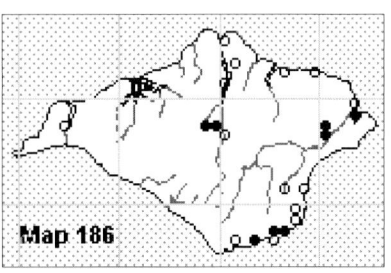

S. verbenaca

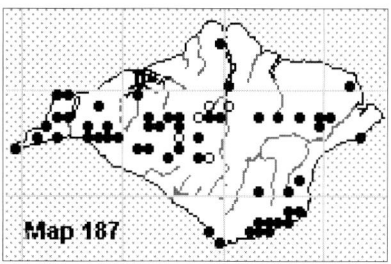

P. media

in many other places, a popular child's game was to fire the flowering heads by wrapping the pliant stems around the neck of the flowerhead and pulling sharply.

[*P. arenaria* Waldst. & Kit.
BRANCHED PLANTAIN
C. Extinct. Bromfield (1856) recorded a few plants found in Ryde Dover in 1843. No other records.]

(*P. afra* L.
GLANDULAR PLANTAIN
C. A plant considered to be this species, but not confirmed, was recorded at Brighstone in 1964, DF.)

126. BUDDLEJACEAE
Butterfly-bush family

Buddleja davidii Franch.
BUTTERFLY-BUSH
E. An increasing and often invasive introduction found on waste ground, walls, old chalk pits and undercliffs. First recorded in 1930 on disturbed ground following a major clifffall in 1928 at Gore Cliff, near Blackgang; C. Marquand found five plants, including two in flower, growing in different parts of the landslide (J. Botany LXVIII: 377).

127. OLEACEAE
Ash family

Forsythia x *intermedia* Zabel
FORSYTHIA
S. A relic of cultivation. Surviving on a slumped cliff at Forelands 6385, AC 1997.

Fraxinus excelsior L.
ASH
N. Common in woods, hedgerows and scrub, regenerating freely. Often dominant in woods on the chalk. Bromfield (1856) wrote 'between Shanklin and Luccombe are many fine trees' and some large trees still survive in this area. The biggest trees are usually old pollards but a few very large maiden trees survive in woods.

A remarkable and strange custom was reported in the *Isle of Wight Observer* for 31 March 1877, as a cure for an illegitimate boy born with a rupture in a cottage near Merrie Gardens at Lake. A young ash tree was planted in the garden and the stem cleft and held open with wedges whilst the ruptured child, stark naked, was passed out of the window, through the open parts of the stem of the tree, and then through the doorway of the house and this was repeated five times. The charm was carefully bound up by splicing the split parts of the stem, and as the ash healed, so supposedly did the rupture. This is the same custom that Gilbert White wrote about in *The Natural History of Selbourne*.

Syringa vulgaris L.
LILAC
E. A relic of cultivation but also sometimes naturalising by suckering in hedgerows and waste ground. Established on cliffs at Ventnor 5677, CP 2000.

Ligustrum vulgare L.
WILD PRIVET
N. Frequent in hedgerows, woodland edges and scrub. It is least common on the sandy soils south of the chalk, including the south-west coastline. Bromfield (1856) recorded that, 'the long straight shoots are used in the island, from their toughness and pliability, in tying small bundles or faggots for firing by the country people'.

L. ovalifolium Hassk.
GARDEN PRIVET
S. Not uncommon as a relic of cultivation but occasionally an established garden escape, such as on Lake Cliffs.

128. SCROPHULARIACEAE
Figwort family

Verbascum blattaria L.
MOTH MULLEIN
C. Rare on waste and rough ground, generally about habitation. Bevis *et al* (1978) give several localities including a cleared site in Carisbrooke Road, Newport (1965), Cowes refuse tip near Medham (1966) and dumped spoil by Shide chalkpit (1967). No recent records. First recorded by Bromfield from Carisbrooke in 1839.

V. virgatum Stokes
TWIGGY MULLEIN
C. A rare introduction on waste ground and road verges. Established on the Medina Way roadside verge, Newport 4989 SB 1995 where it was persisting in 2002. First recorded from Cowes by Drabble & Long (1931).

V. phlomoides L.
ORANGE MULLEIN
C. A rare casual first recorded in 1964. It was found growing by the Newport – Cowes cycleway near Arctic Road 4994, SB 1996 and it has persisted here for a number of years.

V. thapsus L.
GREAT MULLEIN
(local: shepherd's club; flannel plant)
N. Frequent in disturbed grassland and waste ground on dry soil. It is commonest on the chalk.

V. nigrum L.
DARK MULLEIN
N. Found on roadside verges on chalk and other light soils. A rarity on the Island, as it is in southern Hampshire and Purbeck. Since the earliest records in Snooke (1823) it has been almost entirely confined to verges around Downend 5387, Arreton Cross 5386, and Crouchers Cross 5285, and it persists here on sandy soils in small quantity 5285 and 5286. A few plants found on a verge in Kemming Road, Whitwell, 5177, in 1982 and still here in 1990, JS.

[*V. pulverulentum* Vill.
HOARY MULLEIN
C. Extinct casual recorded from Bonchurch Landslip, EHW 1925.]

Scrophularia nodosa L.
COMMON FIGWORT
N. Frequent and widespread in damp woods, hedge banks and waste ground.

S. auriculata L.
WATER FIGWORT
N. Frequent on stream banks and many damp spots. It is particularly widespread in all our river valleys.

M. x *robertsii* Silverside
HYBRID MONKEYFLOWER
E. Established by stream banks. Early records refer to both *M. luteus* and *M. guttatus*, but material has not been examined critically and all Island

records are considered here, under the hybrid. The principal extant site for this hybrid is in the Caul Bourne at Barrington Row 4286 (det. Alan Silverside). The first record from this site was in 1924 when it was considered to be abundant. A report in 1929 referred to the plants here having spread rapidly in recent years. Also recorded from Shalcombe Manor pond banks 3985, IR 1991; Yafford 4481, MB 1999 and at Ventnor Cascade 5677, CP 2000. Present in Brook Green stream 3984 AJ 1967, but since lost.

Historically, Monkeyflower was more frequent. It occurred in the Alverstone stream FS, 1904 and was still present near Alverstone Mill in 1925; it was abundant in the Yar tributary from French Mill to Bobberstone FS, 1891 and EHW, 1904; it was recorded from Gatcombe FS, 1904; and from The Wilderness FS, 1912.

Antirrhinum majus L.
SNAPDRAGON
E. A frequent garden escape, sometimes well established on old walls and quarries. Bevis *et al* (1978) record it as established 'in a veritable cascade on the cliff in Freshwater Bay'. First recorded in Snooke (1823) on Yarmouth Castle walls.

Chaenorhinum minus (L.) Lange
SMALL TOADFLAX
Arch. Local on the chalk as an arable weed and in old quarries. Formerly frequent on clinker of old railways, but declining in this habitat with the loss of old railway tracks. In 2000 appeared in several sites on clay soils on the north of the Island when trenching associated with laying sewerage pipelines brought deep soil to the surface. Map 188

Misopates orontium (L.) Raf.
WEASEL'S-SNOUT
Arch. A much declined weed of cultivated ground on light, gravelly soils. Now very rare as an arable weed, but surviving locally as a weed in gardens and allotments. Frequent at Pan allotments, 5088 BG 1999; gardens in Clatterford Road, Carisbrooke 4888, BS 1995 and elsewhere in the area; Afton Park nursery 3486, PS 1999 and gardens in the general area; gardens at Place Road, Cowes 4895, CP 1998 and turned ground on nearby Cockleton Farm. Pale-flowered plants are found in small quantity in three fields to the north-east of Rookley 5184, RPA 2002.

Cymbalaria muralis P.Gaertn., B.Mey. & Scherb. ssp. *muralis*
IVY-LEAVED TOADFLAX
(local: roving jenny; roving sailor)
E. Common and widespread on old walls, dry banks and waste places. Already well established at the time that Bromfield wrote his *Flora*.

Kickxia elatine (L.) Dumort.
SHARP-LEAVED FLUELLEN
Arch. This plant remains a reasonably frequent arable weed on a range of light soils. Map 189

K. spuria (L.) Dumort.
ROUND-LEAVED FLUELLEN
Arch. An arable weed, sometimes found growing with the previous species, although less frequently. It shows a stronger tendency to favour chalky soils. Map 190

Linaria vulgaris Mill.
COMMON TOADFLAX
N. Fairly widespread and frequent on verges, waste ground and rough grassland but least common on clay soils. Although still common, it has declined over the past one hundred years.

L. x *sepium* G.J. Allman
COMMON TOADFLAX X PALE TOADFLAX
N. Recorded from near Cowes in 1838 by Bromfield. Not uncommon by the old Cowes railway track near Arctic Road 4994, BS 1974. It may be overlooked; both species still grow together in the Cowes area.

L. purpurea (L.) Mill.
PURPLE TOADFLAX
E. Commonly naturalised on rough ground and walls near habitations. Bromfield (1856) knew it as an established escape around Bonchurch.

L. repens (L.) Mill.
PALE TOADFLAX
Arch. Always local but much declined. First recorded by Englefield (1816) 'in great profusion in the hedges, just about West Cowes, but does not extend above a mile from it in any direction'. Cowes has always been known as a stronghold for this species and it still occurs here, in the vicinity of Pallance Four Cross 4894, on verges and as a garden weed. Cranmore appears to be the other main centre of distribution. It has recently been found growing on a grassy bank at Golden Hill, Freshwater 3387, CL 2002. Map 191

[*L. supina* (L.) Chaz.
PROSTRATE TOADFLAX
C. Extinct. Known from a single record from Apes Down by Mrs. Sandwith in 1933 (hb. K).]

L. maroccana Hook. f.
ANNUAL TOADFLAX
C. A garden escape found growing at the base of Lake Cliffs 5882, BS 1995.

Digitalis purpurea L.
FOXGLOVE
N. A common plant of well-drained acid soils in woodland clearings,

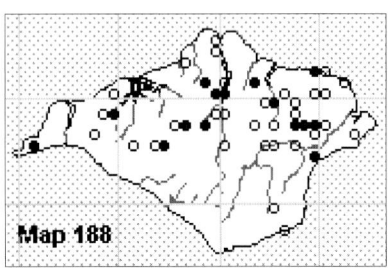

Chaenorhinum minus

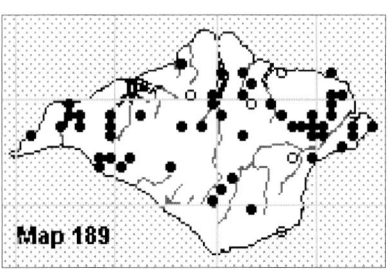

Kickxia elatine

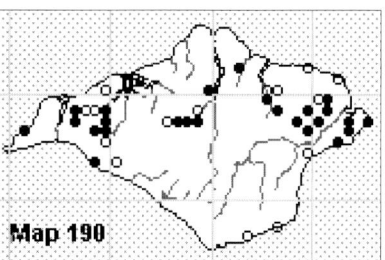

K. spuria

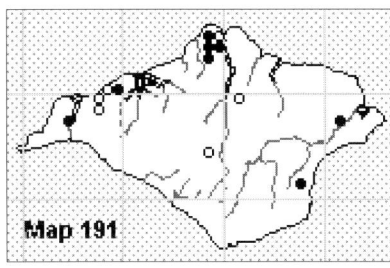
L. repens

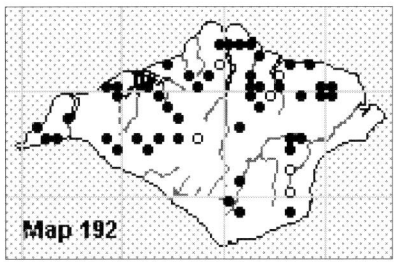
V. officinalis

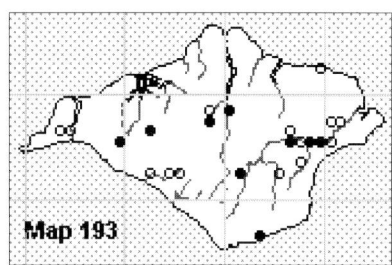
V. anagallis-aquatica

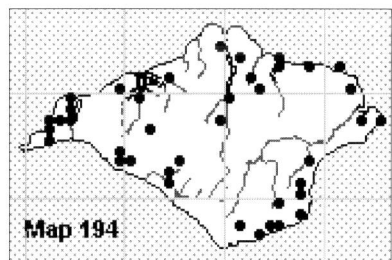
V. filiformis

hedge banks and heathy areas. A children's game used to be popping the unopened flowers. Bevis *et al* (1978) refer to children in the Second World War collecting seed for digitalin. 'It was mainly collected in the Hermitage area, and despite the seed being so fine, the target of 4 lbs. was collected each year from 1941 onwards'.

Veronica serpyllifolia L. ssp. *seryllifolia*
THYME-LEAVED SPEEDWELL
N. Frequent and widespread in short grassland, bare ground in woodland rides, verges and waste ground.

V. officinalis L.
HEATH SPEEDWELL
N. Locally frequent on dry soils on heathland, woodland clearings and dry grassy banks. Restricted to acid and leached soils. Map 192

V. chamaedrys L.
GERMANDER SPEEDWELL
(local: bird's-eyes!)
N. Common and widespread in hedge banks, woods and grasslands.

V. montana L.
WOOD SPEEDWELL
N. Frequent in most of our woods, both ancient and secondary. Also found in verges and shady lanes on the northern half of the Island and occasionally as garden weed.

[*V. scutellata* L.
MARSH SPEEDWELL
N. Extinct. A plant of boggy pastures on acid soils. It persisted longest on Cridmore Bog 4981, where it was last recorded in 1974, FR. Recorded historic sites are Freshwater Marshes, Drabble & Long (1931); Cranmore, Drabble & Long (1931); Bouldnor (possibly the same site as the previous) J. Groves 1923; edges of a pool on Golden Hill, Freshwater, Bromfield 1856; and stream above the mill at Lower Knighton, Wilson Saunders 1856. It is conceivable that the plant was not correctly identified at all of these sites.]

V. beccabunga L.
BROOKLIME
N. Frequent in wet, muddy places in woods, and by streams and ponds right across the Island.

V. anagallis-aquatica L.
BLUE WATER-SPEEDWELL
N. A declining species of streams and ponds, generally only present in small quantity. Long known from a single surviving pool on Bleak Down 5182, CP 1998. Pink Water-speedwell is not recorded from the Island. Map 193

V. arvensis L.
WALL SPEEDWELL
N. Widespread and frequent on walls, short turf on dry banks, anthills, cultivated and waste land.

V. agrestis L.
GREEN FIELD-SPEEDWELL
Arch. A rarity of arable fields and gardens, perhaps under-recorded but certainly scarce. Recent records are sandy ground in a garden centre at Newchurch 5583, PS & EC 1999 and by a path in Yarmouth 3589, PS 1997. First recorded near Godshill in 1930 (Drabble & Long, 1931) and refound at Sainham Farm 5281, BS in 1980. It could not be found here in 1999. Other historic sites, where it may still occur, are Brighstone (1962) and Shepherd's Chine (1958).

V. polita Fr. Grey
FIELD-SPEEDWELL
E. A local but widespread arable and garden weed.

V. persica Poir.
COMMON FIELD-SPEEDWELL
E. Common and widespread in cultivated and waste ground. This plant was first introduced into this country about 1820 and Bromfield (1856) described it as rare.

V. filiformis Sm.
SLENDER SPEEDWELL
E. A locally frequent plant of lawns and churchyards. It must have been overlooked when first introduced here for the first record is in 1962, by which time it is described as spreading in many areas. Map 194

V. hederifolia L.
IVY SPEEDWELL
Arch. Common and widespread on cultivated and waste ground, open woodland, hedge banks and shady lanes. Both ssp. *hederifolia* and ssp. *lucorum* (Klett & Richt.) are recorded, but the majority of records do not distinguish between them.

Hebe salicifolia (G.Forst.) Pennell
KOROMIKO
E. Known only from a single bush, about 1.5m in height, growing in a clearing in Bonchurch Landslip, well away from habitation 5878, CP 1985. The plant has since been lost through ground slumping.

H. x *franciscana* (Eastw.) Souster 'Blue Gem'
HEDGE VERONICA

E. This somewhat frost-tender garden shrub has become established by self-seeding on cliffs at Lake 5882/5983, CP 2000 and Ventnor 5677, CP 2000. It was first noticed as becoming well established at Ventnor, east of The Cascade and below the Winter Gardens in 1974, BS.

Melampyrum arvense L.
FIELD COW-WHEAT
(local: poverty-weed)

N? A rare plant known today from two sites at St Lawrence, a field edge bank, which is a Wildlife Trust reserve, and a stretch of steep, south facing cliff. Field Cow-wheat has not always been so rare. It was first recorded by Snooke (1823), 'by the Inn near Lord Dysart's'; this would have been at the site of Steephill Castle. Bromfield (1856) and Townsend (1883) both give descriptions of its distribution, from which it would appear that it was abundant in cornfields above Steephill and St Lawrence and northwards to Whitwell.

Bromfield also described it as occurring sparingly at Bonchurch. It apparently did not occur below the Undercliff, according to Townsend but he says, 'I saw it in abundance above Steephill and St Lawrence in 1844, in the cornfields and borders of fields and on the cliff banks, descending several yards below the upper edge of the cliff'. Other records also refer to its occurrence in field borders and bushy slopes on the cliff face. At the time, it was clearly a very successful plant within its restricted range.

As an arable weed, Bromfield describes it as a 'grievous nuisance'. On Dean and Ash Farms, near St Lawrence, he described how 'wheat and barley are completely overrun by it, and the crops greatly deteriorated thereby'. The seeds apparently made the flour distasteful and gave it a blue colour, so that it acquired the local name of 'poverty weed', a name still remembered at Dean Farm in 1990. Bromfield's account of this plant is full of delightful anecdotes. One of them is repeated here:

'A respectable shoemaker, named Rabbett, who resided for many years at Whitwell, and has only recently left, told me that when he was employed in harvest on Week Farm they used to pull up the Purple Cow-wheat or Poverty-weed with the greatest care, and carry it off the field to burn it, picking up the very seeds from the ground wherever they could be perceived lying.'

By the late 1800s the plant was apparently less of a problem as a cornfield weed, although still surviving in hedge banks and bushy slopes. A report in the *Proceedings of the Isle of Wight Natural History Society* for 1924 reported, 'In arable field and various odd places between St Lawrence and Niton but now (1923) greatly reduced in numbers. H. Ives.' It continued to survive in one field at St Lawrence for a few more years but was eventually lost from here but persisted as a hedgeside plant in several fields above St Lawrence and eastwards to Coombe Bottom sports ground up until the 1950's (EL). It survives today on a single hedge bank. At this site, numbers fluctuate considerably from one year to the next but, since the bank has been kept clear of invading scrub, numbers have varied between approximately 500 and 4000 since 1990. They do best in dry springs as wet weather early in the year promotes a lush growth of competing vegetation.

In 1984, a second population was found to be thriving nearby on the cliff face. My attention was originally drawn to it by a reddish-purple haze in the grassland, which, through binoculars, was confirmed as Field Cow-wheat. This population comprised several thousand plants and was largely inaccessible. Moreover, the warm, sheltered aspect ensures that the plants flower early and competing vegetation is sparse. These plants are, however, threatened by invasive holm oak. The habitat is similar to some native sites on the continent and the rediscovery of this population renewed suspicions that Field Cow-wheat may be a long-established native species rather than introduced as a cereal contaminant. It is native on the near Continent and it is able to maintain an independent existence away from farmland.

On the other hand, several facts suggest that it may be an introduction:
i) Field cow-wheat has heavy seeds with poor dispersal characteristics, although they can be dispersed by ants attracted to the sweet tasting eliasome attached to one end of the seed. If the inner cliff face were a refugium, then one would have expected the plant to become well established in grassland beneath the cliff, which has not been the case, rather than in fields above.
ii) There are several testaments to workers being employed to pull up and remove the plants from fields and as many of the infested fields lay alongside the top of the cliff, it is very likely that some of these would have been thrown over the cliff.
iii) Early reports refer to the introduction of this plant with contaminated seed. E.H. White writing in 1904 said that, 'some years ago it was suddenly found growing in a cornfield near Ventnor. In a few years it spread to an alarming extent through the cornfields'. This is very much the behaviour of an introduced plant.

In 1993, seed was introduced to a south-facing open chalky bank on Knighton Down, 5587, and this resulted in a small population of up to 46 plants surviving for at least seven years. Field Cow-wheat is protected under the Wildlife and Countryside Act 1981 and collection of seed and introductions are illegal excepting under licence.

The status and conservation of this plant is discussed in detail by Wilson (1993).

M. pratense L. ssp. *pratense*
COMMON COW-WHEAT

N. A local plant principally confined to ancient woodland but also occasionally found on old hedge banks on acid soil. Numbers can fluctuate considerably and it becomes scarce in shady unmanaged woods. Largely confined to the northern half of the Island where it is frequent in Parkhurst Forest and some woods around Havenstreet and Wootton. Much scarcer south of the chalk but still found in several places, including Borthwood Copse, America Woods where it is very rare, and woodland adjoining Sandown Golf Course. A

VASCULAR PLANTS

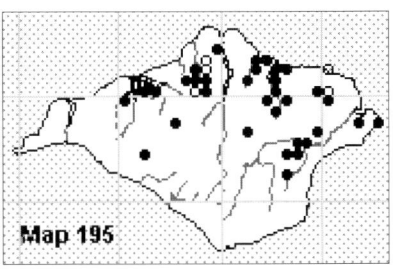

M. pratense

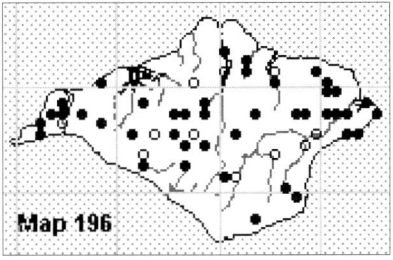

Odontites vernus

small number of rather weak plants were found in Rowridge Copse 4587, AC 1999. These were not determined at the time and plants have not been refound but they are our only record from the chalk and might possibly prove to be ssp. *commutatum* (Tausch ex A. Kern.). Map 195

Euphrasia L.
EYEBRIGHTS
Eyebrights are frequent and widespread across the Island on a range of soil types in short grassland. They are a difficult group and there has been little attempt to differentiate species of late. The following account is acknowledged to be incomplete.

E. anglica Pugsley
N. Status unknown. Occurs in heathland and acid grassland. H.W. Pugsley records this species from Headon Warren in his 1930 publication.

E. arctica Lange ex Rostrup ssp. borealis (F. Towns.) Yeo
N. Status unknown. Recorded from SZ57 by H.W. Pugsley (1930).

E. tetraquetra (Bréb.) Arrond.
N. Believed to be locally frequent in short calcareous grassland around the southern coasts e.g. Tennyson Down 3385, CP 1996; abundant on the undercliff at Brook 3883, PS 2000.

E. tetraquetra x E. pseudokerneri
N. Status unknown. There is a record from Afton Down 1960 conf. P.F. Yeo.

E. nemorosa (Pers.) Wallr.
N. Undoubtedly our commonest species found in pastures on chalk, verges and open woodland rides.

E. nemorosa x E. pseudokerneri
N. Status unknown but presumably present where both parents grow together. Frequent on Arreton Down 5387, CP 1998 (det. Alan Silverside).

E. nemorosa x E. confusa
N. Status unknown. West High Down 3185, CP 1998 and frequent on Headon Warren 3186, CP 1998, both det. Alan Silverside.

E. pseudokerneri Pugsley
N. Believed to be very locally frequent in short calcareous grassland. Arreton Down 5387, CP 1998 (det. Alan Silverside); Knighton Down 5787, BG 1999; Tennyson Down 3385, BG 1999. Old records from Apes Down 1938, Ashey Down 1934 and Brading Down 1960 all conf. P.F. Yeo; Carisbrooke Castle 1930 H. Pugsley.

E. confusa Pugsley
N. Status unknown. H.W. Pugsley records this species from SZ58 in his 1930 publication.

Odontites vernus (Bellardi) Dumort.
RED BARTSIA
N. Frequent in grassland tracks, woodland rides, arable and waste places, particularly on the chalk. Ssp. *serotinus* (Syme) Corbiere is the common form but there are also unconfirmed records for ssp. *vernus*. Map 196

Parentucellia viscosa (L.) Caruel
YELLOW BARTSIA
N. A very rare plant of damp pastures. It is known from a single historic site, a meadow and cornfield margins alongside a lane at Forelands, Bembridge where it was first recorded in 1856 (More, 1871). For many years after 1860 there were no further published records but part of these cliff top fields were lost to erosion. The slumped cliffs below have very sporadically yielded plants in the twentieth century; most recently, 30 plants were found in 1986, 6486, BS and these had declined to just 2 plants by 1987, after which no further specimens have been found. A second site was found in 1997. Over 250 plants were growing in a small donkey grazed pasture at Wootton Common 5391, CP and plants were still present in 2001.

Surprisingly for so rare a plant, there appears to be an early unlocalised record of 1667 by a Mr Cole in Christopher Merret's *Pinax*. This constitutes the first British record.

Rhinanthus minor L.
YELLOW-RATTLE (local: fiddle-cases)
N. Locally frequent in old grassland. Increasingly uncommon away from the chalk but still surviving in some hay meadows and damp grassland. No attempt has been made to separate the subspecies.

[Pedicularis palustris L.
MARSH LOUSEWORT
N. Extinct. Bromfield (1856) knew this plant from several fens and marshy meadows including Freshwater Marshes, Wolverton Marsh at Shorwell, Plaish water meadows and Apse Heath. However, there are very few twentieth century records. E.H. White, writing in 1904, said that 'in the boggy portion of Blackpan Common it throws its head of pink flowers through the furze, often overtopping it'. This suggests that this site was already becoming overgrown. The last record was in 1927 when James Groves led a meeting of the Natural History Society through The Wilderness.]

P. sylvatica L. ssp. sylvatica
LOUSEWORT
N. Much declined but still present locally on damp heaths, acid pastures and flushes, particularly in the Eastern Yar valley. It survives in open plantation rides in Parkhurst Forest and Firestone Copse, and grows in heathy mown grassland in Cowes cemetery. Plants with salmon coloured flowers are sometimes found on St George's Down. Map 197 (*see overleaf*)

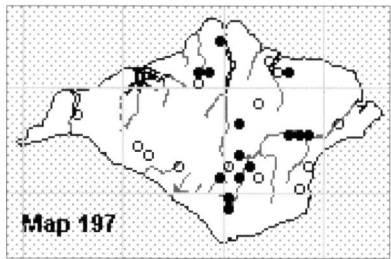

P. sylvatica

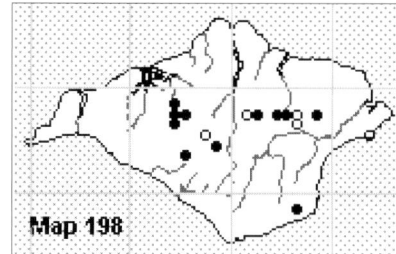

Lathraea squamaria

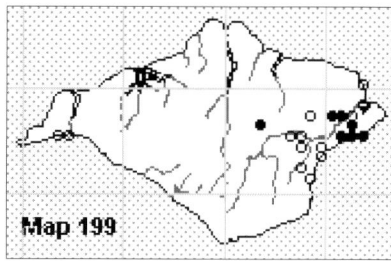

Orobanche purpurea

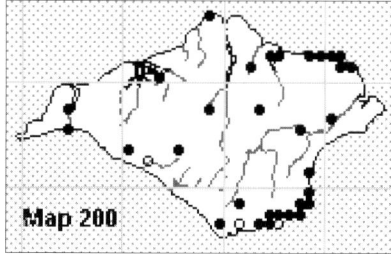

O. hederae

129. OROBANCHACEAE
Broomrape family

Lathraea squamaria L.
TOOTHWORT
N. Locally frequent in woods on chalk or heavy clay soils. Most frequently parasitic on Hazel but also found on Field Maple, Sycamore, Elm, Lime and Cherry Laurel. It is not confined to ancient woodland but also appears in plantation and in secondary woods on the chalk. The number of flowering stems fluctuates from year to year but in favourable years, Eaglehead Copse 5887, a particularly good site, can hold over two thousand plants, in three different colour forms. In addition to the usual flowers, distinctive lemon-yellow and beetroot-red colour forms occur. Its status has probably changed little; it is still present in all of Bromfield's sites. These include Northcourt Manor garden at Shorwell 4583, where Bromfield describes it as occurring abundantly under the shrubs on the terraces. Map 198

Orobanche purpurea Jacq.
YARROW BROOMRAPE
N. A Red Data Book plant occurring very locally in a few sites in dry undisturbed grassland on greensand or chalk. Past records indicate a much wider range, but today it seems to be confined to a narrow belt between Arreton in the west and Culver in the east, with most sites in the Brading and Bembridge/Culver Down areas. One of the largest surviving British populations occurs on the greensand by the cliff edge at the north end of Sandown Bay at Redcliff 6285 and extends onto the chalk on Culver Down 6385. In good years, the size of this extensive population can be numbered in the low hundreds. At most other sites, the number of flowering plants in any one year rarely exceeds twenty. It may still survive, unrecorded, in the Alverstone and Lake areas. First recorded in 1841 near the cliffs opposite the barracks on Royal Heath, Sandown Bay (Bromfield, 1856). Map 199

O. rapum-genistae Thuill.
GREATER BROOMRAPE
N. A rare plant of scrubby areas on rough ground, historically recorded from three sites and surviving in two. It grows on gorse at Little Lynn Common 5389; there were 34 spikes here in 1986 and 11 in 1997, but only 4 in 2002, CP. This is a remnant of a once larger Briddlesford Heath population. Found on Broom in heavily overgrown scrub at Common Copse, Borthwood 5784, MS 1993. There were 8 spikes here in 1997. Borthwood is an historic site, as also is a heathy pasture between Newnham Farm and Quarr, from where it was recorded by Bromfield in 1846. It is probably declining in its surviving sites.

O. hederae Duby
IVY BROOMRAPE
N. Common on Atlantic Ivy in woods in the Undercliff from Shanklin through to Niton, from where it was first recorded by Bromfield (1856). It is also frequent along the coastal belt in Dorset, particularly on Portland, but rare in Hampshire. Increasingly, Ivy Broomrape is being recorded away from the Undercliff, sometimes growing on cultivated ivies in gardens. Map 200

O. artemisae-campestris Vaucher ex Gaudin
OXTONGUE BROOMRAPE
N. This Red Data Book plant has one of its very few British stations on exposed vegetated chalk cliff ledges, or greens, on West High Down, Freshwater, 3285. It was familiar to Bromfield, who recorded it first in 1844 from Rosehall Green, and he provided a very full description. He would have been provided with specimens by local cliffsmen who descended the cliffs on ropes to procure gulls eggs and samphire. This trade ceased around the turn of the century and there are no records from 1890 until 1967, when Phoebe Yule found what she believed to be a specimen of this plant from the top of the cliff. In 1986, 20 plants were found in cliff top grassland above Rosehall Green, CP, and more specimens could be seen through binoculars, flowering on the ledges below. The plant has continued to appear sporadically in clifftop grassland whilst the main population survives on inaccessible cliff ledges. In 1883, it was described by Stratton as plentiful, but is less so today. 59 plants were counted here in 1992, a good year for the plant, but this is likely to be an under-estimate. It is conceivable that the population is currently expanding. Since 1999, plants have appeared in two other well-separated cliff top locations along this stretch of downland where the host plant Hawkweed Oxtongue is frequent, PJ. It is the latest flowering of our species, not usually at its best until well into July. Otherwise only known from undercliffs in East Kent where a larger population survives.

There is a herbarium sheet at Bolton museum of this plant (conf. Fred Rumsey) collected by Miss Livens on 9 June 1875 and labelled as Colwell.

O. minor Sm. var. *minor*
COMMON BROOMRAPE
(local: shepherd's pouches)
N. Occasional but distributed locally right across the Island, in rough grassland and verges and sometimes in flowerbeds. Parasitic on a range of host plants, particularly members of the pea family.
(var. *maritima* (Pugsley) Rumsey & Jury is often recorded from wild carrot along the southern coast of the Island with most records from Ventnor to St Catherine's Point and Compton. Records have not yet been confirmed.)
In 1859, A.G. More recorded a broomrape growing on Sea Holly on St Helen's Spit, which he named as *O. amethystea*. However, H.W. Pugsley (*Journal of Botany*, LI 1913: 336) considered the plant to be *O. minor* f. *procerior* (Rchb.) Beck. ; it is included here as a form of *O. minor*. It was recorded sporadically from this site through the twentieth century. It was found in 1961, after an absence of many years (DF), and then not again until 1987, when three stems appeared on a single Sea Holly clump, CP. There are no further records and as the Sea Holly in the part of St Helen's Duver yielding specimens has now gone, it must be assumed to be extinct.

131. ACANTHACEAE
Bear's-breech family

Acanthus mollis L.
BEAR'S-BREECH
E. A persistent garden escape on cliffs above La Falaise car park at Ventnor 5677, CP 1998.

132. LENTIBULARIACEAE
Bladderwort family

Pinguicula lusitanica L.
PALE BUTTERWORT
N. A very rare plant of spring-fed bogs and flushed clay heaths at the eastern edge of its range. Historically known from three sites, and surviving in one. It is still frequent in the tiny site of Bohemia Bog 5183, CP 2001, where it was first reported in 1908 by F. Stratton. The first Island record was from Cockleton Bog, Northwood in 1839, when it was described as plentiful on boggy ground. By 1913, the date of the last published record from this site (Drabble & Long, 1931), it had probably already been lost through drainage of the site. The third site was Colwell Heath. It was found sparingly here according to Bromfield (1856) and was last recorded in 1869 by F. Stratton, after which time the site was enclosed.

[*Utricularia australis* R.Br.
BLADDERWORT
N. Extinct. Only ever known from the slightly acid waters of ditches at Freshwater Marshes. First recorded by Snooke (1823) as plentiful in the ditches on Easton Marsh and recorded by subsequent botanists but only rarely flowering. First found in flower by F. Stratton in 1867 and the herbarium material at OXF has been determined as this species by CDP. E.H. White found flowering specimens in 1906; he recorded in his diary, 'only three plants were flowering, but under the clear water I could distinguish other of their remarkable bladder-covered roots.' It flowered in profusion in 1927 when J. Groves found it in a recently cleared ditch but after 1936 there are no further records of flowering. It persisted for many more years in a single deteriorating and increasingly murky ditch, but was last recorded in 1970, BS.]

[*U. minor* L.
LESSER BLADDERWORT
N. Extinct. Only ever recorded from ditches in the Yar below Langbridge Farm, Newchurch and westwards to Horringford. Bromfield (1856) described it as abundant and flowering only sparingly in 1842 (hb. K & HCMS) and Townsend recorded it in 1879. There are no further records.]

133. CAMPANULACEAE
Bellflower family

[*Campanula patula* L.
SPREADING BELLFLOWER
C. Known only from a record in Bromfield (1856), referring to a record by A.J. Hamborough from a hedge bank above Shanklin Chine.]

C. medium L.
CANTERBURY-BELLS
E. Established in open plantation woodland in the grounds of Freshwater Court 3386, CP 1989. Bromfield (1856) records this species occurring spontaneously on bushy banks at Brading and Bonchurch.

C. glomerata L.
CLUSTERED BELLFLOWER
N. This plant has an interesting, localised distribution on the chalk, in ancient dry, short turf and sometimes in old quarries. It is locally frequent in the west Wight from the Needles headland westwards through to Newbarn Down, Calbourne but it does not occur eastwards from here except on Bembridge and Culver Downs, where it tends to be scarce. Flowering plants can be extremely dwarfed in coastally exposed situations. Two mapped records are away from its main centres of distribution and may, perhaps, refer to garden escapes. A single plant found west of a footpath on Mersley Down 5587, MS 1993; and a single plant on a roadside verge at Wellow, on limestone 3987, PS 2000. Map 201

C. portenschlagiana Schult.
ADRIA BELLFLOWER
E. A garden escape well established on walls in Cowes, Newport and Ryde and in some churchyards elsewhere. First recorded in the early 1970s, but likely to have been established for much longer than this; it was a popular Victorian and Edwardian edging plant.

C. poscharskyana Degen
TRAILING BELLFLOWER
E. A more recent introduction, also

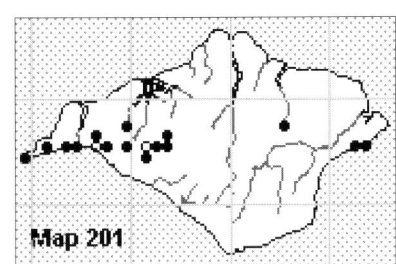

C. glomerata

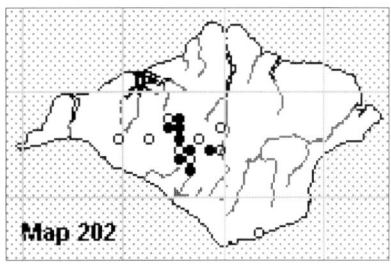
C. trachelium

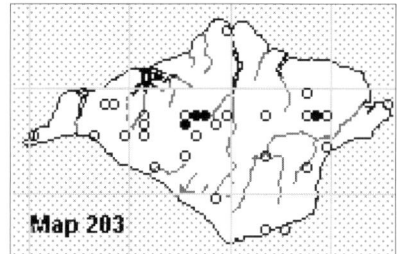
Legousia hybrida

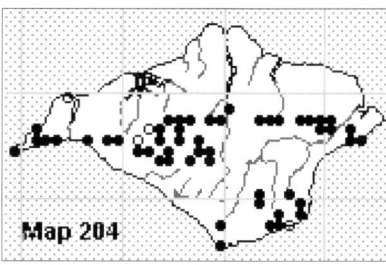
Asperula cynanchica

established on walls and pavement cracks but more widely distributed than the previous species and more invasive.

C. trachelium L.
NETTLE-LEAVED BELLFLOWER
N. Very local in open woods, hedge banks and scrub on chalk. This plant has a remarkably restricted distribution, being confined to the inland chalk outcrop west of the Medina. There has been a contraction in its range and its abundance over the past one hundred years. Its principal extant sites are the Rowridge valley, and other woods at Swainston, Idlecombe Down, Westridge Copse at Shorwell and Tolt Copse at Gatcombe. Map 202

C. rapunculoides L.
CREEPING BELLFLOWER
E. A rare garden escape found growing on a verge at Newbarn Road, Cowes 5195, MB 2000. There are old records from Steephill Cove, J.F. Rayner 1917 and St Lawrence undercliff (Bromfield, 1856).

C. rotundifolia L.
HAREBELL
N. Harebells remain locally frequent in most surviving old calcareous turf but they have declined greatly in dry grasslands off the chalk. Still surviving in sandy turf in a few spots such as Headon Warren, Bleak Down verge and Brook churchyard.

Legousia hybrida (L.) Delarbre
VENUS'S-LOOKING-GLASS
Arch. A much declined arable weed, formerly widespread although most frequent on the chalk. In recent years, recorded from just two areas on chalk, where it remains frequent, West Nunwell Down 5887, CP 1996 and at Bowcombe, west of Carisbrooke. Sites here include 4586, CP & JP 1998; 4687, DD 1998; and 4787, CP 1996. It is likely to be under recorded, and the map probably exaggerates the decline. In 1985, hundreds of plants were growing in a peaty arable field near Great Budbridge Manor 5383, BS. Map 203

[*Wahlenbergia hederacea* (L.) Rchb.
IVY-LEAVED BELLFLOWER
N. Extinct. A former rarity of boggy grasslands alongside stream banks. This plant probably disappeared early in the twentieth century but it was recorded from a surprising number of sites in the nineteenth century. It grew in several places on Bleak Down, including the edges of Leechmore pond by the roadside at the southern end. It grew in the Medina valley in 'spongy meadows' in many spots in The Wilderness and northwards of here by Rookley Farm. E.H. White recorded it in The Wilderness in 1904, but his diary entry for 17 August 1907 reads, 'I also looked for the dainty little bog campanula but owing to the thick vegetation that filled the streams, it could not be found.' A.G. More recorded it from a wet corner of Buck's Heath near Kingston church and between Pyle and Gladice's farm, Chale Green in 1871. F. Stratton recorded it from Budbridge Moor in 1909.]

Jasione montana L.
SHEEP'S-BIT
N. A declining rarity of open, dry sandy ground. Sheep's-bit has always been local although Bromfield (1856) described it as not infrequent and More (1898) said that it 'abounds on every bank' on Pan and Lake Commons. It survives best today in two stretches of coastal sandy cliffs, on Luccombe Chine ledge 5879, to the south and more particularly to the north of the Chine, CP 1999; and, more sparingly, at Blackgang Chine holiday complex 4876, JS 1988 and on the ledge to the north west 4877, CP 1993. A very few plants survive in heathy grassland in one spot on Sandown Golf Course 5885, MB 1998, a former stronghold. Last recorded in 1968 from the roadside verge on Bleak Down 5181, BS, and from sandy ground by Munsley Bog 5282, RK.

Lobelia erinus L.
GARDEN LOBELIA
C. A garden self-sown garden escape on tips and in pavement cracks, not persisting.

134. RUBIACEAE
Bedstraw family

Sherardia arvensis L.
FIELD MADDER
N. This plant is a widespread and frequent arable weed of dry soils and disturbed chalk turf. Bevis *et al* (1978) considered that it was a decreasing species but that is not the case today.

Phuopsis stylosa (Trin.) Benth. & Hook. f. ex. B.D. Jacks.
CAUCASIAN CROSSWORT
E. A persistent garden escape on roadside verges, first recorded from a lane in Carisbrooke by F. Stratton in 1866 (More, 1871). It was recorded from a verge in Whitwell Road, Upper Ventnor 5577 in 1966 TW, and was still present here in 1990. More recently it has not been refound, although there were periods in the past when it was believed to have been lost.

Asperula cynanchica L. ssp. *cynanchica*
SQUINANCYWORT
N. Widespread and frequent on short,

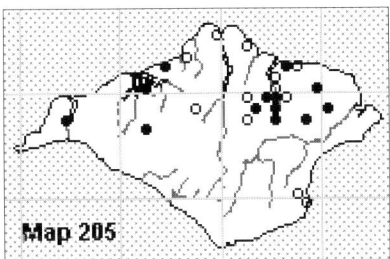
Galium odoratum

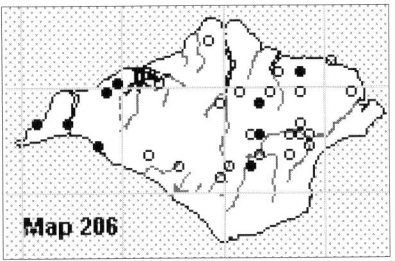
G. uliginosum

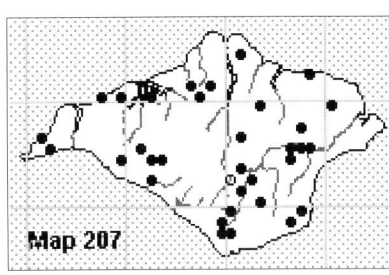
G. saxatile

old chalk grassland turf in exposed situations. It also grows in calcareous grassland outcrops above St Catherine's Point 4975, JO 1998. A more unusual site is slumped lime-rich grassland on Headon Warren undercliffs 3186, CP 1998. It is found in cemeteries on chalk at Lowtherville, Upper Ventnor 5577, and Mount Joy, Carisbrooke 4987, MB.

The first record from the Island was by the Flemish herbalist, Lobelius who settled in this country. The record was published in 1655 in William How's flora (*Phytologia Britannica*). However, Lobelius died in 1616, so this observation must have been made earlier than the first published British record, by John Goodyer from dry chalky grounds of Hampshire in 1619. Map 204

(*A. arvensis* L.
BLUE WOODRUFF
C. Apparently recorded as a casual at Havenstreet by Miss Bullock in 1936 but no herbarium material survives.)

Galium odoratum (L.) Scop.
WOODRUFF
N. Local in old, generally ancient, woods on clay soils. It is also found in ancient woods on the chalk, of which there are very few. Most frequent in ancient woods around Havenstreet. Elsewhere (e.g. Calbourne and Freshwater) it occurs as a garden escape. Map 205

G. uliginosum L.
FEN BEDSTRAW
N. Very local in marshy ditches, fens and wet grassland on neutral or slightly acidic soils. Very much scarcer than Common Marsh-bedstraw but probably under recorded, particularly in the Eastern Yar valley. Map 206

g. palustre L.
COMMON MARSH-BEDSTRAW
N. Frequent and widespread in wet woodland rides, marshes and fens, streamsides and ditches. Not found on the chalk.

G. verum L.
LADY'S BEDSTRAW
N. Frequent in dry grasslands. Most common on the chalk and around the coasts. Less frequent, and probably declining, in dry non-calcareous grasslands inland but surviving well in many cemeteries.

G. x *pomeranicum* Retz.
LADY'S BEDSTRAW X HEDGE BEDSTRAW
N. Sometimes recorded where both parents are growing together on the chalk. The most recent records are from West High Down 3084, CDP & TDD, 1998, an area from where it has been recorded regularly over the past 150 years, and Fairlee cemetery, Newport 5089, MB 2001.

G. mollugo L. ssp. *mollugo*
HEDGE BEDSTRAW
N. Frequent and widespread in hedge banks, scrub, woodland clearings and grasslands. Ssp. *erectum* has not been confirmed but may be overlooked on chalk.

G. saxatile L.
HEATH BEDSTRAW
N. Local but sometimes frequent on acid soils. A good indicator of heathland remnants being found at sites including Headon Warren, the tops of Ventnor Downs, Parkhurst Forest, Bleak Down, St George's Down and Sandown Golf Course. Also sometimes found in chalk heath vegetation. Map 207

G. aparine L.
CLEAVERS (local: cliders)
N. Common and widespread everywhere, especially on nutrient rich soils. Goosegrass derives its name from a once widespread practice described by Bromfield (1856), 'the herb, chopped small, is given to goslings in this island'.

[*G. tricornutum* Dandy
CORN CLEAVERS
Arch. Extinct. Bromfield (1856) considered that this plant was not infrequent as an annual weed on arable ground. He listed fifteen sites, many of them on the chalk. It appears to have been lost around the 1900s. E.H. White (1905) wrote that he had picked a few plants on a bank in Shanklin cemetery and F. Stratton recorded it from the chalk cliff at Culver in 1900. There is also a specimen collected by J. Long (hb. BM) in 1932 by the Medina at Newport. This would have been a casual occurrence.]

Cruciata laevipes Opiz
CROSSWORT
N. Locally frequent in hedge banks, woodland edges and calcareous scrub. Most local floras refer to an inexplicably strange distribution of this plant. On the Island it is particularly frequent in, but not confined to, the south-eastern sector.

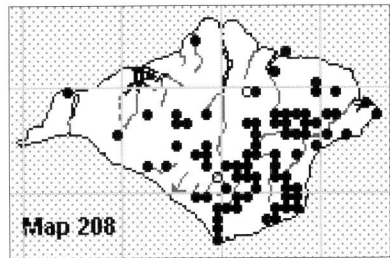
Cruciata laevipes

Map 208
Rubia peregrina L.
WILD MADDER (local: evergreen cliver)
N. A frequent plant with an interesting predominantly coastal distribution found in hedges, woodland edges, scrub and coastal cliffs. It is not found right around the coasts, but restricted to areas of calcareous or base-enriched slumps and undercliffs. In Hampshire it is confined to the southern part of the New Forest. First recorded in William Turner's *New Herbal* of 1562, 'the moste in the Yle of Wyght'. Map 209

135. CAPRIFOLIACEAE
Honeysuckle family

Sambucus nigra L.
ELDER
N. Common and widespread in woods, scrub, hedges and waste ground. Twigs were used by children to make whistle pipes. Both flowers and fruit are still gathered for amateur winemaking.

S. ebulus L.
DWARF ELDER
Arch. A rare plant, established very locally in hedge banks and tracksides. Bromfield recorded it from a cornfield near St Catherine's Point in 1850 and it still grows in hedge banks at Blackgang, below the viewpoint car park 4876, GT 1999. It survives, in decreased quantity, by the side of a footpath near the Brading Roman villa 6086, RHW 2000, from where it was first noted in 1929. Well established by a gateway off the Bowcombe road at Idlecombe 4686, CP 1997. Also recorded from the tip at Ventnor Botanic Gardens, where it is likely to be a garden outcast 5476, PS 2000. It was found in several places by footpaths on heathland between Bierley and Wydcombe in 1987, 5078, BS, but has not been refound here. First recorded in 1770 by R. Waring (*Phil. Transac.*, lxi), between Newport and Carisbrooke Castle.

Bromfield (1856) wrote, ' The plant is, I understand, sought by farriers and horse-doctors as a stimulant and to improve the coats of horses, which may account for its present scarcity in some localities, as between Chine cottage and Rose cliff, where a countryman informed me he had formerly seen it in abundance.'

Viburnum opulus L.
GUELDER-ROSE (local: stink-tree)
N. Locally frequent in ancient woodlands and old hedgerows on the clay soils of the north of the Island. It is much scarcer on the chalk and is rare in the south of the Island. The local name, formerly used by boys, referred to the strong smell when the bark is scraped. Map 210

V. lantana L.
WAYFARING-TREE (local: whip-crop!)
N. Common in scrub on the chalk. More local on lime rich clays in the north of the Island where it is found in woodland borders, coastal slumps and old hedgerows. Extremely rare, or absent as a native, on greensand soils.

Bromfield (1856) wrote, 'the slender stems are used in Russia for whip-handles, a purpose to which they are sometimes applied in this island'. At Ryde, it was used for the bows or benders of shrimp nets (*Proc. IWNHAS* 1924). Map 211

V. tinus L.
LAURISTINUS
E. An established garden escape, naturalised in a few sites on the chalk. Sites include the holm oak wood on St Boniface Down 5778, CP 1990; cliffs at Ventnor 5577, CP 1993; an old chalk quarry on Brading Down 6086, AC 1996; and scrub at Luccombe 5879, GT 1999. Also planted in Victorian shrubberies and occasionally bird-sown from gardens into wasteland.

Symphoricarpos albus (L.) S.F. Blake
SNOWBERRY
E. Formerly planted in woods as pheasant cover and spreading by suckers and also bird-sown. Sometimes found in hedgerows.

Leycesteria formosa Wall.
HIMALAYAN HONEYSUCKLE
E. Found increasingly as a bird-sown escape in woodland, verges waste ground and walls but generally only a few specimens in any one location.

Lonicera nitida E.H. Wilson
WILSON'S HONEYSUCKLE

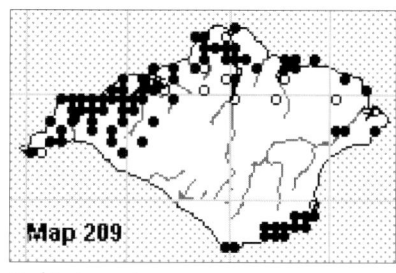

Rubia peregrina

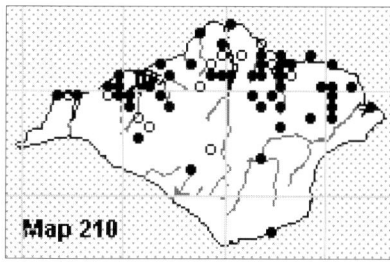

Viburnum opulus

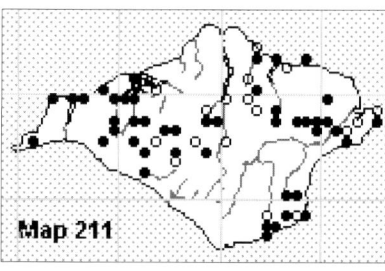

V. lantana

E. Frequently planted in village hedgerows and as pheasant cover in woodlands. It is a frequent relic of habitation.

L. henryi Hemsl.
HENRY'S HONEYSUCKLE
S. Planted around an electricity sub station by Porchfield cemetery and growing vigorously into surrounding streamside woodland 4491, MB (conf. E. J. Clement) 2001.

L. japonica Thunb. ex Murray
JAPANESE HONEYSUCKLE
E. Sometimes becoming rampant when planted alongside boundaries or dumped with garden waste. Occasionally bird-sown.

L. periclymenum L.
HONEYSUCKLE
N. Frequent in woods, hedges, undercliffs and amongst scrub on all soils across the Island. It was popular amongst children to pick the individual flowers to suck for nectar.

136. ADOXACEAE
Moschatel family

Adoxa moschatellina L.
MOSCHATEL
N. Widespread and quite frequent in damp places. Most often found in ancient and old secondary woodlands but also characteristic of stream banks and old sunken lanes. Found on all soil types.

137. VALERIANACEAE
Valerian family

Valerianella locusta (L.) Laterr.
COMMON CORNSALAD
N. A rare plant of cultivated and disturbed ground. Bromfield (1856) described it as common and Bevis *et al* (1978) said, 'frequent, but distribution uncertain owing to confusion with *V. carinata*.' Recent studies have shown it to be a rare plant found in scattered spots, always in very small quantity, but historic records do not allow us to determine whether this has always been the case or whether the species has declined sharply in recent years. Bonchurch coast 5778, JO 1988; arable field at Carisbrooke 4887, CP 1989; St Helen's Duver 6389, CP 1993; Alverstone 5785, JO 1996; Newchurch churchyard 5685, AC 2002; Pound Green, Freshwater 3386, MR 2000; arable field on Bleak Down 5081, PS 2000; Alverstone railway trackbed 5085, AC 2001; Sandy Way, Shorwell 4682, PS 2002.

V. carinata Loisel.
KEELED-FRUITED CORNSALAD
Arch. This is now a common and widespread species of cultivated and disturbed ground. It was first recorded in More (1871) as an occasional weed in the garden of Vectis Lodge, Bembridge. There are a handful of records up to the time of Bevis *et al* (1978) where it is described as quite common in the Newport and Ventnor areas. These were well-botanised spots at the time. It is now by far the commonest cornsalad.

V. rimosa Bastard
BROAD-FRUITED CORNSALAD
Arch. An extremely rare arable plant today, and considered to be critically endangered nationally. Bromfield (1856) described it as 'not infrequent' but there are only two recorded stations post-1900. Drabble (1931) recorded it from a field at Middleton, Freshwater. No further records until 1998 when eleven plants were found in an arable field at Cridmore 5081, PW. A single plant was found here in 2000 (PS).

V. dentata (L.) Pollich
NARROW-FRUITED CORNSALAD
Arch. A rare plant of arable field margins, generally on chalky soil. Although early botanists found this to be a frequent species, it has only been recorded from six stations post 1987. Fields at Cridmore 5080/5081, BG 1999; frequent on Bowcombe Down arable 4687, DD 1998; West Nunwell Down 5887, CP 1996; Newbarn Down arable 4485, CP 1991; Ashey Down 5787, CP 1987; set-aside field east of Warren Farm, Totland 3185, GT 2002.

V. eriocarpa Desv.
HAIRY-FRUITED CORNSALAD
N. A Red Data Book plant which gives all the appearance of a native species in open parched calcareous turf, apparently confined to a narrow band of outcropping Middle and Lower Chalk. Its sites are very similar to many of its native sites in Purbeck. It grows on steep, south-facing slopes at Carisbrooke Castle 4887, where there were an estimated 250 plants in 2000, AM. This is the site where F. Stratton found thousands of plants in 1912 (*J. Botany* 1912: 231-231). It grows on the cliff edge at Culver 6285, CP 2000, where up to 600 plants are found in good years. Up to 20 plants were found in coastal grassland on Afton Down below the road 3685, AM 1999 and in 2002, a population of two to three thousand plants were found in this area, PS. First recorded in 1883 by Rev. G. E. Smith from 'the Isle of Wight'.

Valeriana officinalis L.
COMMON VALERIAN
N. Local in fens and wet woodlands, perhaps declining. All post-1987 records are from the East Wight. The middle stretches of the Eastern Yar valley between Sandown Waterworks

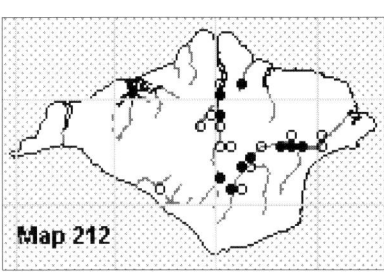

Valeriana officinalis

and Newchurch are a stronghold. Ssp. *collina* Nyman is recorded from Shide chalkpit where a small colony was present on the quarry floor 5088, BS 1996, still present in 2000. Map 212

V. dioica L.
MARSH VALERIAN
N. A rare and declining plant of wet meadows and open carr woodland. Only known today from two sites where it occurs in small quantity: Plaish water meadows, Carisbrooke 4887, BS 1994 and by the Caul Bourne at Newbridge 4187, CP 1998. The Newbridge plants were all female. Four male plants were recorded in 1977, BS, at the edge of alder carr by the Medina between Blackwater and Shide but the site was destroyed during agricultural improvements in 1979. Historic sites, with last recorded dates, include Freshwater Marshes (1931), marsh by Sandown waterworks (1930), streamside at Merrie Gardens, Lake (1904) wet meadows at Thorley (1856) and streamside on Briddlesford Heath (1856).

Centranthus ruber (L.) DC.
RED VALERIAN
(local: ventnor pride!)
E. A common and widely dispersed alien especially around the coast on old walls, chalk quarries and chalk cliffs. Sometimes abundant on south facing chalk cliffs such as at Culver and Freshwater Bay. It is especially prolific around Ventnor and 'Ventnor Pride' is one of the few examples of a modern vernacular name. Bromfield (1856) described it as commonly grown in gardens, but rare as a naturalised species. However, More (1898) writing in 1857 of its occurrence on the inner cliff at St Lawrence found it to be 'everywhere establishing itself' and considered that it 'might well pass for a native to a

stranger'.

138. DIPSACEAE
Teasel family

Dipsacus fullonum L.
WILD TEASEL
N. Widespread and common on waste ground, field margins, verges and woodland clearings.

Knautia arvensis (L.) Coult.
FIELD SCABIOUS
(local: gipsy rose; egyptian rose)
N. A common species in calcareous grassland, widely dispersed on the downs and sometimes found in relic limestone grassland. Also found, although much more sparingly, in unimproved meadows on limey clay soils and in tall grassland on sandy soils.

Succisa pratensis Moench
DEVIL'S-BIT SCABIOUS
N. Locally frequent on low-nutrient soils in species-rich grassland. It is tolerant of a fairly wide range of soil moisture and pH values. Characteristic of clay heath grasslands, damp north-facing chalk grassland, marshy grassland, and damp heathy woodland rides. Map 213

Scabiosa columbaria L.
SMALL SCABIOUS
N. Found in calcareous grassland and particularly frequent and widespread on our downs. Rare away from the chalk but recorded from fixed shingle at Norton Spit and sandy grassland at Brook.

139. ASTERACEAE
Daisy family

Carlina vulgaris L.
CARLINE THISTLE
N. A plant of thin, dry soils which is most characteristic of south-facing chalk grassland slopes and quarries. Also found locally on coastal slumped clay cliffs, in relict limestone grassland and on sandy ground in heathland. Map 214

Arctium lappa L.
GREATER BURDOCK
Arch. Very local and occasional in waste places, river banks, trackside verges and around farmsteads. Bevis *et al* (1978) recorded only two stations, by the railway track between Shide and Blackwater and near the quayside at Shalfleet. It still occurs in both sites, but seems to be more widespread than formerly. Two particular stronghold areas are at Thorness and in the Barnsley Farm and Hill Farm area east of Whitefield Woods. Map 215

A. minus (Hill) Bernh.
LESSER BURDOCK
N. Common and widespread in waste places, woodland clearings, verges and stream banks. Ssp. *minus* is the commonest form; ssp. *nemorosum* (Lej.) Syme is also recorded but there are only nineteenth century records for ssp. *pubens* (Bab.) P.Fourn.

Carduus tenuiflorus Curtis
SLENDER THISTLE
N. Local on dry cliff tops, grassy waste ground by coast, rabbit warrens and disturbed, exposed ground on chalk ridges. In Hampshire this plant is confined to a narrow coastal belt; on the Island it is widespread along the central downland ridge, a reflection of the maritime influence felt right across the Island. Map 216

C. crispus L. ssp. *multiflorus* (Gaudin) Gremli
WELTED THISTLE
N. Widespread but local in disturbed ground and waysides, particularly on the chalk. Perhaps under-recorded. Map 217

C. nutans L.
MUSK THISTLE
N. Frequent and widespread in grassland and disturbed ground on the chalk. Also found less frequently on sandy soils in the southern half of the Island.

Cirsium eriophorum (L.) Scop.
WOOLLY THISTLE
N. A very local plant of chalk grassland and old quarries. It has always been local and its population is probably unchanged over the past

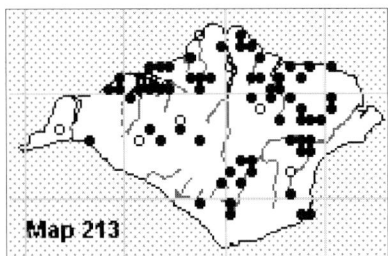
Succisa pratensis

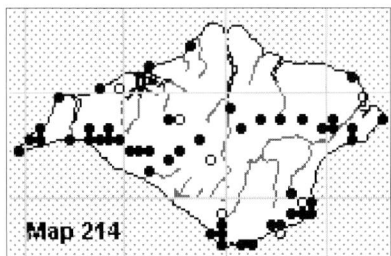
Carlina vulgaris

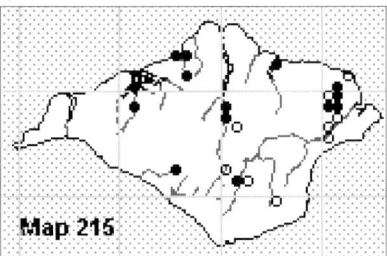
Arctium lappa

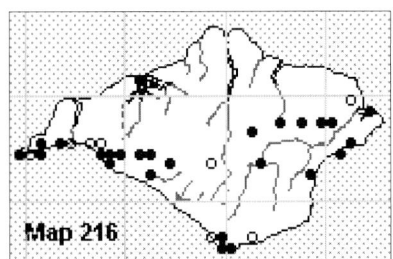
Carduus tenuiflorus

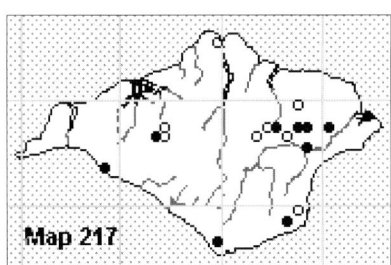
C. crispus

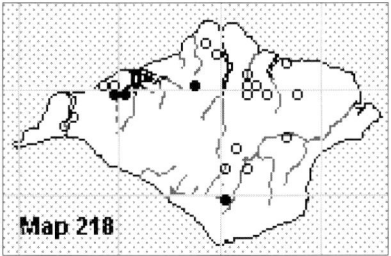
C. dissectum

hundred years. Open, scrubby chalk grassland at Rowridge 4586, MB 1996; Shanklin Down 5779, CP 1997; Nansen Hill, Bonchurch 5778, CP 2001; the 'Leg of Mutton' behind Ventnor industrial estate 5678, CP 1989; two specimens found on Mersley Down 5587, MB 1999.

C. vulgare (Savi) Ten.
SPEAR THISTLE
N. Common everywhere in grassland, verges and waste ground.

C. dissectum (L.) Hill
MEADOW THISTLE
N. A greatly declined plant of unimproved flushed damp acid fen-meadows. Historically recorded from around fifteen sites, the reclamation of moors and heaths resulted in its loss from many of these sites by the twentieth century. Bevis *et al* (1978) knew it from just four sites and it has since disappeared from two of these. Frequent on Smallgains Heath, south of Fattingpark Copse 5290 in 1966 but the site was brought into cultivation in 1972 by which time the plant became confined to the hedge bottom adjoining the copse, BS. There are no further records from this site. Frequent in damp heathy meadow behind Elm Court, Godshill 5282, CC 1986, but the site was destroyed in 1988 to create a playing field. Common in rough grass at Cranmore 3890 in 1967, BS; Bevis *et al* (1978) describe it as occurring in several places at Hamstead and Cranmore and it still persists amongst cleared plantation on the north side of Ningwood Lake 3989 & 4089, GT 2002. The plant still survives along three heathy rides in the south-west of Parkhurst Forest 4790, CP 2000. It has also been discovered growing in quantity in a previously overlooked site, an unimproved wet valley bottom, a tributary of the River Yar, at Fairfields 5079, CP 1997. Map 218

[*C.x forsteri* (Smith) Loudon
MEADOW THISTLE X MARSH THISTLE
N. Described by Bromfield (1856) as 'extremely rare, and now pretty generally considered as a casual mule'.]

C. acaule (L.) Scop.
DWARF THISTLE
N. A frequent and widespread species. It is a characteristic component of unimproved chalk grassland turf and relic limestone grassland. Also found rather locally in neutral clay meadows and short grassland on limey clay slumps along the coast. Occurs rarely in remnant slightly flushed unimproved grassland on greensand soils. In long grassland in meadows off the chalk, it can develop a stem bearing several flowerheads.

C. palustre (L.) Scop.
MARSH THISTLE
N. Frequent and widespread in marshes, damp grassland, verges and clearings in damp woods.

[*C. x celakovskianum* Knaf
MARSH THISTLE X CREEPING THISTLE
N. Bromfield (1856) reports a single plant found by Rev. G.E. Smith between the Needles Hotel and Alum Bay.]

C. arvense (L.) Scop.
CREEPING THISTLE
N. Common everywhere in grassland, verges and waste ground.

Onopordum acanthium L.
COTTON THISTLE
C. A rare plant occurring in small numbers on disturbed ground on verges and chalk banks. Horringford Farm verge 5485, GT 2000; South Down, Chale, eleven rosettes growing from cut down parent plants 4779, GT 1999; Forelands cliff 6587, AC 1999; several in chalk pit on Garstons Down 4785, CL 1991. Bevis *et al* (1978) refer to 3 records from 1960/70 and Bromfield (1856) refers to two records from the 1840s.

Silybum marianum (L.) Gaertn.
MILK THISTLE
Arch. A sporadic casual on waste ground and an established plant along the south coast of the Island. Bromfield (1856) describes this plant as 'truly wild in several places along the Undercliff', where it is still found locally and behaves like a native plant in calcareous grassland and by rabbit warrens along the edges of sunny cliffs. Locally frequent in bare ground above St Catherine's Point 4975, CP 1997; cliff top grassland at St Lawrence above Undercliff Drive 5376, CP 1993.

Serratula tinctoria L.
SAW-WORT
N. A local but sometimes frequent plant of clay heaths and species-rich woodland rides on clay soils across the northern half of the Island. It also occurs in quantity in chalk grassland at the extreme western end of the Island (West High Down, Tennyson Down, Afton Down and Compton Down) and, further to the east on Calbourne Down, 4385. A few plants grow in chalk heath on the top of West High Down. Formerly, it occurred in heathland and bogs on the southern half of the Island, but today it is very rare here. The only recent records from here are Bleak Down 5181, where it was very rare in 1994 and only a single plant could be found in 1998, CP; and Sandown Golf Course 5884, where several plants survived in a small area of heathy grassland in 1993, CP. Map 219

Centaurea scabiosa L.
GREATER KNAPWEED
N. Locally frequent on roadside verges, field borders and in tall grassland. Common in chalk and remnant limestone grassland but also occurs more locally in cliff edge grassland and sandy soils where there is some base enrichment.

C. montana L.
PERENNIAL CORNFLOWER
E. A scarce persistent garden outcast on verges, rough ground and tips. Planted in churchyards.

C. cyanus L.
CORNFLOWER
Arch. Cornflowers have persisted at North Appleford 5081, in the sandy fields west of Bleak Down, for a very

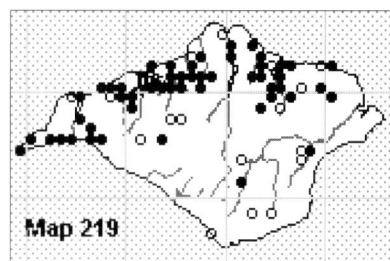

Serratula tinctoria

long time. They were first recorded from here in 1907 by E.H. White but sub-fossil pollen has been found in quantity in peats dated to 300 years ago, with other arable weed pollen just 1.7 km to the north-west of this site. To this day, in good years extensive splashes of blue can be seen amongst the crops in about four fields, often accompanied by the gold of Corn Marigolds. Cornflower populations nationally have crashed and this is now one of only two known persistent sites in the country.

Formerly, this was a widespread cornfield weed, generally on acid sandy soils, but records suggest that it was already becoming scarce by 1900 and by 1960, it was only found very rarely and for a short period when soil was deeply dug. No plants were found in 2000 when deep trenches to lay sewerage pipes were dug across extensive areas of arable land, mostly on the northern half of the Island, but subsequently a few plants appeared in a cereal crop at Tapnell 3786, PS 2002, along the pipeline route. In the same year, a very few plants appeared in scruffy field margins at Idlecombe Down 4585, AB and Nettlestone 6190, L&SS. The plant also occurs as a casual garden escape on waste ground near habitations.

[*C. calcitrapa* L.
RED STAR-THISTLE
C. Extinct. Considered to be a very rare plant of waste places on dry soils in the nineteenth century. There are only two twentieth century records. A few found growing in a field at Rowlands, near Havenstreet, with Yellow Star-thistle in 1932, GB. Recorded from Thorncross Farm, Brighstone in 1962, DF, but not subsequently. There are no further records.]

[*C. solstitialis* L.
YELLOW STAR-THISTLE
C. Extinct. Recorded as an arable weed in a field at Bonchurch in 1836 by Rev. G.E. Smith; from a hedgebank at Cowes in 1922 by J.F. Rayner; at Newport in 1924 by J.W. Long; and in a field at Rowlands, near Havenstreet, with Red Star-thistle in 1932, GB. No further records.]

[*C. melitensis* L.
MALTESE STAR-THISTLE
C. Extinct. A casual of waste tips reported from Cowes and Newport in 1931 by J.W. Long.]

C. nigra L.
COMMON KNAPWEED
N. Frequent and widespread in meadows, verges, woodland rides, scrub and waste places on a wide range of soils. The radiate form is often common but not confined to old grassland sites. Occasionally both rayed and unrayed forms are found growing together.

[*Carthamus tinctorius* L.
SAFFLOWER
C. A very rare casual of waste and rough ground. J.W. Long (1931) described it as frequent on allotment ground at Ryde and on reclaimed land between Cowes and Newport. One further record from deposited soil at Arreton chalk pit, BS 1964.]

Cichorium intybus L.
CHICORY
Arch. A scarce plant, which is usually casual on disturbed ground. Although never common, it was widespread, perhaps mostly as an agricultural relic. It seems to be rare today, but it has appeared consistently in verges and fields in the Bowcombe valley where it was first recorded by Bromfield in 1839. The only post-1987 records are fields between Rowborough and Idlecombe Farms 4685, CP 1999; and in small quantity on Headon Warren undercliffs 3186, RSC 2001.

Lapsana communis L. ssp. *communis*
NIPPLEWORT
N. Common and widespread in a wide range of habitats.

Hypochaeris radicata L.
CAT'S-EAR
N. Common and widespread in grasslands, especially on neutral and acid soils.

H. glabra L.
SMOOTH CAT'S-EAR
N. A rarity of acid sandy soils, historically recorded from five areas. It is believed to survive in three of these but its numbers fluctuate considerably and in some years no plants can be found. At St Helen's Duver, it was recorded in 1879 by Townsend but then not again until 1993 when it was abundant around the car park, 6389. It subsequently declined but there were a few plants here in 1995. Sandy fields at Alverstone were a stronghold of the species and Bevis *et al* (1978) refer to a specific field north east of Hill Farm where it was found reliably up to 1978. Found in a disused sandpit north of Alverstone Farm 5885, PS 1998. Also recorded in small quantity from a sandpit and set-aside field at Hill Heath, Shorwell 4582 & 4682 PS, 1999. Other historic sites, with last recorded dates were a field near Cliff Farm, Shanklin (1909) and Headon Warren (1931).

Leontodon autumnalis L.
AUTUMN HAWKBIT
N. Frequent and widespread in grasslands on a range of soil types.

L. hispidus L.
ROUGH HAWKBIT
N. Frequent and widespread in old grasslands, particularly on calcareous and base enriched soils.

L. saxatilis Lam.
LESSER HAWKBIT
N. Frequent and widespread in dry calcareous and acid soils. Also found on slumping cliffs on the south coast.

Picris echioides L.
BRISTLY OXTONGUE
Arch. Common in disturbed ground and undercliffs especially on heavy clay soils and around the coast. Bromfield (1856) wrote, 'I am informed by Mr. Rawkins, late of Hardingshute Farm, that sheep are very partial to the early radical herbage of this very rough plant, which in that neighbourhood at least is known under the very incorrect name of Borage.'

P. hieracioides L.
HAWKWEED OXTONGUE
N. Occurs locally in dry grassland and disturbed ground, particularly on the chalk and around the coast.

Tragopogon pratensis L.
GOAT'S-BEARD
N. Frequent and widespread in tall, dry grassland. Ssp. *minor* (Mill.) Wahlenb. is the common form. There is a single record for ssp. *pratensis* from Lucket's Farm, Cranmore 3889, PS 2000.

T. x *mirabilis* Rouy
GOAT'S-BEARD X SALSIFY
N. A single plant found on a roadside verge at Bowcombe 4787 in 1997 and again in 2000, SB.

T. porrifolius L.
SALSIFY
E. An occasional casual or established relic of cultivation on waste ground and verges. There are post-1987 records from Brighstone 4282, Thorness 4592, Bowcombe 4687 & 4787, Gurnard 4795, Carisbrooke 4887, Ventnor 5677, Sandown 6084 and Ryde 6092. By 2002 it had increased considerably at Gurnard, JEG.

[*T. hybridus* L.
SLENDER SALSIFY
C. A single record of a plant growing on wasteland at Ventnor, TW 1956 (conf. K).]

Sonchus arvensis L.
PERENNIAL SOW-THISTLE
N. Frequent and widespread in arable fields, verges, undercliffs, marshes and along the drift line on beaches and estuaries.

S. oleraceus L.
SMOOTH SOW-THISTLE
N. Common on bare, disturbed soil everywhere.

S. asper (L.) Hill
PRICKLY SOW-THISTLE
N. Common on bare, disturbed soil everywhere.

Lactuca serriola L.
PRICKLY LETTUCE
Arch. Locally frequent on disturbed ground. Bevis *et al* (1978) only list a small handful of stations, which would suggest that it is increasing. First recorded in 1920, a single specimen at Cowes. By 1929, James Groves described it as abundant near Cowes. Map 220

L. virosa L.
GREAT LETTUCE
N. Very rare, on waste and disturbed ground. Single plants have been recorded from a roadside in Yarmouth 3589, CP 2000; Osborne shore 5295, CP 1999; a car park in Sandown 5883, GT 1997; and waste ground in Newport 5089, DB 1990. The only earlier records were Lynn Bottom refuse site, where it was locally abundant in 1973 (Bevis *et al* 1978) and from a hedgebank near Winford, between Wroxall and Newchurch, where there were a succession of records between 1844 and 1900.

Cicerbita macrophylla (Willd.) Wallr.
COMMON BLUE-SOW-THISTLE
E. A rare but persistent garden throw-out, first recorded in 1979 from a roadside bank at Whitepit Lane, Newport, where it was still present in 1984, BS; Gurnard 4694, JO & PS 1996; chalk pit near Combley Farm 5487, RK 1985.

Mycelis muralis (L.) Dumort.
WALL LETTUCE
N. Very locally frequent in open areas and shady banks in a few woods on the chalk and on old walls in gardens, churchyards and ruins. It still grows on the rocky greensand ledges in Cliff Copse, Shanklin 5680, where Bromfield (1856) found it to be frequent. Other Bromfield sites where it still survives are Quarr Abbey ruins, Westover plantation and woods about Rowridge. Map 221

Genus *Taraxacum*
Dandelions

There has been scarcely any critical recording of Dandelions on the Island. The following species have been confirmed from the county. The majority of records were made between 1960 and 1970:

Section *Erythrosperma*: *T. lacistophyllum*; *T. rubicundum*; *T. oxoniense*; *T. fulviforme*; *T. fulvum*; *T. retzii*. This is the group that was formerly known as *T. laevigatum* (Willd.) DC. Lesser Dandelion. These are species of dry, frequently species-rich calcareous grasslands. *T. retzii* is a very local species of acid sandy heaths in southern Britain. It was found on Norton Spit 3489, ABo 1981 (confirmed Chris Haworth).

Section *Celtica*: *T. bracteatum*; *T. subbracteatum*; *T. nordstedtii*.
T. nordstedtii occurs in wet meadows.

Section *Hamata*: *T. pseudohamatum*.

Section *Ruderalia*: *T. subexpallidum*; *T. stenacrum*; *T. sellandii*.

Recent random collections (1998-2002) by GT from the eastern side of the Island have extended the list significantly, bringing the current total to 38 taxa. Specimens have been determined by A.J. Richards. The following taxa have been recorded:

Section *Erythrosperma*: *T. arenastrum* (A local and rare species recorded from St Helen's Duver, 2001); *T. wallonicum* (Sandown roadside 6084, 2001).

Section *Celtica*: *T. bracteatum*

Section *Hamata*: *T. hamatum*; *T. subhamatum*; *T. pseudohamatum*; *T. atactum*; *T. sahlinianum*.

Section *Ruderalia*: *T. pannucium*; *T. subexpallidum*; *T. corynodes*; *T. undulatum*; *T. dilaceratum*; *T. insigne*; *T. lepidum*; *T. pallidipes*;

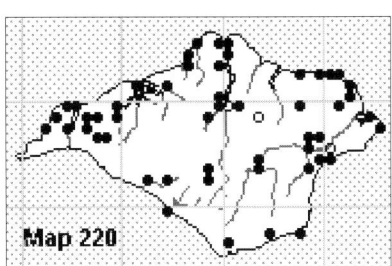

Lactuta serriola

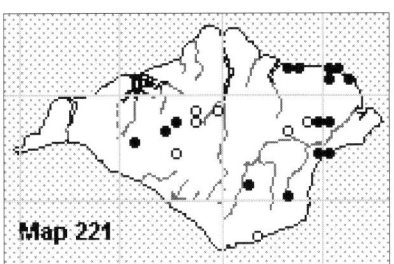

Mycelis muralis

T. undulatiflorum;
T. angustisquameum; *T. exacutum*;
T. leptodon; *T. macranthoides*;
T. oblongatum; *T. cophocentrum*;
T. pachymerum; *T. pulchrifolium*;
T. polyodon; *T. fasciatum*;
T. subxanthostigma.

Crepis biennis L.
ROUGH HAWK'S-BEARD
N? A rare plant of roadside verges and rough ground. There is a scatter of historic records including a number of stations listed by Drabble & Long (1931). There are recent records from three sites. In 1968, it was confirmed from a roadside ditch along the Calbourne Road, west of Swainston, 4387, where it was abundant; it was frequent here in 1994, BS and still surviving in small quantity in 2002. Bevis *et al* (1978) list two other sites: around the old mill at Alverstone, 1971 and by the roadside north of Little Whitefield 5889, 1963 and 1974. It is still present at Alverstone 5785 in small quantity, JO & PS 1996. Also recorded from waste ground by Down Lane, Upper Ventnor 5577, GT 1997.

[*C. nicaeensis* Balb.
FRENCH HAWK'S-BEARD
C. Extinct. Only known from Townsend (1883) where it is listed as growing in sown grass in two sites in 1880, near Shanklin and between Niton and Whitwell.]

C. capillaris (L.) Wallr.
SMOOTH HAWK'S-BEARD
N. Common and widespread in grasslands and disturbed ground.

C. vesicaria L. ssp. *taraxacifolia* (Thuill.) Thell. ex Schinz & R. Keller
BEAKED HAWK'S-BEARD
E. Frequent and widespread in grasslands and disturbed ground. First recorded in 1880 and still considered by F. Stratton to be uncommon in the early 1900s.

[*C. setosa* Haller f.
BRISTLY HAWK'S-BEARD
C. Extinct. A former casual of grass seed and fodder recorded from several sites in the nineteenth century but not since 1879.]

Pilosella peleteriana (Mérat) F.W. Schultz. & Sch. Bip. ssp. *peleteriana*
SHAGGY MOUSE-EAR-HAWKWEED
N. A Red Data Book species restricted to 3km of south-facing chalk cliffs between Freshwater Bay and West High Down, 3084 to 3385, TDD & CDP, 1998. It was first recorded from here by Albert Hamborough in 1849 (Bromfield, 1856). This striking plant is frequent in open ground and this is the largest U.K. population growing on chalk. The range of this subspecies extends westwards into Dorset, where it occurs in similar situations on coastal chalk and on one site on limestone in Purbeck.

P. officinarum F.W. Schultz & Sch. Bip.
MOUSE-EAR-HAWKWEED
N. Frequent and widespread in dry, short grassland particularly on chalk, cliff slopes and sandy soils.

P. aurantiaca (L.) F.W. Schultz & Sch. Bip.
FOX-AND-CUBS
E. An occasional garden escape in dry grassland, perhaps less frequent than formerly. Shorwell churchyard 4583, MB 2000; Afton Road verge, Freshwater 3486, RD 2000; Mackett's Farm, Horringford 5484, LS 2000. First recorded in 1838 from St Lawrence churchyard, where it was suspected of having been planted (Townsend, 1883).

Genus *Hieracium*
Hawkweeds

A difficult group, which is poorly recorded with few confirmed records. Widely distributed but local.

Hieracium sabaudum (sect. *Sabauda* F.N. Williams)
N. Confirmed material has been collected from Borthwood, J. Rayner 1927 (hb. BM); Parkhurst Forest 1958 (hb. CGE); and Quarr Abbey ruins 5692, AWW 1962. Refound at Parkhurst Forest 4790, RCS 1987.

Hieracium umbellatum L. (sect. *Umbellata* F.N. Williams)
N. Not uncommon in Parkhurst Forest 4790, RCS 1987. Bleak Down 5181, 1958 (conf. Sell & West); still present here 1994. Believed to occur in a number of sites on heathland and open woodland rides.

H. eboracense Pugsley
? A record of this taxon from SZ58 is shown in *The Critical Supplement to the Atlas of the British Flora*. F.H. Perring (ed.) 1968.

H. trichocaulon (Dahlst.) Johanss. (sect. *Tridentata* F.N. Williams)
N. Material confirmed by Sell & West collected from Borthwood, 1858 (hb. CGE); Bonchurch, A.G. More 1860 (hb. DBN); Combley Great Wood roadside 5489, AWW 1960; and Parkhurst Forest 4790, RCS 1987.

H. exotericum Jordan ex Boreau (sect. *Vulgata* (Fries) F.N. Williams)
N. Recorded as growing on a bank at Player's Wood, near Ryde 5893, JB 1968. The record was confirmed at the time by C.E.A. Andrews as belonging to this aggregate which suggests that it could belong to a different taxon within this Section.

Gazania rigens (L.) Gaertn.
TREASUREFLOWER
S. Persistent in uncultivated sandy soil on the south-facing bank of Ryde Canoe Lake 6092, CP 2001, having survived here untended for over ten years.

Filago vulgaris Lam.
COMMON CUDWEED
N. Scarce and thinly scattered on nutrient-poor dry sandy and gravelly soils. Common Cudweed was a common plant in Bromfield's time but it has declined considerably over the past one hundred years. Bevis *et al* (1978) only knew it from three hectads, but it appears to be currently increasing. Sites include a sloping roadside verge at Stone Shute and nearby on St George's Down 5186, CP 1997; a set-aside field at Heath Hill, Shorwell 4582, PS 1999; roadside verge by Lake Middle School 5883, GT 2000; sandy field at Hill Farm, Alverstone 5785, PJW 1997;and uncultivated sandy field to east of Newchurch 5685, CP 2001. At most sites it is present in small quantity but a set-aside field below Brook House,

6092, supported many thousands of vigorous plants (many over 23cm in height) in 2000, PS & EC. It is suspected that they were introduced as a seed contaminant. Map 222

[*F. lutescens* Jord.
RED-TIPPED CUDWEED
N. Extinct. A former rarity of dry sandy soils although historically identification has been confused. Two historic records from SZ48: Brighstone, F. Townsend 1904 (conf. T. Rich); and field above Idlecombe Down, F. Stratton 1913 (*J. Botany* 1913).]

[*F. pyramidata* L.
BROAD-LEAVED CUDWEED
Arch. Extinct. A former scarce plant of sandy arable fields, reported from many scattered sites in the nineteenth century, with a concentration in the Kingston area. Only known from two sites in the twentieth century. A garden weed at Totland, W.C. Barton 1912 (hb. BM, NMW, conf. T. Rich). Also recorded from Alverstone, H.S. Redgrove 1918 (hb. BM, conf. T. Rich). Edward Lousley came across 'exceptionally fine erect plants' of this species here in 1947, in a lane near Hill Farm (hb RNG). There are no further records]

F. minima (Sm.) Pers.
N. A rare plant of dry, sandy heaths. Recorded from a number of sites including Heath Hill at Shorwell, Borthwood, Headon Warren and Bleak Down, up to 1931 but no further records for more than fifty years. Bevis *et al* (1978) considered it to be extinct but in 1987 it was discovered growing on a south-facing sandstone slope, much disturbed by rabbits, at Rowdown, Brighstone 4383, CC. This site is now (2002) completely overgrown and the plant has been lost but in 1994, small numbers of plants were found growing on St George's Down 5186, CP, a well-botanised site where it was recorded by Bromfield (1856). Numbers have increased here and by 2001 it was locally frequent.

[*Anaphalis margaritacea* (L.) Benth.
PEARLY EVERLASTING
E. Extinct. A garden outcast found growing on dumped spoil in the chalk

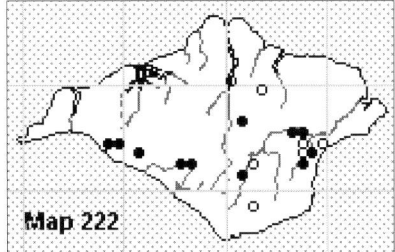

Filago vulgaris

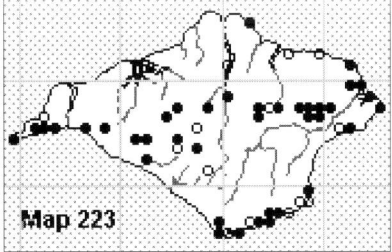

I. conyzae

pit at Downend on Arreton Down, GB 1963. It persisted until 1967 but was eventually lost due to tipping.]

(*Gnaphalium sylvaticum* L.
HEATH CUDWEED
N. Status unknown. Described by J. Woods in *The Botanist's Guide through England and Wales* (1805) as frequent in the south-east parts of the Island. Bromfield (1856) wrote of this species, 'said to inhabit the Isle of Wight, but I have never met with it myself, or seen indigenous specimens from others.')

G. uliginosum L.
MARSH CUDWEED
N. Widespread in damp rutted tracks and fields, gateways and receding margins of ponds, generally avoiding calcareous soils.

Inula helenium L.
ELECAMPANE
(local: velvet dock)
Arch. A garden outcast in hedgebanks, fields and waste ground particularly on clay soils. Bromfield (1856) knew it from a number of sites. It persists in a few of these but has generally declined and there are recent records from just four areas. Well established in Brading churchyard 6087, AC 1996, where it has been known for more than thirty years; recorded from the top of the beach in Thorness Bay in 1967 and still growing in the area near Sticelett Copse 4693, CP 2001; recorded by Bromfield (1856) from a field near Thorley church and still present in roadside verge just south of Thorley 3788, PS 1998. At other long established sites it has not been seen recently. Bromfield (1856) recorded it from within the walls of Quarr Abbey ruins and it was still at Quarr in 1971, JH. It survived in another Bromfield site, a field by the Blackbridge Brook at the bottom of Havenstreet village, until 1986 when the field was ploughed. There is a continuity of records from the upper reaches of King's Quay until 1942. First recorded in 1770, in waste places about Freshwater and elsewhere 'in ye Isle of Wight' by Richard Waring.

I. conyzae (Griess.) Meikle
PLOUGHMAN'S-SPIKENARD
N. Locally frequent on dry, chalk grassland. Also found in a few sites in grassland on lime-rich clay soils on the northern half of the Island. Map 223

I. crithmoides L.
GOLDEN SAMPHIRE
N. Confined to the Newtown and Medina estuaries. It is quite frequent at Newtown in several sites including the east and west spits and on the banks around the old saltpans in the harbour, a Bromfield site. On the Medina, it is confined to a few sites on both banks at the north end of the estuary below Cowes including Werrar and Medham saltmarshes and as scattered plants on the eroding eastern estuary bank. It is recorded from near Yarmouth between 1913 and 1931 (Drabble & Long, 1931) and three plants were established behind the seawall at Bouldnor 3689, BS 1985. These may have originated from the large Hurst Spit population on the adjacent mainland. There are no records from the chalk cliffs although it occurs in this habitat in Purbeck.

Rev. R. Warner (1795) mentions that in his day this plant was gathered 'with little trouble and no danger on all the sea beaches in and near the island' as a poor substitute for Rock Samphire (*Crithmum maritimum*). Bromfield (1856) refers to the plant as having 'an aromatic not ungrateful smell, and a warm, pungent saline

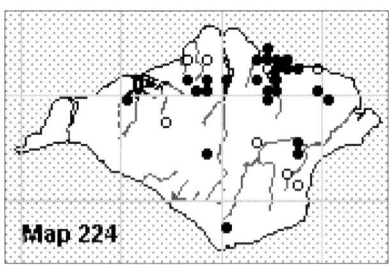
Solidago virgaurea

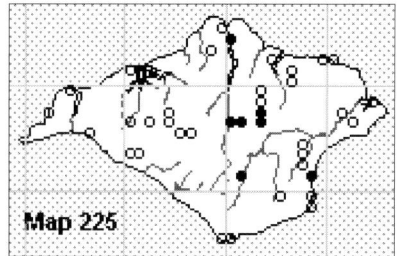
E. acer

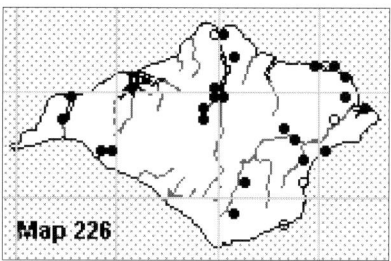
Conyza canadensis

taste, approaching in both respects to the true samphire (*Crithmum*) for which perhaps it would be a good and certainly more accessible substitute.' Surprisingly, neither author refers to gathering Glasswort as a substitute for Rock Samphire.

Pulicaria dysenterica (L.) Bernh.
COMMON FLEABANE
N. Frequent in damp meadows, verges, marshes and woodland rides. Although widespread, it has a tendency to avoid dry chalky soils.

[*P. vulgaris* Gaertn.
SMALL FLEABANE
N. Extinct. Formerly found on greens and about farmyards in wet ground, heavily poached by livestock. Described by Bromfield (1856) as plentiful on parts of St Helens Green that were flooded during the winter. It was still so in 1900 (F. Stratton) but there are no records from here or any other Island site after 1909. Other historic sites were around Walpen Farm, Chale (Bromfield, 1856); west side of The Wilderness (A.J. Hamborough, 1871); by barns at Leesland Farm, Bathingbourne (E.H. White, 1904); Apse Manor farmyard, 1858 (More, 1898); and by Hardingshute pond, near Nunwell (F. Stratton, 1909).]

Solidago virgaurea L.
GOLDENROD
N. A local plant of species-rich rides in ancient woodlands, generally on heavy clay soils; quite frequent in coastal woodlands. An unusual site is around the scrubby edges of chalk grassland on Chillerton Down 4884, CP 1999 and in the adjoining Tolt Copse. Bromfield (1856) described this plant as being abundant in woods, hedgebanks, heaths and dry, hilly bushy places. Much declined since his day, particularly on the sandy soils in the south of the Island. Map 224

S. canadensis L.
CANADIAN GOLDENROD
E. A persistent garden relic on roadsides and waste ground.

Aster novi-belgii L. agg.
MICHAELMAS-DAISIES
E. Frequent and persistent garden relics on waste ground. There has been no systematic recording of this group but the following have been recorded: *A. novi-belgii* L. (Confused Michaelmas-daisy), *A.* x *versicolor* Willd. (Late Michaelmas-daisy), *A.* x *salignus* Willd. (Common Michaelmas-daisy) and *A. lanceolatus* Willd. (Narrow-leaved Michaelmas-daisy).

Erigeron glaucus Ker Gawl.
SEASIDE DAISY
E. Much grown in gardens and increasingly is an outcast naturalised on walls and cliffs around the coast. Although widespread, it is particularly common around the coast from Sandown to Shanklin and at Ventnor. On chalk cliffs below Fort Redoubt, Freshwater Bay.

E. karvinskianus DC.
MEXICAN FLEABANE
E. An increasing garden escape, naturalised on walls and pavements in many of our towns and villages. First recorded by J. Long (1931) from Shide at Newport. Recorded on walls in Ventnor in 1945, TW, where it is still found today. It was not recorded from the West Wight before 1985. It can colonise new sites rapidly. At Fort Victoria, 3389, a single plant was found on the walls in 1994, SC, and it was well established here in 1997.

E. acer L.
BLUE FLEABANE
N. A scarce and declining plant in nutrient-poor dry, open grassland.

Bromfield (1856) described it as 'not infrequent' and Bevis *et al* (1978) described it as 'locally plentiful in a few places'. It has continued to decline with the loss of open, dry heathy grassland and the few recent sites are all in East Wight: several plants on St George's Down 5186, BS 1998; a single plant on Bleak Down 5181, CP 1981; 2 plants at Arreton Cross verge 5386, AC 2000; a few plants in thin grassland on Arreton Down near Michal Morey's hump 5387, CP 1998; boundary wall at East Cowes cemetery 5094, CP 1994, but not seen recently; top of old brick wall at Shanklin 5881, CP 1998. Map 225

Conyza canadensis (L.) Cronquist
CANADIAN FLEABANE
E. A locally frequent and increasing introduction of dry, bare waste ground and disturbed sandy soils, particularly in and around towns. First recorded in 1926 by J.W. Long, from a gravel pit near Newport, where he described it as abundant in some years. Map 226

C. cf. *bilbaoana* Rémy
E/C. A single plant of this recently recognised alien was found on waste ground at Seaclose near Newport 5089, EC, in 2001. It is likely to be found more frequently in the future.

C. sumatrensis (Retz.) E. Walker
GUERNSEY FLEABANE
E. Occurs in similar places to Canadian Fleabane. First noticed in 1999 by PS who found plants at Totland 3287, Ryde 5992, Puckpool 6192 and Horringford 5584. In 2000 it was also found at Brighstone, Newchurch, Sandown and Lake. In 2001 it was recorded from St Helen's Duver and Steephill. This is clearly a rapidly increasing species likely to

Bellis perennis L.
DAISY
N. Widespread and common in short grasslands everywhere.

Tanacetum parthenium (L.) Sch. Bip.
FEVERFEW (local: whitewort)
Arch. A widespread garden escape, particularly found near habitation. Bromfield (1856) described this plant as a common herb in rustic gardens, from whence it readily escapes.

T. vulgare L.
TANSY
N/E. Widespread in small quantity on roadsides, waste places and stream banks. At some sites, for instance along the River Yar, it may be of native origin but at most it is considered to have originated from garden escapes.

Seriphidium maritimum (L.) Polj.
SEA WORMWOOD
N. A rare plant of upper shingly saltmarshes. In Bromfield's time, it grew in all the estuaries on the north coast but today it is only in the Medina and Newtown estuaries and Bembridge Harbour. In the Medina, it is confined to a single saltmarsh spur south of Medham 5093, AM 1996 but at Newtown there are three sites including the east spit and a saltmarsh bank below Walter's Copse 4191, 4390 & 4491. It grew in two sites at Bembridge bordering the Mill pond 6389 but, in 1997, the main colony was lost from a shingle spit due to coastal erosion, CP. Historically recorded from Quarr shore where it was last recorded in 1963, JH.

Artemisia vulgaris L.
MUGWORT
Arch. Common and widespread in waste places and roadsides. E.H. White (1906) said the leaves, 'when dried and rolled form the wild tobacco known to most schoolboys.'

A. verlotiorum Lamotte
CHINESE MUGWORT
N. Not known until 1994, when it was found at the edge of the car park at Osborne House 5194, GK. It has since been found to be locally frequent on the Estate.

A. absinthium L.
WORMWOOD
Arch. A very rare plant of dry hedgebanks on calcareous soils. Bromfield (1856) described it as frequent in hedgebanks and dry waste places about villages but the majority of identified historic sites seem to be between Luccombe and Blackgang, particularly about the inner cliff. More (1898) described it as 'one of the most striking and common species all along the Undercliff'. The only extant site is at the top of the west cliff at Niton where it has declined to about ten plants surviving at the edge of invading scrub 4975, CP 2001; it may be native here. It may still survive at St Lawrence as there are records of two specimens on waste ground opposite Ventnor Botanical Gardens 5477, RK 1973, and a few plants at Seven Sisters Road 5376, RK 1980.

[***Otanthus maritimus*** (L.) Hoffmanns. & Link.
COTTONWEED
N. Extinct. Known only from a record by Snooke (1823), who recorded this plant from the 'shore at Sconce Towe', a site that would be in the vicinity of Fort Victoria at Yarmouth. Bromfield and succeeding botanists were unable to refind it. Cottonweed is now extinct in England but in the late eighteenth and early nineteenth century, it grew on the shore at Mudeford, by Christchurch Harbour, (where it survived until 1891) and on Brownsea Island in Poole Harbour.]

Achillea ptarmica L.
SNEEZEWORT
N. A greatly declined plant of grassy heaths and acid marshes. Although described as 'not common' by Bromfield (1856), from the records he cites, it is clear that this was a locally frequent plant in the northern half of the Island. It has further declined since Bevis *et al* (1978). Extant sites are: an unimproved meadow on the west bank of Palmer's Brook at Blacklands Road 5289, AM 1996; persisting in small quantity at the edges of rides in three locations in Parkhurst Forest 4691, 4789 & 4790, CP 1996; surviving at the edge of a scrubby patch of grassland at Corfheath Firs,

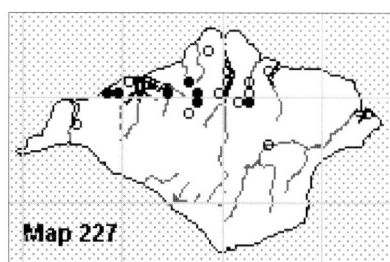

Achillea ptarmica

Porchfield 4490, CP 1998; about 50 plants in a single ride edge location in Bouldnor Copse 3890, CP 1997; and a single patch of heathy grassland at Bouldnor 3990, VS 1998. Map 227

A. millefolium L.
YARROW
N. Common and widespread in grasslands everywhere.

Chamaemelum nobile (L.) All.
CHAMOMILE
N. This plant of short acid turf was widespread in Bromfield's time but has since declined drastically so that Bevis *et al* (1978) believed it to be extinct. In fact it had been overlooked. In 1980, a plant with 'unmistakable scent' was found on St Helen's Duver in short turf near the millpond causeway, MK. It has not been refound here, but the following year it was found in quantity on St Helen's Green 6289, BS, a Bromfield site. It survives in quantity here, but close mowing rarely permits it to flower. Also present in a similar location in mown grass at Puckpool Battery 6192, CP 1996 and in inundation grassland at Seaview Duver 6291, SC 2001; there are historic records from this area. Another historic site where it might survive undetected is Blackpan Common, Sandown.

Anthemis punctata Vahl
SICILIAN CHAMOMILE
E. Abundantly naturalised by the coast at Sandown around Lake Battery and alongside the cliff path 5983, CP 2000. First recorded here in 1972, EC pers. com., but must have been established for much longer.

A. arvensis L.
CORN CHAMOMILE
Arch. An occasional arable weed, which has greatly declined and is now

a very rare casual. Bevis *et al* (1978) described it as very uncommon and not persistent. The only post-1950 records of which I am aware are from Duxmore 5588, RK 1964; Hamstead 4091, RK 1972; growing with poppies in an arable field below Headon Warren 3185, CP 1997; east of Tapnell Farm 3786, PS 2002 and 3787, SB 2002; and near Shalcombe 3886, SB 2002.

A. cotula L.
STINKING CHAMOMILE
(local: morgin! or mavin)
Arch. An occasionally abundant plant of arable fields and waste ground, scattered across the Island. It is nowhere near as prolific as it was at the time of Bromfield (1856). He wrote 'well known to reapers by the name of Morgin, and unanimously accused of blistering the feet, hands and open bosums of those employed in making up the corn into shocks. ...I have been repeatedly assured by the peasantry that they have known men incapacitated for work and laid up from the injurious operation of this noxious weed for days together in harvest time.'

A. tinctoria L.
YELLOW CHAMOMILE
C. Bevis *et al* (1978) described this as a garden escape but the only record of which I am aware is at Ventnor, TW 1966.

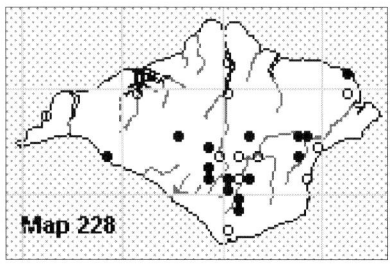

Chrysanthemum segetum

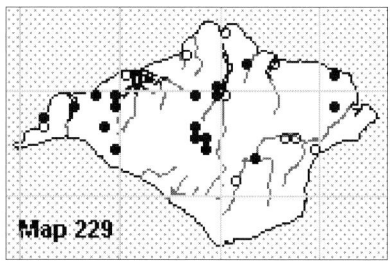

Matricaria recutita

Chrysanthemum segetum L.
CORN MARIGOLD
(local: yellow bozzum; botherum!)
Arch. Although much declined as a plant of arable weeds, it still persists in sandy fields on the south of the Island, sometimes still in great abundance turning whole fields golden. Map 228

Leucanthemum vulgare Lam.
OXEYE DAISY (local: horse daisy!)
N. Frequent and widespread in old meadows, and more recent grasslands. Also found on bare roadside banks.

L. x *superbum* (Bergmans ex J.W. Ingram) D.H. Kent
SHASTA DAISY
E. A persistent and locally well-established garden throw-out. Frequent in cemeteries. Sometimes found some distance from habitation such as Gurnard cliffs 4695, CP 2001; Forelands cliffs 6587, AC 1996; Afton Down roadside 3585, PS 2000.

Matricaria recutita L.
SCENTLESS MAYWEED
Arch. A local arable weed of well-drained soils. Bevis *et al* (1978) only confirmed it from five sites. The map of current known sites suggests it may be increasing. Map 229

M. discoidea DC.
PINEAPPLEWEED
E. A frequent and widespread weed of gateways, trampled verges and farm tracks. No known records before 1900 but, in 1913, F. Stratton wrote, 'the extraordinary rapidity with which this plant has spread itself in the Island is quite unaccountable' (*J. Botany*, 1913).

Tripleurospermum maritimum (L.) W.D.J. Koch
SEA MAYWEED

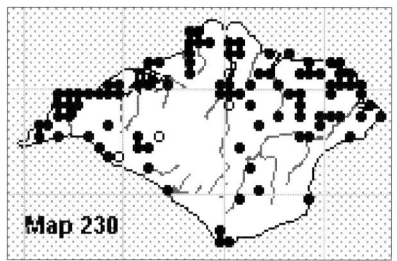

S. erucifolius

N. A frequent plant on shingle and muddy beaches and on cliffs of clay and chalk around the coast.

T. inodorum (L.) Sch. Bip.
SCENTLESS MAYWEED
Arch. A common weed of arable and waste ground everywhere.

Senecio cineraria DC.
SILVER RAGWORT
E. Naturalised on cliffs around the coast at Ventnor, Bonchurch, Steephill, Lake and Forelands. Also sometimes found as a garden escape on walls and freely-drained ground. First recorded at Steephill (Drabble & Long, 1931).

S. x *albescens* Burb. & Colgan
SILVER RAGWORT X COMMON RAGWORT
N. Recorded from a trackside at Wellow 3888, PS 2000. This hybrid was also recorded by Drabble & Long (1931) from Yarmouth and Newport.

S. jacobaea L.
COMMON RAGWORT
N. Frequent and widespread in overgrazed grasslands, rough ground and beaches.

S. aquaticus Hill
MARSH RAGWORT
N. Widespread but local in damp and marshy grasslands, particularly in river valleys.

S. erucifolius L.
HOARY RAGWORT
N. Widespread in unimproved grassland and woodland rides, particularly on heavy clay soils and on chalk. Map 230

S. squalidus L.
OXFORD RAGWORT
E. Surprisingly scarce and not persistent on waste ground and verges. First noted in 1961 from Quarr Abbey ruins and the Osborne estate, JH. From then on, it has been recorded from a scatter of sites, the great majority of which are in north-east Wight. Recent records have been from a roadside verge at Quarr 5692, CP 1990-2002; and the side of a lane by Hermitage Farm, Upper Dolcoppice 5079, GT 1997. This plant is abundant in urban areas on the

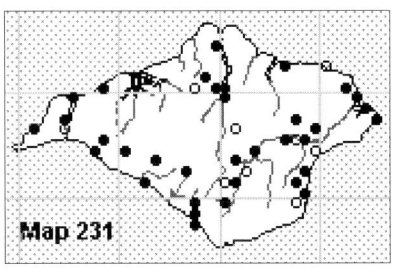
S. sylvaticus

immediate Hampshire mainland.

S. x baxteri Druce
OXFORD RAGWORT X GROUNDSEL
C. Recorded by A.W. Westrup from Ryde in 1964.

S. vulgaris L.
GROUNDSEL
N. A frequent and widespread weed of arable and waste ground. The rayed form occurs sometimes. Both were found growing together on the upper sandy beach on the east spit at Newtown estuary 4191, CP 1997.

S. sylvaticus L.
HEATH GROUNDSEL
N. Local and scattered on disturbed acid soils on heathland, woodland rides and sandy coastal cliffs. Map 231

S. viscosus L.
STICKY GROUNDSEL
E. A rare plant on shingle and disturbed ground. First recorded from Cowes and Newport (Drabble & Long, 1931). Found on shingle at Osborne Bay 5295, MMS 1962; rediscovered here in 1979, BS but not seen since. Recent records from Gurnard 4694, JO & PS 1996; car park at Yarmouth 3589, PS 1998; St John's railway station, Ryde 5992, PS 1999, where several hundred plants were growing on track ballast; and on ballast at Sandown railway station 5984, GT 2002.

[***Tephroseris integrifolia*** (L.) Holub ssp. ***integrifolia***
FIELD FLEAWORT
N. Extinct. A rarity of short calcareous grassland with historic records from two sites. F. Stratton recorded it from Afton Down in 1883 (*J. Botany* 1913) and it was recorded from here in 1930 by a Miss U.K. Smith from Brighton Technical College. At the second site, Westover Down, it persisted much longer. It was first recorded here in 1868 by J. Baker, growing plentifully on the south-east slopes close to the road passing through Calbourne Bottom (More, 1871). It was still plentiful here when Stratton recorded it in 1913, but by 1963, the colony was severely threatened by young conifer plantations over the south-eastern slopes of the down. An area was left unplanted with conifers, Miss Smith grew plants from seed and turf transplants were tried on Brook Down. None of these measures was successful and it was last recorded, in very small numbers, in 1974, RK.]

Doronicum pardalianches L.
LEOPARD'S-BANE
E. Very rare garden escape recorded from a hedgerow bordering a lane between Gatcombe Church and Sheat Manor 4984, BS 1998, where it was first recorded in 1936, GB.

Tussilago farfara L.
COLT'S-FOOT
N. Frequent and widespread on bare and disturbed ground. A characteristic colonising species of unstable undercliffs. Bromfield (1856) described this plant as extremely troublesome in cornfields on stiff clay soils on the north side of the Island.

Petasites japonicus (Siebold & Zucc.) Maxim.
GIANT BUTTERBUR
E. Well established in a few spots around Freshwater: a former duck pond on the south side of School Green Road, Freshwater 3387, CP 1998; wet ground where Blackbridge Road crosses the brook 3486, CP 2000; slipping coastal clay at Fort Victoria 3489, BS 1973. It is believed to have been lost from the latter site by 1998 as a result of development. The Freshwater Marshes plants are said to have been introduced from the School Green Road site sometime during the 1970s. Recorded in error as *P. albus* in Bevis *et al* (1978).

P. fragrans (Vill.) C. Presl
WINTER HELIOTROPE
E. Well established in abundance right across the Island in plantations, verges and on slumping cliffs around the coast, flowering well in sheltered sunny sites. Well known to Bromfield (1856) as a naturalised species.

Calendula officinalis L.
POT MARIGOLD
C. An occasional casual of disturbed ground and waste tips.

Ambrosia artemisiifolia L.
RAGWEED
C. A very rare casual. Appeared in a garden in Carisbrooke 4888 from bird seed, BS (det. J.L. Mason) 1998. There are old records from Newport in 1895 (Rev. E. Linton, hb BM) and Long (1931) who described it as having persisted for several years by a flourmill.

[***Xanthium strumarium*** L.
ROUGH COCKLEBUR
C. Extinct. Long (1931) records this plant from waste ground near Cowes and Newport. No further records.]

X. spinosum L.
SPINY COCKLEBUR
C. Bevis *et al* (1978) reported that four plants appeared in a greenhouse at Watergate nursery, Newport in 1969, growing from wool shoddy used as manure for tomatoes. Bromfield (1856) described this plant as a weed in the garden of St John's, Ryde about 1850.

[***Guizotia abyssinica*** (L. f.) Cass.
NIGER
C. Extinct. Long (1931) describes this plant as occurring frequently on waste ground at Ryde, Cowes and Shide. No further records.]

Helianthus annuus L.
SUNFLOWER
C. An occasional casual on waste ground.

Galinsoga parviflora Cav.
GALLANT-SOLDIER
E. A rare weed in allotments, gardens and urban sites, in all cases clearly spread from plant nurseries. First recorded in 1974 growing amongst newly planted beech hedge at Carisbrooke High School in 1974, BS.

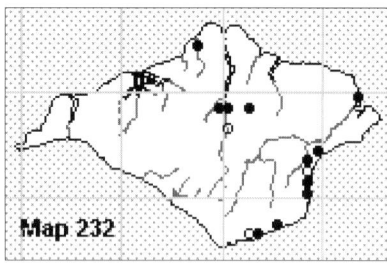

G. quadriradiata

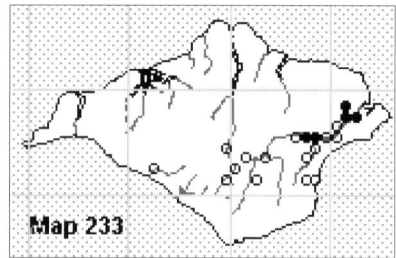

Bidens cernua

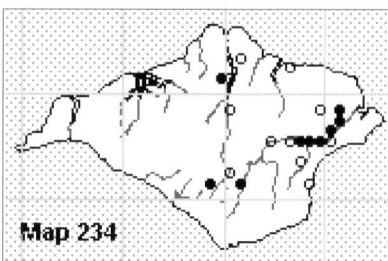

B. tripartita

The only other records are a single plant on disturbed ground on St Helen's Green 6289, AC 1998; and as a weed in Isle of Wight College nursery 4989, CP 2000.

G. quadriradiata Ruiz & Pav.
SHAGGY-SOLDIER
E. A local weed in allotments, gardens and urban sites, mostly in east Wight. First recorded, in error as *G. parviflora*, in 1956 from the garden of Mirables, St Lawrence. Map 232

Bidens cernua L.
NODDING BUR-MARIGOLD
N. An uncommon plant of ponds, streams and ditches. Bromfield (1856) described it as frequent in the Medina and Eastern Yar river valleys but Bevis *et al* (1978) describe it as occasional and mainly in the East Wight. Since that time, its decline has been exacerbated by a general lack of river and ditch management and all modern records are from the Eastern Yar around Alverstone and Carpenters on Brading Marshes. Map 233

B. tripartita L.
TRIFID BUR-MARIGOLD
N. In similar situations to *B cernua* and having suffered a similar decline. It does, however, remain more widespread and more frequently recorded than that species. Abundant by Morton Brook, Sandown 5985, following ground disturbance for pipe-laying, CP 2002. Map 234

Eupatorium cannabinum L.
HEMP-AGRIMONY (local: raspberries and cream)
N. A frequent and widespread plant of marshes, river banks, woodland rides, undercliffs and sometimes disturbed ground.

LILIIDAE
MONOCOTYLEDONS

140. BUTOMACEAE
Flowering-rush family

Butomus umbellatus L.
FLOWERING-RUSH
N/E. Extinct as a native plant. Perhaps the only native station for this plant has been Freshwater Marshes, where it was first recorded in 1842, and in 1879 was described as occurring 'in tolerable plenty in Freshwater Gate marsh' (Townsend, 1883). In the twentieth century, numbers of plants at this site fluctuated but were generally small. By 1972, it was confined to a single patch by the Blackbridge Brook (BS) and it was lost from here soon after. There was a large flowering clump here in 1986, CP, but this may have resulted from an introduction, and it did not persist. Also known from several ponds as an established introduction. It persisted until recent years in a field pond at Kemphill Farm, near Havenstreet, 5790, where it was supposedly introduced by Gladys Bullock around 1950. Other recent sites include Priory Pond, Carisbrooke 4888, where it has been known since 1982, SB 1996; Gunville brickworks pond 4788, where it was first recorded in 1963, TDD & CDP 1996; and near Holden Farm, Roud 5180, BS 1995.

141. ALISMATACEAE
Water-plantain family

Baldellia ranunculoides (L.) Parl.
LESSER WATER-PLANTAIN
N. A rare plant of muddy ponds and ditches on acid soils. It has always been local but has declined considerably over the past hundred years and is now confined to two

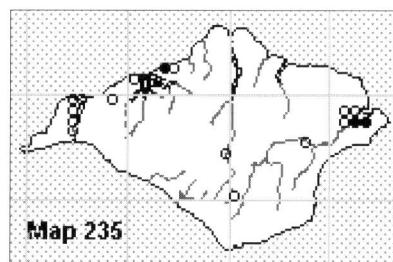

Baldellia ranunculoides

areas where it occurs in small numbers. Long known from Brading Marshes where it survives in three ditches 6287 & 6387, DDa 2001. Also found in a single pond at Elmsworth Farm, Porchfield, where it is very locally frequent in a marshy, cattle-grazed area 4392, MB 2000. Map 235

Alisma plantago-aquatica L.
WATER-PLANTAIN
N. Locally frequent, although generally in small quantity, in ponds, ditches and streams. Map 236

142. HYDROCHARITACEAE
Frogbit family

[*Hydrocharis morsus-ranae* L.
FROGBIT
E. Extinct. Introduced together with Water-soldier (see below) into a pond near Barrett's, a site described as being 'about two miles from Ryde on the Brading Road' around 1856 and apparently still present in 1931

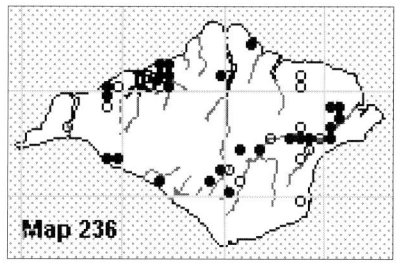

Alisma plantago-aquatica

(Drabble & Long). Also recorded from a pool on a small heath between Freshwater Farm and Norton in 1905, (EHW) and still present here in June 1944, when inspected by the Natural History Society on a walk. No further records.]

Stratiotes aloides L.
WATER-SOLDIER
E. Known from two long-established sites. Albert Bruce Jackson discovered it in abundance in a pond on the borders of Lake Common 5885, near the waterworks, in 1899 when a few years previously there had been none. It has persisted in this site until recently, at least until 1996, MB. In 1995, a few plants were discovered nearby in a ditch by the railway track 5885, BS and it is now dominant here over a short stretch. First recorded in 1856 as naturalised by Dr Thomas Bell Salter in a small pond at Barrett's (see under Frogbit). It persisted in a field pond, a flooded gravel pit, at Kemphill Farm, near Havenstreet, 5790, since at least 1900 and it has been suggested that this may be the site where it was introduced by Salter. It was abundant here in the 1980s but had been lost before 2002.

Elodea canadensis Michx.
CANADIAN WATERWEED
E. Occasional in ponds and streams. The first English record for this introduction was in 1847 and the first Island record is of an introduction into a pond at Spencer Road, Ryde by Dr Bell Salter 'previous to 1850'. It rapidly spread and Stratton (1909) recorded how 'in 1864, and for some years after, it choked mill ponds and streams near Newport, but gradually it decreased, and by 1880 had ceased to be the prevailing plant in the ponds in this locality, and almost disappeared in succeeding years. It is now (1908) making some progress again.' This pattern of rapid increase and gradual decline to a stabilised level was typical throughout the country. Throughout much of the twentieth century it was found in tolerable plenty in many ponds and streams but has once again declined in recent years.

E. nuttallii (Planch.) H. St. John
NUTTALL'S WATERWEED
E. An aquarists' introduction that has probably been overlooked. Recorded from the River Yar at Yarbridge 6085, PDG 1978; and ditches on Brading Marshes in the vicinity of the old seawall 6187, PDG 1978.

Lagarosiphon major (Ridl.) Moss
CURLY WATERWEED
E. An increasing introduction in ponds, often becoming dominant. First recorded in 1987 from a pond at Fairlee, Newport 5090, CP (conf. David Simpson) and since found in many artificial ponds and flooded pits. It is also in the River Medina at Pan Mill 5088, CP 1987.

145. JUNCAGINACEAE
Arrowgrass family

Triglochin palustre L.
MARSH ARROWGRASS
N. A local and declining plant of grazed marshy fields and basic flushes. Still frequent in a tightly sheep-grazed water meadow west of Clatterford near Carisbrooke 4887, CP 1997; frequent in a small marsh at Wydcombe 5078, CP 1998; a cattle-trampled former river meander on Brading Marshes 6187, CP 1991; Compton Marsh 3685, BS 1998. Formerly grew in wet flushes on coastal undercliffs at Forelands (recorded in 1856), Thorness Bay (1856), Compton Chine (1969) and Brook cliffs (1969). It is still present in this habitat at Colwell Bay 3287, JO 1990. Map 237

T. maritimum L.
SEA ARROWGRASS
N. A frequent and characteristic plant of saltmarshes in all our estuaries along the north coast of the Island.

146. POTAMOGETONACEAE
Pondweed family

Potamogeton natans L.
BROAD-LEAVED PONDWEED
N. Scattered distribution in slow-flowing streams, ditches, old pits and ponds, tolerating some eutrophication. Although most frequent in the Eastern Yar valley, it is not so scarce in the West Wight as suggested by Bevis *et al* (1978). It was long known from the main stream in Freshwater Marshes but has not been recorded from here since 1970. Map 238

P. polygonifolius Pourr.
BOG PONDWEED
N. A scarce and declining plant, growing in acid water in ditches and bogs and tolerating shading. The main surviving sites are ditches in Alverstone Marsh 5085, CP 1995; Moor Farm, Godshill 5383, CP 1994; Bohemia Bog 5183, CP 2001; and shaded ditches in The Wilderness 5082, CDP & TDD 1996. Present at Munsley Bog 5282, CP in 1991 but reduced to a single patch in wet woodland in 2002. Map 239

(*P. lucens* L.
SHINING PONDWEED
N? Recorded by Bromfield (1856) from marsh ditches at Sandown, but More later disputed the authenticity of this record. Also recorded from

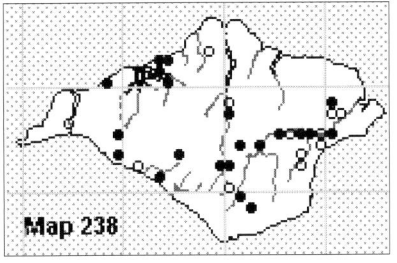

Triglochin palustre

Potamogeton natans

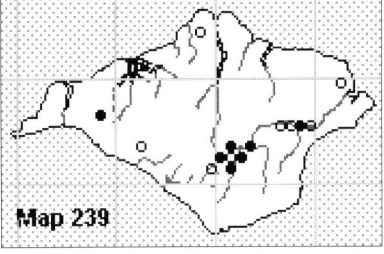

P. polygonifolius

ditches at Langbridge, Newchurch, in 1934 by EHW, but no herbarium material survives.)

[*P. perfoliatus* L.
PERFOLIATE PONDWEED
N. Extinct. Known only from a single record, a pool on slumped cliffs at Compton 3784, BS 1967 (conf. J.E. Dandy). The pond, which dried out in the summer, was subsequently lost due to slumping.]

P. pusillus L.
LESSER PONDWEED
N. Formerly described as occurring in streams and ditches in the main river valleys (Bevis *et al* 1978) but often confused with Small Pondweed. Nevertheless, it is considered to be less frequent than formerly. The only recent records are: pond at Thorncross, near Brighstone 4381, CDP & TDD 1996; reservoir at Cridmore Bog 4981, BS 1999 (conf. Nigel Holmes); reservoir at Newchurch 5686, CP 1997; and pond at Marsh House, Carpenters 6088, CP 2000.

P. obtusifolius Mert. & W.D.J. Koch
BLUNT-LEAVED PONDWEED
N. A rare plant of shallow ponds. The only recent record is from a field pond at Elmsworth Farm, Porchfield 4392 where it is abundantly fertile, CP 2000 (conf. Nigel Holmes). It may have been in this area for a long time, as H.T. Mennell collected this species at Newtown in 1879 (hb. BM). Also found growing in a pond at Spencer Road, Ryde 5892 in 1989 and for a few years after, RK (conf. Clive Jermy).

P. berchtoldii Fieber
SMALL PONDWEED
N. This species remains the most frequently encountered of our smaller linear-leaved pondweeds. It occurs in ditches, ponds and farm reservoirs in widely scattered locations. Probably under-recorded. Map 240

P. crispus L.
CURLED PONDWEED
N. Widespread but scattered in ponds, ditches and streams. Sometimes found in quantity in rivers, such as the Medina at Shide and the Yar at Newchurch and Yar Bridge. Map 241

P. pectinatus L.
FENNEL PONDWEED
N. A characteristic species of eutrophic, frequently brackish, standing water. Bevis *et al* (1978) knew this from three stations but it is actually more widespread than this. Reservoir at Yafford 4580, CP 2001; River Yar at Langbridge 5585, CC 1987; Sandown Canoe Lake 6084, CP 1996; and several sites around Bembridge Harbour. Map 242

Groenlandia densa (L.) Fourr.
OPPOSITE-LEAVED PONDWEED
N. An uncommon and declining species of shallow, clear, base-rich streams. Frequent in the Lukely Brook at Carisbrooke 4888, MB 1998; and in the stream at Ventnor Park 5577, BS 1992. Historically recorded from several sites, especially on Brading Marshes and Freshwater Marshes. Only known from three stations by Bevis *et al* (1978), namely the River Medina at Pan, drainage ditch on Brading Marshes near the railway station, and a pond by the Military Road at Thorncross. Not seen recently at any of these sites.

147. RUPPIACEAE
Tasselweed family

Ruppia maritima L.
BEAKED TASSELWEED
N. Scarce but locally frequent in brackish pools and ditches. Brackish ditch behind Yarmouth Mill 3589, CP 1997; Newtown Quay lagoon 4191, MSh 1987; Ryde Canoe Lake 6092, CP 1988; and Bembridge lagoons 6388, AC 1996. Not recorded recently from brackish ponds by the Medina at Medham, BS 1965, where it may have been overlooked.

R. cirrhosa (Petagna) Grande
SPIRAL TASSELWEED
N. In similar situations to Beaked Tasselweed, but rarer. Both species recorded growing together in old saltworks pools at Newtown by James Groves in 1879; Spiral Tasselweed was refound here in 1980, 4191 BS and was still present in 2000. Found growing in abundance in Sandown Canoe Lake 6084, AC 1999.

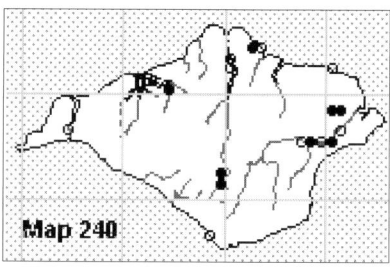

P. berchtoldii

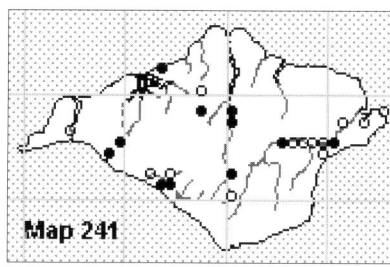

P. crispus

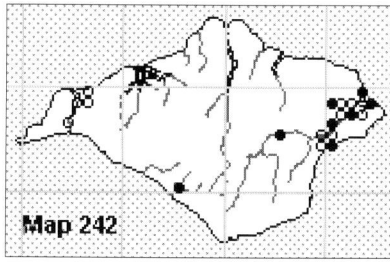

P. pectinatus

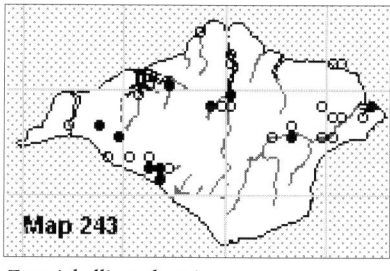

Zannichellia palustris

149. ZANNICHELLIACEAE
Horned Pondweed family

Zannichellia palustris L.
HORNED PONDWEED
N. Rather local in streams, ponds and ditches in fresh and brackish waters. Probably under-recorded. Map 243

150. ZOSTERACEAE
Eelgrass family

Zostera marina L.
EELGRASS (local: grassweed)
N. Locally dominant below low water on sandy or muddy substrates in

numerous places along the north shore of the Island and at Bembridge Ledges. Bromfield (1856) described this plant as 'a troublesome weed of shallow waters, by impeding the passage of boats, as in the Southampton river crossing over to Hyde etc, where it impedes the progress of small craft'. However, a 'wasting disease' caused a catastrophic decline all around the coasts in the early 1930s and by the late 1930s, the Island's *Zostera* beds had gone (Wadham, 1940). More recently there has been a widespread recovery. Tubbs (1999) considered that since about 1965, the three *Zostera* species have spread widely on the Island's shoreline. Recorded growing in the moat at Fort Victoria, Yarmouth in 1968, just before it was filled in, BS.

Z. angustifolia (Hornem.) Rchb.
NARROW-LEAVED EELGRASS
N. Locally frequent on soft muddy substrates at or below mid tide level on the Solent shores of the Island, particularly from King's Quay to Bembridge Ledges, often with Dwarf Eelgrass. Tubbs (1999) described how sub-littoral forms of this species growing at Bembridge Ledges produced generally longer and somewhat more robust leaves than the eulittoral forms.

Z. noltei Hornem.
DWARF EELGRASS
N. Locally frequent on muddy substrates in many places along the north shore of the Island. All three species grow together at the lagoon margins on Bembridge Ledges and Tubbs (1999) recorded how, in 1980, the small patches of Dwarf Eelgrass were longer leaved than their eulittoral counterparts and not easily distinguished from Narrow-leaved Eelgrass on purely vegetative characters. Bromfield (1856) claimed that Dwarf Eelgrass was first detected as British species by a Mr. Sonder, of Hamburgh, who came across it amongst a package of seaweeds sent to him from Ryde shore in 1847.

151. ARACEAE
Lords-and-Ladies family

Arum maculatum L.
LORDS-AND-LADIES
(local: cuckoo-babies)
N. Frequent in woods, scrub and hedgebanks everywhere.

A. italicum Mill.
ITALIAN LORDS-AND-LADIES
N. ssp. *neglectum* (F.Towns.) Prime Confined to the Undercliff between Luccombe and St Catherine's Point, where it is locally frequent in sun or light shade on the calcareous talus slopes beneath the shelter of the cliff. Also grows in the grounds of The Hermitage on St Catherine's Down 4978, CP 1998, where it may possibly have been introduced. The plant was first discovered in this country by A. Hambrough, a naturalist and owner of Steephill Castle. He first noticed it growing near his home in 1854. Steephill Castle was demolished in the 1960s but the plant remains frequent here. Map 244

E. ssp. *italicum* A scarce but increasing garden escape established in widely scattered locations close to gardens.

Dranunculus vulgaris Schott
DRAGON ARUM
C. A plant recorded growing on a roadside verge near Niton Primary School, 5076, 1991, was presumably the result of dumping of garden waste.

152. LEMNACEAE
Duckweed family

Spirodela polyrhiza (L.) Schleid.
GREATER DUCKWEED
N. Extinct or overlooked, in slow flowing ditches. Bromfield (1856) knew this duckweed from ditches in the Freshwater Marshes and, abundantly, in marsh ditches on Sandown Levels, especially between Yaverland and Yarbridge. There are no further records from Freshwater but it persisted in the Eastern Yar valley. It was abundant in ditches between Alverstone Lynch and Sandown Waterworks 5885, RK 1976; and in a ditch below riding stables at Adgestone 5885, JB & RK 1976. There are no further records and RK considered it to have become extinct below Alverstone Lynch by 1982 as a result of drainage and ditching. Today, the ditches at Adgestone are either choked with vegetation through lack of management, or dominated by Parrot's-feather.

Lemna gibba L.
FAT DUCKWEED
N. Very rare in slow-flowing ditches. Recently recorded only from Brading Marshes around Carpenters 6188 where it occurs in ditches near the causeway bridge in fluctuating quantity, AC 1995. There is also a record from a recently formed pond south-east of St Lawrence Church 5376, AJ 1967. Bromfield (1856) also recorded this duckweed from a small pond near Yafford Farm.

L. minor L.
COMMON DUCKWEED
N. Widespread and quite frequent in ditches, ponds, wet woodland rides and water tanks although perhaps less abundant than formerly.

L. trisulca L.
IVY-LEAVED DUCKWEED
N. Scarce, although probably under recorded, in slow-flowing ditches and ponds, generally with Common Duckweed. Many of its former localities are ditches, which are no longer cleared and have become

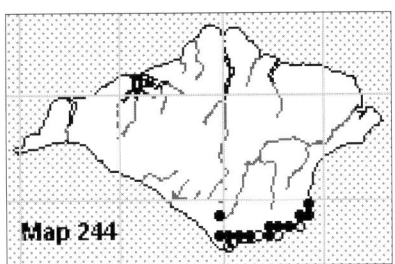

A. italicum ssp. *neglectum*

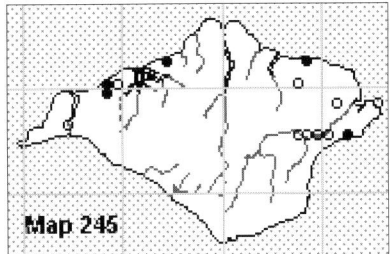

L. trisulca

choked with vegetation. Formerly common in the main stream through Freshwater Marshes, but last recorded here in 1970. Map 245

L. minuta Kunth.
LEAST DUCKWEED
E. Local although rapidly spreading in ponds. First recorded in 1994 from Mottistone Mill pond 4283, BS. Subsequently recorded from ponds at Niton 5076, CP 1995; slumped bowl at Redcliff 6285, AC 1996; Yafford Mill 4482, PS 1999; and Lavender's Farm, near Godshill 5281, AC 2000. Increasing.

155. JUNCACEAE
Rush family

Juncus squarrosus L.
HEATH RUSH
N. Extinct? This plant of damp heathlands is, remarkably, one of our rarest plants. Always rare, historically it has been confined to above Blackgang Chine (last recorded 1856), The Wilderness (1860), Bleak Down and Munsley Bog. It has survived longest on Bleak Down 5181, where it was still common in 1965, BS. In 1997, about twenty clumps were scattered in open ground along the western edge of the site, CP, although it has not been found here since. A very few plants were found on Munsley Bog, 5282, in 1970, BS, and again in 1992, CP, but not since.

J. gerardii Loisel.
SALTMARSH RUSH
N. Locally frequent in saltmarshes in all our estuaries along the north coast. Also occurs in brackish grazing marsh on Brading Marshes. There is an outlying extant site on Sandown Levels, 6084, on the south coast.

J. foliosus Desf.
LEAFY RUSH
N. Likely to be under-recorded. There is a record from bare gravelly soil at Alum Bay 3085, PS 2002 and herbarium material collected from moist ground at Shanklin by H. Trimen in 1860, BM (conf. Clive Stace).

J. bufonius L.
TOAD RUSH
N. Frequent and widespread in damp tracks, pond margins, roadsides and damp arable.

J. ambiguus Guss.
FROG RUSH
N. Local but under-recorded in damp brackish areas near the coast. Ningwood Lake, Newtown estuary 3989, PS 1994; muddy area at Watershoot Bay, St Catherine's 4975, PS 1999; near All Saints' Church, Freshwater 3487, PS 1999; Cranmore 3989, PS 1999.

J. subnodulosus Schrank
BLUNT-FLOWERED RUSH
N. A scarce plant of fens, marshes and coastal landslips on base rich soils. Although there has been some contraction of range over the past one hundred years, it still survives in seven sites. Its extant fen sites, where it is a distinct component of the community, are: on Brading Marshes in a boggy flush below Centurion's Hill 6287, NS 1991; High Grange marsh at Compton 3784, CP 1999; and Freshwater Marshes, where a relict fen community survives beneath reeds in the south marsh 3486, CP 2001. Coastal landslip sites are: streams at St Catherine's 5075, CP 1995; streams at Blackgang landslip 4876, GT 1997; cliff seepages at Colwell Bay 3388, CC & JO 1990; and the landslips below Headon Warren 3286, PS 1999. Map 246

J. articulatus L.
JOINTED RUSH
N. A frequent and widespread species of marshy fields, damp woodland rides, ponds and ditches. Not found on the chalk.

J. acutiflorus Ehrh. ex Hoffm.
SHARP-FLOWERED RUSH
N. Frequent, often in similar situations to Jointed Rush.

J. bulbosus L.
BULBOUS RUSH
N. Widespread but local on acid soils. It is often found in vehicle ruts and ditches in woodland rides and marshes. It is sometimes fully aquatic in suitable conditions. Map 247

J. maritimus Lam.
SEA RUSH
N. A characteristic upper salt marsh

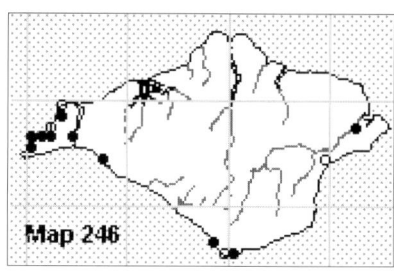
J. subnodulosus

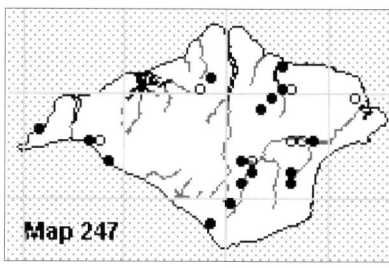
J. bulbosus

plant in all our estuaries on the Solent coast.

J. inflexus L.
HARD RUSH
N. Frequent in grasslands with impeded drainage right across the Island, only absent from very acid soils. It is particularly characteristic of pastures on heavy clay soils.

J. x diffusus Hoppe
HARD RUSH X SOFT RUSH
N. Probably occurs sporadically with the two parents but only known from historic records. 'Very fine in Alum Bay', J.G. Baker 1871; and several records from the roadside on the south side of Parkhurst Forest, W. Borrer 1847, F. Stratton 1900.

J. effusus L.
SOFT RUSH
N. Widespread and frequent in wet pastures, marshes, ditches, damp woodland rides and by ponds.

J. conglomeratus L.
COMPACT RUSH
N. Local in wet meadows and marshes but less common than Hard and Soft Rushes.

Luzula forsteri (Sm.) DC.
SOUTHERN WOOD-RUSH
N. Frequent in ancient woods, old hedge banks, sunken lanes and churchyards. Widespread, excepting

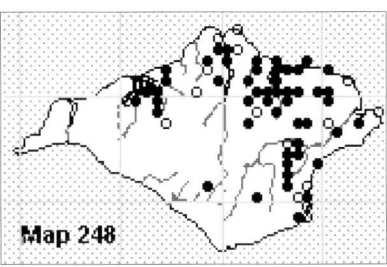
Luzula forsteri

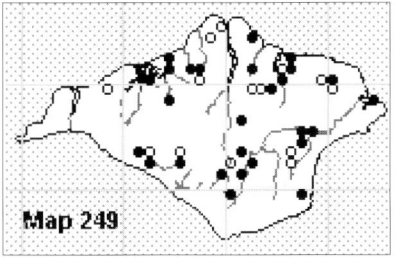
L. multiflora

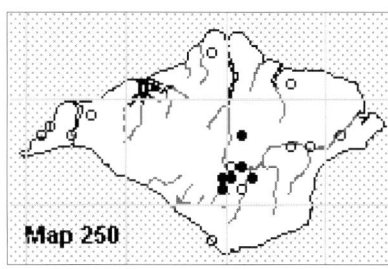
Eriophorum angustifolium

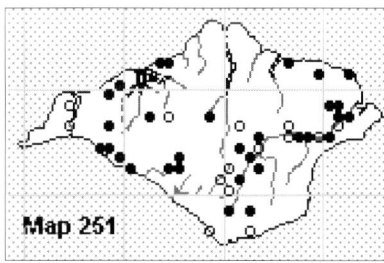
Eleocharis palustris

in south-west Wight, where it is rare. Southern Wood-rush is particularly frequent on the Island. Map 248

L. x *borreri* Bromf. ex Bab.
SOUTHERN WOOD-RUSH X HAIRY WOOD-RUSH
N. Probably overlooked. Known only from historic records: Apse Castle, America Wood, W. Bromfield, who first described this hybrid from here in 1841; and Quarr Copse, A.G. More 1856 and F. Townsend 1883.

L. pilosa (L.) Willd.
HAIRY WOOD-RUSH
N. Occurs in similar situations and with a similar distribution to Southern Wood-rush but less frequent than that species.

L. sylvatica (Huds.) Gaudin
GREAT WOOD-RUSH
N. A very local plant of acid, usually sandy, soils in ancient woodlands in East Wight. Little Standen Wood, near Downend 5287, CP 1997; still frequent in Lynch Copse, Newchurch 5685, CP 1998; western edge of Youngwoods Copse, Alverstone 5785, CP 1998; Borthwood Copse 5784, AB 2002; America Wood 5682, CP 1998; Cliff Copse, Shanklin, particularly at western end 5680, CP 1998. Although often locally dominant in these woods in the past, at most sites it survives in small quantity under heavy shade. Responding to recent management in Borthwood Copse, where plants are springing up in many new clearings.

L. campestris (L.) DC.
FIELD WOOD-RUSH
N. Widespread and frequent in dry grassland and lawns.

L. multiflora (Ehrh.) Lej.
HEATH WOOD-RUSH
N. Widespread but local and generally in small quantity in clearings in heathy woods, and acid grassland. Map 249

156. CYPERACEAE
Sedge family

Eriophorum angustifolium Honck.
COMMON COTTONGRASS
N. A greatly declined species of acid bogs. It is most frequent in the wettest parts of Cridmore Bog 4981, BG 1999 but it flowers poorly; only 17 stems could be found in 2002. Also grows at Bohemia Bog 5183, CP 2001 and in decreasing quantity at Munsley Bog 5282, SC 1997. In 1996, a clump of vigorous plants were found growing at the edge of a flooded gravel pit on Bleak Down 5186, DB, a site from which there are no historic records. Sites from which it has been lost in the last fifty years, with dates of last records, are: Newnham Common between Newnham Farm and Quarr Abbey (1961); Bleak Down (1966); wet meadow between Newchurch and Alverstone by railway track (1966); and relic bog east of Centurion's Copse, near Brading (1985). Map 250

[*E. latifolium* Hoppe
BROAD-LEAVED COTTONGRASS
N. Extinct. Formerly, a rarity of base-rich flushed bogs recorded from Cockleton Bog, near Cowes, by P.D. Radcliffe in 1839 (Drabble & Long, 1931) and from the wettest parts of Colwell Heath, where it was first described as plentiful in 1841 (Bromfield, 1856) and subsequently as sparing (Townsend, 1883). By 1908, Stratton considered that it had become extinct.]

Eleocharis palustris (L.) Roem. & Schult.
COMMON SPIKE-RUSH
N. A fairly frequent and widespread plant in suitable places. It grows in wet meadows, ditches, pond margins and brackish marshes. Map 251

E. uniglumis (Link) Schult.
SLENDER SPIKE-RUSH
N. Very local in coastal brackish grassland, and recently only recorded from Brading Marshes where it is locally frequent 6387, CP 1996 and 6287, NS 1991. Formerly recorded from Freshwater Marshes where it was last recorded in 1966 when it had become very rare (BS). There are also two unusual records: it was found growing around a restored pond at Swainston 4387, CP 1983; and on the landslip between Blackgang and Niton Undercliff in 1967, JB.

E. multicaulis (Sm.) Desv.
MANY-STALKED SPIKE-RUSH
N. A rarity of acid bogs. It occurs in quantity at Bohemia Bog 5183, CP 2001, where, in the past, it has sometimes been mistaken for *Trichophorum caespitosum*, Deer-grass (see Bevis *et al* 1978), a species that does not occur on the Island. Also recorded from around ponds on Bleak Down, but not since 1969 (RK). Other nineteenth century sites included Freshwater Marshes, St Helen's Green, Cockleton Bog and Blackpan at Sandown.

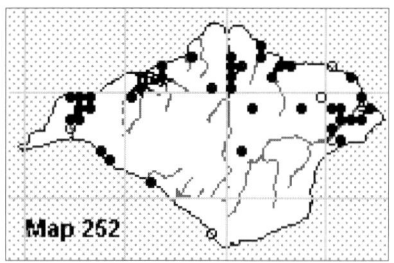

Bolboschoenus maritimus

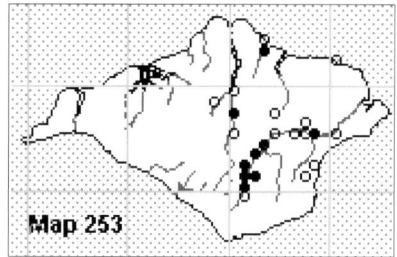

Scirpus sylvaticus

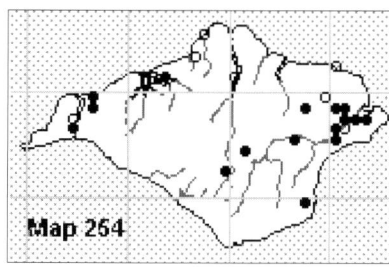

Schoenoplectus tabernaemontani

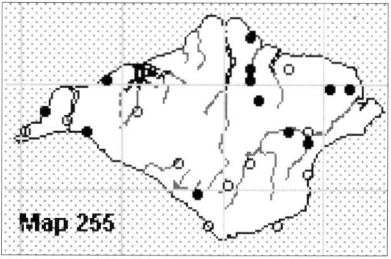

Isolepis setacea

[*E. quinqueflora* (Hartmann) O. Schwarz
FEW-FLOWERED SPIKE-RUSH
N. Extinct. Historically recorded from flushed bog sites at Colwell Heath (1840, W.A. Bromfield), a site on the east bank of the Western Yar (1871, J.G. Baker), and a boggy slope at the western end of St Helen's Green (1871, A.G. More).]

Bolboschoenus maritimus (L.) Palla
SEA CLUB-RUSH
N. Frequent in brackish ditches and upper saltmarshes in estuaries and mouths of streams along the north coast. Widespread in ditches on Brading Marshes and also present in the Yar valley at Sandown Levels, 6185. Also found, more locally, around pools on the south-west undercliffs. There are also a few inland records: invading an old reservoir on the Prison estate 4890, CP 2001; frequent, but non-flowering, in a ditch at Rookley country park 5184, PS & EC 2000; colonising a redug field pond at Beanacre, Ashey 5788, BW 1998; recent pond at Staplers 5288, AM 2002. Also recorded from Whitefield Farm pond, near Ashey in 1961, JH. Map 252

Scirpus sylvaticus L.
WOOD CLUB-RUSH
N. Local in wet, neglected pastures and ditches by streams. Although this plant has declined over the past one hundred years, it remains locally dominant in parts of the Eastern Yar valley. Map 253

Schoenoplectus tabernaemontani (C.C.Gmel.) Palla
GREY CLUB-RUSH
N. Occasionally found in brackish pools and ditches, and fringing the upper tidal reaches of rivers. Grey Club-rush is also recorded from a number of inland sites, some of which are shared with Sea Club-rush. Common Club-rush, *S. lacustris*, is not recorded from the Island. Map 254

Isolepis setacea (L.) R. Br.
BRISTLE CLUB-RUSH
N. A very local plant of bare, wet acid soils in rutted tracks, woodland rides and flushes. Bromfield (1856) considered this to be a frequent plant but Bevis *et al* (1978) only knew it from four stations. It has since been recorded from a number of other sites, including some historic ones, on disturbed ground and is no doubt overlooked in others. It survives, very locally, on Colwell Common 3287, CP 2000, and on St Helen's Green west 6289, CP 1994, both of these being historic sites. Map 255

I. cernua (Vahl) Roem. & Schult.
SLENDER CLUB-RUSH
N. Found on bare, wet soils but distinctly coastal in distribution. Bromfield (1856) considered this plant to be commoner than Bristle Club-rush but there was clearly confusion between the two species. The only known extant sites are: wet spots on Luccombe Chine ledge 5879, CP 2000; scarce by a stream on Blackgang cliffs 4975, CP 1996; and two plants in grazed saltmarsh at Bembridge Harbour 6388, FR 1992.

[*Eleogiton fluitans* (L.) Link
FLOATING CLUB-RUSH
N. Extinct. Bromfield (1856) describes this plant as 'not uncommon' but only lists six stations, all of them flooded pools or ditches with acid water. The only site with specific records in the twentieth century was Bleak Down where, in 1973, it was still plentiful in flooded gravel pits at the southern end of the site, 5181 BS. By 1994, it was confined to a single pond but remained frequent here, CP. The plant was lost sometime between 1994 and 1998, and this is considered to be due to deteriorating water quality.]

Cyperus longus L.
GALINGALE
N/E. Extremely rare as a native plant in a single, reed-dominated pool side near St Catherine's 5075, where only a very small handful of flowering stems could be found in 2000, AB. Scarce, but increasing as an established garden outcast in ponds: Priory pond, Carisbrooke, 4888, SB 1997, where it was already well established in 1980; Barton pond, Osborne estate 5294, CP 2000; field pond at Swanmore, Ryde 5991, RG 1998; Bonchurch pond 5778, CP 2000: Lukely stream in Castle Street, Carisbrooke 4888, BS 1987.

Historically known from three sites: in small quantity in a wet meadow below Carisbrooke Castle (Bromfield, 1856) from where it was lost by 1880; a wet meadow below Apes Down (SZ457878) which is now wet woodland and from where it was last recorded in 1985; and Castle Mead, St Catherine's. The latter is described by Bromfield as follows: 'At St Catherine's Point, the plant was cut for a late hay-crop by the former occupant of the ground, and its sweetness, permanence and ample produce seem to point it out as a valuable object of cultivation on wet meadow-lands. The station is now on the property of my friend George Kirkpatrick, Esq., by whom the meadow has been fenced in for the protection of the *Cyperus*, which, through the zeal of that gentleman, and his love for whatever is rare and beautiful in nature, will henceforth flourish in security from the scythe of the utilitarian farmer'. Ironically, it was probably the scythe which gave rise to ideal conditions for the plant and as the population dwindled in the late twentieth century and was fenced off from cattle and the stream diverted, its tall fen community gradually succeeded to reed bed and willow carr. By 1998, it was believed to have been lost, but conservation work carried out in 1999 revealed a very few surviving plants, which so far have shown no inclination to spread.

C. eragrostis Lam.
PALE GALINGALE
E. A garden escape, establishing in a few places and increasing. First recorded in 1984 invading an old cracked reservoir floor at Los Altos Park, Sandown 5983, CP. Other records from: waste ground behind a garden centre at Lake 5783, PS 1999; abundant in paving cracks on a side road at Seaview 6291, PS 1999; between paving stones at Ryde 6092, PS 2000; garden centre at Apse Heath 5783, RHW 2001; and between paving stones in Newport 4990, BS 2002.

[*Rhynchospora alba* (L.) Vahl
WHITE BEAK-SEDGE
N. Extinct. Only ever recorded from Lake Common (Bromfield, 1856), abundantly in one or two spots on the edges of the common. The only subsequent record is by A.G. More who, in 1858, found it 'in the wetter spots, and where the spongy moss scarcely affords a footing' (More 1898).]

[*Cladium mariscus* (L.) Pohl
GREAT FEN-SEDGE
N. Extinct. Known from just a single enigmatic record in Bromfield (1856), 'in the bog at Easton, Freshwater Gate, G.S. Mill, Esq.'. Neither Bromfield nor any succeeding

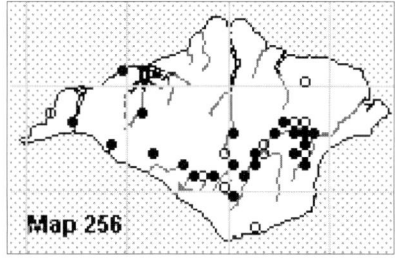

Carex paniculata

botanists were able to find it here, but he did examine non-flowering dried material collected by Mr. Mill, described by him as 'tolerably plentiful' in one or two meadows.]

Carex paniculata L.
GREATER TUSSOCK-SEDGE
N. Local and declining in swampy alder and willow carr by streams and flushes on acid soils. No longer survives in fens. Its distribution has been much affected by drainage and old tussocks often survive as relict features in drying woods. Map 256

[*C.* x *boenninghausiana* Weihe
GREATER TUSSOCK-SEDGE X REMOTE SEDGE
N. Only ever recorded from Lynch Copse, Newchurch, first by A.G. More in 1857 (More, 1871) and finally in 1893 by J.H. Steuart (hb. IPS). It could still be present here.]

[*C. diandra* Schrank
LESSER TUSSOCK-SEDGE
N. Extinct. Historically recorded from wet ground at The Wilderness, 1870, F. Stratton and sparingly in the wettest parts of Freshwater Marshes A.G. More, 1862 (both from More, 1871), but with no further records.]

[*C. vulpina* L.
TRUE FOX-SEDGE
N. Extinct. Historic records for this species are considered to be in error for *Carex otrubae*, with one exception. Material collected by F.J. Hanbury (hb. BM) in 1920 from King's Quay has been confirmed as this species by Clive Jermy. There are no other records.]

C. otrubae Podp.
FALSE FOX-SEDGE
N. Locally frequent in wet places on heavy clay soils and in brackish pastures near the coast. It is widespread north of the chalk, but scarce and with a coastal distribution in the south of the Island.

C. x *pseudoaxillaris* K. Richt.
FALSE FOX-SEDGE X REMOTE SEDGE
N. There were four nineteenth century stations for this hybrid recorded in Bromfield (1856). These were Quarr Wood, Little Smallbrook Wood, Appley Wood and Lynch Copse, Newchurch.

C. spicata Huds.
SPIKED SEDGE
N. Occurs quite frequently in dry grassland on roadside verges and banks in many places, generally on chalk or clay soils. Many earlier records were confused with Prickly Sedge and this is believed to also be the case today. Recent confirmed stations include Freshwater, Wellow, Dodnor, Fattingpark Copse, Ryde and Nansen Hill.

C. muricata L. **ssp.** *lamprocarpa* Celak
PRICKLY SEDGE
N. Occurs quite frequently in dry grassland on hedges and banks and heathy grasslands, generally on acid soils. Some records have been confused with Spiked Sedge but recent confirmed stations include Colwell, Yarmouth, Brook Hill, Niton, Godshill, Great Budbridge, Merstone, Newchurch and Arreton.

C. divulsa Stokes **ssp.** *divulsa*
GREY SEDGE
N. Frequent and widespread in dry grassland on verges, banks and woodland rides.

Ssp. *leersii* (Kneuck.) W. Koch has not been confirmed from the Island but may well have been overlooked. This subspecies was believed to have been found at Thorley 3788 in 1998 and at Lucketts Farm, Bouldnor 3889 in 2000 (PS).

C. arenaria L.
SAND SEDGE
N. Very locally frequent on sandy ground, generally around the coast on sand dunes and sandy shingle and on the tops of sandy cliffs. Well established on the cliff top perched

dunes at Ladder Chine 4778. Also found inland on dry grassland in sandy soils in several places on the greensand. Map 257

C. disticha Huds.
BROWN SEDGE
N. Very local but sometimes frequent in damp unimproved grassland and fens. Recent sites include: Brading Marshes 6287, NS 1991; Sandown Levels 6085, GT 2000; Alverstone Marsh 5785, JO & PS, 1996; Compton Marsh 3685, CP 1998; and High Grange Marsh 3784, CP 1999. Last recorded from Freshwater Marshes in 1970, BS, when it was described as occurring very sparingly. An unusual site is on slumped flushed clay cliffs at Totland Bay 3287, PS 1999. Bromfield (1856) described this plant as 'not rare'. Map 258

C. divisa Huds.
DIVIDED SEDGE
N. Locally frequent in upper saltmarshes around our estuaries and in brackish pastures along the north coast. It is often abundant in depressions and ditch margins on coastal grazing marsh. Before the land was developed, it grew in the meadows behind Ryde Dover, where Bromfield (1856) described it as constituting a large proportion of the hay crop. It sometimes persists in small patches of coastal grassland well away from surviving natural habitats. Map 259

C. remota L.
REMOTE SEDGE
N. A widespread and frequent species of both ancient and secondary damp woodlands, flushes and ditches. Occurs right across the Island, but most frequent north of the chalk.

C. ovalis Gooden.
OVAL SEDGE
N. Locally frequent in unimproved and semi-improved neutral to acid grasslands, particularly in the Eastern Yar valley. Map 260

C. echinata Murray
STAR SEDGE
N. Very local but sometimes frequent in bogs, flushes and acidic wet meadows, largely confined to the middle and upper sections of the Eastern Yar. Recent sites include: Hill Farm drain, Alverstone 5785, CP 2001; Munsley Bog 5282, SC 1995; Kennerley Moor 5283, CP 1993; Bohemia Bog 5183, CP 2001; boggy flush at Upper Dolcoppice 5079, CP 1997; and Cridmore Bog 4981, CP 1995. Described by Bromfield (1856) as frequent, this plant has declined considerably over the past hundred years. Map 261

C. curta Gooen.
WHITE SEDGE
N. A rare and declining plant of bogs. Recently only recorded from two sites, namely a localised area on Cridmore Bog 4981, CP 1995; and Munsley Bog 5282, BS 1985 where it was frequent amongst sphagnum. May now have been lost from Munsley Bog as a result of drainage and growth of willow scrub. Bromfield (1856) described this plant as 'extremely abundant almost everywhere on Rookley moors, and about the Wilderness, 1844'.

C. hirta L.
HAIRY SEDGE
N. A frequent and widespread plant found in unimproved and semi-improved grasslands, marshes, pond edges and verges.

C. acutiformis Ehrh.
LESSER POND-SEDGE
N. Very local in flood-plain marshes, ditches and wet woods. Much declined over the last hundred years but perhaps overlooked, particularly where rarely flowering. Extant sites where it is still locally dominant include Wolverton Marsh, Shorwell 4482, CP 1995; north side of River Yar on Alverstone Marsh 5885, TDD & CDP 1998; and Brading Marshes 6287, NS 1991. Still present at Freshwater Marshes 3486, BG 2000 but much declined here since 1945 when the entomologist, K.G. Blair discovered a new British moth, Blair's Wainscot. At that time, the moth was frequent on the marsh, the larvae feeding on the stems of Lesser Pond-sedge, but it subsequently declined to

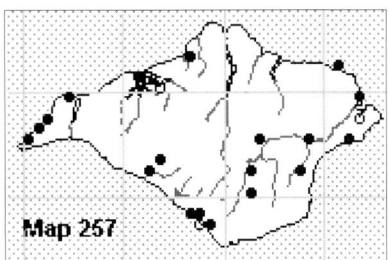

C. arenaria

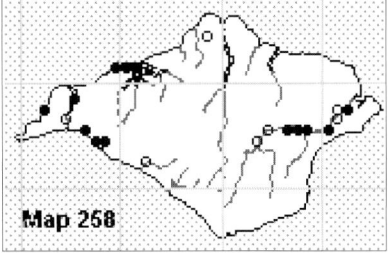

C. disticha

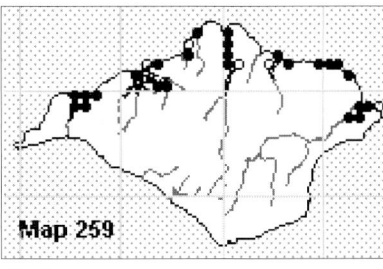

C. divisa

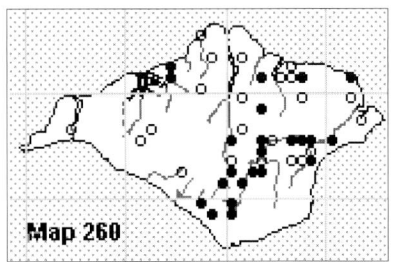

C. ovalis

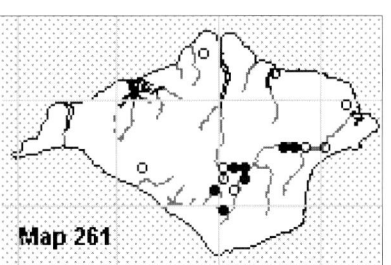

C. echinata

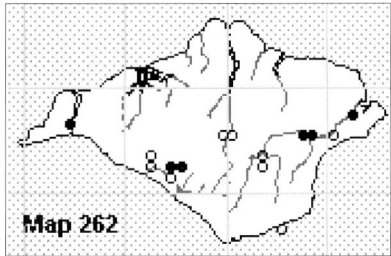

C. acutiformis

VASCULAR PLANTS 177

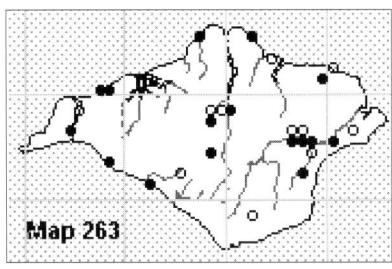

C. riparia

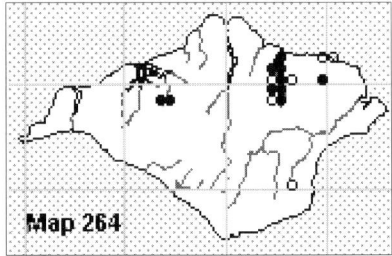

C. strigosa

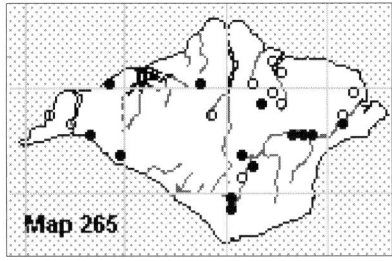

C. panicea

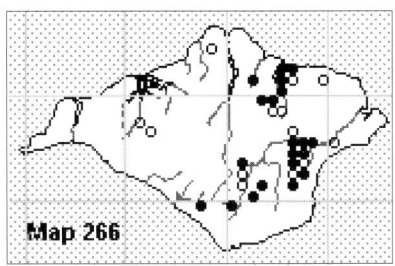

C. laevigata

extinction in the early 1950s with the loss of the food plant resulting from drainage and burning. Map 262

C. riparia Curtis
GREATER POND-SEDGE
N. Fairly widespread and locally frequent on ditch banks, wet meadows and edges of ponds. Map 263

C. pseudocyperus L.
CYPERUS SEDGE
N. A rare plant of freshwater ditches, alder carr and ponds. There has been a continuity of records from Freshwater Marshes since Bromfield's time. It is still present here in small quantity in both the north marsh 3486, RG 1995 and in carr woodland in the south marsh, CP 1994. The only other recent records have been from pond margins well away from historic sites where it has behaved as a pioneer species: two ponds in Bouldnor Copse, one to the north of the Battery 3790, CP 1997 and one to the east 3890, BS 1980, where it was abundant in 2002, GT; and a pond in Lushington Copse, Wootton 5392, JKN 1993. Bevis *et al* also recorded it from two sites on Alverstone Marshes, namely secondary ditches just west of the Mill 575858, and ditches below Alverstone Lynch 583856, both BS & RK 1971. Also recorded from several other sites, but not since the nineteenth century.

C. rostrata Stokes
BOTTLE SEDGE
N. A declining species of very wet peaty swamps and fens. It survives in a few spots on Alverstone Marshes 583857, CP 1990; 585854, CP 1990 and 579853, JO & PS 1996. Also recorded from Wolverton Marsh, Shorwell 4582, CP 1990; in small quantity at Munsley Bog 5282, SC 1995; and Cridmore Bog 5082, CP 1995, where it has declined since 1983. Recorded at Bohemia Bog 5183 in 1971, BS, but not since; and from a pond at the southern extremity of Bleak Down 5181, BS & JB 1971, which was subsequently infilled. In Bromfield's time (1856) this plant was considered to be abundant in several of the sites referred to above and also at Freshwater Marshes and marshes adjoining Lake Common.

C. pendula Huds.
PENDULOUS SEDGE
N. A common and often abundant plant in rides and clearings in wet woods, especially on clay soils and along spring lines. Also found commonly on moist talus of undercliffs and sometimes as a garden weed.

C. sylvatica Huds.
WOOD-SEDGE
N. Frequent in ancient and secondary woods and on hedgebanks, particularly on clay and chalk soils. It is widespread, excepting in the south west of the Island where it is rather scarce.

C. strigosa Huds.
THIN-SPIKED WOOD-SEDGE
N. Local in damp ancient woodlands on base-rich soils, often in low-lying wet areas. It is however, not as scarce as suggested by Bevis *et al* (1978). Its headquarters are in woods at Swainston and at the head of Wootton Creek, where it is recorded from Briddlesford Copse, Combley Great Wood, Firestone Copse, Hoglease Copse, Hurst Copse and New Copse. Also present in Swanpond Copse, 5990, south of Ryde. Map 264

C. flacca Schreb.
GLAUCOUS SEDGE
N. Frequent and widespread in dry and damp grasslands on a range of soil types right across the Island.

C. panicea L.
CARNATION SEDGE
N. A local and declining species of flushed, sometimes base-enriched grasslands. Recent sites include: Compton Marsh 3685, CP 1998; Cranmore grassland 3890, CP 1998; Roughland Cliff, Brook 3983, PS 2000; Parkhurst Forest ride 4790, CP 1994; Bohemia Bog 5183, CP 2001; Munsley Bog 5282, CP 2002; Borthwood Lynch, Alverstone 5785, CP 1994; and Yarbridge Marsh 6186, CC 1989. Map 265

C. laevigata Sm.
SMOOTH-STALKED SEDGE
N. Local but not infrequent in the East Wight growing in acidic flushes in ancient woodlands and in bogs and acid marshland. It has probably declined over the past hundred years. Map 266

C. binervis Sm.
GREEN-RIBBED SEDGE
N. Very locally frequent in both dry and damp heathy grasslands. Extant sites include Colwell Heath 3287, CP 2001; rides in Parkhurst Forest 4691/ 4790/ 4891, CP 2000; Combley Great Wood 5489, CP 1995; Bohemia Bog

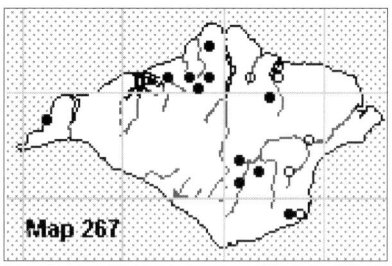
C. binervis

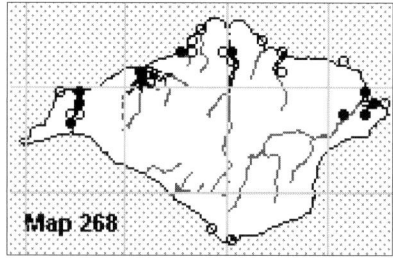
C. distans

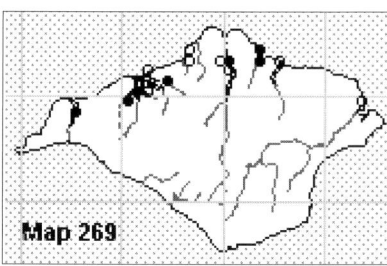
C. extensa

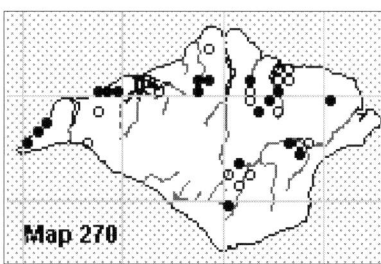
C. viridula

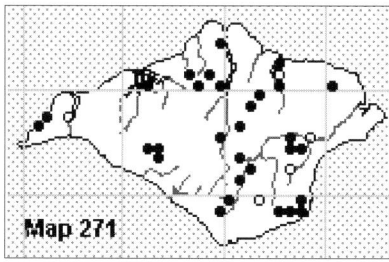
C. pilulifera

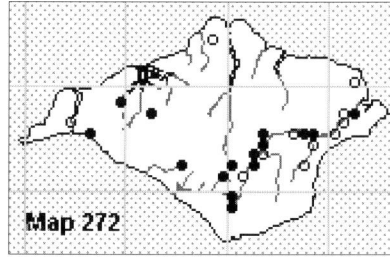
C. nigra

5183, CP 20001; Bleak Down 5181, CP 1996; and heath on top of Ventnor Downs 5678, CP 2001. Map 267

C. distans L.
DISTANT SEDGE
N. Local and often in small quantity in brackish grassland along the north coast of the Island. Extant sites include several spots by Western Yar marshes 3588, PS 1999; Thorness marsh 4593, CP 2002; grassland at Medham by the Medina 5093, CP 1999; rare in dune slacks on St Helen's Duver 6389, CP 2000; and grassland by Bembridge ponds 6488, AC 1996, and other places on Brading Marshes. Map 268

[*C. punctata* Gaudin
DOTTED SEDGE
N. Extinct. Known only from a specimen collected by A.G. More in 1862 from Colwell Heath (CGE) and mis-identified by him at the time as *C. distans* (conf. R.W. David).]

C. extensa Gooden.
LONG-BRACTED SEDGE
N. Scarce in the upper saltmarshes of our estuaries, although very locally common. It is most frequent in the Newtown Estuary, where it occurs in several sites. Map 269

[*C. hostiana* DC.
TAWNY SEDGE
N. Extinct. Formerly a rarity of flushed fens. Recorded by Bromfield (1856) from Briddlesford Heath, boggy ground at the upper end of Colwell Heath and from Freshwater Marshes, the latter two sites in abundance. Townsend (1883) questioned the authenticity of Bromfield's identification but A.G. More collected material from Briddlesford Heath and Colwell Heath and agreed with Bromfield's determination. Material collected by More from Colwell Heath in 1862 is deposited at CGE (conf. R.W. David) and BM.]

C. viridula Michx. ssp. oedocarpa (Andersson) B. Schmid
COMMON YELLOW-SEDGE
N. A local species of wet tracks on heathland and woodland rides, always on acid soils. Although not common, it is less rare than suggested by Bevis *et al* (1978). Frequent in rides in Parkhurst Forest. Map 270

C. pallescens L.
PALE SEDGE
N. This is a scarce sedge of damp ancient woodlands, often in newly opened rides. It is also found in neutral grasslands, sometimes in older secondary woods. Bevis *et al* (1978) knew this plant from three sites; it is now recorded from eight. These are: Walter's Copse, Newtown 4390, CP 1999; Lock's Copse, Porchfield 4491, BS 1996; Fattingpark Copse 5291, CP 1997; Briddlesford Copse 5590, CP 1999; Pucker's Copse, Quarr 5659, CP 1989; Dame Anthony's Common, Ryde 5791, CP 1997; Rowlands Wood 5689, CP 1997; and Peakyclose Copse, Nunwell 6088, CP 1995. It probably still occurs in Parkhurst Forest 4790, BS 1966; Combley Great Wood 5489, BS 1974 and Firestone Copse 5591, BS 1975.

C. caryophyllea Latourr.
SPRING-SEDGE
N. A widespread and frequent species of unimproved calcareous grasslands and neutral grasslands on sandy or clay soils. It is often characteristic of cemetery grasslands on a range of soil types.

[*C. montana* L.
SOFT-LEAVED SEDGE
N. Extinct. Known only from collections made by G.C. Druce in Parkhurst Forest in 1920 (BM & OXF). Druce wrote, 'I noticed this sedge last in the spring of 1920, and had just before predicted it as a likely plant to occur.' Successive visiting botanists have predicted that it is likely to survive at Parkhurst or elsewhere on the calcareous Headon beds, but it has not been refound.]

C. pilulifera L.
PILL SEDGE
N. A local plant of heathland and woodland rides on acid soils.

Bromfield (1856) described this plant as rare and Bevis *et al* (1978) described it as uncommon, occurring in just eight tetrads. The number of recent records suggests that it may have been overlooked. Map 271

[*C. acuta* L.
SLENDER TUFTED-SEDGE
N. Extinct. Known only from nineteenth century records from Sandown Marshes where Bromfield (1856) found it abundantly in several wet meadows. There is a specimen collected in 1858 at Sandown by A.G. More in BM (conf. A.C. Jermy).]

C. nigra (L.) Reichard
COMMON SEDGE
N. A local and probably declining plant of damp grassland and flushes, principally in the Eastern Yar valley. Sites include Compton Marsh 3685, CP 1998; Wolverton Marsh, Shorwell 4582, CP 1990; Wydcombe Marsh 5078 CP 1998; Cridmore Bog 4981, CP 1995; Munsley Bog 5282, SC 1997; Alverstone Marsh 5785, JO & PS 1996 and 5885, CDP & TDD 1998; Bembridge Marshes 6287, NS 1991. Map 272

[*C. elata* All.
TUFTED-SEDGE
N. Extinct. Bromfield (1856) recorded this sedge from Freshwater Marshes but both More and Townsend subsequently dismissed the record as a misidentification for tall growing *Carex nigra*. However, material collected by Bromfield in 1840 and deposited at K has been confirmed as being this species by David Simpson. The label reads 'Marshy meadow near the sluice at the head of Brading harbour', a locality not referred to in *Flora Vectensis*.]

C. pulicaris L.
FLEA SEDGE
N. A scarce plant of marshes and flushes, principally today on remnant clay heath sites on the north of the Island. Bromfield (1856) describes it as 'not infrequent' but Bevis *et al* (1978) considered it to be a rarity recorded from just two sites. It must have been overlooked at that time. Recent sites are: recently cleared heathy vegetation at Golden Hill 3387, CP 1998; heathy

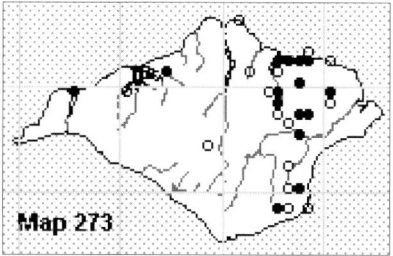

Milium effusum

area at Bouldnor cliff edge 3790, CDP & TDD 1996; grassland at Cranmore 3890, VS 1998 and 3990, CP 1994; heathy field at Brickfields, East Spit, Newtown 4292, RG 1997; newly cleared ride in Walter's Copse, Newport 4390, CP 1999; Bohemia Bog 5183, CP 1995, the only recently recorded site in the south of the Island. Last recorded on Bleak Down in 1969, RK.

157. POACEAE
Grass family

Sasa palmata (Burb.) E.G.Camus
BROAD-LEAVED BAMBOO
E. Well-established and forming thickets in derelict woodland gardens. Sites include grounds of Ryde House 5892, BS 1982 (conf. D. McClintock); spreading beneath trees at the head of Luccombe Chine 5879, BS 1993; and Player's Wood, Ryde 5893, CP 1998.

Pseudosasa japonica (Siebold & Zucc. ex Steud.) Makino ex Nakai
ARROW BAMBOO
E. Many well-established clumps in wet woodland at Landguard Manor, Shanklin 5782, CP 1999.

Nardus stricta L.
MAT-GRASS
N. A rare upland plant of unimproved dry acid grasslands, known today from just two sites. Scarce and very local in heathy grassland on Sandown Golf Course 5885, MB 1996; locally dominating an area of dry acid grassland on Brading Marshes 6287, NS 1991 forming a community which is possibly unique in lowland England. Formerly found on Bleak Down but not recorded here since 1969, RK. Bromfield (1856) found this plant to be local on heaths and moors, recording it from The Wilderness, Rookley Farm, Bleak Down and Headon Warren.

Milium effusum L.
WOOD MILLET
N. Local and generally in small quantity in old woods, particularly in East Wight. Perhaps under recorded. Map 273

Festuca pratensis Huds.
MEADOW FESCUE
N. Widespread, in unimproved grasslands on heavy soils and on verges and marshes.

F. arundinacea Schreb.
TALL FESCUE
N. Frequent on verges, rough grassland and waste land. Nineteenth century botanists referred to this as a plant of undercliffs and clay banks around the coast but it is widespread today.

F. gigantea (L.) Vill.
GIANT FESCUE
N. A widespread and frequent grass of woods, hedgerows, stream sides and scrub.

F. rubra L.
RED FESCUE
N. Very common in grassy places throughout the Island on all soil types. The species has been split into several subspecies but no attempt has been made to differentiate between them. Ssp. *juncea* (Hack.) K. Richt. probably occurs frequently around the coast but has rarely been confirmed.

F. ovina L.
SHEEP'S-FESCUE
N. Locally frequent in unimproved grasslands in short calcareous turf and neutral grassland. The different subspecies have not been differentiated.

F. filiformis Pourr.
FINE-LEAVED SHEEP'S-FESCUE
N. Very local on short dry acid grassland but probably under recorded, or recorded as Sheep's-fescue.

X *Festulolium* hybrid grasses are very under recorded. Bevis *et al* (1978) considered that they were not infrequent.

X *Festulolium loliaceum* (Huds.) P.Fourn.
MEADOW FESCUE X PERENNIAL RYE-GRASS
N. Probably local in old grassland. Pathside at Nettlestone 6190. AC 1999 (conf. T.A. Cope). Bevis *et al* (1978) considered this hybrid to be frequent in marshes such as at Clatterford, Alverstone and in marshy ground south of Kingston Manor.

X *F. braunii* (K.Richt.) A. Camus
MEADOW FESCUE X ITALIAN RYE-GRASS
N/C. Known from two records. Recorded growing in a clover field at St Helen's 6189, AC 1999 (conf. T.A. Cope); and from Quarr JH 1971.

F. arundinacea x *L. multiflorum*
TALL FESCUE X ITALIAN RYE-GRASS
N/C. Known only from material gathered near Havenstreet JH 1972 (conf. C.E. Hubbard).

X *F. brinkmannii* (A. Braun) Asch. & Graebn.
GIANT FESCUE X PERENNIAL RYE-GRASS
N. Known only from Rowlands Lane, Havenstreet, both on roadside verges and in the adjoining cleared wood, 5689 JH 1971 (conf. C.E. Hubbard).

Lolium perenne L.
PERENNIAL RYE-GRASS
N/E. Very common in grasslands throughout the Island.

L. x *boucheanum* Kunth
PERENNIAL RYE-GRASS X ITALIAN RYE-GRASS
C. An escape from cultivation, which is poorly recorded but probably increasing. By Sutton Farm pond, Yafford 4481, PS 2001.

L. multiflorum Lam.
ITALIAN RYE-GRASS
E. A widespread introduction in grasslands, verges and waste ground.

L. temulentum L.
DARNEL (local: cheat)
C. Not recently recorded but Bevis *et al* (1978) refer to this as a bird seed alien, referring to its occurrence on refuse tips at West Medina Mill in 1974 and Nettlestone tip in 1975. It was a frequent cornfield weed in the nineteenth century. The local name was derived from the resemblance of the seeds to that of the crop.

L. remotum Schrank
FLAXFIELD RYE-GRASS
C. Known only from a record from Newport, Drabble & Long, 1931.

Vulpia fasciculata (Forssk.) Fritsch
DUNE FESCUE
N. The classic site for this species is St Helen's Duver where it grows abundantly today and was first recorded by William Borrer in 1839. In recent years it has expanded its range, extending to Bembridge Point 6488, AC 1996; Norton Spit 3589, BG 1999, where it was first found in 1993; Thorness Bay 4593, SB 2002; and the upper beach below Lake Cliffs 5880, AC 1999.

V. bromoides (L.) Gray
SQUIRRELTAIL FESCUE
N. Widespread and locally frequent in short dry grassland and bare ground on acid to neutral soils. Map 274

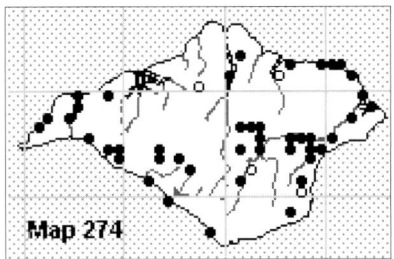
V. bromoides

V. myuros (L.) C.C. Gmel.
RAT'S-TAIL FESCUE
Arch. Locally frequent in scattered locations on waste ground, walls, railway trackbeds and sandy ground around the coast. Map 275

V. ciliata Dumort. ssp. *ambigua* (Le Gall) Stace & Auquier
BEARDED FESCUE
N. A very local grass of coastal sands. First recorded by A.G. More in 1838 from Ryde Dover, but by the time of Bevis *et al* (1978), it was only known from St Helen's Duver, where it still occurs. More recently, there has been a modest spread from this site. It now grows along the old railway track behind Bembridge Embankment 6488, AC 1999. In 1996, it was found growing on the sandy banks of Ryde Canoe Lake 6092, CP, on the site of the former Ryde Dover.

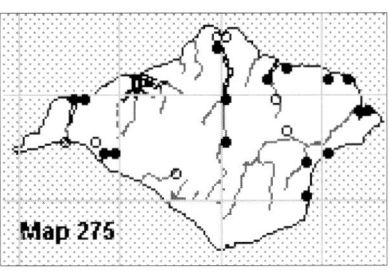
V. myuros

Cynosurus cristatus L.
CRESTED DOG'S-TAIL
N. Frequent in unimproved and semi-improved dry grasslands right across the Island.

C. echinatus L.
ROUGH DOG'S-TAIL
E. A rare casual on wasteland and sandy ground. Recently only recorded from sandy ground at Bembridge Point, 6488, PS 2001, where it has been seen regularly for the past eight years. It seems to be well established in the Bembridge Harbour area, having been first recorded here in 1965 and appearing in various spots around the harbour ever since. First recorded in 1920 by J.T. Knight near the sea at Springvale, near Seaview.

Puccinellia maritima (Huds.) Parl.
COMMON SALTMARSH-GRASS
N. A common plant on the north coast of the Island found in saltmarshes in all our estuaries and on muddy shores.

P. distans (Jacq.) Parl.
REFLEXED SALTMARSH-GRASS
N. An uncommon plant of brackish mud, in a variety of situations. It is less scarce than suggested by Bevis *et al* (1978) who knew it from a single site on the River Medina, where it still occurs. Back of seawall at Forelands 6587, GK 2001; coastal grassland at Gurnard 4795, CP 2001; Seaview 6291, PS 1999; Cowes seafront 4896, JEG 1999; coastal grassland at Sandown Levels 6084, CP 1997; Medina estuary in vicinity of Island Harbour 5091, CP 1996; Yarmouth Promenade 3989, CP 1995.

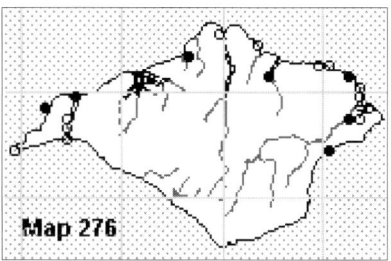
P. fasciculata

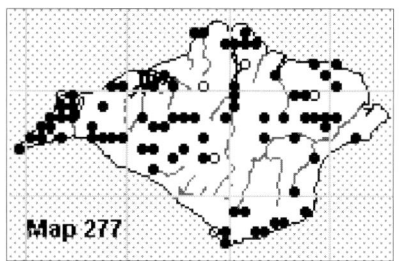
Briza media

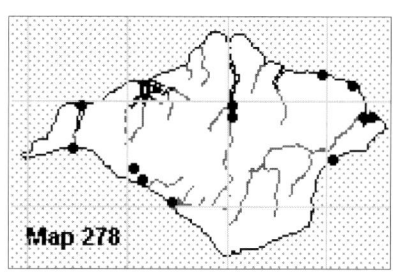
Poa infirma

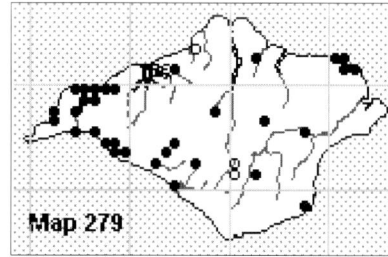
P. humilis

P. fasciculata (Torr.) E.P. Bicknell
BORRER'S SALTMARSH-GRASS
N. Scarce in bare, dry ground around the coast, at the edge of saltmarshes and behind seawalls. It may have declined over the past hundred years but it is less scarce than suggested by Bevis *et al* (1978) who knew it from a single site. Map 276

P. rupestris (With.) Fernald & Weath.
STIFF SALTMARSH-GRASS
N. Uncommon in similar situations to the above species and very occasionally growing with it. Freshwater Bay 3475, RHW 2001; Seaview seafront 6291, PS 1999; by old railway track by Western Yar 3587, PS 1999; Colwell Bay seafront 3288, PS 1994; near the boathouse at Newtown 4191, MMS 1980.

Briza media L.
QUAKING-GRASS
(local: rattle-grass; maiden's-hair; totter-grass; wiggle-waggles!)
N. Locally frequent in agriculturally unimproved grasslands. Particularly characteristic of downland turf, but also found in remnant limestone grassland, in neutral meadows on clay soils and in damp, flushed grasslands. Probably declining off the chalk, as habitats become unsuitable for it. Map 277

B. minor L.
LESSER QUAKING-GRASS
Arch. A rare plant of neutral to acid gravelly or clay soils in the north of the Island. It does appear to be remarkably persistent in its sites although there is rarely a continuity of records. Abundant in several fields at Alverstone Farm, Whippingham 5292, BS 1998 and 5293, PS 2000, from where it was first recorded by James Pristo in 1871. One specimen seen in a field at Palmer's Farm, near Whippingham 5393, CP 2002. Arable field on the west side of The Old Millpond, Wootton 5491, JC 2001; it was recorded from fields on the east bank of Wootton Creek in 1871 by A.G. More. There has been a continuity of records from one or two fields near Quarr Abbey, 5692, where it was first recorded by Bromfield in 1836 and most frequently in 1996, BS. Bromfield found it growing with *Anthemis cotula* and *Gastridium ventricosum*. There are old records from the Freshwater area, where it was described as an introduction, and from Thorley in 1918. There is a casual record of a plant on wasteland near Newport Quay 5089, BS 1996.

B. maxima L.
GREATER QUAKING-GRASS
E. Generally recorded as a non-persistent casual, but it is well established and increasing (2002) in cliff top sandy grassland at Lake Cliffs 5983, where it was first recorded in 1981, JS. It has persisted for several years on a dry sandy bank at Sandford, near Godshill 5481, CP 1996; and between paving stones on a housing estate at Brighstone 4282, MB 1996.

Poa infirma Kunth
EARLY MEADOW-GRASS
N. First recorded from the Island growing at the edge of a car park in Freshwater Bay 3485, PS 1999 (conf. J.R. Edmondson). Targeted searching of its niche, sheltered sunny spots in well-drained soil, often by roadsides, early in the spring has shown this plant to occur widely. The population in Mottistone car park 4083, PS 2002, comprises several thousand plants. Populations are very localised and discrete, and it is unclear at this stage whether the plant is a recent colonist or has been overlooked. Until quite recently, this grass was confined to the coastline of the extreme south-west of England but it has now been found right along the south coast and into East Anglia. Map 278

P. annua L.
ANNUAL MEADOW-GRASS
N. Common everywhere, especially where the soil has been trampled or disturbed.

P. trivialis L.
ROUGH MEADOW-GRASS
N. Frequent and widespread in damp grasslands, verges and woodland rides.

P. humilis Ehrh. ex Hoffm.
SPREADING MEADOW-GRASS
N. Local, but under recorded, in damp meadows, chalk grassland, coastal grassland and dunes. Map 279

P. pratensis L.
SMOOTH MEADOW-GRASS
N. Frequent and widespread in a wide range of habitats across the Island.

P. angustifolia L.
NARROW-LEAVED MEADOW-GRASS
N. Status unknown but probably scarce in dry grassland. Needles headland 2984, DD 1998; roadside verge near Totland 3185, PS 2002; and at Brighstone Down, above road 4284, PS 2002. Bevis *et al* (1978) record this species from a roadside at

Plaish, Bleak Down and Parkhurst Forest paths.

P. compressa L.
FLATTENED MEADOW-GRASS
N. Local on old walls with crumbling mortar and probably in dry grassland. Under recorded. Wall near village stores at Shorwell 4583, PS 1999; wall at Mount Joy cemetery, Carisbrooke 4987, BS 1992; Bonchurch 5778, JO 1988. Bevis *et al* (1978) recorded this species from walls at Apes Farm and Rowridge Farm, Lower Shide Mill and Puckpool Park, all sites where it probably still survives.

P. nemoralis L.
WOOD MEADOW-GRASS
N. An infrequent and local grass in woodlands and shady hedge banks, but probably under recorded. Also believed to be occasionally sown in woodland rides to provide food for pheasants. Map 280

P. bulbosa L.
BULBOUS MEADOW-GRASS
N. A very local grass found in short coastal grassland on well-drained soils. Frequent on St Helen's Duver 6389, CP 1993, where it was first recorded in 1850; until 1968, this was the only known Island site for this plant. Frequent along the cliff edge at Freshwater from Tennyson Down to the Needles headland 2984, 3185, 3285 and 3385, CP 1997, where it was first noticed in 1983. At the eastern end of the Island, it grows on Culver Cliff edge 6385 and in sandy cliff top grassland at Redcliff 6285, CP 2002. Other sites are sandy banks around Ryde Canoe Lake 6092, PS 1999; Norton Spit 3589, JKN 1996, where it was first recorded in 1968; and sandy ground at Fort Victoria 3389, PS 1998. Bevis *et al* (1978) only knew this grass from St Helen's Duver and Norton Spit and it is now considered to be an increasing species.

Dactylis glomerata L.
COCK'S-FOOT
N. Frequent and widespread in coarse grasslands everywhere.

Catabrosa aquatica (L.) P. Beauv.
WHORL-GRASS
N. Very local and scarce in shallow

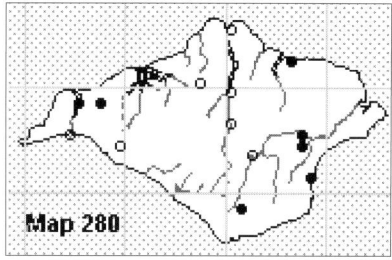

P. nemoralis

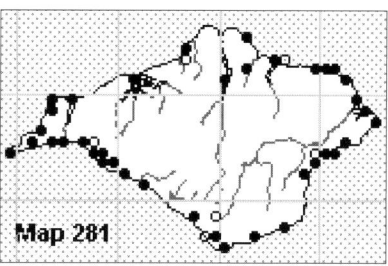

C. marinum

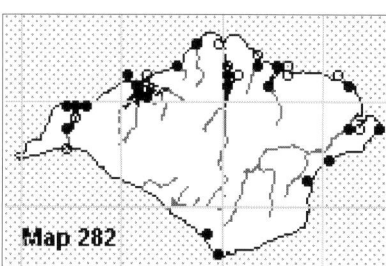

Parapholis strigosa

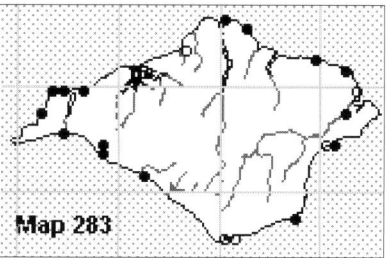

P. incurva

streams and ditches, particularly in the Eastern Yar valley. Bevis *et al* (1978) described this plant as frequent; that is no longer the case although it is probably under recorded. Recently recorded from drainage ditches on Alverstone Marsh 5785, JO & PS, 1996 and from a feeder stream at Wacklands Farm near Newchurch 5585, ST 1991. Preston *et al* (2002) suggest that this grass is declining throughout England.

Catapodium rigidum (L.) C.E. Hubb.
FERN-GRASS
N. Widespread but local on walls, pavement cracks, dry grasslands, anthills and sandy and shingly beaches.

C. marinum (L.) C.E. Hubb.
SEA FERN-GRASS
N. Local in bare ground around the coast in clifftop grassland, shingle, dunes and old walls. Map 281

Parapholis strigosa (Dumort.) C.E. Hubb.
HARD-GRASS
N. Local but widespread around the coast in upper saltmarshes and behind seawalls. It is principally found along the north coast including all our estuaries, but there are a few records from muddy cliffs on the south coast. Map 282

P. incurva (L.) C.E. Hubb.
CURVED HARD-GRASS
N. Found very locally around the coastal fringe on compacted clay slopes, on sandy shingle and behind seawalls. Although local, it is less scarce than suggested by Bevis *et al* (1978). The recent number of records suggests that it may be increasing. Map 283

Glyceria maxima (Hartm.) Holmb.
REED SWEET-GRASS
N. In contrast to its abundance in marshes and streams on the mainland, this plant has always been a rarity on the Island and is today only known from rivers around Newport. There are large patches in the middle of the town by the Lukely Brook near Towngate Mill 4989, BS 2001, a site where it was first noticed in 1971. Also found by the Medina between Shide and Blackwater 5087, SB 2001, where it was first noticed in 1978. First recorded by A.G. More in 1861 from the chine at Colwell Bay, but it had been lost from here by 1913.

G. fluitans (L.) R. Br.
FLOATING SWEET-GRASS
N. Frequent and widespread in shallow ponds, ditches, wet woodland rides, marshes and wet meadows.

G. x *pedicellata* F. Towns.
HYBRID SWEET-GRASS
N. Uncommon or overlooked in

marshes and water meadows. Recorded from a drainage ditch on Brading Marsh 6086, AC 1999; and from Plaish water meadows 4887, BS 1969. Stratton (1909) recorded this hybrid from seven sites and Townsend (1883) recorded it from six sites.

G. declinata Bréb.
SMALL SWEET-GRASS

N. Occasionally recorded in wetland sites, although probably overlooked. Newtown Brook at Fleetlands Farm 4289, PS 2000; ditch at Sainham, near Godshill 5280, CP 2002; Brook Chine 3883, CP 1999; wet peaty ground at Bohemia Bog 5183, CP 1995.

G. notata Chevall.
PLICATE SWEET-GRASS

N. Under recorded from similar wetland sites to *G. fluitans*. Recorded from Compton Marsh 3685, CP 1998; and Wolverton Marsh, Shorwell 4582, CP 2001. Bevis *et al* (1978) recorded it from a stream at Shalfleet, water meadow at Plaish, wet ride in Fattingpark Copse, marshy field at Quarr and from Alverstone Marsh.

Melica uniflora Retz.
WOOD MELICK

N. A local grass of ancient woodlands and species-rich shady woodland banks on chalk or clay soils. Map 284

Helictotrichon pubescens (Huds.) Pilg.
DOWNY OAT-GRASS

N. Frequent on unimproved chalk grassland but also found, more locally, in dry or damp neutral meadows. Map 285

H. pratense (L.) Besser
MEADOW OAT-GRASS

N. Widespread, with a similar distribution to Downy Oat-grass.

Arrhenatherum elatius (L.) P.Beauv. ex J.&C. Presl
FALSE OAT-GRASS

N. A frequent and widespread coarse grass throughout the Island in a range of habitats. It is a frequent coloniser of ungrazed meadows.

var. *bulbosum* (Willd.) St-Amans, This variety, called Onion Couch, has been recorded frequently both as an arable weed and in hedgebanks, old grassland and within woodlands. Bromfield (1856) described it as frequent, referring to the local name of Knot-grass.

[*Avena strigosa* Schreb.
BRISTLE OAT

C. Extinct. Bromfield (1856) referred to this plant as an occasional weed of cornfields and cultivated ground, but only cites one record of Rev. G.E. Smith from amongst potatoes at St John's Turnpike, Ryde in 1838. Townsend (1883) quotes an additional record from a field at Steephill.]

A. fatua L.
WILD-OAT

Arch. A frequent and pernicious weed of crops on most soils. Also found on waste ground.

A. sterilis L. ssp. *ludoviciana* (Durieu) Gillet & Magne.
WINTER WILD-OAT

C. A rare weed of arable land. Fields below Bleak Down 5081, PJW 1998; field margin east of Tapnell Cottages 3786, CDP & TDD 1998; and at Wootton on disturbed arable soil from pipeline trench 5391, PS 2000. First recorded from Havenstreet, GB 1962.

A. sativa L.
OAT

C. An occasional casual on waste ground, rubbish tips and verges.

Gaudinia fragilis (L.) P. Beauv.
FRENCH OAT-GRASS

E? This grass is a frequent component of unimproved and semi-improved neutral meadows on heavy soils. There is some doubt as to whether it is a native or introduced species but it is strange that none of the Victorian botanists came across it, which suggests that it may have been introduced with grass seed imported from Europe around the time of the First World War. The first non-casual record of this species in Britain was made by J.W. Long in July 1917, 'growing freely on meadow land near Ryde' (BM). In 1938, he saw it growing at Havenstreet and wrote, 'A meadow from 8 to 10 acres seemed to contain quite as much *Gaudinia* as any other grass. In a meadow on the other side of the village it seemed to be growing plentifully. From its abundance and the considerable area in which it has proved its existence, I am convinced that it must have been established for many years'. Gladys Bullock had discovered it here and she had established that the meadows had been under permanent grassland for at least twenty years.

Although its stronghold is on the clay soils on the north of the Island, it has been increasingly found on the central chalk (first in 1975) and to the south, particularly since 1998. It is unclear whether these sites had previously been overlooked or whether the records represent an increase in range. It has been suggested that the plant may be being spread by the current practice of using contractors' machinery for hay cutting; with this equipment operating across the Island. *Gaudinia* has its British strongholds in meadows in Somerset and the Island. Map 286

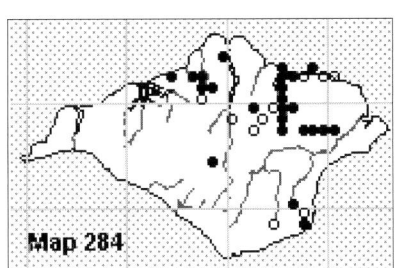

Melica uniflora

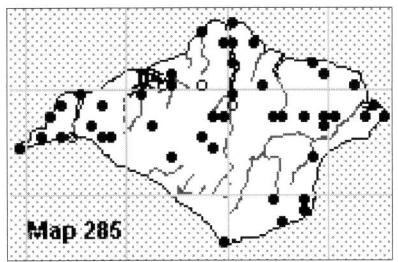

Helictotrichon pubescens

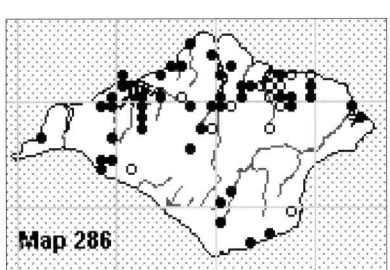

Gaudinia fragilis

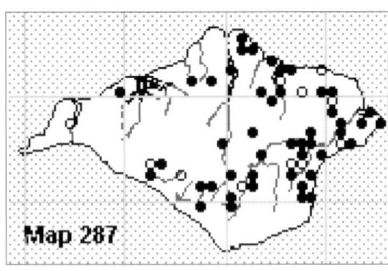

H. mollis

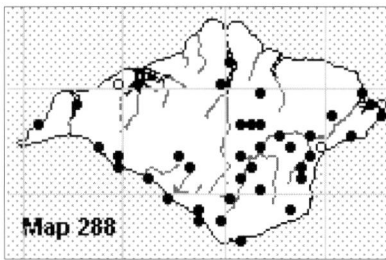

Aira caryophyllea

Trisetum flavescens (L.) P. Beauv.
YELLOW OAT-GRASS
N. Frequent in unimproved calcareous and neutral dry grasslands and verges, particularly common on the chalk.

Koeleria macrantha (Ledeb.) Schult.
CRESTED HAIR-GRASS
N. Widespread, but not abundant, in unimproved chalk grassland. There are few records off the chalk: cliff top sandy grassland at Redcliff 6285, GT 1999; St Helen's Duver 6389, BS 1966; and calcareous slumped ground at Headon Warren undercliff 3186, CP 1998.

Deschampsia cespitosa (L.) P. Beauv.
TUFTED HAIR-GRASS
N. A frequent and widespread grass in rough pastures and marshes across the Island, away from the chalk. Particularly frequent on the clay soils on the north of the Island, where it is regularly found in damp woodland rides. Bromfield (1856) wrote, referring to the texture of its leaves, 'might not this roughness be employed in polishing turnery and metal wares, instead of the imported Dutch Rushes (*Equisetum hyemale*)?'

D. flexuosa (L.) Trin.
WAVY HAIR-GRASS
N. An uncommon species on dry acid soils on heaths and sandy banks. Very local at the eastern end of Headon Warren 3186, CP 1998; several patches in meadow in front of Osborne House 5194, CP 2001; locally frequent on western slopes of Luccombe Down 5779, CP 1998; in small quantity on Sandown Golf Course 3186, CP 1998; and locally frequent in grassland on Brading Marshes 6287 & 6388, NS 1991. A record from the eastward slope of Tennyson Down, above Freshwater Bay 3485, BG 1999 must be on leached soil.

Holcus lanatus L.
YORKSHIRE FOG
N. A frequent and widespread species in grasslands, waste ground and woodland rides across the Island.

H. mollis L.
CREEPING SOFT-GRASS
N. A widespread but local grass on acid soils, particularly on the greensand, often growing in woods. Map 287

Aira caryophyllea L.
SILVER HAIR-GRASS
N. Widespread but local in dry, bare sandy or gravelly soil such as on heaths, cliff top vegetation along the southwest coastline and sandy coastal grassland. Map 288

A. praecox L.
EARLY HAIR-GRASS
N. Found in similar situations to Silver Hair-grass, but more widespread and generally in larger quantity. The majority of records are from the sandy soils on the south of the Island.

Anthoxanthum odoratum L.
SWEET VERNAL-GRASS
N. Frequent and widespread in dry grasslands across the Island.

Phalaris arundinacea L.
REED CANARY-GRASS
N. Widely distributed on riversides, ditches and pond margins. Bevis *et al* (1978) noted that this grass was especially common along the Eastern Yar between Newchurch and Brading, and this remains true today.

P. aquatica L.
BULBOUS CANARY-GRASS
C. A casual on a roadside verge in Newport 4989, BS & SB 2002 (conf. B. Ryves).

P. canariensis L.
CANARY-GRASS
C. Known since Bromfield's time as a casual of waste ground and rubbish tips. Most records probably originate from birdseed.

P. minor Retz.
LESSER CANARY-GRASS
C. Known from a handful of records from waste or arable sandy ground: Atherfield Farm 4779, PJS 1998; Hulverstone 3983, PS 2000; and Wellow 3988, PS 2000. Otherwise only recorded from waste ground at Newport in 1928, (Drabble & Long 1931).

P. paradoxa L.
AWNED CANARY-GRASS
C. Found growing in great quantity on trench spoil from pipeline excavation in fields near Wootton 5390, PS 2000 and Wellow 3988, PS 2000; amongst maize crop at Gatcombe 4985, SB 2002. Otherwise only recorded from waste ground at Newport 'in fair quantity' in 1917, (Drabble & Long 1931).

Agrostis capillaris L.
COMMON BENT
N. Frequent and widespread in dry grasslands.

A. gigantea Roth
BLACK BENT
Arch. Under-recorded but widespread on rough cultivated and waste ground. Sometimes found as an abundant weed of arable fields on light soils. Bevis *et al* (1978) write 'distribution unknown owing to confusion with other *Agrostis* species', but there have been many records since then.

A. stolonifera L.
CREEPING BENT
N. Frequent in damp and brackish grasslands, undercliffs, overgrown ponds and improved chalk grassland.

A. curtsii Kerguélen
BRISTLE BENT (local: rabbit-grass)
N. A local plant of dry heaths and hedgebanks on gravelly and sandy soils. It can be locally abundant as, for instance, on the top of Ventnor Downs 5678, and around the covered

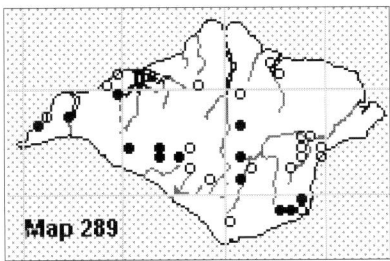

A. curtsii

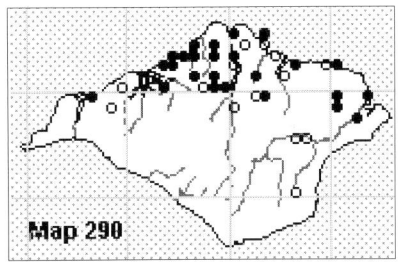

Calamagrostis epigejos

reservoir on Brighstone Down 4384. Much declined over the past hundred years, with the loss of suitable heathy grassland sites to cultivation and woodland. This south-western grass approaches the eastern limits of its distribution on the Island. Map 289

A. canina L.
VELVET BENT
N. Locally frequent in damp grassland on acid soils.

A. vinealis Schreb.
BROWN BENT
N. Locally frequent on dry grassland on acidic, often sandy, soils, but under recorded. Older records were confused with *A. canina*.

Calamagrostis epigejos (L.) Roth
WOOD SMALL-REED
N. Locally frequent on heavy clay soils on the north of the Island, forming patches in damp woodland rides, scrubby areas and slumped coastal cliffs. Bromfield (1856) described it as 'abundant in the northern part of the island' but Bevis *et al* (1978) considered it to be 'much rarer today possibly as a result of drainage'. Recent records suggest that this conspicuous grass may have increased. Map 290

[*C. canescens* (F.H. Wigg.) Roth
PURPLE SMALL-REED
N. Extinct. Known only from a small boggy alder carr near Knighton, where it was recorded by A.G. More in 1857 (More, 1898). He described it as growing plentifully with Marsh Fern, but there are no further records.]

Ammophila arenaria (L.) Link
MARRAM
N. A very locally frequent grass of sandy shores. Long known from St Helen's Duver 6389, TCR 1996,
Norton Spit 3589, JKN 1996 and sparingly at Thorness Bay 4693, JEG 1997. Also recorded from Bembridge Point 6488, AC 1996; sparingly on Binstead beach 5792, BS 1992; a single patch at the mouth of King's Quay 5394, CP 2001; and a patch in sandy ground at the foot of Lake Cliffs 5880, AC 1999. Reported as common on Hamstead shore in 1965, BS. Bromfield (1856) wrote, 'I cannot learn that there exists any peculiar name for this plant in the island, where it is known only as Spire, a term applied by the Islanders to all the larger-spiked grasses, Carices and Typhae.'

Gastridium ventricosum (Gouan) Schinz & Thell.
NIT-GRASS
N. Formerly a widespread arable weed, particularly on the northern half of the Island, but last recorded in this habitat before 1931 and considered to have become extinct until rediscovered in 1982. About 20 plants were found on an eroded shoreline cliff at Watershoot Bay, St Catherine's 4975, BS 1982 (conf. P.J. Trist); it was found again here in almost the same spot in 1999, BS. About 40 plants were found growing on a roadside wall opposite Shorwell Church 4583, MB 1998 and it was still growing here in 2002. It is likely that populations survive on dry calcareous grassland sites, as elsewhere in south and south-west England. The only record to date suggestive of this habitat was in 1908, when it was found growing amongst gorse on Rew Down by A.B. Jackson (OXF). Two plants were found on a sandy grass bank adjoining an arable field west of Leechmore Farm, to the west of Bleak Down 5080, GT 2002. There is a casual record from Quarr in 1963, JH.

Apera spica-venti (L.) P. Beauv.
LOOSE SILKY-BENT
Arch. A very rare grass of arable land. Several plants were found in a sandy field by Burnt House Lane, Alverstone 5885, CDP & TDD 1998. The field had been turf stripped three years previously. Plants were again found in this field in 2001. Otherwise, only known from waste ground at Newport, 1930 and Cowes, 1929 (Drabble & Long, 1931).

[*Polypogon monspeliensis* (L.) Desf.
ANNUAL BEARD-GRASS
C. The only modern records are from Bleak Down 5182, BS & SB 2002, where it was abundant on the landfill, and on disturbed ground at Seaclose, Newport, a few plants, 5089, PS 2002. There are one or two historic casual records but also records suggesting it occurred as a native of saline coastal marshes: very sparingly on the saltmarsh above the bridge at Norton, Yarmouth, 1868 G.R. Tate (More, 1871) and by the Medina estuary, on several occasions, Drabble & Long, 1931.]

P. viridis (Gouan) Breistr.
WATER BENT
E. A rare but increasing casual grass of waste ground but the first record, in 1989, was of a well established population on landslipped cliffs at Grange Chine 4281, BS. There are casual records, from 1994 onwards, from Newport, Shanklin, Newchurch, Niton and Freshwater.

Alopecurus pratensis L.
MEADOW FOXTAIL
N. Frequent and widespread in unimproved and semi-improved grasslands across the Island, excepting on the most acid soils.

A. geniculatus L.
MARSH FOXTAIL
N. Frequent in permanently and seasonally wet grasslands. Widespread but not recorded from the chalk.

A. bulbosus Gouan
BULBOUS FOXTAIL
N. A rather rare grass of brackish pastures along the north coast. The principal extant sites are the pastures to the south of Yarmouth Mill, 3688

and 3589 RHW, 2000, where it is locally frequent. Also found east of Ryde at Springvale Marsh 6192, CP 1999 and Seaview Duver 6291, CC 1985; brackish areas of Brading Marshes at Home Farm 6387, CC 1985; Marsh Farm, Brading 6188, CC 1985 (site since destroyed) and Quarr Marsh 5692, CC 1985. Bromfield (1856) knew it from the marshes behind Ryde Dover, where it was last recorded in 1900.

A. aequalis Sobol.
ORANGE FOXTAIL
N. Known only from around a small field pond at Fleetlands Farm, Newtown, where it was plentiful 4190, MS, 1976 (conf. C. E. Hubbard). It has not proved possible to revisit this site.

A. myosuroides Huds.
BLACK-GRASS
Arch. Locally frequent as a weed of dry arable soils. It may have increased since Bevis *et al* (1978), but was a pernicious weed in Bromfield's time. He writes, 'this is the Black-grass of Isle of Wight farmers, and is no doubt so called from its injurious qualities, and not from its colour.'

Phleum pratense L.
TIMOTHY
N. Frequent and widespread in grasslands on a range of soils.

P. bertolonii DC.
SMALLER CAT'S-TAIL
N. Frequent in unimproved grasslands, particularly on calcareous soils but also in short turf on sandy or clay soils.

P. arenarium L.
SAND CAT'S-TAIL
N. Only ever known from sandy ground at Norton Spit 3589, BS 2000 where it can be frequent in good years. It was first recorded here in 1846 (Bromfield, 1856).

[*Bromus arvensis* L.
FIELD BROME
C. Extinct. This was a rare cornfield weed in the nineteenth century. The last record was from Bleak Down (Drabble & Long, 1931).]

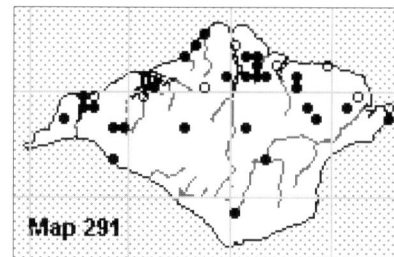
B. commutatus

B. commutatus Schrad.
MEADOW BROME
N. Very locally frequent in unimproved and semi-improved meadows, particularly on clay soils, and on verges and trackways. Occasionally found in cornfields. It is under recorded but is more frequent than suggested by Bevis *et al* (1978). Map 291

B. racemosus L.
SMOOTH BROME
N. Apparently scarce in semi-improved meadows, although likely to have been overlooked. Scattered plants on trackways north of Hamstead Farm 4091, PS 1994; meadow at Rew Street 4693, CP 2001; meadow at Springhill, East Cowes 4693, CP 2001; and Perreton Farm, Arreton 5385, CP 1996. Records suggest that it has never been particularly frequent.

B. hordaceus L.
SOFT BROME
N. Frequent and widespread in dry grasslands and rough ground. This is a very variable grass and there has been confusion over the identity of some forms.
ssp. *hordaceus* is the commonest form. Dwarf specimens growing near the cliff edge on West High Down 3285 were identified as this subspecies by P.J. Trist in 1980.
ssp. *ferronii* (Mabille) P.M. Sm. is likely to occur in thin dry clifftop grassland along the south-west coastline but has yet to be confirmed. It is frequent in this situation on the Dorset coast.
ssp. *thominei* (Hardouin) Braun-Blanq. is a small plant which grows in sandy ground and bare places on cliffs: amongst blown sand on Rowdown 4383, RHW 2000; sandy bank at Ryde Canoe Lake 6092, BS 1980 (conf. P.J. Trist); Culver Cliff edge growing as a 5cm high sward 6385, BS 1983 (conf. P.J. Trist); Afton Down 3685, EAP 1981 (conf. P.J. Trist).

A recently recognised taxon, provisionally called *longipedicillatus*, was found on a grassy roadside verge in Newport 5088, CP 2000 (conf. L. Spalton). This tall, early flowering grass has subsequently been found in a number of grassland sites on the clay and is probably widespread, as has proved to be the case on much of the mainland.

B. x *pseudothominei* P.M. Sm.
LESSER SOFT-BROME
E. Rare but under recorded in meadows. Gathered in meadows around The Wilderness 5082, BS 1980 and on a sandy bank at Ryde Canoe Lake 6092, BS 1980 (both conf. P.J. Trist); grazing marsh by seawall at Brading 6187, CP 1998 (conf. R.M. Payne).

B. lepidus Holmb.
SLENDER SOFT-BROME
E? This grass is only known from three records, most recently from a field to the east of The Wilderness 5082, FR 1974. Also recorded from near Freshwater, JH 1966 (confirmed BM) and waste ground by the river at Newport in 1932, J.W. Long (K). Preston *et al* (2002) have noted a national paucity of modern records for this taxon.

B. secalinus L.
RYE BROME
Arch. Recently recorded from waste ground at Wellow 3987, PS 2000 where it was growing in quantity around a large grain silo, and subsequently from other arable sites in the vicinity 3786, 3787, 3788 & 3886, PS 2002. Bromfield (1856) described Rye Brome as a locally abundant cornfield weed.

Bevis *et al* (1978) refer to this plant as occurring abundantly in permanent grassland at Palmer's Brook Farm 5291, an unusual habitat. Material collected from here in 1974 was identified by C.E. Hubbard as *B. pseudosecalinus*. It has been searched for without success in recent years.

Bromopsis ramosa (Huds.) Holub
HAIRY-BROME
N. Frequent in woods, hedgebanks and shady lanes. Widespread but most frequent on clay and chalk soils.

B. erecta (Huds.) Fourr.
UPRIGHT BROME
N. Locally frequent in tall calcareous grassland, particularly where ungrazed. First recorded by Turner at Luccombe in 1823. Bromfield (1856) only knew it from this area and it was not until the second quarter of the twentieth century that records started to appear from other downland sites. Also recorded rarely from neutral grassland e.g. Osborne estate grassland 5294, NS 1996; Medham by old railway track 5093, BS 1972; and near Ryde 5992, JH 1965.

B. inermis (Leyss.) Holub
HUNGARIAN BROME
E. A seed contaminant established in dry roadside grassland at Newport 4887, BS 1989 (conf. P.J. Trist) and Carisbrooke 4989, SB 1998.

Anisantha diandra (Roth) Tutin ex Tzvelev
GREAT BROME
E. Rarely well established in warm, dry waste ground, but probably increasing. Recorded from sandy ground at Bembridge Point 6488, PJS 2000, where it was first noticed in 1982; Ventnor cliff east 5677, CP 2000; headland of an arable field at Thorley 3688, PS 2000; waste ground on Forest Road, Parkhurst 4789, CP 2002; and waste ground at Whippance Farm, Rew Street 4693, JEG 1997. Prior to this, the only records were from Ventnor and Newport in Drabble & Long (1931).

A. sterilis (L.) Nevski
BARREN BROME
Arch. Frequent and widespread on verges, walls, waste ground and as a weed of cultivated ground.

[*A. tectorum* (L.) Nevski
DROOPING BROME
C. Only ever recorded as a casual weed of a grass crop at Bembridge in 1858 (More, 1871).]

[*A. madritensis* (L.) Nevski
COMPACT BROME
C. Only known from two old records: a casual at Newport (Drabble & Long, 1931) and from Quarr GB, 1965.]

Ceratochloa cathartica (Vahl) Herter
RESCUE BROME
E. Well established by roadside verges at the north end of Alverstone village 5785 & 5885, AC 1999, where it has been known since at least 1972. Also recorded as a casual from a field at Bowcombe 4486, MB 1998; car park at Brighstone 4282, GK 1994; railway embankment at Lake 5983, ST 1993; Bembridge shore east of the Point 6488, JEG 1993; and Nettlestone tip 6290, RK 1974. First recorded as established on waste land at Cowes between 1928 and 1931 (Drabble & Long, 1931).

Brachypodium pinnatum (L.) P. Beauv.
TOR-GRASS
N. A remarkable absentee from the Island's flora. However, a single patch, which was found on the west face of Tennyson Down 3385, BS 1996, was confirmed as this species by T.A. Cope.

B. sylvaticum (Huds.) P. Beauv.
FALSE BROME
N. Frequent and widespread in woods, scrub, chalk grassland and hedgebanks.

Elytrigia repens (L.) Desv. ex Nevski
COMMON COUCH
N. Frequent and widespread in disturbed grasslands, hedgebanks and cultivated ground.

E. atherica (Link) Kerguélen ex Carreras Mart.
SEA COUCH
N. Frequent in upper saltmarshes and low clay cliffs along the Solent coastline. Also found on low cliffs in suitable spots between Lake and Steephill. Recorded from the Needles headland 2984, DD 1998. var. *setigera*, the form with long awns, has been noted at King's Quay saltmarsh 5393 CP, 2002.

E. x obtusiuscula (Lange) Hyl.
SEA COUCH X SAND COUCH
N. Status unknown. There are old records of this hybrid from the upper shore where both parents occurred together at Norton Spit, St Helen's Duver, Newtown estuary and Sandown Bay (More, 1871).

E. juncea (L.) Nevski
SAND COUCH
N. Very local on sand dunes and sandy foreshores. Bembridge Point 6488, AC 1996; St Helen's Duver 6389, CP 1993; King's Quay east spit 5394, JC 2000; Thorness Bay 4593, CP 2000; Norton Spit 3589, JKN 1996.

Leymus arenarius (L.) Hochst.
LYME-GRASS
N? Extremely locally frequent on sand dunes. Bembridge Point 6488, AC 1996; St Helen's Duver 6389, CP 1993; slowly increasing at Norton Spit 3589, JKN 1996; a single clump on Thorness beach 4693, CP 2000. Presumably native but only recorded for the first time in 1927, at Shanklin.

Hordeum distichon L.
TWO-ROWED BARLEY
C. A casual of cultivation.

H. murinum L. ssp. *murinum*
WALL BARLEY
Arch. Frequent in dry grassland and waste ground. It is widespread, but least so in inland rural areas.

H. jubatum L.
FOXTAIL BARLEY
C. A rare casual in sown grassland or waste ground. Recorded from the car park above Bonchurch beach 5777, BS 1992. First recorded at Ventnor in 1972, TW (conf. C.E. Hubbard).

H. secalinum Schreb.
MEADOW BARLEY
N. Frequent in unimproved and semi-improved meadows, particularly on clay soils. Not recorded on the chalk. Map 292 (*see overleaf*)

[*H. marinum* Huds.
SEA BARLEY
N. Extinct. Bromfield (1856) knew this as a frequent grass of brackish meadows at Newtown, Yarmouth,

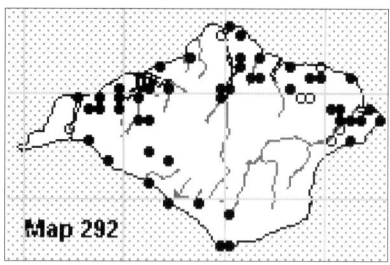

H. secalinum

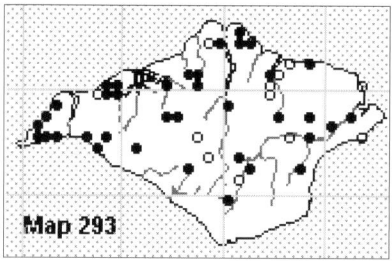

Danthonia decumbens

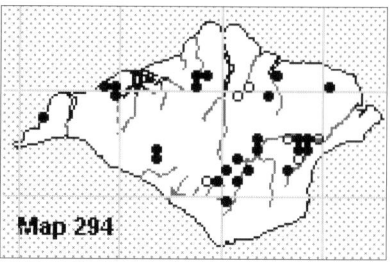

Molinia caerulea

Springfield and Brading Harbour. Last recorded in 1879, when it was still abundant near Yarmouth and St Helens (Townsend, 1883). It has not been refound in recent years, despite many searches and the presence of apparently suitable habitat.]

Triticum aestivum L.
BREAD WHEAT
C. A frequent casual of waste ground, rubbish tips and pavement cracks.

Danthonia decumbens (L.) DC.
HEATH-GRASS
N. A local, but probably under recorded, grass found in small quantities in unimproved grasslands. It occurs in chalk grassland as well as a few neutral meadows, but it is most frequent in wet and dry grassy heaths on acid soils. Map 293

Cortaderia selloana (Schult. & Schult.) Asch. & Graebn.
PAMPAS-GRASS
E. Persisting in a few places on slumped cliffs, perhaps as a garden throw-out. One plant well away from habitation on Roughland Cliff, Brook 3882, PS 2000; Binnel Bay 5075, CP 2000.

Molinia caerulea (L.) Moench
PURPLE MOOR-GRASS
N. A local grass of wet heathy grasslands, bogs and woods on acid soils. The subspecies have not been investigated. In the absence of grazing or if sites dry out, it can become dominant, excluding more delicate species. Bevis *et al* (1978) recorded it from 17 tetrads and its status has probably changed little since then. Bromfield (1856) wrote, 'a large proportion of the grass in Parkhurst forest consists of this species, under the trees in the extensive plantations (principally fir) in that enclosure.' Map 294

Phragmites australis (Cav.) Trin. ex Steud.
COMMON REED
N. A frequent, often dominant plant in still and slow-moving water courses, tolerating brackish conditions. It is also frequent on slumped undercliffs around the coast, where it can produce impressively long rhizomes across the surface of the ground. Map 295

Cynodon dactylon (L.) Pers.
BERMUDA-GRASS
C. A very rare introduced grass, surviving for a number of years but not known for its long term persistence. A weed in a hotel garden in Victoria Avenue, Sandown 6084, GT 2000; persisted for several years on Quarry Road allotments, Ryde 5991, CP 1989, until the site was destroyed by pipe laying. Previously recorded as persisting for a number of years in rough pasture at Newport just after the 1914-18 War (Drabble & Long, 1931).

Spartina maritima (Curtis) Fernald
SMALL CORD-GRASS
N. Known only from saltmarsh bordering Clamerkin Creek, Newtown 4290, 4291, 4390 & 4391, CP & AM 2000, where it is abundant over a large area. It grows as scattered tufts in mixed saltmarsh and as a dense sward on pool margins, excepting where cattle have access. This is almost the last surviving site for this declining species along the South coast of England, although it remains frequent in many East Anglian saltmarshes.

It was probably once widespread in all our estuaries. Bromfield (1856), writing of this plant said, 'this rank-smelling grass is quite destitute of beauty; nor does it recommend itself by any known use, unless by its creeping and fibrous roots serving to consolidate the soft fluctuating soil on which it grows, and affording a safe, if not a dry footing over the dreary waste of flat salt-marsh.' However, by the beginning of the twentieth century, it was in decline. Drabble & Long, 1931, recorded it from Norton Spit in 1913 but added that it had not been seen recently. Stratton said that although it was persisting longer here than on the Hampshire mainland, by 1913 it was becoming increasingly difficult to find in creeks on the eastern side of the Medina (Rayner, 1929). In fact, it persisted longer here than in most of our estuaries but when P.J. Goodman collected seed here, from the saltmarsh at Whippingham (508933) in 1953 he found only one surviving clump about two feet square. Map 296

S. x townsendii H. & J. Groves
TOWNSEND'S CORD-GRASS
N. This is the original hybrid between the native Small Cord-grass and the American introduction, Smooth Cord-grass, which arose spontaneously in Southampton Water around 1870. It was named in honour of Frederick Townsend, author of *Flora of Hampshire* (1883), by the brothers Henry and James Groves. This sterile hybrid was more vigorous than either parent and it spread on saltmarshes around the Solent. It was first recorded from the Island in 1893, by James Groves, on the banks of the Medina and at Norton, Yarmouth (but see also under Common Cord-grass below). It was subsequently recorded from other estuaries but later declined. Bevis *et al* (1978) described it as a decreasing species occurring in small patches amongst the dominant *S. anglica*. It survives in small quantity on the saltmarshes at Newtown 4290, AJG 1999.

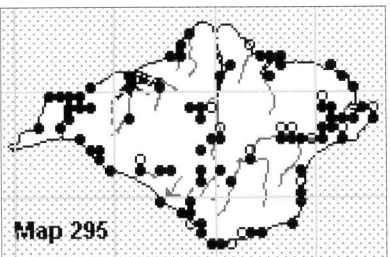
Phragmites australis

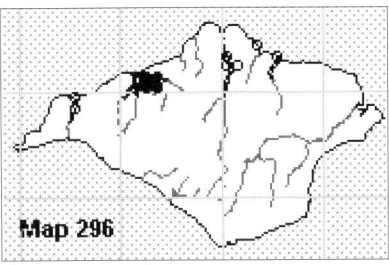
Spartina maritima

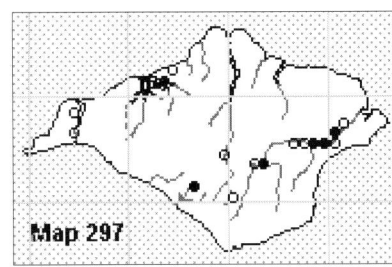
S. emersum

S. anglica C.E. Hubb.
COMMON CORD-GRASS
N. This is the fertile polyploid hybrid derived from Townsend's Cord-grass. It has colonised all our saltmarshes along the Solent coast, sometimes to the detriment of other saltmarsh species. The taxon was not recognised until 1957 by C.E. Hubbard and it is not clear when it first appeared on the Island. Cord-grass is virtually confined to estuaries today but 'around 1885, an invasion of *Spartina* grass began in front of Quarr and proceeded and extended in subsequent years' (Colenutt, 1939). In 1942, it covered extensive mudflats on the shore at Quarr. More recently, it has virtually been lost from here although two or three small patches survive.

Panicum miliaceum L.
COMMON MILLET
C. A casual from bird seed and on tips. Occasionally planted in arable for pheasants. First recorded in 1931 by J.W. Long.

Echinochloa crus-galli (L.) P. Beauv.
COCKSPUR
C/E. This is a casual of waste ground and a persistent weed of maize crops. Recorded as an arable weed near Shalfleet 4289, PS 2000; Champion Farm, Rookley 5085, SB 2002; and especially in sandy fields at Newchurch 5685 & 5686 CP 1997. Persisting over many years around dismantled aviaries at Puckpool Park near Ryde 6192, AC 1999. Recorded as a casual from a number of scattered sites across the Island, involving both long and short awned plants. First recorded from rough ground near the shore at Freshwater in 1869 by F. Stratton (More, 1871).

Setaria pumila (Poir.) Kerguélen
KNOTROOT BRISTLE-GRASS
C. A very rare casual of waste ground but persisting over many years around dismantled aviaries at Puckpool Park near Ryde 6192, AC 1999. Several plants on disturbed soil from pipeline excavations at Lake Common 5884, CP 1999; amongst maize crop at Champion Farm, Rookley 5085, SB 2002; Newchurch churchyard 5685, MB 2002; Nettlestone tip 6290, BS 1973. First recorded in 1869 on waste ground at Whippingham by James Pristo (More, 1871).

S. verticillata (L) P. Beauv.
ROUGH BRISTLE-GRASS
C. Only known from old records: Quarr Abbey 1971, JH; Medina valley below Newport 1928, (Drabble & Long, 1931) and Cowes, 1925 (Rayner, 1929).

S. viridis (L.) P. Beauv.
GREEN BRISTLE-GRASS
C. An increasing casual in warm dry urban places such as in gardens and between paving stones. Recorded recently from Ryde, Sandown, Carisbrooke and Freshwater.

S. italica (L.) P. Beauv.
FOXTAIL BRISTLE-GRASS
C. Known only from a single record from Cowes river-bank, Long, 1931.

Digitaria sanguinalis (L.) Scop.
HAIRY FINGER-GRASS
C/E. This is an increasing casual, occurring as a pavement weed and sometimes persisting for a few years. Also recorded as a weed of maize crops on sandy soils at Newchurch 5584, PS 1999, and Apse Heath 5483, DB 1998.

Sorghum halepense (L.) Pers.
JOHNSON-GRASS
C. An occasional casual from bird seed and on waste ground.

158. SPARGANIACEAE
Bur-reed family

Sparganium erectum L.
BRANCHED BUR-REED
N. Frequent in ponds, streams and drainage ditches, often choking water courses. Bevis *et al* (1978) records it as being commonest in the East Wight and, whilst it is a common feature of the Eastern Yar and its tributaries, it is also found in many streams and ponds in West Wight.
ssp. **erectum** and ssp. **neglectum** (Beeby) K. Richt. have both been recorded but more work needs to be done to accurately reflect their distribution.

S. emersum Rehmann
UNBRANCHED BUR-REED
N. Occasional in slow flowing streams and ponds. Much less frequent than Branched Bur-reed and, although likely to be under recorded, it has probably decreased with lack of ditch clearances. Map 297

[*S. natans* L.
LEAST BUR-REED
N. Extinct. Recorded in some plenty by Bromfield from old clay pits on a small common a little to the east of Cranmore Farm in 1840. Subsequently recorded by More (1871) but considered to have become extinct by 1900. It is still found in a few pools with similar chemistry in the New Forest.]

159. TYPHACEAE
Bulrush family

Typha latifolia L.
BULRUSH
(local: black-puddings, blackamoors, blackheads, bacco-bolts)
N. Widespread although local in streams, ponds, ditches and old clay

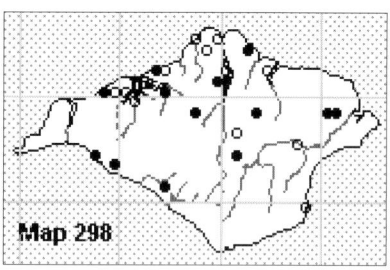

T. angustifolia

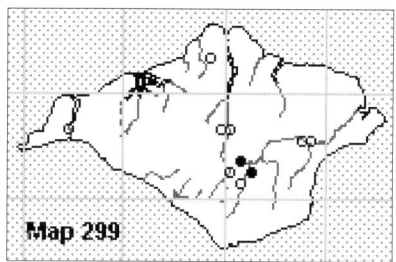

Narthecium ossifragum

pits across the Island. Often found in temporary ponds on the cliffs of the south-west coast. Bromfield (1856) refers to a greater number of local names for this plant than any other; Bacco-bolts apparently referred to a resemblance in the spikes to a roll of tobacco.

T. x *glauca* Godr.
BULRUSH X LESSER BULRUSH
N. Rare but probably occurs wherever both parents grow together. Pond at Rookley Country Park 5184, PS & EC 2000; pond on Rifle Ranges, Porchfield 4490, CP 2001.

T. angustifolia L.
LESSER BULRUSH
N. Occurs in ponds but has a thinly scattered distribution and is much more local than Bulrush. Bromfield (1856) found this plant to be more common than the Bulrush, and especially characteristic of old clay pits. He notes that it was particularly abundant in pools and clay pits on heathy ground to the west of Newtown estuary. Map 298

162. LILIACEAE
Lily family

Narthecium ossifragum (L.) Huds.
BOG ASPHODEL
N. Very rare and declining, but still frequent in Bohemia Bog 5148, CP 2001. Much declined at Munsley Bog 5282, where it was still frequent in 1988. Only 44 flowering clumps were counted in 1992, with just 3 clumps and a scatter of other plants in 2002, CP. Historic sites, with last recorded dates, are as follows: Bleak Down 1975; Blackpan Bog by Sandown Waterworks 1943; The Wilderness 1927; Blackwater Marsh & Marvel Marsh 1909; Freshwater Marshes 1900; and Alverstone Lynch 1841. Map 299

Hemerocallis fulva (L.) L.
ORANGE DAY-LILY
E. An occasional persistent garden outcast, sometimes found on cliffs.

Kniphofia uvaria group
RED-HOT-POKER
E. A persistent escape on the cliffs to the east of Ventnor 5677, CP 2000; and at Orchard Bay, Steephill 5476, CP 1996.

[*Colchicum autumnale* L.
MEADOW SAFFRON
E. Extinct. Stratton writing in the *Journal of Botany* in 1913, said that it had been well established in the grassy borders at Blackwater House, Shide for more than 15 years. G. Kirkpatrick recorded it before 1850 from a field by the Medina above Shide Bridge and this may well have been the same site, and raises the question as to whether it was native here. There is also a record from a field at Bembridge, EHW 1947.]

[*Tulipa sylvestris* L.
WILD TULIP
E. Extinct. In 1846 it was persistent in a clayey meadow near Hardingshute Farm (Bromfield, 1856), the site of a former cottage. By 1871, A.G. More said the site had been destroyed when a new farmhouse was built on the same spot.]

Lilium martagon L.
MARTAGON LILY
C. Only known from a single, purple flowered specimen persisting in a damp shady wood by the Buddle Brook at Brighstone 4283, MB 1996. Still present here in 1998.

Polygonatum x *hybridum* Brügger
GARDEN SOLOMON'S-SEAL
E. A very rare garden escape persisting for over forty years in a hedge bank at Hollow Lane, Godshill 5281, CP 1989. Several plants at the edge of gorse scrub at Brook Hill 3984, CP 1999.

Ornithogalum angustifolium Boreau
STAR-OF-BETHLEHEM
E. Occasionally naturalised in dry grassland in cemeteries, roadside verges and beside footpaths. Recent records are from sandy roadside verge at Kingston 4681, CP 1986; sandy pathside at Hulverstone 4084, PS 1993; sandy pathside in Beech Copse, Godshill 5381, CP 1987; south facing chalk bank at Carisbrooke Castle 4887, CP 2000; beside a ride near car park in Parkhurst Forest 4890, SB 2001; and churchyards at Newchurch 5685, BS 1985 and Fairlee cemetery 5089, MB 2001. Bromfield (1856) described it as occurring in meadows about Steephill and appearing truly wild and in 1973, a few specimens were found here in a meadow near Orchard Bay, TW.

Scilla peruviana L.
PORTUGUESE SQUILL
E. Frequently grown in gardens and sometimes in churchyards and very persistent. Large flowering clump at the edge of the cliff at Blackgang on the site of a former garden 4976, CP 1999.

S. autumnalis L.
AUTUMN SQUILL
N. Occurs in quantity in dry sandy grassland on St Helen's Duver, where it was first recorded in 1823. White flowered plants occur occasionally. This is the only sand dune locality in this country. It is principally a south-western species, occurring in cliff top grassland in Devon and Cornwall, but it still survives in one or two dry acid grassland sites in the Thames valley.

Hyacinthoides non-scripta (L.) Chouard ex Rothm.
BLUEBELL (local: blue-bottles)
N. A frequent and widespread plant, often dominant in woodlands, particularly ancient woods. It also occurs frequently in old hedge banks. Occasional white flowered plants can

be found in all large populations. Bluebells are also characteristic of open ground beneath bracken, particularly but not exclusively on sandy soils. Perhaps the most spectacular show of open-grown Bluebells is on the north slopes of Wroxall Down.

H. non-scripta* x *H. hispanica
BLUEBELL X SPANISH BLUEBELL
E. This hybrid is not infrequent but almost always occurs in the vicinity of habitations or as a result of garden outcasts. It is widespread in secondary woodland in the Undercliff.

H. hispanica (Mill.) Rothm.
SPANISH BLUEBELL
E. A not uncommon garden escape, although less frequent than the hybrid.

Muscari armeniacum Leichtlin ex Baker
GARDEN GRAPE-HYACINTH
E. An occasional persistent garden outcast.

Allium roseum L.
ROSY GARLIC
E. A garden escape, first recorded in 1960 in Ducie Avenue, Bembridge, GB. In 1964, it was found in the chalk pit by Clerken Lane by Carisbrooke Castle 4888, BS, and it has persisted and increased here. Although it remains scarce, Rosy Garlic has been recorded from a further seven sites in recent years.

A. neapolitanum Cirillo
NEAPOLITAN GARLIC
E. A scarce but increasing garden escape. It grows on Ventnor east cliffs 5677, CP 1995, and on the cliff below St Catherine's lighthouse 4975, SB 1997, having been first recorded from both sites in 1949, TW. Also recorded recently from roadside verges at Bouldnor, Whippingham, Bonchurch Shute and Undercliff Drive at St Lawrence.

A. subhirsutum L.
HAIRY GARLIC
E. A rare garden escape, perhaps overlooked. Established at Christchurch, Totland 3286, MB 2000 and on the roadside in Weston Lane, nearby.

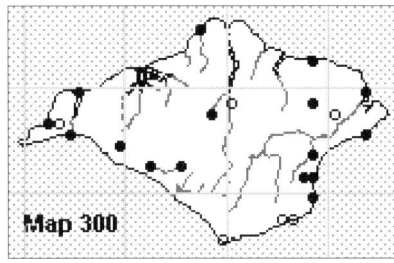

A. triquetrum

A. moly L.
YELLOW GARLIC
E. There is a single record as an established garden escape on a roadside at Bowcombe 4787, DF 1963.

A. triquetrum L.
THREE-CORNERED GARLIC
E. An established and increasing garden escape, increasingly seen on roadside verges. First recorded in 1924 as naturalised on the cliffs at Ventnor. It was recorded from a short stretch of roadside at East Ashey 5888 in 1967, RK; it still survives here but has not spread. Since Bevis *et al* (1978), it has been recorded from an increasing number of sites across the Island but, although locally dominant, it has not become invasive to the extent that it is in south west England. Map 300

A. paradoxum (M.Bieb.) G.Don
FEW-FLOWERED GARLIC
E. A very locally established garden escape. By the brook at Binstead 5792, CP 1993; Westhill Park, Ryde 6092, CP 1991; roadside at Whitefield 5988, BS 1978; known from Pump Lane, Bembridge 6488, since at least 1962.

A. ursinum L.
RAMSONS (local: gypsy onion)
N. Locally dominant in old woods but also found in moist hollows and shady hedge banks. It is particularly abundant in woods on the chalk where it dominates the ground flora to the exclusion of most other species. In Wroxall Copse 5678, where it is abundant, a large clump is growing in the crutch of an old ash pollard at a height of 2m, CP 2002. Ramsdown Farm, near Chillerton, is said to have a name deriving from Old English meaning 'the hill where wild garlic

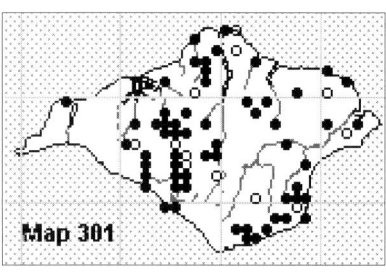

A. ursinum

grows' (Mills, 1996). Map 301

A. oleraceum L.
FIELD GARLIC
N. A rarity of dry grassland only known from one localised stretch of moat at Carisbrooke Castle 4887, CP 1991; 130 flowering stems were counted here in 1992, BS. It may still survive in cliff top grassland; there are old records in this habitat from Ventnor (1932) and from St Lawrence and Steephill inner cliff (1871).

A. vineale L.
WILD ONION
N. Frequent and widespread in verges, rough grassland, borders of arable fields and cliffs. It is commonest in the northern half of the Island. Considering its abundance today, it is surprising that Bromfield (1856) describes this plant as being 'not common, and seldom if ever flowering in this island'.

A. nigrum L.
BROAD-LEAVED LEEK
C. Apparently established at St Helen's in 1952, 6289 GB, and confirmed by Kew.

Nothoscordum borbonicum Kunth
HONEYBELLS
C. Found on waste ground at Luccombe in 1970, TW (conf. BM). There is an unconfirmed record from the cliffs between Ventnor and Flowers Brook in 1980.

Tristagma uniflorum (Lindl.) Traub
SPRING STARFLOWER
E. An occasional garden escape which can become established on sandy ground. Well established on the south facing bank of Ryde Canoe Lake 6092, CP 1998; spreading in mown grass at Yarmouth sailing club 3589, CP 2000, Castle Court, Steephill

5477, CP 2000 and Shore Road, Bonchurch 5778, SB 2002.

Leucojum aestivum L.
SUMMER SNOWFLAKE
E. A persistent garden outcast, sometimes established in plantations and churchyards.

Galanthus nivalis L.
SNOWDROP
E. Well established in many damp woods, lanesides, churchyards and cemeteries. Small numbers survive in Snowdrop Lane, Gatcombe, where Bromfield (1856) described it as occurring 'in great profusion on the steep bushy sides'. Map 302

Other *Galanthus* species and varieties are sometimes established in churchyards and cemeteries.

Narcissus x *medioluteus* Mill.
PRIMROSE PEERLESS
E. In Bromfield's time, this cultivar was a frequent escape from gardens into fields with light soils. Today it is not often grown in gardens and it is no longer found in meadows. It still survives in chalk grassland around the old buildings above Sun Corner, Needles headland 2984, where it was first noticed in 1983, BS (conf. D. McClintock).

N. pseudonarcissus L. ssp. *pseudonarcissus*
WILD DAFFODIL
(local: lent or lenten lily)
N. A locally frequent plant of ancient woodlands, scattered across the Island but with a concentration of sites in woods around Havenstreet. It was also formerly frequent in hedge banks and old meadows, but is rare today in these habitats.

The map shows largely native sites but it is difficult to be sure whether or not it has been introduced into some woodland sites. Bevis *et al* (1978) suggest that this plant was rapidly declining but this is not borne out by recent records.

The showy, early flowers of daffodils used to be popular, picked in bunches or dug up to plant in cottage gardens. This probably explains their frequent occurrence in churchyards and in cemeteries today. Wild Daffodils were exploited as a source of income and huge quantities were collected and sold in the early part of the twentieth century from certain woods. Daffodil Valley off Redhill Lane, Sandford was a celebrated venue. In the 1920s, people would travel out in buses and pay six pence at the kiosk for the privilege of seeing the wild daffodils and picking a bunch to take away.

At Shorwell, boys gathered bunches from Troopers Withybed behind Wolverton Manor and sold them for a penny a bunch to coach passengers arriving at the Crown Inn (recollection of Sid Riddett to Bill Shepard). However, this developed into such an industry that in 1939, two 17 year old lads were each fined £1 for being in possession of large sacks strapped to their bicycles containing daffodils collected from this wood. Three large baths filled with the daffodils were displayed as evidence when the case was heard in Newport. (*Isle of Wight County Press*, 8 April 1939). As late as the 1950s, gypsies would gather large bunches in Firestone Copse to sell in towns on the Island and in Portsmouth.

Bromfield (1856) refers to how the local name had become corrupted to Lantern Lilies. Map 303

Narcissus agg.
GARDEN DAFFODIL
E. A frequent introduction on roadside verges, churchyards and cemeteries but the many cultivars have not been distinguished.

Asparagus officinalis L. ssp. *officinalis*
GARDEN ASPARAGUS
E. Well-established on sandy ground by the coast. Norton Spit 3589, JKN 1996, where it has been known since 1823; long established by the old trackside at Bembridge Harbour 6488, AC 1999. It grew on St Helen's Duver, where it was first recorded by More in 1871 and last noted in 1969, BS. It has also been recorded more recently from a number of other coastal sites, where it may become permanent. Asparagus was one of the crops growing on Brading Marshes in the early years following the reclamation from the sea of Brading Haven in the 1880s. Bromfield (1856) mentions the 'vulgar name' for this plant as Speerage Sparrow-grass. Young spears of plants from Norton Spit have sometimes been collected in recent years.

Ruscus aculeatus L.
BUTCHER'S-BROOM
(local: box holly!; french holly!)
N. Locally frequent, particularly in ancient woodland on heavy clay soils, but also found in old hedgerows and scrubby thickets on acid sandy soils; not recorded from the chalk. Butcher's-broom was also sometimes planted in churchyards and in plantations and a few of the mapped sites will not be native. It is still known by local names to some woodsmen. Bromfield (1856) wrote that local butchers 'deck their mighty Christmas sirloins with the berry-bearing twigs'. Map 304

Galanthus nivalis

N. pseudonarcissus

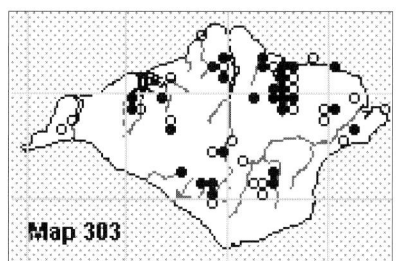

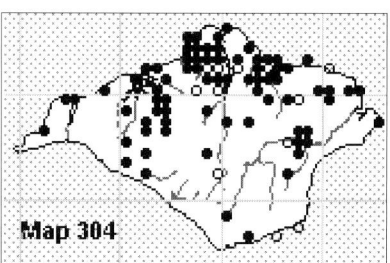

Ruscus aculeatus

163. IRIDACEAE
Iris family

Sisyrinchium bermudiana L.
BLUE-EYED-GRASS
E. The only record is of seven specimens established in grassland at the foot of Rew Down in 1972, TW & RK.

S. striatum Sm.
PALE YELLOW-EYED-GRASS
E. An occasional persistent garden escape or outcast. Slumped cliff below Reeth Lodge, St Lawrence 5075, CP 2000; at edge of scrub on St Helen's Duver 6388, CP 1993; Puckpool Park 6192, DB 1999.

Iris germanica group
BEARDED IRIS
E. Very rarely surviving as a garden outcast. Growing in the old quarry at Tapnell 3886, CP 1996.

I. pseudacorus L.
YELLOW IRIS
N. Locally frequent beside streams and in marshes along all our watercourses. It is also rather frequent in ponds. Well distributed across the Island.

(***I. orientalis*** Mill.
TURKISH IRIS
There is a record of a large stand growing in wet ground on a grassy bank just above the beach at Old Park, St Lawrence 5275, ABo 1981, but this has not been confirmed.)

I. foetidissima L.
STINKING IRIS
N. Frequent and widespread in woods, hedge banks, scrub and dry, rough grasslands, particularly on chalk and clay soils. Especially abundant in the Underliff and in the open on coastal cliffs where it sometimes flowers prolifically. Bromfield (1856) said that a variety with variegated leaves was grown in gardens at Ryde. He also refers to var. *citrina* Bromf. with lemon-yellow petals, found in a wood near Yarmouth in 1847. There are no modern records of this form outwith gardens. Map 305

Crocus vernus (L.) Hill
SPRING CROCUS
E. Sometimes established in churchyards and cemeteries. The Glebe field at Freshwater, 340868 was a famous crocus field where local people gathered bunches for either Valentine's Day or Mothering Sunday, depending upon the season, until 1967 when the site was developed for housing. Stratton (1916) described the field as 'a most lovely sight, the flowers even then constituting a purple haze when seen from some distance away.' He added, 'Lord Tennyson says he has known the plant growing there for fifty years past'. When the field was eventually sold by the Church Commissioners, members of the Isle of Wight Natural History and Archaeological Society dug up many of the corms and transplanted them in several suitable sites across the Island. They have established successfully at some of these sites such as Fairlee cemetery and Marsh Farmhouse, Newtown. Although the Glebe field is now occupied by housing, some crocuses still survive in the hedgebank on Victoria Road and in some garden lawns 3486, CP 1999.

C. tommasinianus Herb.
EARLY CROCUS
E. Well established in Newchurch churchyard, from where it is spreading to the adjoining footpath verges 5685, CP 1997.

Other Crocus species and cultivars are well established in churchyards and cemeteries.

[***Gladiolus illyricus*** W.D.J. Koch
WILD GLADIOLUS
N. Extinct. It was formerly an extremely rare plant of heathy grassland beneath bracken. A Mrs. Phillips collected a single specimen in bud 'in the midst of a wild tract of copse and heath called Apse or America woods' in July 1855 (More, 1861-62). She took it home to make a drawing and the specimen is preserved in Bromfield's herbarium at HCMS. The date is a year earlier than the generally accepted first British record of W.H. Lucas from the New Forest. More says that 'there is a tradition on the spot; it has long been known to the inhabitants of a neighbouring farmhouse that a wild Gladiolus grows in the woods at Shanklin'. In 1872, a plant was gathered amongst bracken on Lake Common, by a lady who found several specimens growing there, and this record was reported by Stratton (OXF). A further specimen was gathered from nearby Blackpan Common in 1897. Last recorded in 1931, without a locality, 'Mr. R.H. Fox has sent Mr. Long a small flower spike of *G. illyricus*'. Lousley wrote in 1947 that many hours had been spent searching for it at its four Island localities. The historic native distribution of this plant outside the New Forest, where it still occurs, included both the Island and eastern Dorset, although it was recorded only once from Dorset, in 1874.]

G. communis L. ssp. ***byzantinus*** (Mill.) R.C.V. Douin.
EASTERN GLADIOLUS
E. Very locally well established garden escape, found growing on waste ground, coastal cliffs, roadside banks and cemeteries. First recorded in 1931 from Ventnor east cliffs 5677, TW, a site where it still grows.

Crocosmia paniculata (Klatt) Goldblatt
AUNT-ELIZA
E. Established in scrubby ground behind the seawall at Bouldnor 3689, EC & PS 1999.

C. x crocosmiiflora (Lemoine) N.E. Br.
MONTBRETIA
N. Locally well established as a garden outcast on waste ground, verges, woodland edges and coastal cliffs. First recorded by J.E. Lousley in 1947 as thoroughly established on Lake Cliffs 5882, a site where it still grows abundantly.

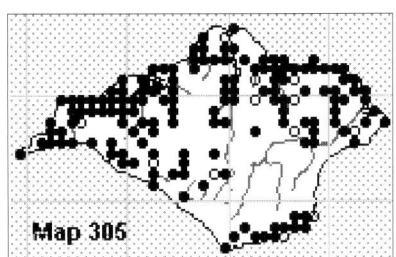
I. foetidissima

164. AGAVACEAE
Centuryplant family

Cordyline australis (G.Forst.) Endl.
CABBAGE-PALM
S. Widely planted in gardens where it sometimes self-seeds. Plants continue to survive in derelict gardens. A wind-battered plant by the cliff edge at Blackgang 4976, CP 1999.

Phormium tenax J.R. & G. Forst.
NEW ZEALAND FLAX
S. Frequently grown in gardens and persistent where planted. Three large clumps growing on the cliff below Reeth Lodge, St Lawrence 5075, MB 1999.

165. DISCOREACEAE
Black Bryony family

Tamus communis L.
BLACK BRYONY
(local: wild vine; murrain berries)
N. Frequent and widespread right across the Island in woods, old hedges and in scrub. Bromfield (1856) said that this plant was 'of universal occurrence with us, often seen twining, even around the stalks of corn and herbaceous plants, in open fields and pastures'. The berries, steeped in gin, were used, according to him, as a popular local remedy for chilblains, acting as a counter-irritant.

166. ORCHIDACEAE
Orchid family

Cephalanthera damasonium (Mill.) Druce
WHITE HELLEBORINE
N. A rare plant of secondary woods on the chalk, usually under beech, but probably more frequent than at any time in the recorded past. Recorded from Lynch Lane Rew at Calbourne Bottom 4285, ST 1996 where numbers in recent years have ranged between 0 and 3; and nearby in Brighstone Forest 4285, where 30 fine flowering specimens were recorded in 1998, MS. The most prolific site is beneath Holm Oak woodland on St Boniface Down 5678, AB 1999 where it was first recorded in 1989 and around 150 plants are counted in good years. It has been known since 1847 from beneath beeches in The Shrubbery by Clerken Lane, Carisbrooke Castle, 4888. Since 1966, a maximum of ten specimens has been seen in any year but it was last found here in 1984.

Epipactis palustris (L.) Crantz
MARSH HELLEBORINE
N. A greatly declined species now restricted to two base-rich flushed fens on coastal undercliffs. Present on Headon Warren undercliff 3085 & 3186, CP 1998 where about two dozen plants survive in two localised sites; around 200 plants were counted here in 1969, RK but the site has since become largely overgrown. Also present in two localised sites near Luccombe Chine 5878 & 5879, JS 2000 where numbers occasionally reach the low hundreds but are generally lower and the sites are highly susceptible to cliff erosion. Two plants were moved from here to a calcareous flush in Shide chalk quarry 5088, in 1995 and flowered for the first time in 2001, MS. Formerly also recorded from calcareous fens and wet meadows but not since 1935. These historic sites included Freshwater Marshes, Colwell Heath, claypits near Cranmore Farm and Compton Marsh. Map 306

E. purpurata Sm.
VIOLET HELLEBORINE
N. Not known from the Island until three flowering plants were discovered in an ancient wood on the Briddlesford Estate 5489 in 1980, BS (conf. J.T.H. Knight). Six good flowering plants were found in 1982 but it subsequently declined and was last recorded in 1994.

E. helleborine (L.) Crantz
BROAD-LEAVED HELLEBORINE
N. An uncommon plant of shady woods on chalk and base-rich clay soils, generally found in small quantity but possibly under recorded, particularly from woods on the chalk in the centre of the Island. The greatest concentration of sites is in the ancient woodlands around Havenstreet. Map 307

[*E. phyllanthes* G.E. Sm.
GREEN-FLOWERED HELLEBORINE
N. Extinct. A small, green-flowered

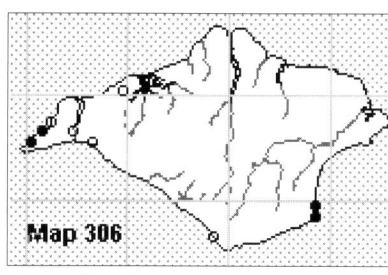
Epipactis palustris

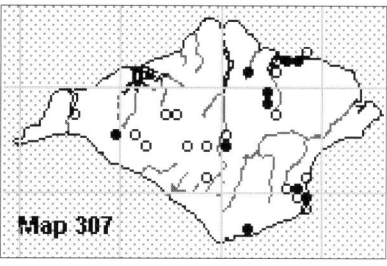
E. helleborine

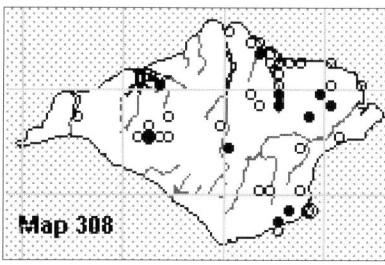
Neottia nidus-avis

helleborine was discovered by E.W. Hunnybun in a wood on the chalk near Ventnor in 1913 (CGE). At the time it had not been found elsewhere in this country, and it was named the Isle of Wight Helleborine, *E. vectensis* (T. & T.A. Steph.) Brooke & Rose. In July 1930, a similar plant was found in Bonchurch landslip, EHW, and further specimens were found here by a Miss Newnham in the 1940s. In 1951, about twelve plants were found on a wooded bank to the rear of St Lawrence Church by Dr F. Laidlaw (conf. P. Young). It survived here for several years, where it was last recorded in 1959, TW. There have been no further records but this inconspicuous species may yet reappear in the Undercliff.]

Neottia nidus-avis (L.) Rich.
BIRD'S-NEST ORCHID
N. A rare plant found in small numbers in shady woods on the chalk and on heavy clays. The map is probably an under-estimate of its

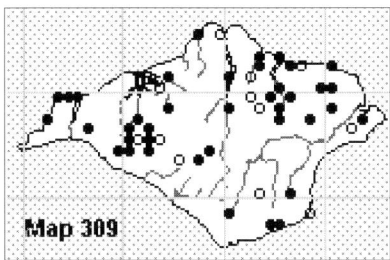

Listera ovata

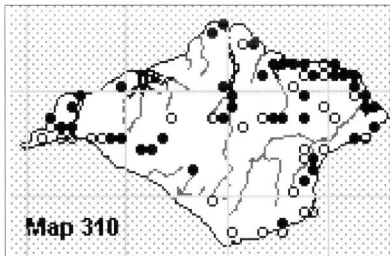

Spiranthes spiralis

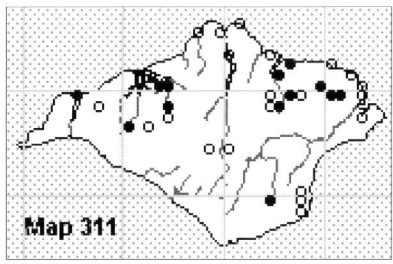

Platanthera chlorantha

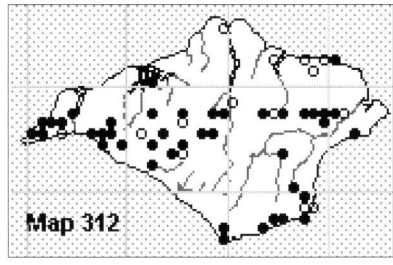

Anacamptis pyramidalis

current distribution. An unusual site is in deep leaf litter in the Holm Oak wood on St Boniface Down 5678, BS 1989. Map 308

Listera ovata (L.) R. Br.
COMMON TWAYBLADE
N. A widespread and sometimes locally frequent species of woodland, scrub and chalk grassland. It occurs principally on chalk and clay soils. Map 309

Spiranthes spiralis (L.) Chevall.
AUTUMN LADY'S-TRESSES
N. A grassland species which is widespread and more frequent than Bevis *et al* (1978) suggest. It occurs on many of our downs although often in small quantity. It can be abundant in unimproved neutral meadows both on clay and greensand soils, and on older garden lawns which have been left unmown in late summer. Some of these lawns are known to have been derived from turfs cut from downland, with Arreton Down in particular being a prolific source in the first half of the twentieth century. Map 310

Platanthera chlorantha (Custer) Rchb.
GREATER BUTTERFLY-ORCHID
N. Sparsely distributed in woods, principally on clay soils on the north of the Island. This orchid declines rapidly when woods cease being actively coppiced. It was once a common sight in quantity in woods around Havenstreet and woodsmen were keen to have a small bunch in a glass in their living room so that they could enjoy the powerful scent. Not known as a downland species on the Island but very rarely found in scrubby grassland on clay. The only modern record from south Wight is at the edge of Appuldurcombe Wood 5479, JS 1998. Bromfield (1856) described this species as very frequent in woods, thickets and grassy slopes. Map 311

[*P. bifolia* (L.) Rich.
LESSER BUTTERFLY-ORCHID
N. Extinct. This orchid has always been rare although the identity of some older records of butterfly orchids were confused. It survived on clay heath pasture at Cranmore until at least 1973, RK 3890 but a maximum of only three plants were seen in any year. Historic sites are Colwell Heath (two specimens in 1841); Stroud Wood coppice, near Ryde, 1838; and near Luccombe, 1871.]

Anacamptis pyramidalis (L.) Rich.
PYRAMIDAL ORCHID
N. A frequent orchid of chalk grassland across the Island, and locally on lime-rich undercliffs and remnant limestone grassland. It used to be found in meadows on sandy or clay soils but is of rare occurrence here today. Known sites on clay are: Appley Park, Ryde 6092, CP 1999; field at Newtown 4290, JP 1999; and Golden Hill, Freshwater 3487, CP 1995. It grows in greensand grassland at Branstone Cross verge, Apse Heath 5583, CP 2002. Map 312

Gymnadenia conopsea (L.) R. Br.
FRAGRANT ORCHID
N. ssp. *conopsea* Despite the frequent occurrence of this plant on the chalk in Hampshire, it has always been rare with us with, at most, a handful of plants occurring in any one site. It has suffered declines over the past fifty years but recently, one or two plants have been recorded from the following sites: Carisbrooke Castle moat 4887, MS 1998; north facing downland above Duxmore Farm 5587, MS 2000; Five Barrows earthworks, Chillerton Down 4883 RDP 1987; and Idlecombe Down 4585, CC 1987. Sites where it has not been seen recently, with last dates, include: Binnel Bay (1984); Ashey Down (1983); clearing in Walter's Copse, Newtown (1981); and Alvington chalk pit (1975).

ssp. *densiflora* (Wahlenb.) E.G. Camus, Bergon & A. Camus. A rare plant only ever known from three sites. Its stronghold was around flushes on the Headon Warren undercliff 3085, AM 1998 and 3186, CP 1998. The number of plants today is about 30 whereas in the 1970s there were several hundreds. Bromfield (1856) knew it from very wet ground on Colwell Heath. The second recent site is on north-facing chalk grassland at Nansen Hill 5778. It was recognised from here in 1982, MJ, but this was probably the Fragrant Orchid recorded by E.H. White in 1904 from the down slopes above Luccombe. A maximum of 28 plants was found and it was last seen here in 1986, CP.

Coeloglossum viride (L.) Hartm.
FROG ORCHID
N. This has always been a rare orchid with us and it has recently only been recorded from the north side of Afton Down 3685, RK 1992. In 1981, there were several hundred specimens but it has subsequently declined to very few. A small number of plants were found on Tennyson Down in 1963 but it has not been found here since 1979. It

used to grow off the chalk in a neutral clay meadow at Duxmore, Havenstreet (548886) where it was first recorded in 1916 and survived until 1962 after which the field was sprayed.

Dactylorhiza fuchsii (Druce) Soó
COMMON SPOTTED-ORCHID

N. Our commonest orchid found in woodland, scrub, chalk grassland, roadside verges and disturbed ground. It is frequent and widespread apart from on sandy soils.

D. x *grandis* (Druce) P.F. Hunt
COMMON SPOTTED-ORCHID X SOUTHERN MARSH-ORCHID

N. Our most frequent *Dactylorhiza* hybrid, occurring as single plants or swarms in marshy places, particularly on coastal landslips such as Luccombe Chine ledge and Totland Cliff.

D. maculata (L.) Soó
HEATH SPOTTED-ORCHID

N. Very local in bogs and marshy acidic grassland, having declined in the past one hundred years. Sites include: Munsley Bog 5282, SC 1995; Bohemia Bog 5183, CP 2001; Alverstone Marsh 5785 JO 1996; Cridmore Bog 4981, CP 1998; and Wolverton Marsh, Shorwell 4582, BG 2001. Map 313

D. x *hallii* (Druce) Soó
HEATH SPOTTED-ORCHID X SOUTHERN MARSH-ORCHID

N. Occasionally found in acidic marshy spots. Upper Dolcoppice 5079, CP 1997; Munsley Bog 5282, SC 1995; The Wilderness 5082, CP 1983; Alverstone Marsh 5785 BS 1974.

D. praetermissa (Druce) Soó
SOUTHERN MARSH-ORCHID

N. A locally frequent orchid of fens and wet marshy places, particularly in the Eastern Yar and Medina river valleys and on coastal undercliffs. Also found occasionally in chalk grassland, generally on north facing slopes. Map 314

var. *junialis* (Verm.) Senghas is a variant with unbroken purple loops on the lip, which can be confused with *D.* x *grandis*. According to Jenkinson (1995), 'Some populations are unusually distinctive: three populations on the Isle of Wight were visited by Eric Nelson in the 1970s (D.C. Lang pers. comm. 1993) where he saw plants which he described as very close to the European nominate variety of *D. majalis*, which is not recognised to occur in the British Isles. When I visited one of those sites in 1994, however, I could only find large numbers of robust hybrids'.

Orchis mascula (L.) L.
EARLY-PURPLE ORCHID
(local: kettle-cases!)

N. A widespread and locally frequent species, principally found in old woodland but also sometimes occurring in unimproved grasslands on all soil types. Most frequent on the north side of the Island. The local name, for which there seems to be no satisfactory explanation, is still sometimes used.

O. morio L.
GREEN-WINGED ORCHID

N. A local species of unimproved grasslands particularly in neutral meadows on clay soils. Although this orchid has undoubtedly declined over the past one hundred years, it is still remarkably frequent, turning up regularly in cemeteries and old garden lawns. There are several good sites but the best of these is the Rifle Ranges hay meadow at Porchfield, 4490, where numbers have been estimated at 250,000 in good years. Local extinctions have been many. It is no longer found around Havenstreet where it was formerly abundant in several meadows and referred to as 'kettle-cases' (see above). One local inhabitant remembered picking a posy each year which he arranged by flower colour with a white specimen in the middle surrounded by pale pink flowers and with the darkest colours to the outside. Map 315

[*O. ustulata* L.
BURNT ORCHID

N. Extinct. Present on Garstons Down 4785, in 1991, when five specimens were found, BS. In 1973, over 300 specimens were counted at this site but numbers subsequently declined although it was seen in most succeeding years up to 1991. Not found since, despite searching. It was formerly very local on several chalk downs. Recorded sites over the last fifty years, with last dates, are: Rew Down (2 blooms in 1972, TW); north slopes of Ashey Down (one in 1969, RK); and Nansen Hill (3 plants in 1967, TW & LS).]

[*Aceras anthropophorum* (L.) W.T. Aiton
MAN ORCHID

N. Extinct. Known only from a north facing chalk grassland slope on Shanklin Down, 571799, where two stunted specimens were last seen in

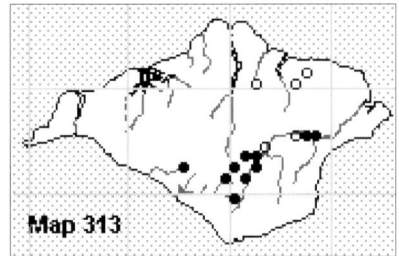
D. *maculata*

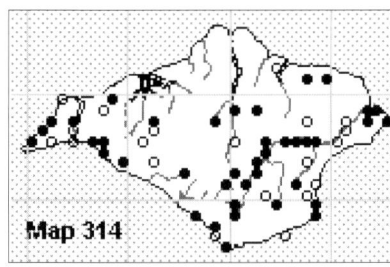
D. *praetermissa*

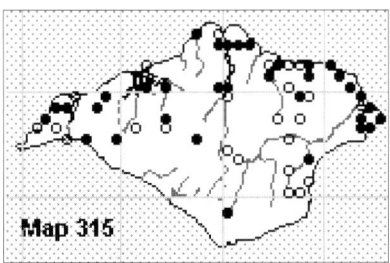
O. *morio*

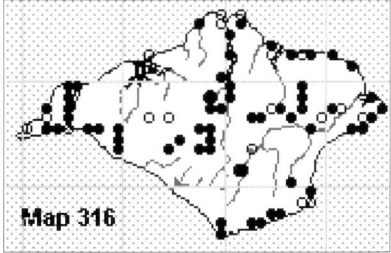
O. *apifera*

1983, RK. The vegetation here has since become somewhat rank. It was first found here in 1905 by E.H. White; in 1907 he observed that it was increasing. In 1945, 33 flowering plants were counted but then the population declined. It was refound in 1960 by Miss Newnham after an absence of several years. Subsequently, numbers fluctuated between 0 and 2.]

[*Himantoglossum hircinum* (L.) Spreng.
LIZARD ORCHID
N. Extinct. Known only from two records at a time when the English population was at its peak. In 1933, a fine specimen in bud was gathered by a school child and taken to her biology mistress, Miss Bullock. It had been found near the railway track by Rowlands Lane, just south of Havenstreet. The following year, a discarded bloom was found nearby and again taken to Miss Bullock.]

[*Serapias neglecta* De Not.
SCARCE TONGUE-ORCHID
E. Extinct. Known from a single remarkable record. 'In a cornfield, Binstead, Isle of Wight, June 2, 1918, noticed for several years, Mrs Wedgewood. Unfortunately, the roots were removed by the tenant to a garden' *BEC Rep.* 5: 308. The specimen is at OXF. The origin of this plant is obscure. There are no other British records for this species of Tongue-orchid and it has been tentatively suggested that it may have been accidentally introduced with imported cereal seed.]

[*Ophrys insectifera* L.
FLY ORCHID
N. Extinct? An enigmatic species, which has always been rare with us and generally present in very small numbers. Despite rumours to the contrary, there have been no confirmed reports since 1982. The most reliable site was Alvington Copse and chalk pit 4788, where Bromfield found it 'in considerable abundance' in 1848. It was recorded regularly here between 1967 and 1975, with a maximum of 31 in 1972, RK. By 1975, the pit was being infilled and the only subsequent record is of a single specimen in 1982, BS. There is also a reliable record of three specimens at the edge of Kelly's Copse, Brading Down 6086, BW 1981. Other twentieth century sites, with last dates are: north side of Arreton Down (5 specimens in 1966, RK); Duxmore chalk pit, Mersley Down (2 specimens in 1966, RK); Westover Down (single specimen in 1965, DF); Luccombe Copse (1961, Miss Newnham); Woodside (1937, GB); north of Long Copse, Gatcombe (1930); Eaglehead Copse (1916); and Combley (1909).]

O. sphegodes Mill.
EARLY SPIDER-ORCHID
N. Despite the proximity to large populations on the coast in Purbeck, this orchid has always been a rarity on the Island. A single plant flowered on a lawn by Saltern Wood, Norton 3589 in 2003, Helen Danby. In 1992, a single specimen was seen and photographed growing by the cliff edge on Tennyson Down 3385, AMu. Despite searching, no further plants have been found here. Prior to this, the most recent record was of one gathered on Brading Down in 1945, GB. In the nineteenth century, it occurred sporadically on the chalk on Shanklin and Ventnor Downs and Luccombe Landslip.

O. apifera Huds.
BEE ORCHID (local: bee flower)
N. Despite having declined considerably, this plant remains frequent on many of our downs and also occurs in meadows on clay and sandy soils, including garden lawns, and rarely in woodland clearings. It will also colonise disturbed ground, sometimes appearing in large numbers in the early years before coarser vegetation takes over. Numbers fluctuate considerably from year to year. var. *chlorantha* Godfrey, the cream coloured form occurs locally. Very rarely, forms approaching var. *trollii* (Hegetschweiler) Nelson, the 'wasp orchid', are found. Bee orchids used to be regularly picked by gypsies for sale in the towns. They also were sent to London where, in 1964, they purportedly fetched 1s 6d (7 pence) per bloom. Map 316

REFERENCES

Anon. (2002) *Forgotten Forelands*. Desktop Studio. Bembridge, I.W.

Bamford, F. (ed.) (1936) *A Royalists' Notebook: the Commonplace book of Sir John Oglander Kt*. London.

Basford, V. & Smout, R. (2000) The Landscape and History of Wydcombe, Isle of Wight. *Proc. Isle Wight nat. Hist. archaeol.Soc.***16**: 9-20.

Bevis, J., Kettell, R. & Shepard, B. (1978) *Flora of the Isle of Wight*. Isle of Wight Natural History and Archaeological Society. Newport, Isle of Wight.

Bowen, H. (2000) *The Flora of Dorset*. Pisces Publications. Newbury, Berks.

Briggs, M. (ed.) (2001) *The Sussex Rare Plant Register*. Sussex Wildlife Trust.

Bromfield, W.A. (1843) Notice of a new British *Calamintha*, discovered in the Isle of Wight. *The Phytologist* CLXXVII: 768-770.

Bromfield, W.A. (1856) *Flora Vectensis*. London. William Pamplin.

Brewis, A., Bowman P. & Rose, F. (1996) *The Flora of Hampshire*. Harley Books. Colchester.

Chatters, C. (1991) A brief ecological history of Parkhurst forest, Isle of Wight. *Proc. Isle Wight nat. Hist. Archaeol. Soc.* **11**: 43-60.

Clement, E.J. & Foster, M.C. (1994) *Alien Plants of the British Isles*. B.S.B.I. London

Colenutt, G.W. (1939). Fifty years of Island coastal erosion. *Proc. Isle Wight nat.Hist. archaeol.Soc.***III** (I): 50-57.

Cottrell, J.E., Forrest, G.I. & White, I.M.S. (1997) The use of RAPD analysis to study diversity in British black poplar (*Populus nigra* L. subsp. *betulifolia* (Pursh) W. Wettst. (Salicaceae)) in Great Britain. *Watsonia* **21**: 305-312.

Drabble, E. & Long, J.W. (1931) A list of plants from the Isle of Wight. *Rep. Bot. Exch. Club* **9**: 734-757.

Englefield, Sir Henry (1816) *A History of the Isle of Wight*.

Gunther, R.W.T. (1922) *Early British Botanists and their gardens, based on unpublished writings of Goodyer, Tradescant and others*. Oxford University Press, Oxford.

Hants Repos. (1799) *The Annual Hampshire Repository*, by Dean

Garnier & Revd. Poulter, Vol.1

Harrison, John (1630) *The 1630 Survey of Swainston*. IWPRO.

Jenkinson, M.N. (1995) *Wild Orchids of Hampshire & the Isle of Wight*. Orchid Sundries Ltd. Gillingham, Dorset.

Jones, J.D. (1979) *The Isle of Wight 1558-1642*. Thesis. Pp. 204-216. University of Southampton

Jones, M.J. (2003) A survey of the manors of Swainston and Brighstone 1630 - farm buildings and farm lands. *Proc. Isle Wight nat. Hist. Archaeol. Soc.* **19**

Long, W.H. (1886) *A Dictionary of the Isle of Wight dialect*. Newport, I.W. County Press.

Long, J.W. (1931) Some alien plants of the Isle of Wight. *Rep. Bot. Exch. Club*: **9**:758-760.

Lousley, J.E. (1948) Notes on plants observed in the Isle of Wight in July 1947. *Proc. Isle Wight nat. Hist. Archaeol. Soc.* **IV** (II): 54-56.

Lousley, J.E. (1950) *Wild Flowers of Chalk & Limestone*. The New Naturalist. Collins, London.

Marshall, W. (1798) *Rural Economy of the Southern Counties*

Merrett, C. (1666) *Pinax Rerum, naturalium Britannicarum, etc.*

Mills, A.D. (1996) *The Place-names of the Isle of Wight*. Paul Watkins, Stamford

More, A.G. (1858) *A Catalogue of Flowering Plants and Ferns growing wild in the Isle of Wight.*

More, A.G. (1861-62) On the discovery of *Gladiolus Illyricus* (Koch) in the Isle of Wight. *J. Proc. Linn. Soc.* 5-6: 177-182.

More, A.G. (1871) *A Supplement to Flora Vectensis*. London. Taylor & Co.

More, A.G. (1898) *Life and letters of Alexander Goodman More, with selections from his zoological and botanical writings*. C.B. Moffat (ed.) Dublin.

Perring, W.H. & Farrell, L. (1977) *British Red Data Books: 1 Vascular Plants* SPNC, Lincoln.

Preston, C.D., Pearman D.A. & Dines T.D. (eds.) (2002) *New Atlas of the British & Irish Flora*. Oxford University Press.

Rayner, J.F. (1929) *Supplement to Flora of Hampshire*

Rich, T.C.G., Holyoak, D.T., Margetts, L.G. & Murphy, R.J. (1997) Hybridisation between *Gentianella amarella* (L.) Boerner and *G. anglica* (Pugsley) E.F. Warb. (*Gentianaceae*). *Watsonia* **21**: 313-325.

Snooke, W.D. (1823) *Flora Vectiana*

Stace, C.A. (ed.), 1975. *Hybridisation and the Flora of the British Isles*, xiii, 626pp. Academic Press, London, in collaboration with the Botanical Society of the British Isles.

Stace, C.A. (1997) *New Flora of the British Isles*. 2nd ed. Cambridge University Press

Stratton, F. (1878) On an Isle of Wight gentian. *Journal of Botany* : **16**: 263-265

Stratton, F. (1900) *Wild Flowers of the Isle of Wight*. Isle of Wight County Press, Newport.

Stratton, F. (1909) Flowering Plants and ferns and their allies. In *A Guide to the Natural History of the Isle of Wight*, F. Morey (ed.) Newport, Isle of Wight County Press

Stratton, F. (1913) Isle of Wight plants. *Journal of Botany*: 285-294.

Stratton, F. (1916) *Crocus vernus* in the Isle of Wight. *Journal of Botany*: **LIV**: 114

Telfer, Sue (1994) *A survey of early gentian (Gentianella anglica). The Isle of Wight*. Unpublished report to English Nature. Wight Wildlife, Shanklin.

Townsend, F. (1883) *Flora of Hampshire including the Isle of Wight*. L. Reeve & Co. London.

Tubbs, C. (1999) *The Ecology, Conservation and History of the Solent*. Packard Publishing Ltd. Chichester.

Turner, Robert (1664) *Botonologia: The British Physician*.

Wadham, p. (1940) The impoverishment of marine life round the Isle of Wight. *Proc. Isle Wight nat. Hist. Arch. Soc.* **3**:119-121.

Warner, Richard (1795) *The History of the Isle of Wight*. Southampton, Baker.

White, E.H. (1905) *The Wild Flora of Shanklin and District*.

Wilson, P, J. (1993) *The ecology and conservation of field cow-wheat (Melampyrum arvense)*. Eastleigh. Hampshire & Isle of Wight Wildlife Trust.

Womens Institute (WI) 1971 *Niton Calling*. Ryde, I.W.

Bryophytes

Liverworts & Mosses

LORNA SNOW

The bryophyte flora comprises 74 liverworts, of which eight are considered to have become extinct, and 271 mosses, of which perhaps sixteen are extinct. There are, in addition, one unconfirmed liverwort and twelve unconfirmed moss species.

There has been little published work on bryophytes on the Isle of Wight. Apart from a few scattered records in Venables (1860), the first list for the Island was by A. Bruce Jackson, in Proceedings of the Hampshire Field Club for 1907. These records were mainly from Rev. Livens and J. Groves. An extended list of mosses was compiled by Livens, and of hepatics by W. Ingham, and published in Morey (1909). This has been the most complete list of bryophytes published to date. In 1926, Livens published a list of additional records in Proceedings of the Isle of Wight Natural History Society. Percy Long collected and recorded bryophytes on the Island, aided by H.H. Knight, until his death in 1944. All his remaining herbaria and notes are held by the Natural History Society.

There were very few records made subsequently until the British Bryological Society held their annual meeting on the Island in the spring of 1964. Six days in the field produced a great deal of information, including several new vice county records, but as recording was carried out on a 10 kilometre square basis, some records are difficult to locate. There was no further recording until Francis Rose visited the Island around 1977 - 1980 and later. During the 1980s and 1990s, Rod Stern was working on the Island on Forestry Commission sites, and made comprehensive records of bryophytes in these areas. Distribution maps of bryophytes were published by Snow (1989, 1992 and 1997).

The most recent bryological records included in this account are from the Spring Field Meeting of the British Bryological Society in March 2002, where some 1200 records were made including the confirmation of some lapsed and doubtful records and the addition of some 12 new species to the Island list.

The following notes refer to the bryophytes recorded on the Island, as the situation is known at the end of 2002. The systematic list and nomenclature is in accordance with Blockeel & Long (1998). Where the name has changed since Hill *et al* (1991, 1992 & 1994), the older name is included in parenthesis in light face type. Species believed to be extinct are enclosed in square parentheses [ ]. Unchecked records and records believed to be in error are enclosed in round parentheses ( ). A few highly dubious unconfirmed records have been omitted.

The species name is followed by the total number of 1km squares in which there are records for the species. Four figure grid references are given for most sites mentioned; all are within the 100 km square SZ. The recording has been on a site basis where possible, otherwise on a 1 km square basis. Not every square has been surveyed, but the most interesting areas have been covered. Abbreviations used for recorders are listed in Appendix 1. Standard abbreviations for herbaria are listed at the start of the Vascular Plants chapter. The earliest known record for each species is given at the end of the relevant notes. The distribution of a few species is illustrated by maps. Open circles are pre-1970 records; closed circles are post-1970.

HEPATICAE
LIVERWORTS

Kurzia pauciflora (2 sq.) Found in boggy ground, often growing amongst sphagnum. Very rare; it is only recorded from Whale Chine 4678, LS & JAP 1985 and Bohemia Bog 5183, BBS 2002. Frequent in New Forest bogs. Bohemia Bog (as Lake Farm, Rookley) PL 1938.

Lepidozia reptans (7 sq.) A woodland species, uncommon on the Island, although nationally very common in the south and west. The most recent record is from the south-west corner of Headon Warren 3085, DGL 2002. Found in The Wilderness 4987, JAP 1964 but this area has changed and the species is unlikely to survive here.

Calypogeia fissa (31 sq.) A fairly common species, found on a variety of damp substrates, heath, streams, tracks and old logs. Parkhurst, HML 1906.

Calypogeia muelleriana (15 sq.) Distribution similar to *C. fissa*, but found on more acidic soil in woodlands. Bohemia Bog (as Lake Farm, Rookley), HHK 1938.

Calypogeia arguta (19 sq.) A calcifuge, found on clay banks in woodland. Common in the south and west of the British Isles, and mainly found in the north and east of the Island. Parkhurst, HML 1906.

Cephalozia bicuspidata (19 sq.) A common calcifuge species, found in a variety of sites, from sphagnum bogs, to woodland and chines. Apparently absent from some areas where it was found 70 years ago, such as Borthwood Copse 5784. Whale Chine, HML 1906.

Cephalozia macrostachya var. *macrostachya* (1 sq.) A species of wet acid sites, usually amongst sphagnum. The only Island site is at Bohemia Bog 5183, FR & JAP 1975; refound there BBS 2002.

Cephalozia connivens (4 sq.) A rare species of damp heath and bogs. The most recent records are from The Wilderness 4981, and Bohemia Bog 5183, both BBS 2002. The Wilderness HHK 1908.

Cladopodiella fluitans (1 sq.) A species of wet bogs and peat. The only Island record is from Bohemia Bog 5183, among *Sphagnum fallax* subsp. *fallax* FR 1975. Neither species were relocated by BBS, 2002.

Cephaloziella hampeana (3 sq.) A rare acid-loving species which can grow in a variety of habitats, but is recorded only from Chilton Chine 4182, BBS 1964 and St. Helen's Duver 6389, JAP & DTH 2002. The Wilderness HHK 1909.

Cephaloziella baumgartneri (1 sq.) A Red Data Book, southern European, calcicolous species at the northern edge of its range. The only recorded site is St. Catherine's Point 4975, on Upper Greensand BBS 1964, and again in the same site in 1983 RCS, FR & ECW, and TLB 2001.

[*Cephaloziella divaricata* (5 sq.) Thought to be extinct on the Island, this species was found in acid, gravelly sites - such as on Bleak Down and St. George's Down. The last record was from Bouldnor, JG 1925. St. Boniface Down, Ventnor HPR 1903.]

Cephaloziella stellulifera (2 sq.) A rare western plant found in turf on cliff tops. There is an early record from Whitecliff Bay by HHK 1912, the only other record being St.Catherine's Point 4975, BBS 1964.

Cephaloziella turneri (1 sq.) A rare species only found in Parkhurst Forest, on banks of ditches by HHK in 1908 and 1920 and most recently in 1964 by BBS.

(*Cephaloziella* sp. Records from Headon Warren, LS 1977 and Alum Bay Chine RCS & HM 1999 were not identifiable to species.)

Lophozia ventricosa (3 sq.) A nationally common calcifuge species found in a variety of habitats, but apparently rare on the Island. The only recent records are from Headon Warren 3085, TLB 2001 and Munsley Bog 5382, FR 1976. The Wilderness, HML 1906 Hb. BON.

[*Lophozia excisa* (1 sq.) Widespread on the mainland, but probably extinct on the Island. The only record is from Headon Warren HHK Hb. HML, BON 1908.]

[*Lophozia bicrenata* (2 sq.) A pioneer of sandy places, possibly extinct on the Island. The only records are from the north-east corner of Parkhurst Forest 4891, UKD 1964; and Bleak Down 5181, HHK 1928.]

Leiocolea turbinata (30 sq.) A common species on limestone and chalk, often on boulders in woodland. Scattered across the chalk ridge and the south-east. The Landslip, Bonchurch HML 1909.

Gymnocolea inflata (3 sq.) A common species on a variety of acid substrata, but scarce on the Island. The only post-1970 record is from Headon Warren 3085, TLB 2001. Bleak Down 5181, HHK 1908.

(*Tritomaria exsectiformis* (1 sq.) A species of damp woodland, decorticate logs and stumps. The only Island record, from Ventnor in 1909 by Rev. Reader, has not been traced.)

Mylia anomala (1 sq.) Grows amongst sphagnum hummocks. The only Island record is from Munsley Bog, Godshill 5382, FR & ECW 1976.

Jungermannia gracillima (10 sq.) A scarce pioneer on wet acid soil of woodland tracks, cliffs, stream sides and ditches, always in areas which are

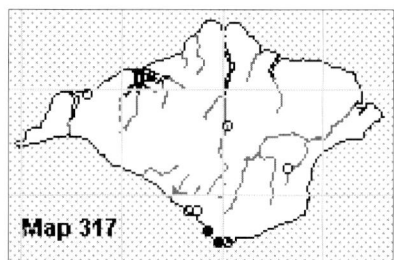

Jungermannia gracillima

kept open. Blackgang 4976, and Whale Chine 4678, HML 1909. Map 317

[*Jungermannia caespiticia* (1 sq.) Extinct. The only record is from the sides of a ditch in The Wilderness in 1908 and 1912 by HHK Hb. BON.]

[*Jungermannia hyalina* (1 sq.) Extinct. A species of damp shady rocks, clay and loamy banks in woods and ditches. The only Island record was from shady rocks, Bonchurch Landslip, HHK 1908. Hb. BON.]

Nardia scalaris (6 sq.) An uncommon species of base poor sandy banks and paths. Headon Warren, HML 1906.

Diplophyllum albicans (14 sq.) A calcifuge species on soil, logs, boulders, etc. Parkhurst and Westover, HML 1909.

Scapania compacta (7 sq.) On dry sandy areas, recorded from Bleak Down 5081, Beacon Alley near Godshill 5181, and a few other similar sites. Headon Warren, Pearson 1906.

Scapania nemorea (2 sq.) The only records are from Rowlands Copse 5689, HHK 1923 and Parkhurst Forest 4790, RCS 1980.

[*Scapania irrigua* (1 sq.) Extinct. The only record is from Parkhurst Forest, PL & HHK 1908; the specimen is in NMW.]

Scapania undulata (1 sq.) Although nationally a common species of mainly western and Wealden distribution in damp calcifuge sites, the only Island record is from the north-east corner of Parkhurst Forest 4891, ACC 1964.

Scapania aspera (1 sq.) Found in calcareous turf. Only recorded from a north-facing slope near The Needles 2984, AJES 1964.

Lophocolea bidentata (including *L. cuspidata*) (84 sq.) A very common species of damp woodland on the ground, rotting wood, mossy stones, etc. Apes Down, HML 1909.

Lophocolea heterophylla (86 sq.) The commonest liverwort in Southern England, found on shaded tree boles and rotting wood. It has a distinctive smell of rotting wet wood. The Wilderness and Carisbrooke, HML 1906.

Lophocolea semiteres (1 sq.) An introduced species, first recorded in this country in 1955. Found on rotting logs, tree stumps, and peaty banks. The first Island record is from a peaty slope beneath heather at Headon Warren 3085, DGL 2002.

Lophocolea fragrans (1 sq.) An Atlantic species of damp shaded rocks and ravines, often near the coast. The only Island site, a distant outlier from its western strongholds, is in a gully in Cliff Copse, Shanklin 5680, LS 1981 and HM 1987.

Chiloscyphus polyanthos (12 sq.) There are scattered records from soil and decaying wood in damp places. Also known from the stream in Ventnor Park 5577, since 1926, JG. Map 318.

Chiloscyphus pallescens (*C. polyanthos* var. *pallescens*) (5 sq.) Occasionally found in damp marshy woods and willow carr, generally in north-east Wight e.g. Briddlesford Copse 5490 and Fattingpark Copse 5291, BBS 2002. Rowlands Wood, Ryde 5689, HHK 1910.

Plagiochila porelloides (28 sq.) This is a common species of the western British Isles, found on rocks and banks in basic woodlands. On the Island, it occurs mainly in East Wight. Bonchurch Landslip, HML 1909.

Plagiochila asplenioides (22 sq.) A larger version of the previous species,

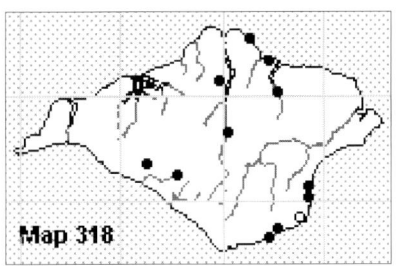

Chiloscyphus polyanthos

found mixed with other bryophytes in shaded woods. Westridge Plantation, HML 1909.

Southbya nigrella (1 sq.) A Red Data Book species known only from Portland in Dorset and St Catherine's Point 4975. The Island's site is on shaded calcareous rocks by a stream, TLB 2001, where it was first discovered in 1979 by CCT. Its open habitat is maintained by natural erosion, but the site is threatened by invasive *Buddleja*. In dry weather, the plant becomes almost invisible.

Radula complanata (29 sq.) Epiphytic on trunks and roots of a variety of trees. On the Island, it is largely confined to the area north of the chalk ridge. Newbarn Down, HML 1909.

Ptilidium pulcherrimum (1 sq.) A rare corticolous species, colonising a variety of trees. The only record is from elder in Rowborough Bottom, Brighstone 4584, RCS 1983.

Porella platyphylla (13 sq.) Fairly common on beech and ash in chalk woodlands, occasionally on calcareous walls. Carisbrooke, HML 1909.

Porella arboris-vitae (5 sq.) An uncommon species, preferring shaded calcareous boulders and ravines. The most recent Island records are from East Afton Down 3485, and Alum Bay 3085, BBS 1964. Bonchurch Landslip, ABl 1864.

Porella obtusata (1 sq.) A rare Atlantic species on dry rocks exposed to maritime winds. Recorded from Tolt Rocks, above Blackgang 4977, CP 1995, which is the only site on the Island. It is at the eastern end of its range, apart from one outlier in Kent.

Frullania tamarisci (15 sq.) Found as an epiphyte in ancient woodlands, and occasionally on rocks in chalk grassland. It also grows with *Porella obtusata* on greensand outcrops at Tolt Rocks 4977. Westover, HML 1909.

Frullania dilatata (19 sq.) A common epiphyte on elder, willow, hazel, oak and similar shrubs and trees, in dry shaded sites. Combley Great Wood, HML 1909.

Marchesinia mackaii (1 sq.) An oceanic calcicolous species, often growing on vertical rock faces. The first Island record was in 1964 at St Catherine's Point, DFC. Seen again in 1971 on a north facing rock alongside the Undercliff Road near Niton, FR. It has not been refound.

Microlejeunea ulicina (*Lejeunea ulcina*) (68 sq.) A common epiphyte on bushes and shrubs, often with *Frullania dilatata*, scattered widely over the Island. Parkhurst, HHK 1908.

Lejeunea lamacerina (8 sq.) An Atlantic species of wet wooded valleys and rocks. The most recent records are from Bonchurch Landslip 5878 and Briddlesford Copse 5490, BBS 2002. Bonchurch Landslip, HHK 1908.

Cololejeunea rossettiana (1 sq.) A rare shade-loving species, growing on calcareous boulders. The only Island site is the eastern end of Bonchurch Landslip 5878, last seen there by HM in 1987, epiphyllous on *Thamnobryum alopecurum*. Bonchurch Landslip, HHK 1920.

Cololejeunea minutissima (40 sq.) Found on the trunks, branches and twigs of a variety of trees and bushes, particularly along the south and west coasts of Britain, and scattered widely across the Island. Calbourne, HHK 1908. Map 319

Fossombronia pusilla (9 sq.) A pioneer species of damp road and tracksides, ditches and disturbed ground, not common on the Island. Bouldnor, JG 1919.

[*Fossombronia wondraczekii* (2 sq.) Extinct. Occurs in similar sites to *F. pusilla*. There are two early records for the Island, HHK 1908 from a damp wood near Sandown, and Parkhurst Forest, 1908 hb. HML, BON]

Pellia epiphylla (40 sq.) A widespread calcifuge species, often growing in sheets on rock faces, and the banks of streams and ditches. Shanklin Chine 5880, HML 1909.

Pellia endiviifolia (47 sq.) Calcicole, widespread on wet rocks and ground in woods by water. Marvel 5087, HML 1909.

[*Pallavicinia lyellii* (1 sq.) A rare species of wet ditches and rocks. The only Island record was in The Wilderness 4981, EMH 1964. This site has deteriorated in recent years, and the species was not refound during a search in 2002.]

Blasia pusilla (3 sq.) A rare plant on bare, moist sandy soil around the coast. Recently found at Blackgang Chine Ledge 4876, CP 1998 and Shanklin Chine 5880, BBS 2002. Known from the cliff face at Hope Point, Shanklin 5881, since 1908, HML.

Aneura pinguis (25 sq.) Frequently found in damp woods, on slightly basic turf and ditches, and also on damp brickwork. Sometimes found submerged, as in a stream at Pelham Woods at St Lawrence 5476. Bonchurch Landslip, HML 1909.

Riccardia multifida (7 sq.) A scarce species of boggy flushes, damp banks and tracks. The most recent record is from Bohemia Bog 5183, BBS 2002. The Wilderness 4980, HML 1909.

Riccardia chamedryfolia (16 sq.) Found in damp woodland rides, mainly in the east of the Island but the most recent record is from Headon Warren 3085, DL & GR 2002. Blackwater 5086, HML 1908.

Riccardia latifrons (1 sq.) Generally found growing on *Sphagnum* and *Molinia* tussocks. The only confirmed record is from Bohemia Bog 5183, BBS 2002.

Metzgeria fruticulosa (13 sq.) An occasional epiphyte, mainly on elder. Tinkers Hole, America Wood 5681, LS 1983.

Metzgeria temperata (14 sq.) Found in similar sites to *M. fruticulosa*, in woodland sites scattered around the Island. The Wilderness 4981, AJES 1964.

Metzgeria furcata (92 sq.) By far the most common of the three species, found on a wider variety of trees throughout the Island. Apes Down Wood 4587, HML 1906.

Lunularia cruciata (49 sq.) A common liverwort often found in gardens, along the base of walls, in flowerpots and greenhouses; also in woods, along the sides of ditches and rides. Most Island records are from the East Wight. Bonchurch Landslip, HML 1909.

Conocephalum conicum (39 sq.) One of the largest of the liverworts, common in damp shady places like the sides of streams and ditches in woods, or wet vertical brickwork under bridges and similar places. Blackwater, HML 1909.

Reboulia hemisphaerica (3 sq.) An uncommon species on the Island, found on sandy banks and rocks. Recorded from Shanklin Chine in 1978 but not seen there since; it may have been destroyed by a falling tree. The only other modern record is the railway line at Cliff Bridge, Shanklin 5681, LS 1978. Ashey Down 5787, HML 1909 BON.

Marchantia polymorpha ssp. *ruderalis* (8 sq.) Known from artificial habitats, and commonly found in gardens and increasing. The only known native site was Shanklin Beach, JR 1854.

Riccia fluitans (3 sq.) A rare aquatic floating on the surface of ponds, which can also occur on mud. It has been found on clumps of tussock sedge in a wet meadow near America

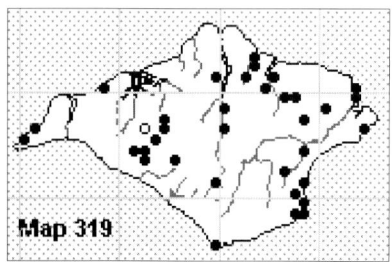

Cololejeunea minutissima

Wood 5882, LS 1986, and in a pond at Elmsworth, near Porchfield 4492, BS 1986. Burnt Wood pond 4492, RCS 1981.

Riccia sorocarpa (10 sq.) An ephemeral found in gardens and arable fields, most recently at an organic farm at Lower Yard near Godshill 5283, RDP 2002. Totland HML 1909.

Riccia glauca (10 sq.) Another ephemeral found in arable fields on acid soils. The Wilderness, HHK 1909.

Anthoceros punctatus (1 sq.) Only known from Hope Cliff, Shanklin 5880, RDP 1995, JDS & RCS 2002, Shanklin Chine 5880, BBS 2002 and the cliffs below the Channel View Hotel, Shanklin 5880, JDS 2002.

Anthoceros agrestis (3 sq.) A rare annual species found in damp arable fields, and sandstone cliffs. No records since 1978, when it was found at Upper Hyde Farm, Shanklin 5781, LS. Sandown JG 1926.

Phaeoceros laevis (5 sq.) Found growing on wet banks and sandstone. The most recent records are from Bonchurch Landslip 5878, RCS 2002, and Shanklin Chine 5880, BBS 2002. There are earlier records from Shanklin Chine, PL 1923 and the adjacent Chine Hollow LS 1978. Also recorded from the clay bank of the garden of Alverstone Manor Hotel, Shanklin 5880, LS 1982/3. Sandown, HHK 1908.

MUSCI
MOSSES

Sphagnum papillosum (3 sq.) A rare species preferring open peaty bogs and marshland. The only records are from Sandown Golf Course (Lake Common) 5885, 1942; Bohemia Bog 5183, RAF & RCS 2002; and Munsley Bog 5282, FR 1998. Lost from Lake Common and declining at Munsley Bog. Lake Common, PL 1942.

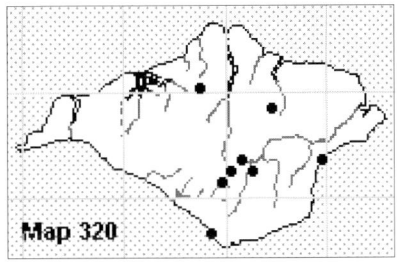

Sphagnum fimbriatum

Sphagnum palustre (9 sq.) A rather more frequent, shade tolerant species found in wet woods and ditches. Recorded from the same areas as *S. papillosum*, but also from Parkhurst Forest and Combley Great Wood. The Wilderness, HML 1907.

Sphagnum squarrosum (7 sq.) This species was found in the same localities as the previous two, but has greatly declined. The most recent records are from Munsley Bog 5282, LS 1980; a wet wood by Sandown Waterworks 5885, LS 1991; and a poor specimen from The Wilderness 5082, CP 1996. The Wilderness, HML 1907.

Sphagnum fimbriatum (8 sq.) Occasionally found in peaty bogs and wet acid woodland. Also recorded from Blackgang Chine Ledge 4877, CP 1999, and in a seepage area on the vertical cliff face above Lake beach 5882, LS 1984. The Wilderness, FMM 1909. Map 320.

Sphagnum capillifolium subsp. *rubellum* (3 sq.) A rare bog species recorded from Bohemia Bog 5183, RAF & RCS 2002; Munsley Bog 5282, LS 1980; and The Wilderness 5082, FR 1974.

Sphagnum subnitens (9 sq.) A less demanding species, found in boggy grass, wet woods and ditches, although most of the records come from The Wilderness, Munsley Bog and Bohemia Bog. The most recent records are from Cridmore/The Wilderness 4981, LS 1995; and Bohemia Bog 5183, RAF & RCS 2002. Recorded from Newchurch Marshes PL & WRS 1924 and Lake Common PL 1942, but has not been seen since in either of these locations.

[*Sphagnum compactum* (1 sq.) Believed extinct. The only record is from Bleak Down in 1909, HHK. The site was later used as a rubbish dump.]

Sphagnum denticulatum (*S. auriculatum*) (14 sq.) Found in a wide range of wet acid habitats. On the Island it is recorded from the same localities as other sphagnum species. It also occurs in woods - Briddlesford, Combley Great Wood and Parkhurst Forest. Sainham Copse, Godshill, PL 1926.

(*Sphagnum contortum* The records in Morey (1909) are not supported by specimens, and so the species is not included on the current Isle of Wight list.)

Sphagnum cuspidatum (3 sq.) A rare and declining species. The only records are from The Wilderness 5082, BBS 1964, Munsley Bog 5282, LS 1980 and Bohemia Bog 5183, RAF & RCS 2002.

Sphagnum fallax ssp. *fallax* (*S. recurvum*) (5 sq.) Another rare and declining bog-moss. Found in The Wilderness between 1908 and 1964; Lake Common 5885, DGL 1942; Bohemia Bog 5183, FR 1975 and LS 1980; and Munsley Bog 5282 FR 1998. The Wilderness, HML 1909.

Pogonatum nanum (7 sq.) A colonist of acidic, comparatively bare soil in woods and heaths. The most recent records are from Headon Warren 3186, FR 1998 and TLB 2001. Can be confused with *P. aloides*. Hamstead, HML 1909.

Pogonatum aloides (14 sq.) Very similar to the last species in habitat, but commoner and often on vertical banks. Found in chines and sand pits. Parkhurst, HML 1909.

[*Pogonatum urnigerum* (1 sq.) Probably extinct. A calcifuge species of gravel quarries and walls. The only Island record is from sandy ground in Parkhurst Forest, HHK 1909 (BON).]

[*Polytrichum longisetum* (1 sq.) Probably extinct. The only record is from Marvel Copse, HML 1909. The post 1950 record in Hill *et al.* (1992)

from the Sandown area has not been traced.]

Polytrichum formosum (47 sq.) A common plant of slightly acid soil in woods under birch, oak and pine. Parkhurst Forest, HML 1907.

Polytrichum commune (8 sq.) The largest of the genus, an acid-loving plant that is now almost extinct in the Isle of Wight. Previously found in Newchurch Marshes, Bleak Down, and similar sites, but now only known from The Wilderness 5082, BBS 2002; Munsley Bog 5282, FR 1998 (this area is declining); and Cridmore Bog 4981, CP 1995. The Wilderness, HML 1907.

Polytrichum piliferum (20 sq.) A species of dry gravelly or sandy areas in woods, heaths and dunes. It is widespread in suitable sites. St. George's Down, HML 1907.

Polytrichum juniperinum (42 sq.) An acid-loving species of heaths and dry shingle banks, often where the ground has been burnt. It is common on the Island. Brighstone Down, HML 1907.

Atrichum undulatum (74 sq.) A common plant of shaded fairly open ground in woods and grassland. It is widely distributed on the Island. Carisbrooke, HML 1907.
(var. *minus* is recorded by FM from Combley Wood in 1907, but is no longer recognised as a variety (Blockeel & Long, 1998).

Tetraphis pellucida (7 sq.) Found in damp shady conditions, often on rotten logs and tree stumps in woods. Rarely recorded recently, but this could be through lack of searching. The most recent records are from a wet wood by Sandown Waterworks 5885, LS 1991; Fattingpark Copse 5291, BBS 2002, and The Wilderness 5082, BBS 2002. Willow carr in The Wilderness, PW et al 1958.

Archidium alternifolium (7 sq.) An uncommon plant. There are a few early records from Whippingham, Parkhurst and King's Quay, but the only record since 1964 is from Headon Warren, TLB 2001. Firestone Copse & Parkhurst Forest, HML 1909.

Pleuridium acuminatum (18 sq.) Grows in bare patches of sandy soil in woods and gravel pits. The plants frequently have capsules. Almost confined to the north of the chalk ridge. Also found colonising bare soil in flowerpots. Parkhurst, HML 1907

Pleuridium subulatum (5 sq.) A scarce plant, of similar habitats to the last species. The most recent Island record is from Locks Copse, Porchfield 4491, LS 1987. St. George's Down PL 1926.

Pseudephemerum nitidum (7 sq.) A colonist of bare patches in damp acid areas in woods, arable fields and by streams. The most recent record is from The Wilderness 5082, TLB & DGL 2002. The Wilderness, HML 1906.

Ditrichum cylindricum (4 sq.) Although a widely distributed plant in Britain on sandy banks and roadsides, it is rare on the Island. The most recent record is from Brook, RDP & DTH 2002. S.E.Wight, BBS 1964.

Ditrichum heteromallum (2 sq.) Associated with gravelly soils in sandpits and on tracks in woodland. It is recorded from only two Island sites, namely Blackgang Chine 4876, BBS 1964 and Blackwater 5086, JAP 1959.

Ditrichum gracile (including ***Ditrichum flexicaule***) (13 sq.) A plant of chalk grassland, found the length of the chalk ridge. The most recent records are from Tennyson Down 3385, LS 1987 (det. GPR), and Shide chalk pit 5088, RCS 2002. Ashey Down, HML 1907.

Ceratodon purpureus (79 sq.) This is one of our commonest bryophytes, occurring in a wide variety of generally dry habitats. Found throughout the East Wight, but much less frequent west of the River Medina. Brighstone Down & St. George's Down, HML 1909.

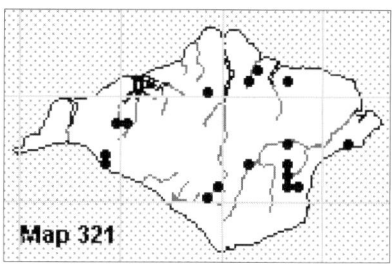

Dicranella staphylina

Dichodontium pellucidum (4 sq.) There are currently just four known sites for this species of gravelly banks of streams: Shanklin Chine 5881, BBS 2002; Fattingpark Copse 5291, RCS & DJH 2002; Chillingwood Copse 5589, RCS 1983; and Whale Chine 4678 BBS 1964. Brighstone Down HML 1909.

Dicranella schreberana (17 sq.) This is a frequent species of arable fields, ditches and disturbed ground. Chilton Chine, BBS 1964.

Dicranella varia (68 sq.) Widespread over the Island, in woods, chalk pits and arable fields. Garstons, HML 1906.

Dicranella staphylina (17 sq.) A fairly frequent plant of arable fields and disturbed ground; rare on chalky soils. It was first recorded in an arable field at Brook, ACC 1964. Map 321.

Dicranella cerviculata (5 sq.) Rarely found in damp gravelly or peaty sites. The most recent record is from Luccombe Chine Ledge 5879, CP 1982. The Wilderness, HML 1909.

Dicranella heteromalla (84 sq.) Widespread and common in a wide variety of habitats, particularly woodland, growing on banks. tree stumps and the sides of ditches. Combley Great Wood, HML 1909.

Dicranoweisia cirrata (79 sq.) A common species on tree trunks and particularly the upper surfaces of horizontal branches, widespread on the Island. Godshill & Blackwater, HML 1909.

Dicranum bonjeanii (7 sq.) A scarce and declining plant of damp turf in marshy ground and heaths, and

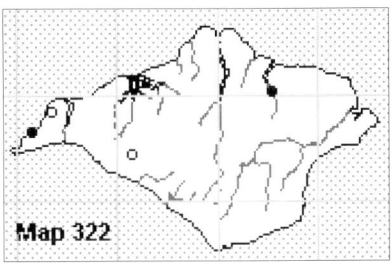

Dicranum majus

around ponds. Munsley Bog 5282, LS 1980; East Cowes cemetery 5094, LS 2001.The Wilderness HML 1909.

Dicranum montanum (1 sq.) Found on the exposed roots and branches of deciduous trees. The only Island record is from Burnt Wood 4492, RCS 2002, but this species is increasing its range in this country.

Dicranum scoparium (67 sq.) A common calcifuge species with a wide range of habitats from leached calcareous and boggy grasslands to tree stumps and trunks, or often on the ground in woods. Garstons Down, HML 1909.

Dicranum majus (5 sq.) This rare species is found on the ground in sheltered woods, particularly with *Quercus petraea*, and on heathy banks. The only recorded extant sites are Briddlesford Copse 5490, BBS 2002; and Headon Warren 3186, TDB 2001. Mottistone Down, JG 1926. Map 322.

Campylopus fragilis (4 sq.) This is a plant of heaths and turf cliff tops near the sea; it is uncommon. The only recent records are Headon Warren 3186, TLB 2001; and path by the Caul Bourne stream 4188, LS 1987. The Longstone, Mottistone HML 1909.

Campylopus pyriformis (23 sq.) Found on bare sandy soil in conifer woods, or on heaths and the sides of ditches, scattered across the Island. St. George's Down, HML 1909.

var. *azoricus* (4 sq.) Prefers damper sites than the previous species, especially *Molinia* and fern stocks; not recorded on the Island until 1979, from The Wilderness 5082, CCT.

Campylopus flexuosus (*C. paradoxus*) (15 sq.) This is a common species of acidic heaths, decaying tree stumps and bare sandy places in woodland; scattered across the Island. Headon Hill, HML 1909.

Campylopus introflexus (39 sq.) Growing in dense mats on acid sandy, gritty ground on heaths, or on bare patches at the sides of ditches. It was first recorded in England in 1941 and is widespread on the Island. Parkhurst Forest, BBS 1964.

Campylopus brevipilus (6 sq.) A localised species of heaths, often near seasonal pools. There are no records since 1979 and it is possibly extinct. Near The Longstone, Mottistone PL 1928.

Leucobryum glaucum (12 sq.) Very local, forming hummocks at the base of trees in acid woodland and on mature heath. Recently recorded from Bohemia Bog 5183, BBS 2002; Parkhurst Forest 4791, CP 2000; Borthwood Copse 5684, LS 1998; Nunneys Wood 4089, RCS 1982; Combley Great Wood 5489, FR 1981; and Godshill Park Wood 5381, FR 1976. Previously recorded from Bleak Down 5181, PL 1930. Headon Warren, HML 1909.

Fissidens viridulus (32 sq.) A widespread species of woodland banks, among rocks and beside streams. Combley Great Wood, HML 1907.

Fissidens pusillus (6 sq.) Found on rocks in shaded limestone and chalky areas. Bonchurch Landslip, HHK 1926.

Fissidens gracilifolius (*F. pusillis* var. *tenuifolius*) (5 sq.) A very small plant found on dry chalk rubble in woods and quarries; uncommon on the Island. Shanklin Great Wood 5780, HM 1994; Tapnell Furze, Wilmingham 3687, RCS 1981; and Battery Gardens, Sandown 5882, LS 1980. Mersley Down, RDF 1964.

Fissidens incurvus (35 sq.) Widespread on the Island, growing on bare patches in woods, turf and banks, in slightly acid areas. Downend, HML 1907.

Fissidens bryoides (73 sq.) A widespread and common species on the Island, found on soil in woods, gardens, on banks and by ditches. Yarmouth, HML 1909.

Fissidens crassipes (1 sq.) A semi-aquatic species, usually on stones at about water level in streams carrying calcareous waters. There are very few suitable sites on the Island. The only record is from the waterfall in Luccombe Chine 5879, RDF 1964.

Fissidens exilis (7 sq.) Ephemeral, occurring on damp clay or loamy sites in woods. Recently only recorded from woods in the north-east of the Island, at Fattingpark, Firestone, Combley and Briddlesford Copses. It may be under-recorded. There are early records from Hamstead, HML 1906 and Bouldnor, JG 1926. Apse Heath, HML 1906.

Fissidens celticus (5 sq.) An Atlantic species of shaded banks, by streams and in woods, largely confined to the west of Britain. The only recorded sites are from Parkhurst Forest and Briddlesford Copse, BBS 1992. Parkhurst Forest, JAP 1964.

Fissidens taxifolius (122 sq.) The commonest of the genus on the Island, found on damp soil on banks in woods and hollow lanes, in fields and gardens. America Wood, HML 1909.

Fissidens dubius (*F. cristatus*) (35 sq.) A plant of chalk grassland, found right along the central chalk ridge and on the downs at Ventnor. Bonchurch Landslip, HML 1907. Map 323.

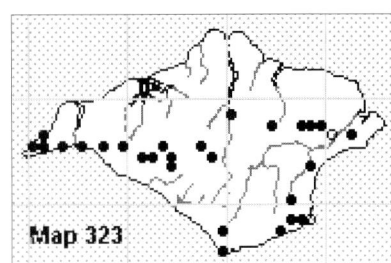

Fissidens dubius

Fissidens adianthoides (15 sq.) A larger plant than others of the genus, found in damp chalk grassland, on wet rocks and ledges, scattered across the Island. Ashey Down, HML 1907.

Encalypta streptocarpa (13 sq.) A calcicole species, scattered across the chalk ridge in old pits and grassland; also found on old walls where the mortar is crumbling. Apes Down, HML 1907.

Encalypta vulgaris (6 sq.) A calcicole species found in similar sites to the previous species, but very much less common. Brighstone chalk pit 4284, LS 1987. Mottistone Down, PL 1927.

Eucladium verticillatum (16 sq.) Found growing on wet calcareous rocks, often forming tufa, on the south-east coast of the Island. Also recorded from flushes at Headon Warren and Shide chalk pit. Luccombe Chine, PL 1927.

Weissia controversa (25 sq.) Scattered across the Island, in a range of habitats on bare soil. It grows in woods and on the downs, on banks and in open areas. Chillerton, HML 1907.

var *crispata* (3 sq.) Strictly a calcicole, growing in chalk grassland. There are only 3 records on the Island, all before 1964. Ashey Down, HML 1909.

Weissia rutilans (2 sq.) Rare on damp, earthy banks. There are only 2 records for the Island: Firestone Copse 5591, RCS 1983 and in a field at Ryde, MSa 1937.

Weissia condensa (*W. tortilis*) (6 sq.?) A nationally scarce species of open, dry chalk habitats. Older records have not been confirmed and may have been confused with *W. controversa*. The two most recent are from Compton Down 3685, and Arreton Down 5387, FR 1986 and 1987. Rowborough Down, HML 1907.

Weissia brachycarpa var. *brachycarpa* (*W. microstoma* var. *brachycarpa*) (1 sq.) A species of damp fields, woods and banks of ditches on non-calcareous soils. The only Island

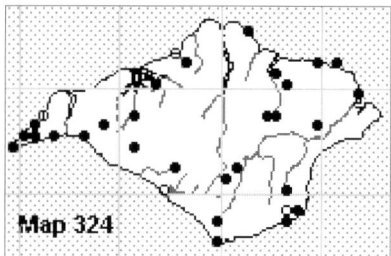

Trichostomum brachydontium

record is from a field near Combley Great Wood, JA 1964.

var *obliqua* (*W. microstoma* var. *microstoma*) (23 sq.) A species of chalk grassland and broken ground, found right along the length of the Island's chalk ridge. Ashey Down, HHK 1910.

Weissia squarrosa (3 sq.) A nationally rare species of damp non-calcareous soil in fields. The most recent record is from an old mustard field near Compton Farm 3784, GE 1964. Gurnard Bay, E.S. Salmon 1894 (NMW).

Weissia sterilis (2 sq.) A very rare species of dry, open chalky ground. Recorded from chalk grassland on Afton Down 3685, RCS 2002. First seen in Bonchurch Landslip, HCN 1912.

Weissia longifolia var *angustifolia* (22 sq.) A chalk-loving plant found in short turf and on bare chalk. It is more common in West Wight, with only a few records east of the River Medina. Carisbrooke, HML 1906.

(*Weissia sp.* (11 sq.) Found in chalk grassland, some plants are not identifiable without capsules. Brook Undercliff LS 1980).

Tortella inflexa (3 sq.) Scarce in sheltered calcareous turf, often on rubble. The most recent record is from a chalk stone (or perhaps Bembridge limestone) in a field at Dodpits, Newbridge 4087, MOH 2002. Afton Down EFW 1964, Brading Down, BBS 1964.

Tortella flavovirens (15 sq.) A coastal species of calcareous dunes and banks, scattered in suitable coastal locations. Freshwater cliffs, HHK 1910.

Trichostomum brachydontium (34 sq.) Scattered distribution, occurs in a variety of generally basic habitats - chalk pits, downs, walls, sand, and gravel. This species has a predominantly western distribution in this country. Totland, HND 1889. Map 324.

Trichostomum crispulum (29 sq.) Confined to chalk and limestone. It occurs along the length of the Downs, and at St.Catherine's Point and Ventnor. Mottistone Down, PL 1923.

Pleurochaete squarrosa (8 sq.) A species of open chalk grassland and dunes. Rare on the Island. Recorded from Mottistone Down chalk pit 4184, BBS 2002, the first record since 1964. Frequently found at Downend, Arreton in the late 1920s, but in 1932, PL reported that it was 'not found - ? the turf removed'. St. Helens, HHK 1909.

Dialytrichia mucronata (*Cinclidotus mucronatus*) (1 sq.) Growing on the base of trees and the banks of streams liable to flood. The only Island record is by the Palmer's Brook, JA 1964.

Pseudocrossidium hornschuchianum (*Barbula hornschuchiana*) (16 sq.) Found on gravely paths, chalk downs and in chalk pits, widely scattered across the Island but not very common. Wight (no locality given), EV 1907.

Pseudocrossidium revolutum (*Barbula revoluta*) (15 sq.) Found in crevices and the mortar of walls, scattered across the Island, but not common. The most recent records are from Shide chalk pit 5088, BBS 2002 and from the south coast of the Island. Mottistone churchyard, HML 1909.

Bryoerythrophyllum recurvirostrum (*Barbula recurvirostra*) (20 sq.) Grows in woods, on banks, in crevices, shady places in parks and gardens, particularly in the south-east of the Island. Cridmore, HML 1909. Map 325.

[*Leptodontium gemmascens* (1 sq.) Extinct? A spring ephemeral species found on old thatch. Modern thatch is covered by wire netting, which seems to inhibit growth. Only recorded from a cottage at Calbourne and not seen there since 1964, but the likely sites are not readily accessible. Calbourne, ECW 1964.]

Gyroweisia tenuis (14 sq.) Found occasional on limestone rocks and mortar in walls, usually in damp and shaded places. Scattered records in East Wight. Luccombe, HHK 1907.

Gymnostomum viridulum (1 sq.) A rare species of dry chalky soil. Only one record for the Island, namely Shide chalk pit 5088, RCS 2002.

(*Gymnostomum calcareum* The only two records, from BBS, 1964, have not been confirmed, as no specimens have been located.)

Barbula convoluta (including var. *commutata*) (66 sq.) A very common species throughout the Island. It occurs on soil on paths, in arable fields, gardens, in chalk grassland and in woods. Godshill, HML 1906.

Barbula unguiculata (98 sq.) Very abundant throughout the Island, found on soil in a great variety of sites, usually on disturbed ground. Newport, HML 1906.

Didymodon acutus (*Barbula acutus*) (1 sq.) Rare, but overlooked, on dry calcareous banks. The only record is by Lynch Lane on Brighstone Down 4284, MOH 2002.

Didymodon rigidulus (*Barbula rigidula*) (22 sq.) Found on rocks and walls, scattered throughout the Island, but not very commonly. Ashey Down, HML 1907.

Didymodon umbrosus (*Trichostomopsis umbrosa*) (1 sq.) Found on damp mortar of shaded walls, such as under railway arches. The only record was from a bridge over the old railway line at Shanklin 5681, CCT 1979, but the site was cleaned off and re-pointed later.

Didymodon vinealis (*Barbula vinealis*) (25 sq.) This species is found on dry limestone rocks, walls, concrete and other stony sites. It has a scattered distribution over the eastern half of the Island. Carisbrooke Castle, HML 1906.

Didymodon insulanus (*Barbula cylindrica*) (54 sq.) Common in damp sites on silt in woods, on walls, in churchyards, tarmac and tree bases, across the Island. Whitwell, HML 1907.

Didymodon luridus (*Barbula trifaria*) (34 sq.) Growing on hard chalky ground, on stones or walls, in churchyards and chines, scattered over the Island. Newbridge, HML 1907.

Didymodon sinuosus (*Barbula sinuosa*) (19 sq.) Found on damp rocks by streams, often in the flood zone. Although the majority of Island records are prior to 1964, it has probably been overlooked and it was recorded from Shanklin and Luccombe Chines, Bonchurch Landslip, St Catherine's Point and Fattingpark Copse in 2002, BBS. Ventnor, JFR 1907.

Didymodon tophaceus (*Barbula tophacea*) (31 sq.) Found in wet areas by streams, on chalk cliffs and also in quarries and chines. Sometimes forming tufa in lime-rich springs. Recent records are clumped. There are also a few records to the north at Firestone Copse 5591, TLB 2001 and Barton Wood, Osborne 5294, RCS 1981. Headon Warren (as Headon Hill), HML 1907. Map 326.

Didymodon fallax (*Barbula fallax*) (84 sq.) A common moss of soil in chalk grassland, also found on tracks, in woods, chines, chalk pits and by ditches and ponds. Bonchurch Landslip, HML 1906.

Pterygoneurum ovatum (3 sq.) A nationally scarce plant of open chalk grassland, which could be overlooked in the turf on cliff tops. Recorded from south-facing chalky banks above St. Catherine's Point 4975, TLB 2001. There are older records from Afton Down, JG 1923, and Culver Cliff, HHK 1909.

(*Aloina rigida* The only record, near Ventnor ABl 1906, published in Morey (1909), has not been substantiated.)

Aloina aloides (10 sq.) Found on bare ground in chalk pits and banks, and the mortar of walls, but not common on the Island. Bonchurch Landslip, HML 1907.

(*Aloina ambigua* (*A. aloides* var. *ambigua*) The records from HML, Ashey Station 1907 and Garstons Down, held in BON, have been reclassified as *Aloina aloides* (TLB 1997). The records from the BBS in 1964 are not confirmed by specimens and are therefore not accepted. The species is currently not on the Island list.)

Leptobarbula berica (4 sq.) A shade-loving species found on limestone, brickwork, and the mortar of walls. First recorded in 1986, and most recently from boulders in Bonchurch Landslip 5878, RCS & ML 2002. Shanklin, RCS 1986.

Tortula subulata (8 sq.) Found on sandy soils around tree bases, on banks and in chines. Not common on the Island. Whale Chine, HHK 1906.

Tortula marginata (46 sq.) A shade-loving plant found in woods on damp limestone, sandstone rocks and low down on walls. Common in south-east

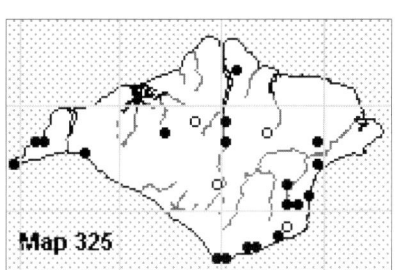

Bryoerythrophyllum recurvirostrum

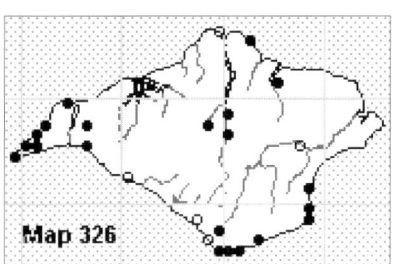

Didymodon tophaceus

Wight, but more scattered over the rest of the Island. Steephill, HML 1907.

Tortula muralis (107 sq.) An abundant species found on walls, both brick and stone, tiles, concrete and similar sites across the Island. Probably under recorded in urban areas. 'Very common', HML 1909.

(***Tortula atrovirens*** (*Desmatodon convolutus*) The only record is from Newtown seawall, PL 1930.)

Tortula lanceola (*Pottia lanceolata*) (12 sq.) Found in chalk grassland and on soil in open places usually chalky, not common. Down End, Arreton, PL 1925.

Tortula viridifolia (*Pottia crinita*) (4 sq.) An annual plant of south-western coastal areas, growing on thin soil in rocky areas. The only known Island sites are: Rocken End 4975, ML & RCS 2002; Brighstone chalk pit 4284, LS 1987; Whale Chine 4678, BBS 1964; and Whitecliff Bay, HHK 1909.

Tortula modica (*Pottia intermedia*) (10 sq.) A coastal ephemeral on disturbed soil, arable fields and banks, uncommon or under recorded on the Island. The most recent record is from Rocken End 4975, LS 1977. Apes Farm, Swainston PL 1924.

Tortula truncata (*Pottia truncata*) (34 sq.) Found in arable fields, gardens, on ant hills, in tracks, gravel pits and similar sites. It has a scattered distribution across the Island. Wootton, HML 1906.

Tortula protobryoides (*Pottia bryoides*) (3 sq.) A rare plant found on chalk soil and in quarries. Most recently recorded from The Needles Battery 2984, RAF & JJG 2002. Near Shalcombe, Chessell, JG 1919.

Tortula acaulon (*Phascum cuspidatum*) (32 sq.) Found in a variety of habitats, in gardens, stubble fields, sparse grassland and in ditches. Scattered across the Island. Hamstead, HML 1909. Map 327.

var ***pilifera*** (*P. cuspidatum* var. *piliferum*) (2 sq.) Only recorded from clay soil in short turf at cliff top at St Catherine's Point 4975, DH 2002, and from East Afton Down 3685, ECW 1964.

Microbryum starckeanum (*Pottia starkeana*) (6 sq.) An annual found in stubble fields, chalk grassland, banks and quarries. It is rare on the Island. Garstons Farm, near Gatcombe, PL 1906.

Microbryum davallianum (*Pottia starckeana* ssp. *conica*) (15 sq.) Found in similar habitats to the previous species, but probably under-recorded. Bowcombe Down, HML 1909.

Microbryum rectum (*Pottia recta*) (12 sq.) Grows in chalk grassland on cliffs, and on banks and tracks. Probably under-recorded. Garstons, HML 1907.

Microbryum curvicolle (*Phascum curvicolle*) (4 sq.) Occasionally found in chalk grassland and by paths. Afton Down 3585, BBS 2002; Needles Battery 2984, RAF & JJG 2002. Recorded regularly from Downend between 1925 and 1932. Freshwater Downs, HHK 1909.

Microbryum floerkeanum (*Phascum floerkeanum*) (3 sq.) An ephemeral found in arable fields and chalky soil, probably overlooked due to its minute size. There are no records since 1986, LS Shalcombe, near Chessell, 3985. Culver Cliff, HPR 1909.

Hennediella macrophylla (2 sq.) First recorded in Britain in 1965, probably from New Zealand, Hill *et al* (1992). The only Island records are from a shaded lawn at Tower Gardens, Shanklin 5881,TLB 2001; and a shady path above Alum Bay 3085, DGL 2002.

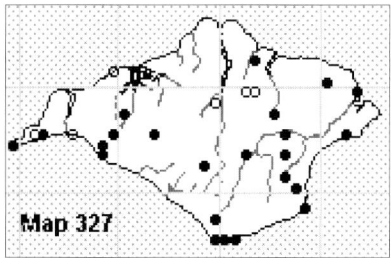

Tortula acaulon

Hennediella heimii (*Pottia heimii*) (9 sq.) An annual confined to saltmarsh edges, and cliff top sites subject to sea spray inundation. There are recent records from The Needles Battery 2984, RAF & JJG 2002; St Helen's Duver 6389, JB 2001; and Newtown seawall 4191, LS 1979. St. Helens, HHK 1909.

Acaulon muticum (3 sq.) Rare, but possibly overlooked. Recorded from sandy ground on Headon Warren and near Calbourne, but no record since 1980, Freshwater downs 3385, FR. Headon Warren (as Headon Hill), HHK 1909.

Acaulon triquetrum (2 sq.) A minute ephemeral of coastal chalk grassland and bare patches on cliffs. This endangered species is only found in six known sites along the south coast of England from Dorset to East Sussex. The national stronghold is in Dorset. On the Island, it has been recorded from 2 sites. First recorded at Afton Down, JG 1923, but could not be refound here in 2002. Found on south and south-west facing calcareous hummocks at Rocken End 4975 TLB 2001 & BBS 2002 after a lapse of almost 40 years.

Chenia leptophylla (*Tortula rhizophylla*) (1 sq.) An introduced species of bare arable ground. When first found, in a ploughed field on the Island in 1964, it was thought to be a new species and named *Tortula vectensis*. Subsequently it was recognised as *T. rhizophylla*. Both these names are now relegated to synonymy. Only three sites are known in Britain. Still present (2002) at the original site. Brook 3883, EFW & ACC 1964.

Syntrichia ruralis (*Tortula ruralis*) (28 sq.) A common species, often found on roof tiles, on sandy ground and dunes and on mortar of walls. Arreton Manor, HML 1907.

Syntrichia ruraliformis (*Tortula ruralis* ssp. *ruraliformis*) (11 sq.) More common on dunes and loose sand than the previous species and generally confined to coastal sites. Thorness Bay, PL 1924.

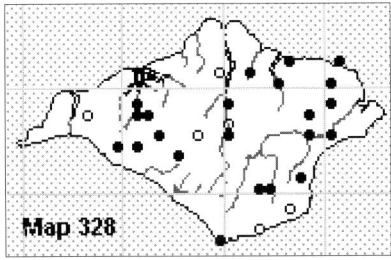

Syntrichia laevipila

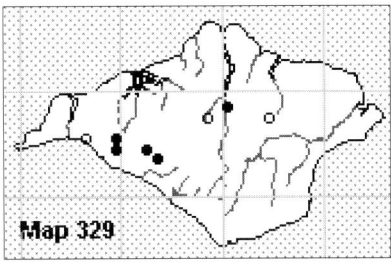

Seligeria calcarea

Syntrichia intermedia (*Tortula intermedia*) (43 sq.) A fairly common species found on walls, roofs and rocks scattered across the Island. Headon Warren (as Headon Hill), HML 1907.

Syntrichia laevipila (*Tortula laevipila*) (30 sq.) Epiphytic on a wide range of trees. Parkhurst, HML 1907. Map 328.

[var. ***laevipilaeformis*** (*T. laevipila* var. *laevipilaeformis*) (3 sq.) Extinct? There are no modern records, the most recent being from Shanklin, CHB 1927].

Syntrichia papillosa (*Tortula papillosa*) (6 sq.) Possibly under-recorded; found along roads, both on tarmac and the boles of trees along the hedges. Sandrock Road, Niton, HM 1987.

Syntrichia latifolia (*Tortula latifolia*) (3 sq.) Uncommon on the Island. Found on the banks of streams liable to flood, and also on tree bases and damp stonework. Most recently recorded from near St Catherine's Point 4975, TDB 2001. Sandown, HCN 1912.

Cinclidotus fontinaloides (1 sq.) A semi-aquatic species of walls, rocks and wood in, and by, running water. The only Island record is on the water wheel at Yafford Mill 4482, LS 1980 and 1997.

Schistidium apocarpum (15 sq.) Found on calcareous stone on roofs and walls, limestone rocks and similar sites, scattered across the Island. Blackgang, HML 1907.

Grimmia ovalis (1 sq.) A comparatively rare species in the British Isles, found on dry rocks and, as here, on roof tiles. It has been recorded from Wales, the Lake District and East Scotland, and seems to be on the decline. Recorded from roof tiles on a cottage near Woodhouse Farm, Briddlesford 5490, JDS 2002.

Grimmia pulvinata (39 sq.) A widespread and common species of man-made habitats such as walls, bridges, tombstones and old concrete, but probably under-recorded. Carisbrooke, HML 1907.

Grimmia trichophylla (1 sq.) Found on dry acid rocks, gravestones and similar sites, particularly in northern and western Britain. The only Island record is from the west wall of Godshill Church 5281, TLB 2001.

(***Seligeria recurvata*** The only record is from a Shanklin wood by Mrs. Venables, mentioned by Bruce Jackson in the Hants. Field Club, 1906 Vol. 6. It has not been substantiated.)

Seligeria calycina (*S. paucifolia*) (15 sq.) Not common on the Island, and found mainly in the west on chalk including on rubble in pits, on the downs and tracks across the ridge. Monkham Copse, Rowridge, PL 1923.

Seligeria calcarea (10 sq.) A scarce species of calcareous rock, mostly recorded from chalk pits. Mount Joy Quarry, Carisbrooke HML 1907. Map 329.

Funaria hygrometrica (61 sq.) Common and widespread in a variety of sites, but often where there has been a bonfire. In 1983, it was abundant between the rows of vines at Barton Manor vineyard, Osborne. Week Down, HML 1907.

Entosthodon fascicularis (13 sq.) Uncommon on the Isle of Wight, but occasionally found in stubble fields and plantations and in waste ground, on non-calcareous soil. St. George's Down, PL 1924.

Entosthodon obtusus (*Funaria obtusa*) (8 sq.) Found in small quantity on damp gravelly soils and on ditch banks, mostly in the north of the Island. Westover, HML 1907.

Physcomitrium pyriforme (14 sq.) Found in damp meadows, banks, ditches and mud by rivers. Recorded locally across the Island. Most recently recorded from Fattingpark Copse 5291, Bohemia Bog 5183 and The Wilderness 5082 (all BBS 2002); and Norton Spit 3489, LS 1990. The Wilderness, HML 1907.

[***Ephemerum recurvifolium*** (1 sq.) Extinct or over-looked. The only record is from a cornfield at Whitecliff Bay, HHK 1909 (BON).]

(***Ephemerum sessile*** (1 sq.) The only record is from New Barn Down, HML 1909 (BON). The specimen has been re-determined as depauperate *Campylopus pyriformis*. TLB 1998.)

Ephemerum serratum var. ***minutissimum*** (3 sq.) A rare ephemeral of damp woodland rides or shady soils in non-calcareous arable fields. Recorded from Briddlesford Copse 5490, BBS 1992; and field near Combley Great Wood 5489, UKD 1964. Field near The Wilderness, HHK 1909.

Orthodontium lineare (15 sq.) Found on shaded banks, tree bases and logs, widely scattered on the Island. It produces abundant capsules, which may be grazed by slugs. Headon Warren, BG 1964.

Leptobryum pyriforme (13 sq.) A garden weed of flower-pots; also found in shaded places by streams and on sandy banks, mostly in south-east Wight. Appuldurcombe, HHK 1907. This moss has been found well preserved in the post-glacial peat deposits at Brook (Clifford 1936). Map 330 (*see overleaf*).

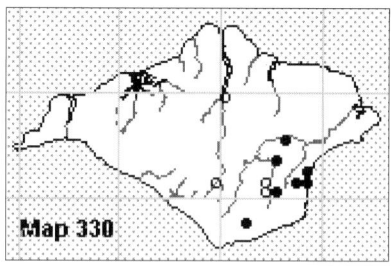
Leptobryum pyriforme

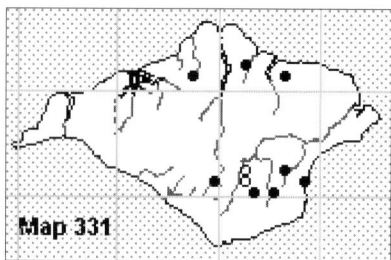
Epipterygium tozeri

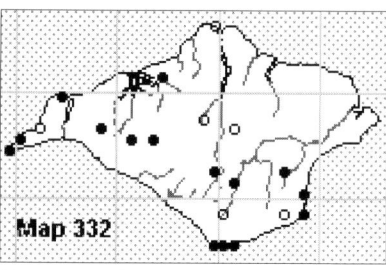
Bryum caespiticium

Pohlia nutans (20 sq.) A common calcifuge species found in a variety of habitats, from sandy banks, peat, rocks and walls to acid grassland and tree stumps. Widely distributed, but more frequent in the south-east. Ladder Chine, HML 1907.

Pohlia annotina (7 sq.) Uncommon on the Island. Found in acid gravely soils, on stream banks and waste ground. Only 2 records since 1964, both at Shanklin 5881, RDP 1995. Bleak Down, HML 1907.

[***Pohlia camptotrachela*** (1 sq.) The only Island record is from Bohemia Bog 5183, LS 1980. Whether it still survives in this site is doubtful; it was not found by BBS in 2002.]

Pohlia lutescens (1 sq.) Prefers open areas on sandy loam, by paths and rabbit holes. Rare or overlooked. The only Island record is from Briddlesford Copse 5490, RCS 1992.

Pohlia lescuriana (2 sq.) Found rarely on damp clay in woodland rides and by streams. Recorded from River Medina by The Wilderness 5082, ACC, JAP & JA 1964 and again in 1979 CCT. The only other record is Parkhurst Forest 4790, RCS & FR 1988.

Pohlia melanodon (*P. carnea*) (49 sq.) The commonest species of the genus on the Island, found in chines, damp woodland rides and by streams, widely scattered across the Island. Whale Chine, HML 1907.

Pohlia wahlenbergii (12 sq.) A species of wet gritty soils and cracks in rocks. Apparently fairly widespread at the start of the twentieth century, but there have been only 4 records since 1964. It may be under-recorded.

Ningwood, HML 1907.

[var. *calcarea* (2 sq.) The only records are from Shanklin and Luccombe Chine, CHB 1927.]

Epipterygium tozeri (10 sq.) Uncommon on the Island. A species of warm, sheltered sandy lanes and chines, on banks. Recent records include Shanklin Chine 5880, The Wilderness 5082 and Woodhouse Copse 5292, all BBS 2002. Apse Heath, HML 1907. Map 331.

Bryum pallens (14 sq.) Occasional, usually in chalk pits and in chalk grassland, but also recorded from damp heath on Headon Warren. Bleak Down, HML 1909.

[***Bryum algovicum*** var. *rutheanum* (2 sq.) Extinct? Recorded from sandy ground but only known from two old records. Near The Longstone, Mottistone, PL 1934 and Shorwell, HHK 1910.]

Bryum imbricatum (*B. inclinatum*) (4 sq.) An uncommon species on the Island, and possibly under-recorded. Godshill railway track, HML 1909.

[***Bryum intermedium*** (4 sq.) Found on damp soil in chines and on dunes; the only record since 1912 is from St. Helens Duver 6389, LS 1979, and is regarded as doubtful. Bonchurch, HML 1907]

Bryum donianum (8 sq.) Found in woods and on banks in calcareous areas. Two records from Blackwater and Apse in 1907 are in the Livens herbarium (BON). The only record since 1941 is from Luccombe Chine 5879, RAF 2002. Blackwater, HML 1907.

Bryum capillare (116 sq.) Common throughout the Island on trees, walls fences, rocks, in ditches and by tracks and streams. Steephill, HML 1907.

Bryum torquescens (4 sq.) Recently recorded from Shide Chalk Pit, 5088, and the Needles Headland 2984, both BBS 2002. The only other records are from Bonchurch Landslip 5878, ARP 1964 and Ashey Down, HHK 1910 (BON). It could be under recorded, as it may be confused with *B.capillare*.

Bryum canariense (1 sq.) The only record is on a rocky outcrop east of St. Catherine's Point 4975, FR & ECW 1971.

Bryum pseudotriquetrum (9 sq.) Prefers damp areas in dunes, by streams and on rocks, and is lime tolerant. The most recent record is from Shide chalk pit, 5088, RCS 2002. Three records by HML in 1906 for var. *bimum* from Whale Chine, below Ladder Chine, and the Undercliff are now included in the species. Bonchurch Landslip EVW & PW 1964.

Bryum caespiticium (21 sq.) Found on basic soils on waste ground, gravel pits and chalk pits. Occasional, mainly in the south and west. St. Georges Down, HML 1907. Map 332.

Bryum argenteum (50 sq.) A common species, including var. *lanatum*, of dry sandy sites, public paths, between paving stones and similar places, also found on walls, mortar and roofs. Undercliff top, HML 1909.

Bryum gemmiferum (3 sq.) Recorded from sandy coastal cliffs at Shanklin 5881, RAF 2002 and Luccombe Chine 5879, BBS 2002. Previously recorded from Chale Bay, 4777, BBS

Bryum bicolor (61 sq.) A common species, widespread on the Island and found in a variety of sites, including roadsides, in gardens and quarries, by streams and muddy banks. Parkhurst, HML 1907.

Bryum dunense (6 sq.) A species of gravely soil, and sometimes calcareous soils, often near the coast. St Helen's Duver 6389, DTH & JAP 2002; Afton Down 3685, BBS 2002; Mottistone Down chalk pit 4184, BBS 2002. Christchurch churchyard, Sandown 5983, LS 1984.

Bryum radiculosum (18 sq.) Found on the mortar of old walls, hard chalky soil and limestone. Not common in the Island, but it may be under recorded. Gatcombe, HML 1907

Bryum ruderale (3 sq.) Found on roadsides, field gateways and paths. Only two records since 1964, at Parkhurst Forest 4790, RCS 1991 and by Dodpits Lane, Newbridge 4087, MOH & JJG 2002. East Afton Down 3685, ACC, ECW & GE 1964.

Bryum violaceum (1 sq.) An arable species, which has only been recorded once. Field at Wellow 3987, MOH 2002.

Bryum klinggraeffii (10 sq.) A species of bare ground and in arable fields, which is not common on the Island. St.Catherine's Point 4975, ACC 1964.

Bryum subapiculatum (*B. microerythrocarpum*) (9 sq.) A species of non-calcareous arable fields, cliffs and heaths. St. George's Down, HML 1907.

(*Bryum bornholmense* The only records from the Island were from Bonchurch Landslip, ACC & FR 1977. These have now been re-determined as *B.rubens* (ACC & HLKW 2001). The species is no longer on the Island list.)

Bryum rubens (24 sq.) A common species of arable fields, molehills, turf,

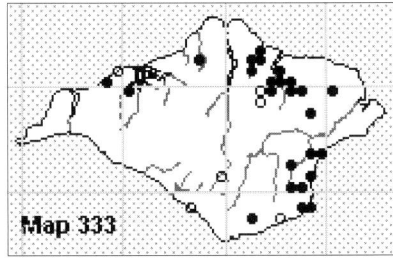

Rhizomnium punctatum

woodland rides and roadside banks. Brighstone Down, HML 1907.

Bryum alpinum (3 sq.) In Britain, a species with a mainly Atlantic distribution. Rare, on heathy tracks. Bagwich gravel pit, Bleak Down 5182, LS 1989; Lake cliffs 5882, LS 1995. Bleak Down, HML 1909.

(*Bryum* sp. (36 sq.) A proportion of *Bryum* records are not identifiable to species, due to lack of ripe capsules. Such records are scattered across the Island in a variety of habitats - downs, chalk pits, churchyards and woods.)

[*Rhodobryum roseum* (4 sq.) Extinct? A large, distinctive species, found on anthills in old chalk grassland. The most recent Island record was from East Afton Down 3685, BBS 1964.]

Mnium hornum (106 sq.) A very common moss of woodland banks, tree bases, rotting logs and sunken lanes. Hamstead, HML 1907.

Mnium stellare (3 sq.) Still present at Bonchurch Landslip 5878, DGL 2002, where it was first recorded in 1907. The Landslip, HML 1907.

Rhizomnium punctatum (37 sq.) Growing in damp woods and on rotting logs. It is fairly common in east Wight, with a few records from west of the River Medina. Steephill, HML 1907. Map 333.

(*Rhizomnium pseudopunctatum* Confined to marshes and fens. The record from Whitwell, HML 1906 is not backed by a voucher specimen, and The Wilderness specimen, HML 1907 (BON) is *R. punctatum*. The species is excluded from the Island list.)

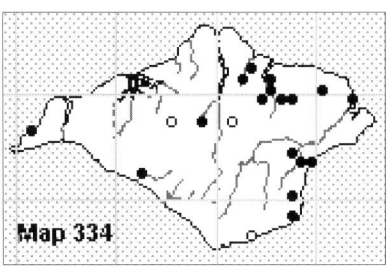

Plagiomnium rostratum

Plagiomnium affine (13 sq.) On damp ground in woods, sides of tracks and in turf, and not very common on the Island. The most recent record is from Brocks Copse 5292, RCS 2002. The Undercliff, HML 1907.

Plagiomnium elatum (2 sq.) In wet calcareous sites, but not recorded since 1964, Combley Great Wood 5489, PW. Arreton, HML 1909.

Plagiomnium undulatum (76 sq.) An abundant species of damp woods, shady lanes and grasslands such as lawns and churchyards. Tolt Copse, Gatcombe HML 1909.

Plagiomnium rostratum (21 sq.) In small quantity on shaded rocks and soil in woods and churchyards. Most records are from East Wight; there are only 3 recent records west of River Medina. Combley, HML 1909. Map 334.

Aulacomnium palustre (11 sq.) Restricted to wet, acidic sites on the Island. The most recent records are from Munsley Bog 5282, 1980 LS; and The Wilderness 5082 and Bohemia Bog 5183, both BBS 2002. Sullens, near Arreton HML 1907.

Aulacomnium androgynum (11 sq.) Growing on rotting tree stumps, hedge banks and old grass tussocks, and fences. Standen Elms, PL 1923.

[*Bartramia pomiformis* (9 sq.) Extinct? An acid-loving species of shady banks, gravel and sandy areas. The most recent record is from Apse Heath 5683, BBS 1964. Lynn Common, FMM 1907.]

(*Philonotis rigida* Originally recorded from Shanklin Chine in 1906 by HML and in 1965 by JA. The material was re-determined as *P. marchica* by J.H. Field in 1981, and deleted from the Island list.)

Philonotis marchica (3 sq.) Very rare. Grows in coastal sites on vertical sandy cliffs and loose damp scree, in sun and shade. First found at Shanklin Chine 5881 in 1869 (WW), but not confirmed until 1974, by AJES. It was re-found on a cliff at Shanklin 5882, in 1978 by LS (conf. J.H. Field). When the Chine site was visited by BBS in 2002, a substantial reduction in the population since 1995 was recorded. The only other site in England was in north Yorkshire, where it has not been recorded recently. Shanklin Chine, WW 1869.

[***Philonotis arnellii*** (1 sq.) The only record is Bleak Down, HHK 1909 (BON). It may be extinct here - a doubtful specimen was found by LS in 1999, but not confirmed.]

[***Philonotis caespitosa*** (1 sq.) Probably extinct. Growing on damp basic soil. The only records are from a track at Rookley PL 1923-1935, det. H.H. Knight & J.H. Field.]

Philonotis fontana (1 sq.) Found in acidic flushes. Still present at Bohemia Bog 5183, JS 2002, which is the only recorded site. Bohemia Bog (as Lake Farm, Rookley) PL 1933.

Zygodon viridissimus (67sq.) Common in woods. Epiphytic on the bark of trees, often on ash, elder and sallows; also on rocks and walls. Westover, HML 1907.

var. *stirtonii* (4 sq.) Usually on rocks, brick walls or mortar, sometimes on elder bushes, but not common on the Island. The most recent record was from St. Catherine's Point 4975, RCS, FR & ECW 1983. Carisbrooke Castle, HML 1909.

Zygodon rupestris (*Z. baumgartneri*) (4 sq.) A rare epiphyte on old trees in ancient woodland sites. Briddlesford Copse 5490, BBS 2002; Parkhurst Forest 4890, FR 1988; and on elder at Shanklin 5781, LS 1998. Briddlesford Copse, FR 1980.

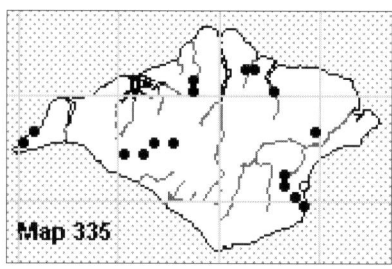

Zygodon conoideus

Zygodon conoideus (20 sq.) Usually on elder, but not confined to this species of tree. It is fairly widespread. Luccombe Chine, JMB 1927. Map 335.

Orthotrichum lyellii (29 sq.) Found on ash, sycamore and similar trees in woods, and on elder in scrubby copses. Appuldurcombe Wood, ABl 1860.

Orthotrichum striatum (5 sq.) Epiphytic on shrubs and trees. Rare and decreasing on the Island. The most recent record is from Barton's Withy Bed, Apse Heath 5783, LS 1979. Bonchurch Landslip, HHK 1909.

Orthotrichum affine (53 sq.) Common on the Island, growing on the bark of trees in woods and by roads. Hamstead, HML 1907

Orthotrichum anomalum (10 sq.) Growing on limestone rocks and walls, and in quarries. Not very common on the Island. Hamstead, HML 1907.

Orthotrichum stramineum (1 sq.) Rare, but possibly under-recorded, on the bark of trees such as ash, sycamore and others. The only Island record is from willows near Bagwich 5182, JAP & JA 1964.

Orthotrichum tenellum (9 sq.) Grows on trees along hedgerows and roadsides, rather than in woods. Not common on Island, but could be under recorded. Bonchurch Landslip, HHK 1910.

Orthotrichum diaphanum (56 sq.) Although an epiphyte on trunks of trees such as elder, ash and hazel, it also found on tarmac and rocks. It is a widespread species. Gatcombe, HML 1907.

Orthotrichum pulchellum (4 sq.) Epiphytic on elder and other trees. First found here in 2001 and apparently spreading its range in England. Blackwater 5086, TLB 2001.

Ulota crispa (56 sq.) A common species throughout the Island, growing on twigs, branches of shrubs and trees such as elder, willow, and hazel. Ventnor, ABl 1860.

Ulota bruchii (*U. crispa* var. *norvegica*) (33 sq.) Found in similar sites to the previous species, but most frequent in the north of the Island. Arreton Down, HML 1907.

Ulota phyllantha (27 sq.) Although it grows on twigs and branches of trees and shrubs, it is also found on rocks near the sea. Commonest in the south of the Island, where it is exposed to maritime conditions. Arreton Down, HML 1907.

Fontinalis antipyretica (8 sq.) A locally occurring aquatic species found in clear water in a few ponds, streams and ditches and water tanks. Carisbrooke, HML 1907.

Cryphaea heteromalla (34 sq.) Epiphytic on shrubs, particularly elders. It is found the length of the chalk ridge, and the south-east coast of the Island. Dark Lane, Whitcombe, HML 1907.

Leucodon sciuroides (14 sq.) Grows on trees, more usually in hedges and along roadsides than in woods. A declining species which was often found on ash in 1920s and 1930s (PL), but not common recently. The most recent record is Eaglehead & Bloodstone Copses, Ashey 5887, RCS 1982. Blackwater, HML 1907.

(***Antitrichia curtipendula*** The only record, Luccombe Down, Mres. V. 1850, listed in the Census Catalogue 2, has not been traced.)

Pterogonium gracile (2 sq.) Very rare on tree trunks in ancient woodlands. Only recorded from Swainston 4486,

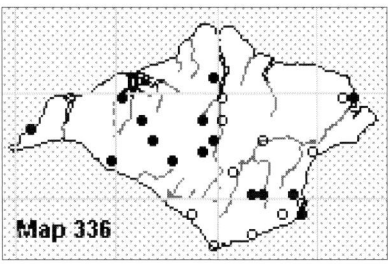
Leptodon smithii

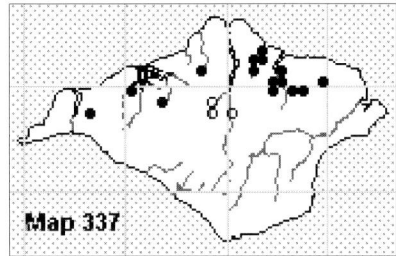
Homalia trichomanoides

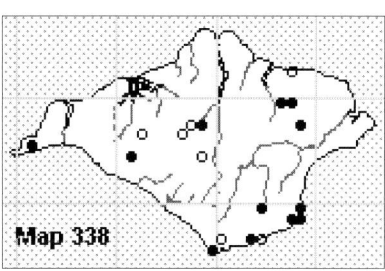
Anomodon viticulosus

on oaks at High Wood 4487, and an ash tree (subsequently felled) by Mudless Copse 4486, FR & ECW 1976/77.

Leptodon smithii (30 sq.) Almost confined to the southern counties of Britain. Widespread on the Island, but not common. Grows on trees such as elder, beech and field maple, and also colonises concrete on walls and gravestones. Gatcombe, HML 1907. Map 336.

Neckera crispa (30 sq.) Found on base rich rocks and walls, in turf along the length of the chalk ridge, and on limestone. Tolt, near Gatcombe HML 1907.

Neckera pumila (24 sq.) Epiphytic on a wide range of trees in woods, and more frequent north of the chalk ridge. Appuldurcombe Woods, ABl 1860.

Neckera complanata (66 sq.) Common and widespread in woods, growing on tree bases and roots; also on calcareous rocks, and in chalk grassland. Combley Great Wood, HML 1907. A well preserved specimen has been discovered in the post-glacial peat deposits at Brook. (Clifford, 1936).

Homalia trichomanoides (20 sq.) Grows on the exposed roots and bases of trees in copses, often just above the waterline of ditches. Confined to the northern half of the Island. Dark Lane, Whitcombe, HML 1907. Map 337.

Thamnobryum alopecurum (66 sq.) A common moss of shady woods, growing on the ground and around tree bases. Distributed particularly on the chalk, but also in woods to the north. Steephill, ABl date unknown.

Hookeria lucens (12 sq.) Found on damp, shady, slightly acid soils, on banks, ditches and stream-sides in woods. It is uncommon on the Island. Recorded from Luccombe Chine 5879, Briddlesford Copse 5490, and Buddle Brook at Brighstone 4285, all BBS 2002; Combley Great Wood 5489, RCS 1990; Parkhurst Forest 4791, RCS 1988; Chillingwood Copse 5689, RCS 1983; and New Copse, Fishbourne 5592, RCS 1979. Parkhurst Forest PW 1958.

Leskea polycarpa (5 sq.) Uncommon. Usually confined to the exposed roots of trees liable to seasonal inundation, alongside ditches and streams. It also occurs on the framework of the waterwheel at Yafford Mill, 4482. Woodhouse Copse, RCS 1983.

Anomodon viticulosus (20 sq.) Occasional in grassland and woods on the chalk, and in rocky sites and walls along the south coast. Tolt, HML 1907. Map 338.

Thuidium tamariscinum (54 sq.) Frequent on damp ground in woods, particularly north of the chalk ridge. Parkhurst Forest, HML 1909.

[*Thuidium philibertii* (2 sq.) Rare in old chalk grassland, and not recorded since 1964 (BBS). Ashey Down, HHK 1923]

Palustriella commutata (*Cratoneuron commutatum*) (8 sq.) Confined to chalk springs, and often forming tufa. Largely restricted to the Undercliff area. Lisle Coombe, St Lawrence ABl 1860. A record from Yarmouth (HML 1909) is doubtful.

Cratoneuron filicinum (54 sq.) Grows around chalk springs and in damp turf in woods, on damp rocks. Also recorded from the waterwheel at Yafford Mill, 4482. Whitwell, HML 1907.

Campylium stellatum var *protensum* (5 sq.) Rare on the Island; found in chalk grassland, old chalk pits and woodland rides. The only record since 1964 is from Briddlesford Copse 5490, RCS 2002. Alvington chalk pit, Carisbrooke PL 1923.

Campyliadelphus chrysophyllus (*Campylium chrysophyllum*) (18 sq.) Found in chalk grassland and chalk pits, all along the central chalk ridge and near Ventnor. Ashey Down, HML 1907.

[*Campylophyllum calcareum* (*Campylium calcareum*) (1sq.) Extinct? A strong calcicole, growing on chalk and hard basic earth. The most recent record was from Bonchurch Landslip, CHB 1927.]

Amblystegium serpens (80 sq.) Common across the Island, in a wide variety of sites. It grows on soil, shady rocks (chalk, sandstone and brick) and on dead and live wood, especially willows and elder. Arreton, HML 1907.

Amblystegium tenax (6 sq.) A scarce aquatic species found in streams, or on stones and masonry liable to flooding. Almost confined to the south-west of the Island. Gatcombe Mill 4985, JAP & JA 1964.

[*Amblystegium varium* (1 sq.) Extinct? Found on wood and stones in damp sites by streams and in willow carr. Only known from a single old record. Whitecroft, near Gatcombe HML 1906.]

Leptodictyum riparium (*Amblystegium riparium*) (22 sq.) Occasional on wet wood, stones and masonry near and in streams, ditches and ponds. Steephill, HML 1907.

Conardia compacta (*Amblystegium compactum*) (1 sq.) A rare moss of damp, shaded soil and rocks in calcareous areas. It is more characteristic of the north and west of Britain. Only recorded from Bonchurch Landslip 5878, EFW 1964 & FR 1977.

Drepanocladus aduncus (4 sq.) A mainly aquatic species floating in marshes and shallow pools. Not common on the Island. Pond at Lambsleaze, near Porchfield 4391, BS 1986; edge of shallow pool on St Helen's Duver 6389, DTH & JAP 2002. St. Helens, HHK 1909.

[*Calliergon stramineum* (2 sq.) Extinct? Found in boggy ground in marshes. The only records are from Newchurch Marshes, PL 1931 and Sandown Golf Course PL1943.]

Calliergon cordifolium (4 sq.) Uncommon. Grows in boggy ground in marshes. Recent records are from Cridmore Bog 4981, CP 1995; and Alverstone Marshes 5885, NS 1992. The Wilderness, HML 1906.

Calliergonella cuspidata (*Calliergon cuspidatum*) (94 sq.) A common species found in chalk grassland, wet meadows, lawns, in woods and churchyards and chalk paths. The Wilderness, HML 1907.

Isothecium myosuroides (79 sq.) Common in woods throughout the Island, growing on tree bases and trunks in shade. Hamstead, HML 1907.

Isothecium alopecuroides (*I. myurum*) (36 sq.) Found in similar sites to the previous species, but not so widespread. It also grows on rocks in shady woods. Swainston, HML 1907.

Scorpiurium circinatum (30 sq.) Grows on chalk and limestone rocks, walls, shaded gravestones and on wood. Most frequent along the Undercliff in south-east Wight. Steephill, HML 1909. Map 339.

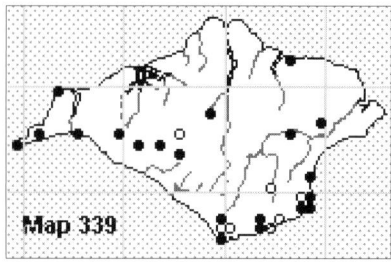

Scorpiurium circinatum

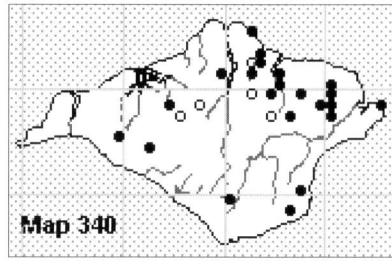

Cirriphyllum piliferum

Homalothecium sericeum (85 sq.) A common species found in woods, chalk pits, graveyards and around houses, growing mainly on limestone rocks, mortar of walls, asbestos roofs and gravestones. Ashey, HML 1909.

Homalothecium lutescens (60 sq.) Characteristic of chalk grasslands, it is found the length of the chalk ridge, and in the Ventnor-St. Catherine's area. Ashey, HML 1907.

Brachythecium albicans (28 sq.) Found in sandy soil and slightly acid grassland and dunes, mainly coastal. Headon Warren, HML 1907.

Brachythecium glareosum (6 sq.) A rare species of dry chalk turf and undercliffs. Recently recorded from Bonchurch Landslip 5878, BBS 2002; and Arreton Down 5387, RDP 1987. Ashey Down, HML 1907.

Brachythecium mildeanum (9 sq.) Uncommon in wet calcareous grasslands. Ladder Chine, HML 1907.

Brachythecium rutabulum (158 sq.) A ubiquitous species of woods, on tree stumps, soil, rocks, in hedges, grassland, quarries and churchyards. Ventnor, ABl 1860.

Brachythecium rivulare (30 sq.) A widespread species of wet woods and carr; also found in and by streams, on rocks, tree bases and rotting logs. Shanklin, CHB 1929.

Brachythecium velutinum (52 sq.) Frequent in woods on tree bases, stumps, logs and the bark of elders. Also found on rocks and soil in lanes, and by streams. Ventnor, ABl 1860.

(*Brachythecium populeum* The only record is from Appuldurcombe, AL 1907, but this is not supported by a specimen, so the species has not been included in the Isle of Wight list.)

Scleropodium purum (*Pseudoscleropodium purum*) (85 sq.) A common species, found in grassland on downs, in woods, churchyards roadsides and on banks. Parkhurst, HML 1907.

Scleropodium cespitans (5 sq.) Rare. Found on tree bases, stones and on masonry liable to occasional immersion by silty water. There are no records since 1982. Godshill, AL 1907.

Scleropodium tourettii (11 sq.) Grows on well-drained soils and on rocks on warm cliffs, banks, paths, and in sparse turf. Not common on the Island. Headon Hill, PL 1930.

Cirriphyllum piliferum (27 sq.) Prefers shady sites, often among grass on stream-sides, in damp woods and churchyards. Largely confined to the north of the Island. Near Ventnor, ABl 1900. Map 340.

Rhynchostegium riparioides (26 sq.) On tree roots, concrete and brickwork, submerged in streams and on weirs and sluices, scattered across the Island. Known since 1930 from the stream at Winkle Street, Calbourne. Alverstone Mill, HML 1907.

Rhynchostegium murale (10 sq.) Found on shady limestone rocks and walls; also on tree bases and damp brickwork. The only records since 1964 are from Shanklin Chine 5881, HM 1990; Brocks Copse, Whippingham 5292, LS 1982; and Springhill Wood, East Cowes 5096, LS 1994. Whitwell, HML 1907.

Rhynchostegium confertum (89 sq.) A common species of shady damp places in woods, on elder bark, on banks, stones, and roofs. Bonchurch Landslip, AL 1907.

Rhynchostegium megapolitanum (8 sq.) Rare on well-drained soils on chalky and sandy banks and in chalk pits. There are no records since 1980 Sandown 5984, LS. Bleak Down, HML 1907.

Eurhynchium striatum (70 sq.) A common species of shady rocks, both sandstone and limestone, and soil in woods, hedges and roadsides, banks and cliff top turf. Ventnor, ABl 1860.

Eurhynchium striatulum (*Isothecium striatulum*) (1 sq.) A calcicole of dry rocks and walls. The most recent record is from St. Catherine's Point 4975, 1964 BBS. Ventnor, AB (date unknown).

Eurhynchium pumilum (56 sq.) A fairly common species of deep shade in woods on banks, ditch sides, silt covered brickwork and chalkpits. Most frequent in the east of the Island. Bonchurch, HML 1906.

Eurhynchium praelongum (incl. var. *stokesii*) (160 sq.) Widespread and common in a wide range of sites, damp or dry, shaded or not, in woods, on logs, tree bases, rocks and soil, in scrub, hedges and banks. Parkhurst Forest, PL 1923. A well-preserved specimen has been reported from the post-glacial peat deposits at Brook (Clifford, 1936).

Eurhynchium hians (*E. swartzii*) (82 sq.) Found across the Island in mainly, but not exclusively, chalk and limestone areas; frequent in woods, scrub and hedges, and chalk grassland. Westover, HML 1907.

Eurhynchium schleicheri (7 sq.) On shady banks, in woods and sunken lanes. Not common on the Island. Freshwater CHB 1928.

Eurhynchium speciosum (6 sq.) Occasionally found in wet shady sites, on soil and tree roots in wet woods and carr, and ditches and wet rocks. Perhaps under-recorded. The Wilderness, HML 1906

Eurhynchium crassinervium (*Cirriphyllum crassinervium*) (32 sq.) Found on soil in woods on the chalk, and also on old mortar and walls. This species has apparently declined since the 1940's. Ventnor, ABl 1860.

Rhynchostegiella tenella (38 sq.) Found on chalk and limestone rocks and pebbles, often in shady woods, also on gravestones and walls. Most frequent in the south-east of the Island. The Undercliff, HML 1907.

Rhynchostegiella litorea (*R. tenella* var. *litorea*) (1 sq.) The only record is from St. Catherine's Point 4975, TLB 2001.

Rhynchostegiella curviseta (5 sq.) Grows on damp stones and tree roots. Not common on the Island. Shanklin Chine, HML 1907.

Entodon concinnus (2 sq.) A moss of species-rich short chalk turf. It is rare on the Island. The only modern record is from a chalk pit on Mottistone Down 4184, RAF & RCS 2002. Carisbrooke, HML 1907.

Pleurozium schreberi (6 sq.) Found on heaths and moors, often under heather. The most recent records are from Munsley Bog 5282, LS 1980; and Headon Warren 3985, FR 1982. Headon Warren, HML 1907.

Plagiothecium latebricola (9 sq.) Grows on decomposing, damp vegetable matter, such as fern and Tussock Sedge stocks, in damp, shady places. Recorded from The Wilderness 5082 and Briddlesford Copse 5490, by BBS 2002. The Wilderness, HHK 1906.

Plagiothecium denticulatum (17 sq.) Frequently found in damp woods and shady sites, growing on tree stumps, logs and banks; also walls and stream sides. Brighstone Chine, HML 1907.

Plagiothecium ruthei (4 sq.) Apparently rare in wet woodlands. The most recent record is from the Buddle Brook, Brighstone 4283, RCS 1984. The Wilderness, EFW 1964.

Plagiothecium curvifolium (13 sq.) Occasionally found on tree stumps, logs, damp litter and soil in conifer woods, including decaying pine needles. The Wilderness 5082, JAP 1964. A well-preserved specimen has been reported from the post-glacial peat deposits at Brook (Clifford, 1936).

Plagiothecium succulentum (16 sq.) Found on banks and tree bases in woods and hedges, and confined to the East Wight. The Wilderness, BBS 1964.

Plagiothecium nemorale (68 sq.) A common species of tree trunks and banks in woods, hedges and near streams. Found across the Island. Gatcombe, HML 1907.

Plagiothecium undulatum (9 sq.) A calcifuge, growing in conifer woods and heaths. It is uncommon on the Island, but it could be under-recorded. Marvel Copse, HML 1907.

Pseudotaxiphyllum elegans (*Isopterygium elegans*) (29 sq.) An acid loving species of sandy soil and banks in woods, and on tree stumps and logs. The majority of records are from the East Wight. The Wilderness, HML 1907.

Herzogiella seligeri (1 sq.) The only record is on a tree stump in Combley Great Wood 5489, FR 1981. It could be under-recorded.

Taxiphyllum wissgrillii (7 sq.) Found in shady woods and gullies in chalk and limestone areas. Very local on the Island; the most recent record is from Cliff Copse, Shanklin 5680, HM 1987. The Undercliff (HND Vict. County Hist.)

Platygyrium repens (1 sq.) An invading species, which grows as an epiphyte on the trunks and branches of a variety of trees and logs. The only record is from The Wilderness 5082, DGL 2002.

Hypnum cupressiforme (114 sq.) Said to be the commonest pleurocarpous moss in Britain (AJES 1994), growing on almost any type of substrate except soil - trees, logs, walls, roofs, gravestones, in woods and towns. Ventnor, ABl 1860.

Hypnum lacunosum (*H. cupressiforme* var. *lacunosum*) (38 sq.) Found in chalk turf, soil and in woods. Hamstead, HML 1907.

var *tectorum* (*H. cupressiforme* var. *tectorum*) (3 sq.) The only records are from Alum Bay 3085, RCS & HM 1999 and from Headon Warren 3085, and St. Catherine's Point 4975, TLB 2001; and St Helen's Duver 6389 JMB 2001.

Hypnum resupinatum (*H. cupressiforme* var. *resupinatum*) (105 sq.) Commonly found on tree trunks in woods and copses, widespread across the Island. Combley Great Wood, HML 1908.

Hypnum andoi (*H. mammillatum*) (42 sq.) Found in similar sites to the previous species, and also on conifers. Apparently less common, but it may be under recorded. Parkhurst, HML 1907.

Hypnum jutlandicum (28 sq.) Found on heathland, and also on trees and stumps in woods and conifer plantations. Mainly found in the north on the Island. Westover, HML 1907.

Hypnum lindbergii (2 sq.) Found along tracks in woodland. The only two records are from Woodhouse Copse 5393, RCS 1982 and Carisbrooke Castle Shrubbery 4887, LS 1983.

Ctenidium molluscum (33 sq.) Grows among turf on cliffs and by streams, on rocks and soil in chalk areas. Fairly common along the chalk ridge. Ashey Down, HML 1907. Map 341.

Rhytidiadelphus triquetrus (24 sq.) Found in chalk grassland and woods, on banks and in open places, but not common on the Island. Ventnor, ABl 1860.

Rhytidiadelphus squarrosus (42 sq.) Often abundant in chalk grassland, lawns, pastures and marshes, woodland rides. Forms a mat under thick grass by Scotchells Brook, Alverstone 5884. Westover, HML 1907.

Hylocomium splendens (13 sq.) Rare and greatly declined. Usually found on heaths and in woods on slightly chalky sites, also in north-facing chalk grassland. Since 1930, it has only been recorded from Bouldnor Cliff 3790, RCS 1990 and from south-east Wight (SZ58, possibly on top of Ventnor Downs), BBS 1964. Ventnor, ABl 1860.

REFERENCES

Blockeel, T.L. & Long, D.G. (1998) *A Check-list and Census Catalogue of British and Irish Bryophytes.* British Bryological Society, Cardiff.

Bloxam, A. (1860) Mosses. In *A Guide to the Isle of Wight*, E. Venables. Lond.

Crundell, A.C. & Whitehouse, H.L.K. (2001) A revision of *Bryum bornholmense* Wink & R. Ruthe. *J. Bryol.* 23, 171-176.

Duncan, J.B. (1926) *A Census Catalogue of British Mosses.* 2nd edition. British Bryological Society.

Edwards, B. (1996) *Southbya nigrella* on the Isle of Portland. *Recording Dorset.* No.6.

Field, J.H. (1979) A modern British specimen of *Philonotis marchica* (Hedw.) Brid. *Proc. Birmingham Nat. Hist. Soc.* 24 (1): 39-40.

Field, J.H. (1963) Notes on the taxonomy of the genus *Philonotis* by means of vegetative characters. *Trans. Br. Bryol. Soc.* 4, 3.

Hill, M.O., Preston, C.D. & Smith, A.J.E. (1991-1994) *Atlas of the Bryophytes of Britain and Ireland,* Vols. 1-3. Harley Books. Colchester, Essex.

Ingham, W. (1909) Hepatics. In *A Guide to the Natural History of the Isle of Wight.* F. Morey. Isle of Wight County Press, Newport.

Jackson, A. B. (1907) The moss flora of Hampshire and the Isle of Wight. *Proc. Hants. Fld. Club* 6: 29-40.

Livens, H.M. (1909) Mosses. In *A Guide to the Natural History of the Isle of Wight.* F. Morey. Isle of Wight County Press, Newport.

Livens, H.M. (1926) A supplementary list of additional species of mosses. *Proc. Isle Wight nat. Hist. Archaeol. Soc.* 1 (VII): 454

Livens, H.M. (1926) Supplementary list of hepatics. *Proc. Isle Wight nat. Hist. Archaeol. Soc.* 1 (VII): 452

Martin, P. (1997) Castle Haven ecological survey, 1997. Appendix A Vegetation, Bryophytes. Unpublished report.

Paton, J.A. & Blackstock, T. (1996) *Cephalozia macrostachya* Kaal. var. *spiniflora* (Schiffn.) Mull. Frib. in Britain and Ireland. *Trans. Br. Bryol. Soc.* 19, 333-339.

Paton, J.A. (1961) *Southbya tophacea* Spruce new to the British Isles. *Trans. Br. Bryol. Soc.* 4, 98-101.

Paton, J.A. (1971) *Southbya tophacea* Spruce in Anglesey. *Trans. Br. Bryol. Soc.* 6, 328-330.

Paton, J.A. (1965) A new British moss, *Fissidens celticus* sp. nova *Trans. Br. Bryol. Soc.* 4, 5.

Paton, J.A. (1977) *Metzgeria temperata* Kuwah. in the British Isles, and *M. fruticulosa* (Dicks.) Evans with sporophytes. *J. Bryol.* 9, 441-449.

Perry, A.R. (1965) The annual meeting 1964. *Trans. Br. Bryol. Soc.* 4, 893-895.

Reese, W.D. (1967) The discovery of *Tortula vectensis* in North America. *Bryologist* 70, 112-114.

Rose, F. & Swinscow, T.D. (1956) *Southbya nigrella* (De Not.) Spruce in fruit. *Trans. Br. Bryol. Soc.* 3, 124.

Smith, A.J.E. (1974) *Philonotis marchica* (Hedw.) Brid. in Britain. *J.Bryol.* 8, 5-8.

Smith, A.J.E. & Whitehouse, H.L.K. (1978) An account of the British species of the *Bryum bicolor* complex, including *B. dunense*, sp. nova Extract - *Bryum gemmiferum* Wilcz. & Dem. *Trans. Br. Bryol. Soc.* 10, 36-39.

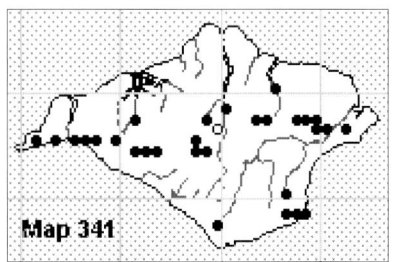

Ctenidium molluscum

Snow, L. (1989) Provisional atlas of the bryophytes of the Isle of Wight: Liverworts. *Proc. Isle Wight nat. Hist. Archaeol. Soc.* **9**, 121-134.

Snow, L. (1992) Provisional atlas of the bryophytes of the Isle of Wight: Mosses. Part 1 *Sphagna - Grimmiales*. *Proc. Isle Wight nat. Hist. Archaeol. Soc.* **12**, 51-70.

Snow, L. (1997) Provisional atlas of the bryophytes of the Isle of Wight: Mosses. Part 2 *Funariales - Hynobryales*. *Proc. Isle Wight nat. Hist. Archaeol. Soc.* **13**, 11-32.

Sollman, F. (1979) *Tortula rhizophylla* (Sak.) Iwats and Saito in Italy. *Lindenbergia* **5**, 109-110.

Syed, H. (1973) Taxonomic studies of *Bryum capillare* Hedw. and related species. Extract 10 - *Bryum torquescens* Bruch. *Trans. Br. Bryol. Soc.* **7**, 307-314.

Venables, Rev. E. (1860) *A Guide to the Isle of Wight, its approaches and places of resort*. London.

Wallace, E.C. (1972) *Tortella inflexa* (Bruch.) Broth. in England. *J. Bryol.* **7**, 153-156.

Warburg, E. (1965) *Pohlia pulchella* in Britain. *Trans. Br. Bryol. Soc.* **4**, 760-762.

Warburg, E.F. & Crundwell, A.C. (1965) *Tortula vectensis*, a new species from the Isle of Wight. *Trans. Br. Bryol. Soc.* **4**, 763-766.

Whitehouse, H.L.K. (1960) *Dicranella staphylina* a new European species. *Trans. Br. Bryol. Soc.* **5**, 757-776.

Lichens

COLIN POPE

The current lichen flora of the Isle of Wight comprises a total of 393 taxa. A further 33 species, supported by herbarium specimens, are now considered extinct. There are also some 12 unconfirmed or dubious records from the 19th century. The surrounding counties have significantly higher totals (published totals, which are now likely to be exceeded, are Dorset 652, Hampshire 564, Sussex 586) but these are substantially larger counties and they have been the subjects of greater scrutiny. The Island's lichen flora is, for its area, comparatively rich, but would undoubtedly benefit from more thorough and systematic surveys. The great bulk of the survey work was carried out in the 1970s and 1980s, and there have been considerable advances in field lichenology since then.

In the mid-nineteenth century, the Island was a fashionable place for botanists to visit, and some visitors recorded lichens. Most famous was William Borrer, who discovered *Fulgensia fulgens* at Freshwater in the early 1800s and spent some time collecting during a visit in 1852. Bloxam (1860) compiled the first published list of Island lichens. He recorded many interesting species in the neighbourhood of Ventnor, Bonchurch and Appuldurcombe, including *Pseudocyphellaria aurata* at the tops of trees near Shanklin. This list also has a short addendum of 17 lichens provided by Rev. T. Salwey. Subsequently, Salwey (1860) reported *Roccella phycopsis* for the first time from several churches on the Island. A more substantial list of lichens of the Isle of Wight was prepared by J. A. Wheldon (1909), vice-president of the Liverpool Botanic Society, together with Rev. H.M. Livens for Morey's Natural History of the Isle of Wight. This list consisted of previously published records from various sources, together with specimens collected by H. M. Livens. Livens' herbarium was eventually donated to Bolton Museum in 1945, during his 85th year, and it includes collections from other members of his family, F. Morey, H.P. Reader and J.H. Wheldon (Seaward, 1978; Coppins, 1978). A revised and enlarged list of lichens of the Isle of Wight was prepared by H.H. Knight (1932); his herbarium is housed at the National Museum of Wales (Cardiff).

Little recording of lichens was carried out subsequently until Brian Coppins, Alan Pentecost and Francis Rose spent a week on the Island in 1971. These records are referred to as 1971* in this list. In 1982, the British Lichen Society held its spring meeting at Newport (Pope, 1984). From that date, contributions to field work on the Island have been made, amongst others by Oliver Gilbert, Peter James, Francis Rose, Neil Sanderson and the author. The current list is a revision and update of the most recently published lichen flora (Pope, 1983).

Nomenclature follows that of Coppins (2002). Where the name has changed since Purvis *et al* (1992), the older name is also included in parenthesis. Species believed to be extinct are enclosed in square parentheses [ ]. Unchecked records and records believed to be in error are enclosed in round parentheses ( ). A list of 10 km grid squares from which each species has been recorded on the Island since 1970 appears at the end of each entry.

The Isle of Wight is covered by eleven 10 km squares but SZ28 covers a very small area at the western-most tip of the Island and has not been listed except in exceptional circumstances. The squares are as follows: SZ38, 39, 47, 48, 49, 57, 58, 59, 68 and 69.

Acarospora fuscata On nutrient-enriched acid stones; reasonably frequent. All except 39, 58, 59, 69.

A. smaragdula On stone influenced by heavy metals; rare. Godshill Church, both brown and yellow-green morphotypes beneath copper window frames, PWJ 1991. 58.

Acrocordia conoidea On hard limestone, Shalfleet Church, KAS 1998; Mottistone Church chest tomb, FR 1998. 48.

A. gemmata On bark of trees, particularly ash and formerly elm; widespread although rather infrequent. All squares.

A. salweyi On limestone rocks and walls; scarce. Recorded from Godshill Church and a crag east of St. Catherine's Point, both 1971*; Carisbrooke Castle 1995 (conf. BJC); St Edmund's Church, Wootton FR 1995. 47, 48, 58, 59.

Agonimia allobata A very rare ancient woodland species. Eastern end of Whitefield Woods 1971*; Hurst Copse, Wootton CP 1991 (conf. PWJ). 59, 68.

A. gelatinosa (*Polyblastia gelatinosa*) On calcareous soils; very rare, perhaps overlooked. A component of the *Fulgensia* community on Tennyson Down CP 1990; north facing chalk bank, West High Down BJC 1973; frequent on Bouldnor cliffs VJG & OLG 2002. 38, 39.

A. octospora A scarce ancient woodland species. One Island record from a single old ash with *Lobaria pulmonaria* at the edge of Mudless Copse FR 1976 (hb. FR). This tree was felled in 1981. 48.

A. tristicula On old stonework; rare, but probably overlooked. Fertile on old walls at Quarr Abbey, BJC 1971 (E); Brading Church wall FR, KAS 1993; Carisbrooke Castle FR 1995; Newtown Church L&SS, CH 2002. 48, 49, 59, 68.

Amandinea punctata (*Buellia punctata*) On bark and fence posts; common and widespread. All squares.

Anaptychia ciliaris Formerly local but widespread on elms: Nunwell Park, Appuldurcombe Park and Quarr (all 1971*); south of Stonesteps, Calbourne FR 1976; near St. Helen's Old Church CP 1979. Has declined dramatically with the demise of elms and now very rare. On ash at Swainston CP 1980, but not seen since. Abundant, but not fertile, on a single ancient parkland oak at Swainston CP 2001. 48, 49, 58, 59, 68.

A. runcinata Recorded by Bloxam (Knight, 1932) from coastal rocks at Old Park and still there today, CP 1999. Sugar Loaf rock by coast at St Lawrence 1999. Extends westwards along the coast to St. Catherine's Point, where it was particularly well developed on a large rock close to the lighthouse, 1980 but in reduced quantity, CP 2000. A relict species, perhaps in decline, its occurrence on the Island constitutes the eastern-most locality on the South coast. 47, 57.

Anisomeridium biforme Occasional on bark of ash, elm and elder. America Wood, Quarr, Bonchurch Landslip, north of Brading, all 1971*; Long Copse, Gatcombe and Mudless Copse, both FR 1977; Parkhurst Forest FR 1988; Hurst Copse, Wootton PWJ 1991. 48, 49, 57, 58, 59, 68.

A. polypori (*A. nyssaegenum*) Probably overlooked. Hurst Copse, Wootton, on hazel PWJ 1991; old oak at Newtown L&SS, CH 2002. 49, 59.

A. ranunculosporum (*Arthopyrenia ranununculospora*) A rare ancient woodland species. Recorded on oak in Parkhurst Forest FR 1989. 49.

Arthonia cinnabarina An old woodland species, usually on shaded bark; infrequent. Borthwood Copse 1971*, Mudless Copse FR 1976; Harelane Plantation, Swainston FR 1981; Parkhurst Forest FR 1988; Hurst Copse, Wootton CP 1991; Bonchurch Landslip CP 1992. 48, 49, 57, 58, 59.

A. didyma Rare on oak and hazel. Recorded from Borthwood Copse FR 1971(BM); Parkhurst Forest FR 1988; St Lawrence CP 1994; and on hazel at Newtown, CH 2002. 49, 57, 58.

A. elegans On oak in Parkhurst Forest FR 1989. 49.

A. lapidicola On copper stained windowsill, Godshill Church PWJ 1991. 58.

A. pruinata (*A. impolita*) On dry bark of old parkland oaks, ash and field maple, formerly on elms; occasional. 48, 49, 57, 58, 59, 68, 69.

A. punctiformis On young twigs; probably under-recorded. Borthwood Copse and Bonchurch Landslip, both 1971*; elms north of Brading BJC 1973; Northpark Copse, Swainston FR 1981; Parkhurst Forest FR 1989; Hurst Copse, Wootton FR 1991; America Wood L&SS, CH 2002. 48, 49, 57, 58, 59, 68.

A. radiata On smooth-barked trees in woods; frequent. All except 38, 69.

A. spadicea In established woodland, particularly on oak; widespread and frequent. 39, 48, 49, 57, 58, 59.

A. vinosa An ancient woodland species on old oak and sometimes ash; very locally frequent. Northpark Copse CP 1980; Mark's Corner, Parkhurst CP 1976; Walter's Copse, Newtown FR 1977; Briddlesford Copse CP 1980; Bonchurch Landslip CP 1992; Gatcombe Withy Bed CP 1995. 48, 49, 57, 58, 59.

Arthopyrenia analepta (*A. lapponina*) On smooth bark; uncommon. Bonchurch Landslip 1971*; Briddlesford Copse CP 1993; America Wood C&SS, CH 2002. 38, 57, 59.

A. punctiformis On smooth bark; occasional but probably overlooked. Bonchurch Landslip and Appuldurcombe Park, both 1971*; on blackthorn at Alum Bay BJC 1973; Parkhurst Forest FR 1988; America Wood C&SS, CH 2002; Newtown L&SS, CH 2002. 38, 49, 57, 58.

Aspicilia caesiocinerea Remarkably frequent on stones on Afton Down OLG 1987 (conf. BJC). 38.

A. calcarea On hard, calcareous rocks; common. All squares.

A. contorta Occasional on limestone. Brading churchyard 1971*; east wall of Carisbrooke Castle BJC 1973; Freshwater All Saints Church FR 1995; calcareous pebbles on Afton Down CP 1986; St Edmund's Church, Wootton FR 1995; Newtown town hall C&SS, CH 2002. 38, 48, 49, 59, 68.

A. radiosa (*A. subcircinata*) Rare. Carisbrooke Castle bridge BJC 1973. 48.

[*A. tuberculosa* Extinct. Ryde beach on flints, single specimen collected c. 1830, C. Lyell (BM). Habitat destroyed by development. This endemic species, which was otherwise known only from the South Downs in Sussex, is considered to be extinct in Britain.]

Bacidia arceutina On ash; rare or overlooked. Long Copse, Gatcombe FR 1977 (hb. FR). Old records from St. Lawrence and Brading. 48.

B. bagliettoana On moss in short turf on chalk; rare. Tennyson and Afton Downs OLG 1990. 38.

B. biatorina An old forest species on oak and occasionally on ash; scarce. Long Copse, Gatcombe on oak and ash FR 1977; Briddlesford Copse FR 1980; Northpark Copse FR, CP 1981; Barton Wood, Osborne on oak CP 1984 (det FR). 48, 59.

B. friesiana Only recorded from ash at Bonchurch Landslip. BJC 1973. 57.

[*B. incompta* Extinct This nationally vulnerable species was scarce on elms but is now considered to be extinct. Nunwell Park and Undercliff, east of Niton, both 1971*; Godshill Park FR 1976. 47, 57, 58.]

B. laurocerasi On ash and sycamore, formerly also on elder; local. All except 39, 59, 69.

B. phacodes Widespread but scarce on nutrient-rich bark. Less frequent today with the disappearance of elms. All except 39, 47, 49, 59.

B. rubella Once frequent on elm; now much less frequent on basic bark. On oak in Northpark Copse FR 1981. All except 39, 69.

B. viridifarinosa On oak in Parkhurst Forest FR 1988. 49.

Baeomyces rufus In heathy places; very locally frequent. Bleak Down 1971*; St. George's Down CP 1981; Headon Warren CP 1976; gravel pit at Downend CP 1982; Sibden Hill sandpit CP 1993 (fruiting well at the last three stations); Parkhurst Forest FR 1988; Luccombe Down CP 1989. 38, 49, 57, 58.

Belonia nidarosiensis On sheltered, north-facing walls of several medieval church walls. Under-recorded. FR 1995.

Biatora epixanthoides (*Bacidia epixanthoides*) A rare old woodland species. Sterile on ash in Northpark Copse FR 1981 (det BJC); on oak in Hurst Copse, Wootton PWJ 1991. 48, 59.

Buellia aethalea On acid stones, particularly in churchyards; widespread, but under recorded. 38, 47, 48, 49, 58, 59, 68.

B. griseovirens On ash; rare, perhaps under recorded. Long Copse, Gatcombe and Mudless Copse FR 1977 (hb. FR). 48.

B. ocellata On acid stones; under-recorded. Carisbrooke Castle BJC 1973; Freshwater All Saints Church BLS 1982; Godshill Church OLG 1990; Brading Church FR, KAS 1993; St Edmund's Church, Wootton FR 1995; Newtown Church L&SS, CH 2002. 38, 48, 49, 58, 59, 68.

B. stellulata On stones and pebbles; scarce or overlooked. Headon Warren and St. Catherine's Point BLS 1982. 38, 47.

Byssoloma leucoblepharum On a Sessile Oak in Parkhurst Forest NS 1997. Very rare; currently (2002) only four other sites known in British Isles, including the New Forest. 49.

Calicium glaucellum On decorticated wood; rare. South of Seaview 1971* (BM); old oak at edge of Walter's Copse, Newtown L&SS, CH 2002. 49, 69.

C. salicinum On old bark; rare. On an old oak at Quarr 1971*. 59.

C. viride On dry bark; occasional. On old oaks at Quarr 1971* and 1987; Barton Wood, Osborne CP 1984; Northpark Copse on oak CP 1987; Pucker's Copse, Quarr, on ash CP 1987. 48, 59.

[*Caloplaca atroflava* A single 19th century record from Ryde (BM).]

C. aurantia On sunny calcareous rocks and tombstones; common and widespread. All except 39.

C. ceracea Rare on greensand. Godshill Church 1971*; Carisbrooke Castle and north side of Whitwell Church, FR 1995. These are the easternmost records on the south coast of Britain of this largely maritime species. 48, 57, 58.

C. cerina On parkland trees; very rare, perhaps extinct. Appuldurcombe Park 1971*. 47, 58.

[*C. cerinella* Recorded 19th century on elder at Niton (JHB). Possibly still extant as this is an easily overlooked species.]

[*C. chlorina* (*C. isidiigera*) Extinct. Fertile crust on elm at Appuldurcombe Park FR 1971, but not recognised until 1989 and now extinct. 58.]

C. citrina s.lat. On calcareous substrates and nutrient-rich bark; abundant and widespread. On vertical sandstone cliff faces at Luccombe and Redcliff, Sandown. All squares.

C. crenularia On church walls and headstones; frequent and sometimes abundant. On rocks at St. Catherine's Point. All except 39, 69.

C. dalmatica Impressively abundant on vertical south-facing stonework of medieval churches and Carisbrooke Castle and on old walls at St

Catherine's Point. All except 39, 49, 69.

C. flavescens On calcareous substrates; widespread and abundant, more so than *C. aurantia*. Corticolous form recorded from elm in Swainston Park 1971 *. All squares.

C. flavovirescens On calcareous rocks; occasional. Brading and Godshill churchyards 1971*; on concrete at Alum Bay and mortar coping on Cowes sea front, BJC 1973; Chale Church, Carisbrooke Castle and St Lawrence Church, FR 1995. All except 39, 59, 69.

C. holocarpa On hard calcareous rocks and concrete; widespread but under-recorded. All except 39, 49.

C. lactea Recorded from chalk pebbles on the cliff edge at Tennyson Down, CP 1982, and Afton Down CP 1984; and from gravestones in Yarmouth churchyard VJG 1982. 38.

[*C. luteoalba* Extinct. This species, classified as nationally vulnerable, was found on large elms but is believed to be extinct. There are old records from near Ventnor (ABl, BM) and from Shanklin in 1874 (JMC, BM).]

C. marina On hard coastal rocks. Most rocks are unsuitable for this species but it may be under-recorded. Tennyson Down OLG 1976. 38.

C. obscurella A single record from elm at Whitefield Woods 1971* (E). 68.

C. saxicola On limestone and mortar; widespread. All except 39, 49, 69.

C. teicholyta On dry, sunny calcareous surfaces; widespread and common. All except 39, 49, 69.

C. ulcerosa A scarce species which was formerly on elms west of Niton 1971* and in Bonchurch churchyard PWJ 1977; only recently recorded from a native black poplar at Flowers Brook, Ventnor NS 1992. 57.

C. variabilis Rare. Brading churchyard chest-tomb FR, KAS 1993. 68.

Candelaria concolor On nutrient-enriched bark; increasingly scarce. On elm in Nunwell Park and Brading churchyard 1971*; Cliff Copse, Shanklin BLS 1982. 58, 68.

Candelariella aurella On limestone, concrete and asbestos-cement; frequent. All except 39, 59, 69.

C. medians On calcareous substrates; widespread. All except 39, 47.

C. reflexa On nutrient-rich bark, formerly on elm. Appuldurcombe Park, elms just west of Niton, Chale Church, Whitefield Woods, all 1971*; Long Copse, Gatcombe FR 1977; Hurst Copse, Wootton PWJ 1991. All except 38, 39, 49, 69.

C. vitellina On man-made substrates and stone; very common. All squares.

C. xanthostigma On old parkland trees; rare. On ash in Nunwell Park and roadside elm north of Brading, both 1971*. 58, 68.

Catapyrenium rufescens Very rare. On a single chest-tomb in Mottistone churchyard, epiphytic on *Collema fuscovirens*, FR 1998 (det. OLG). 48.

C. squamulosum On shallow calcareous soils; occasional. Very locally on Tennyson Down near the cliff edge OLG 1976; cliff top on Afton Down NS 1995; Headon Warren CP 1976; St Lawrence cliff top FR 1986 (still present at these sites); St Catherine's CP 2001. 38, 47, 57.

Catillaria aphana On hard chalk. Only recorded from Afton Down, OLG 1990. 38.

C. chalybeia On calcareous rock; widespread but infrequent or overlooked. All except 39, 59.

C. lenticularis On limestone, especially in churchyards; occasional or overlooked. All except 39, 47, 69.

Catinaria atropurpurea (*Catillaria atropurpurea*) This old woodland species is known only from oak in Northpark Copse FR, CP 1981 and Barton Wood, Osborne CP 1983 (det FR). 48, 59.

Celothelium ischnobelum On hazel in ancient woods; rare. Borthwood Copse 1971*; Cliff Copse, Shanklin CP 1991 (det. FR). 58.

Cetraria aculeata (*Coelocaulon aculeatum*) On acidic peaty or sandy soils; very locally frequent. Headon Warren BJC 1973 and St. Boniface Down FR, CP 1981; St Helen's Duver CP 2001. 38, 57, 68.

C. muricata (*Coelocaulon. muricatum*) Much rarer than *C. aculeata*. Only recorded from heathland on Bonchurch Down NS 1992. 57.

Chaenotheca ferruginea On dry bark of old oak trees; rare, although perhaps overlooked. Records suggest that this species may be a recent colonist. Bouldnor Cliff NS 1990; abundantly fertile at Swanmore near Ryde CP 1994; Osborne Estate CP 1994; Withybed Copse, Westover CP 1994; Walter's Copse, Newtown L&SS, CH 2002. 39, 48, 49, 59.

[*C. furfuracea* (*Coniocybe furfuracea*) Extinct. Only recorded from Shanklin in the 19th century (Borrer, BM).]

Chromatochlamys muscorum On thin calcareous soil; very rare. Tennyson Down BLS 1982, OLG 1992. 38.

Chrysothrix candelaris In dry crevices of bark on oak, ash and occasionally elm and beech; Widespread and fairly frequent. All except 39, 57.

C. chrysophthalma On oak and pine. There are no pre-1989 records for this lichen but it may be an increasing species. Parkhurst Forest FR 1989; Briddlesford Copse FR 1993; Osborne Estate CP 1994. 49, 59.

Cladonia arbuscula Rare on heaths. Headon Warren BLS 1982; Bleak Down 1971*; St. Boniface Down FR 1981. 38, 48, 57.

C. caespiticia Terricolous at Headon Warren BLS 1982; on oak at Borthwood Copse 1971*. 38, 58.

(*C. cariosa* Material collected from Lynn Gravel Pits, Downend 1980, was considered by BJC as likely to be this species but poor material. This site has now been lost.)

C. cervicornis Still present at Headon Warren BJC 1973; Afton Down OLG 1990. 38.

C. chlorophaea s. *lat*. On logs, stumps and trees; common. All except 39, 68, 69.

C. ciliata var. *tenuis* Rare; Headon Warren BJC 1973. 38.

C. coccifera s. *lat*. On peat; very local. Headon Warren CP 1976; Bleak Down 1971*; abundant on Luccombe Down CP 1989; Brading Marshes in Mat-grass (*Nardus*) grassland NS 1991. 38, 58, 68.

C. coniocraea On bark, especially oak and around the bases of trees in woods; common. All squares.

C. crispata var. *cetrariiformis* Headon Warren BJC 1973; Luccombe Down NS 1992; Brading Marshes in Mat-grass (*Nardus*) grassland NS 1991. 38, 57, 68.

C. digitata Headon Warren BLS 1982; Parkhurst Forest FR 1989. 38, 49.

C. fimbriata On old wood and peaty soil; widespread but scarce. All except 39, 48, 69.

C. floerkeana On heathland and rotting wood; local. Headon Warren BJC 1973; Row Down, Brighstone, CP 1984; Luccombe Down CP 1989; Peacock Hill, Bembridge CP 1985; Brading Marshes NS 1991. 38, 48, 57, 68.

C. foliacea Frequent on Tennyson Down, as a component of the calcareous lichen-rich turf, OLG 1976; St Helen's Duver CP 1992. 38, 68.

C. furcata On heaths; locally frequent. All except 39, 47, 57, 69.

C. furcata subsp. *subrangiformis* (*C. subrangiformis*) In short chalk grassland; very local. Tennyson Down OLG 1976; Afton Down 1980; above Scratchell's Bay 1980; and on leached soil edge of chalkpit on south side of Brading Down 1981; top of cliff at St. Lawrence 1984; Brighstone Down 1984 (all CP). 28, 38, 48, 57, 68.

C. glauca Headon Warren CP 1976 (det. JRL). Still present here, 2001. 38.

C. gracilis Rare. Headon Warren BLS 1982; Luccombe Down CP 1992. 38, 57.

C. macilenta On peaty soil and wood; local. Headon Warren CP 1976; Player's Wood, Ryde CP 1976; Appuldurcombe and Borthwood Copse 1971*; Parkhurst Forest FR 1989. 38, 49, 58, 59.

C. ochrochlora Occasional on old stumps. Northpark Copse 1981, Walter's Copse, Newtown 1977 and Gatcombe Withy Bed 1977 (all FR); Borthwood Copse CP 1990. 48, 49, 58.

C. parasitica Local on lignum. Parkhurst Forest CP 1983 (det JRL); Bouldnor NS 1990; Northpark Copse CP 1994. 39, 48, 49.

C. pocillum On chalk; occasional. Tennyson Down OLG 1976; above Scratchell's Bay CP 1981; St. Lawrence cliff CP 1984. 28, 38, 57.

C. polydactyla On peaty soil and decaying wood; very local. Headon Warren BJC 1973; Walter's Copse, Newtown FR 1977; Parkhurst Forest CP 1983. 38, 49.

C. portentosa On peaty soil; locally frequent. Headon Warren BJC 1973; St. Boniface Down FR, CP 1981; Bleak Down 1971*; Brighstone Down reservoir CP 1991; Northwood cemetery, West Cowes CP 1986. 38, 48, 49, 57, 58, 68.

C. pyxidata On walltops and dry soil; frequent. All except 39, 58, 68.

C. ramulosa Headon Warren CP 1976 (det JRL); foreshore at Osborne Bay CP 1983 (det FR); Luccombe & St. Boniface Down CP 1989. 38, 57, 59.

C. rangiformis In base-rich grassland; locally frequent. All except 39, 47, 49, 69.

C. scabriuscula Very rare. Headon Warren BLS 1982. 38.

C. squamosa s. *lat*. On sandy or peaty soils; locally frequent. The dominant species on Headon Warren CP 1976; St. Boniface Down FR, CP 1981; Lynn Gravel Pits, Downend CP 1980; Whale Chine CP 1993. 38, 47, 57, 58. Var. *subsquamosa* is believed to be much less common on the Island. Alum Bay BJC 1973; Parkhurst Forest FR 1989. 38, 49.

C. subcervicornis Brading Marshes amongst Mat-grass (*Nardus*) grassland NS 1991. This is predominantly a species of west and north upland Britain. 68.

C. subulata Rare. Headon Warren CP 1976 (det. OLG); Luccombe Down CP 1992. 38, 57.

C. uncialis subsp. *biuncialis* Very local on peaty soils. Headon Warren BJC 1973, where it is frequent; St. Boniface Down FR, CP 1981. 38, 57.

Clauzadea immersa Frequent on calcareous pebbles on Afton Down CP 1989 and Tennyson Down OLG 1990; Quarr Abbey ruins CP 1987; Brading Church FR & KAS 1993. 38, 59, 68.

C. metzleri A nationally scarce species, recorded from chalk pebbles on West High Down BJC 1973; Tennyson Down OLG 1990; and Afton Down CP 1976 (conf. FR). 38.

C. monticola On chalk, limestone and mortar; rather common. 47, 48, 58, 59, 69.

Cliostomum griffithii Common on dry bark. Also saxicolous on Whitwell Church BJC 1973. All squares.

Collema auriforme On chalky soils and on limestone walls and monuments; frequent. 38, 48, 58, 68.

C. crispum Common on limestone and mortar. All except 39, 49, 69. var.

metzleri A single Island record, AEW 1958.

(*C. cristatum* Known only from 19th century records. On rocks in the Undercliff, Shanklin (BM).)

(*C. flaccidum* Recorded from St. John's, Ryde (BM) and from rocks in the Landslip (ABl) in the 19th century).

[*C. furfuraceum* Extinct; formerly very rare. On elm at Appuldurcombe Park 1971*; on a single ash at the edge of Mudless Copse FR. This tree was felled in 1981 and the lichen continued to survive for a number of years on the fallen trunk. 48, 58.]

C. fuscovirens Very rare. On a single chest-tomb in Mottistone churchyard FR 1998. 48.

C. limosum Very rare, or overlooked. On chalk pebbles on Afton Down FR 1986 (det. PWJ). 38.

(*C. nigrescens* Recorded from early in 20th century on mossy trunks at Ryde, Carisbrooke Castle, Shanklin (HML) and Combley Great Wood, Havenstreet (ABl).)

C. tenax On shallow calcareous soils and on mortar; frequent. All except 49, 69. var. *ceranoides* Local on chalk. 38, 59, 68.

Cresponea premnea (*Lecanactis premnea*) Occasional on dry bark of ancient oaks. Quarr (see also *Lecanographa lyncea*); Nunwell Park 1971*; Binstead and Lakeside, Wootton 1991; Appley Park, Osborne and Bembridge windmill 1994; on plane tree at Swainston 1995 (all CP); America Wood and Newtown L&SS, CH 2002. 48, 49, 58, 59, 68, 69.

Cryptolechia carneolutea On old trees with nutrient-rich bark; a rare and decreasing species nationally with an extreme southern distribution. Recorded on elms at Brook and Swainston Park 1971* but now lost. Survives on a few other trees along the Undercliff. On ash in woods at St. Lawrence 1971*, but tree felled in 1986; still on the root of an old ash at Bonchurch Landslip 1971*, CP 2002; on an old ash at Woolverton House, St Lawrence and a field maple at the foot of the cliff CP 1994. 38, 48, 57.

Cyphelium inquinans On acid bark of old oaks; very rare. Standen Heath and Briddlesford Copse CP 1992. Also very occasionally on old posts. 58, 59.

C. sessile A rare parasite of *Pertusaria* spp. Appuldurcombe Park 1971*. 58.

Cyrtidula quercus (*Mycoporum quercus*) Occasional on oak twigs. Bonchurch Landslip 1971*; Parkhurst Forest FR 1988; Briddlesford Copse FR, KAS 1993; Hurst Copse, Wootton FR 1991. 49, 57, 59.

[*Degelia plumbea* Extinct. An old woodland species. Recorded by Rev. Bloxam from Appuldurcombe in the 19th century (BM).]

Dimerella lutea A very local ancient woodland species. Bonchurch Landslip, on oak 1971*; wood west of Appuldurcombe Park OLG 1976; Cliff Copse, Shanklin on ash CP 1977, but could not be refound here 2000; Mudless Copse on ash FR 1976; Northpark Copse on oak CP 1980 and luxuriantly on ash, 2000. 48, 57, 58.

D. pineti On shaded bark, generally on oak or ash; widespread and fairly frequent. All except 38, 39, 47, 69.

Diploicia canescens On nutrient-rich bark and on limestone; common, frequently fertile. All squares.

Diploschistes muscorum Recorded only from open chalk grassland on Tennyson Down OLG 1976 and old trackway on Afton Down CP 1981; and on mossy soil at Redcliff CP 1989. 38, 68.

D. scruposus On rocks and walls; almost certainly under-recorded. On stones at Carisbrooke Castle BLS 1982 and a greensand outcrop at Limerstone CP 1991. 48.

Diplotomma alboatrum On dry, vertical limestone; frequent. Formerly on elm. All except 39.

D. chlorophaeum A maritime, western species recorded from Blackgang BJC 1973. 47.

Dirina massiliensis f. *sorediata* A mauve crust, frequent on north and east church walls. All except 39, 69.

[*Endocarpon pusillum* Extinct. Known only from material collected in 1886 at Alum Bay (HBH, BM). It may survive, as yet undetected, on Tertiary soft cliffs.]

Enterographa crassa On shaded bark of old trees, most commonly on oak and ash, and formerly on elm; frequent. All squares.

[*E. elaborata* Extinct. This critically endangered species was recorded in the 19th century from Quarr Abbey wood (ABl, BM). Otherwise, only known from the New Forest in this country.]

Evernia prunastri On well-illuminated bark; common and widespread. Fertile on beech at St. Martin's Down CP 1984. Occasionally growing on greensand walls and soil: St. Catherine's Oratory BJC 1973, St George's Down CP 1993 and Redcliff CP 1997. All squares.

Fellhaneropsis vezdae (*Bacidia vezdae*) On shaded bark; rare or overlooked. Whitefield Woods and on oak in Borthwood Copse both 1971*; on ash in Long Copse, Gatcombe FR 1977. 48, 58, 68.

Flavoparmelia caperata (*Parmelia caperata*) On well-illuminated tree trunks; common and widespread. Sometimes on church walls and on roofs. Fertile in America Wood 2002. All squares.

F. soredians (*Parmelia soredians*) On dry bark, fences and monuments; scarce. Swainston Park on oak, Nunwell Park on oak, Appuldurcombe Park, Wroxall Church, near Bullen outside Ryde (all 1971*); on beech in Godshill FR 1976; Freshwater Parish Church FR 1995; Lakeside, Wootton on lignum PWJ 1991. All except 39, 49.

Fulgensia fulgens On parched, base-enriched soil; rare. Material collected at Freshwater Bay by W. Borrer and R.D. Turner was used for the hand-coloured illustrations in Smith and Sowerby's English Botany (1806). This near threatened species was refound on Tennyson Down OLG 1976, where it is dominant over a limited area of lichen-rich grazed open turf (Gilbert, 1978). It also occurs scantily by the cliff edge right along this stretch of chalk westwards to Scratchell's Bay. Growing on a single sandy bluff on the cliff north of Alum Bay Chine CP 1998. 28, 38.

Fuscidea lightfootii Very rare. Fertile on willow in a carr southwest of Alverstone bridge. 1971*; on oak twig, America Wood and Newtown L&SS, CH 2002. 49, 58.

Graphina anguina On smooth bark; rare. St. Catherine's, Chale Church and Bonchurch Landslip, 1971*; on beech in Borthwood Copse CP 1998. 47, 57, 58.

Graphis elegans On moderately shaded smooth bark of hazel, young oak and hornbeam; frequent and widespread. All except 47, 57.

G. scripta In similar habitats to the previous species; frequent, indeed slightly commoner. Also on beech and ash. All except 47, 68, 69.

Gyalecta flotowii On ash; very rare. Long Copse, Gatcombe and Shanklin Great Wood, both FR 1977. 48, 58.

G. truncigena In rain channels of old trees; once frequent on elm, but now scarce. Brook, Appuldurcombe Park, Bonchurch Landslip and between Niton and St. Lawrence, all on elm (all 1971*). Apes Down on ash FR 1976; Mudless Copse on ash FR 1977; Shanklin Great Wood FR 1977; Briddlesford Copse on oak FR, CP 1980; St Lawrence on field maple CP 1994. 38, 48, 57, 58, 59.

Haematomma ochroleucum var. *porphyrium* On vertical limestone and nutrient-enriched bark; rare. Swainston Park and gravestone in Wroxall churchyard, both 1971*; St.

Catherine's oratory BJC 1973; Shalfleet Church KAS 1998; Newtown Church L&SS, CH 2002. 47, 48, 49, 57.

Hymenelia prevostii On hard calcareous pebbles, Afton Down OLG 1990. 38.

Hyperphyscia adglutinata On nutrient-enriched bark, especially ash and elm; occasional but widespread. On rock at St. Catherine's BJC 1973. All except 38, 39, 49.

Hypocenomyce scalaris On old wood; occasional. Brook Hill House CP 1980; King's Quay CP 1980; coastal rocks at St. Catherine's Point BLS 1982; peat bank at Headon Warren BLS 1982. 38, 47, 57, 59.

Hypogymnia physodes On bark, twigs and wood; common and widespread. Also on peat and heather on Headon Warren, St. Boniface Down and St. George's Down. All squares.

H. tubulosa On bark; widespread but much less frequent than the previous species. 38, 47, 48, 49, 58, 59.

Hypotrachyna revoluta (*Parmelia revoluta*) On bark, particularly on trees in woods; frequent. Fertile in willow carr at Godshill CP 1993. All except 38, 39, 69.

Lecanactis abietina On dry bark; extremely local. Lady Wood, Swainston 1980; Cliff Copse, Shanklin 1977, fertile here on a single ash 1992; Briddlesford Copse 1980 (all CP); Parkhurst Forest FR 1988. 48, 49, 58, 59.

L. subabietina On dry, shaded bark, generally on oak and ivy in old woods; increasing. 38, 48, 57, 58, 59, 68.

(*L. sp.* Recorded from sheltered, maritime rocks on I.O.W. See James & Coppins (1979).)

(*Lecania chlorotiza* A single 19th century record from elms near Shanklin (HBH).)

L. cyrtella On smooth nutrient-enriched bark; infrequent but probably under-recorded. Also reported from thyme stalks on Tennyson Down BLS 1982; and St Lawrence cliff CP 1984 (conf. OLG). 38, 48, 57, 58.

L. erysibe Frequent on limestone and mortar. The sorediate form is recorded from chalk pebbles on Afton Down OLG 1990. All except 39.

L. naegelii (*Bacidia naegelii*) On nutrient-enriched bark; local, perhaps overlooked. On elders at St. Catherine's BJC 1971 (E); on elm at Wilmingham CP 1979 (det BJC); on elders between Niton and St Lawrence CP 1995; old field maple on bank of sunken lane, Carisbrooke NS 1997. 38, 47, 48, 57, 68.

L. turicensis Rare on old stonework. Carisbrooke Castle and St Edmund's Church, Wootton FR 1995; rare on flints on Tennyson Down OLG 1990. 38, 48, 59.

Lecanographa lyncea (*Lecanactis lyncea*) Very rare on ancient oaks. On a single maiden oak in a field at the foot of Quarr Hill with *Cresponea premnea* 1971*, still present 1987; by Quarr Abbey ruins CP 1987; Nunwell Park CP 1991; old oak in America Wood and at Newtown L&SS, CH 2002. 49, 58, 59.

Lecanora actophila Very rare. Recorded only from Tennyson Down, on chalk OLG 1976; and Afton Down CP 1984. 38.

L. albella (*L. pallida*) On smooth bark, usually on young oak; uncommon. Borthwood Copse 1971*; Lock's Copse, Porchfield CP 1979; Briddlesford Copse CP 1980; Firestone Copse FR 1981; Parkhurst Forest CP 1988. 49, 58, 59, 69.

L. albescens On calcareous stonework. Probably very common but under-recorded. All except 39, 49, 58, 69.

L. argentata On tree bark at Newtown CH 2002. 49

L. campestris On limestone, sandstone and brick; common. All except 39.

L. carpinea On smooth bark, usually on ash, but also recorded from hornbeam; widespread but local. 38, 47, 49, 58, 68.

L. chlarotera On smooth bark, recorded from oak, ash, hazel and elm; common and widespread. All squares.

L. conferta On hard limestone; very rare. Carisbrooke Castle FR 1995. 48.

L. confusa Frequent on smooth bark. All except 59, 69.

L. conizaeoides On bark and wood and on stone and brick; until recently, abundant, but now greatly declined. All squares.

L. crenulata On chalk and limestone, particularly in old churchyards; local. All except 39, 58, 69.

L. dispersa On rocks and walls and rarely on nutrient-enriched bark; very common. All squares.

L. expallens Common on smooth bark and wood. Also saxicolous at Brading Church and Calbourne Church BJC 1973. All squares.

L. fugiens Very rare. Godshill Church PWJ 1991. One of several maritime lichens occurring in this churchyard. 58.

L. gangaleoides On stone; rare or overlooked. Godshill and Chale Churches, both 1971*; Freshwater All Saints Church BLS 1982. 38, 47, 58.

(*L. helicopsis* Known from a single record from coastal rocks at Niton, made in the 19th century (JHB).)

L. muralis On nutrient-enriched man-made substrates; infrequent and uncommon but increasing. 48, 58, 59.

L. orosthea On acid rock; rare. Shorwell, Wroxall and Godshill Churches, all 1971*; west wall of Carisbrooke Castle BJC 1973. 48, 57, 58.

L. polytropa On acid rocks; infrequent. Cowes BJC 1973; Freshwater All Saints' Church BLS 1982; Redcliff CP 1989; Brading churchyard FR, KAS 1993. 38, 49, 68.

(*L. pulicaris* Old records from trees near Ryde, Shanklin and Brading (HBH, BM).)

(*L. rupicola* Known only from an unlocalised record in the 19th century (ABl).)

L. sulphurea On acid stone, flint and old brick; local. 47, 48, 57, 58.

L. symmicta Occasional on lignum. On wooden stile and on heather at Headon Warren and on gate at Bembridge windmill, all BJC 1973; Lakeside, Wootton PWJ 1991; Forelands CP 1991. 38, 59, 68.

L. varia Rare on old fence posts. Tennyson Down OLG 1976; Lakeside, Wootton PWJ 1991. 38, 59.

Lecidea fuscoatra On acid stone and old brick; scarce. Wroxall Church 1971*; Arreton Church BJC 1973. 57, 58.

L. lichenicola Rare on chalk pebbles. Tennyson and Afton Downs OLG 1990. 38.

(*L. turgidula* A single 19th century record from Shanklin (BM).)

Lecidella elaeochroma On smooth bark; widespread and common. All squares. f. *soralifera* is occasional. All except 59, 68, 69.

L. scabra On hard sandstone and old brick; probably frequent but overlooked. Fertile in Brighstone churchyard CP 1994. All except 39.

L. stigmatea On limestone, concrete and mortar; frequent. On rocks at Totland Bay FR 1980. All except 39, 59, 69.

Lempholemma chalazanum On bare calcareous soil and mortar; very rare. Recorded from the *Fulgensia* community on Tennyson Down OLG 1976; Godshill Church PWJ 1991. 38, 58.

Lepraria incana s. lat. On shaded bark, wood and old walls; abundant. Unfortunately, the closely related taxa have not been differentiated. All squares.

Leproloma vouauxii On limestone; under-recorded but probably not uncommon. On mortar of old walls in Ryde CP 1997 (conf. JRL).

Leptogium biatorinum Rarely recorded. Bonchurch New Church on natural rock outcrop CP 1994 (det. OLG); Bouldnor Cliffs VJG & OLG 2002. 39, 57. Probably this species, recorded as *L. cretaceum*, in the 19th century on chalk and flint (BM).

L. gelatinosum, On sheltered mossy walls and shallow calcareous soil; rare, or overlooked. Afton Down OLG 1990; St Lawrence CP 1994; Bonchurch Landslip CP 1995. 38, 57.

L. lichenoides On old mossy bark; very rare. Bonchurch Landslip 1971*; luxuriant on a single ash in Northpark Copse FR, CP 1981. 48, 57.

L. plicatile Recorded from the walls of Carisbrooke Castle FR 1995. 48.

L. schraderi On thin calcareous soils and on old walls; not infrequent. All except 39, 49, 69.

L. teretiusculum Formerly on elms, where it was believed to have been scarce. Now rare on nutrient-enriched bark but also on calcareous soil and old mortar. Bonchurch New Church 1993 and old wall at north end of Landslip 2001, CP; Godshill churchyard table tombs PWJ 1991; Gatcombe Withy Bed on old ash pollard CP 1995. 38, 57, 58.

L. turgidum Very rare or overlooked. On mortar of old wall at Carisbrooke Castle BJC 1973. 48.

Lichina pygmaea On intertidal rocks; currently known only from a single rock in Freshwater Bay, the easternmost site along the south coast of Britain, CP 1990. Recorded in the 19th century as frequent on coastal rocks at Ventnor (ABl, BM) and also at St. Lawrence shore and Freshwater Bay (BM). 38.

[*Lobaria amplissima* Extinct. An old woodland species, known only from a single, very fertile, unlocalised collection, I.O.W. (ABl, BM).]

L. pulmonaria Rare in ancient woodland. Reported in the 19th century as being common at Appuldurcombe, Shanklin Great Wood and a plantation beneath Cook's Castle (ABl, BM); also from Quarr (BM). Frequently fertile. Today, confined to a declining handful of trees. On four ash at the edge of Mudless Copse CP 1975, of which only 1 tree remained by 1990; on a single ash at the edge of Cliff Copse, Shanklin CP 1976 until 1988; on a single old oak in Briddlesford Copse HM 1992, still present. The best site by far is Northpark Copse where it was found growing luxuriantly on about 13 ash in 1980 (CP). In 2000, it was still luxuriant on about 7 trees and was fertile on one tree. 48, 58, 59.

[*L. scrobiculata* Extinct. An old woodland species, formerly recorded from Quarr Wood (BM), Shanklin Great Wood and Appuldurcombe (ABl, BM). Also from a wood on Apes Down (FM).]

[*L. virens* Extinct. An old woodland species only known from luxuriant 19th century specimens on oak and ash at Shanklin (HBH, BM) and Appuldurcombe (BM). Last collected in 1892.]

Loxospora elatina (*Haematomma elatinum*) An ancient woodland indicator species, recorded from Parkhurst Forest, on oak FR 1989. 49.

[*Megalaria grossa* Extinct. Material of this rare old woodland species was collected in 1886 from Bembridge (HBH, BM).]

Megaspora verrucosa On base-rich soil. This nationally scarce species has been recorded from the *Fulgensia* community on Tennyson Down OLG 1990. This is currently (2002) the second British station on chalk. 38.

Melanelia elegantula (*Parmelia elegantula*) A corticolous species, first collected in this country in 1965. Recorded from Appuldurcombe Park 1971*. 58.

M. exasperata (*Parmelia exasperata*) On branches and twigs; rare and in small quantity, but may be overlooked. Above Bonchurch Landslip, Alum Bay and Appuldurcombe Park, all 1971*; Tennyson Down OLG 1976; St Lawrence Shute CP 1990; Hurst Copse, Wootton PWJ 1991; America Wood CH, L&SS 2002. 38, 57, 58, 59.

M. fuliginosa (*Parmelia glabratula* subsp. *fuliginosa*) Frequent on acid stone. Subsp. **glabratula** (*P. glabratula* subsp. *glabratula*) is common and widespread on bark and trees. All squares.

M. laciniatula (*Parmelia laciniatula*) Recorded only from Appuldurcombe Park 1971*. 58.

M. subaurifera (*Parmelia subaurifera*) On bark and twigs, usually on horizontal branches of elder or willow; occasional. All except 59, 68, 69.

Melaspilea ochrothalamia Only recorded from old oaks in Borthwood Copse 1971*. 58.

[*Micarea bauschiana* Extinct. An old record from Shanklin (HBH, BM).]

M. denigrata On worked timber; overlooked. Lakeside, Wootton PWJ 1991. 59.

M. erratica On flints; occasional or overlooked. Bleak Down 1971*; Headon Warren BLS 1982; Tennyson and Afton Downs OLG 1990. 38, 48.

M. melaena Very rare on wood. Headon Warren BJC 1973. 38.

M. nitschkeana Probably overlooked. Found growing on high impact polystyrene container washed up on Shanklin beach BJC 1973. 58.

M. peliocarpa Very rare. Headon Warren BJC 1973; Briddlesford Copse on oak CP 1980 (det BJC); Parkhurst Forest on oak FR 1989. 38, 49, 59.

M. prasina s. lat. On acid bark and lignum, often in old woods; occasional. 38, 49, 57, 58, 59.

[*Moelleropsis nebulosa* Extinct. A species formerly found on mud-capped walls. The Landslip (ABl, BM).]

Mycoblastus caesius (*Haematomma caesium*) Recorded on oak in Parkhurst Forest, FR 1989. 49.

M. fucatus (*M. sterilis*) On acid bark; probably overlooked. On oak in Parkhurst Forest FR 1989. 49.

Myxobilimbia lobulata (*Toninia lobulata*) On parched, basic soil. This rare species was recorded from near the cliff edge on Tennyson Down OLG 1976, but not seen subsequently. 38.

M. sabuletorum (*Bacidia sabuletorum*) On mosses, bark and thin calcareous soil; frequent. All except 39, 49, 69.

Neofuscelia verruculifera (*Parmelia verruculifera*) Rare on acid stones. Godshill Church 1971*; Calbourne and Bembridge Churches BJC 1973; Freshwater All Saints' churchyard and St Edmund's churchyard, Wootton FR 1995. 38, 48, 58, 59, 68.

Normandina pulchella On mossy bark of oak, ash, elm, elder and Norway maple; widespread although local. All except 38, 39, 69.

Ochrolechia androgyna Rare. On elm SW of Seaview 1971*; saxicolous at Whitwell Church BJC 1973; Parkhurst Forest on oak FR 1989; old oak at Newtown L&SS, CH 2002. 49, 57, 69.

O. parella On old walls and brick; widespread and locally frequent. On natural rock outcrops at St. Catherine's Point and Cliff Copse, Shanklin. Also on sycamore in Brighstone churchyard 1983 and on ash at Carisbrooke Castle 1992 (CP). All except 39, 49, 69.

O. subviridis On large isolated trees of oak, ash, field maple and formerly elm; frequent but sterile. All except 47, 57, 69.

O. turneri s. lat. On bark and wood; rare. Appuldurcombe Park 1971* and Mudless Copse on a single old ash FR 1977, but lost from both these sites as a result of tree felling; America Wood CH 2002. On a wooden church gate at Shorwell 1971* and saxicolous at Carisbrooke Castle FR 1995. 48, 58.

Opegrapha atra On smooth bark; widespread and fairly frequent. All except 39, 59, 69.

O. calcarea (inc. *O. chevallieri*, *O. conferta* and *O. saxatilis*) On dry sheltered limestone church walls and natural rock outcrops; locally abundant. Most of the material from church walls is referable to *O. chevallieri*. This form also occurs on rocks at St. Catherine's Point together with *O. conferta*. All except 39, 49, 59, 68.

O. corticola On old trees; rare. Bonchurch Landslip, Nunwell Park on oak, Brook on elm, Appuldurcombe on elm, all 1971*. 38, 57, 58.

O. gyrocarpa On vertical acid rock; rare. Freshwater All Saints Church and Whitwell Church FR 1995. 38, 57.

O. herbarum On dry shaded bark; rare. Whitefield Woods 1971*; Northpark Copse FR 1981; Parkhurst Forest FR 1988; Newtown CH 2002. 48, 49, 68.

O. mougeotii On sheltered limestone; rare. Godshill Church 1971*; rocks west of St.Catherine's Point BLS 1982; Gatcliff rock outcrop CP 1990; Quarr Abbey ruins CP 1987. 47, 57, 58, 59.

O. multipuncta Rare. On hornbeam in Hurst Copse, Wootton FR, PWJ 1991. 59.

O. ochrocheila On rough bark; scarce. On oak in Parkhurst Forest FR 1988; on sycamore at St Lawrence NS 1992; on ash in Bonchurch Landslip CP 1995. 49, 57.

O. parasitica A rare lichen parasite. Carisbrooke Castle FR 1995. 48.

O. prosodea On parkland oaks; a nationally scarce species, with an extreme southern distribution in this country. Whitefield Woods 1971*; Quarr CP 1987; Nunwell Park on oak CP 1986, formerly also on elm. 58, 59, 68.

O. sorediifera On oak and ash in old woods. Bonchurch Landslip, America Wood, Borthwood Copse, Whitefield Woods (all 1971*); Northpark Copse FR 1981; Parkhurst Forest FR 1989. 48, 49, 57, 58, 68.

O. varia On dry bark of old trees, formerly often on elm; fairly frequent. 38, 48, 57, 58, 68.

O. vermicellifera On large, shaded trees; occasional. Nunwell Park on elm 1971*; Long Copse, Gatcombe, on ash FR 1977; Player's Woods, Ryde on ash CP 1982. 48, 58, 59.

O. vulgata Common on dry bark. All squares.

Pachyphiale carneola On oak; a very rare ancient woodland species. Briddlesford Copse FR, CP 1980; Hurst Copse, Wootton CP 1991. 59.

[*Pannaria rubiginosa* Extinct. An ancient woodland species, known from luxuriant material collected at Appuldurcombe (ABl, BM).]

Parmelia saxatilis Widespread but local on bark; also found on acid stone. On rocks at St. Catherine's Point BLS 1982; terricolous at Headon Warren, where it is particularly frequent CP 1976, St George's Down CP 1993 and Redcliff 1997. Fertile on ash at Newchurch churchyard 1994 and Bembridge Down 1991(all CP). All except 39, 69.

P. sulcata On bark and wood; very common. Less frequent on acid stone and old brick. Fertile at Calbourne Church BJC 1973, on sycamore at Old Park, St. Lawrence 1984 and in Dickson's Copse, Dodnor CP 1986. All squares.

Parmelina pastillifera (*Parmelia pastillifera*) Rare on nutrient-enriched substrates. An old record from Quarr refers to this species (BM). Swainston Park on beech and Appuldurcombe Park, both 1971*; Carisbrooke Castle FR 1995. 48, 58.

[*P. quercina* (*Parmelia quercina*) Extinct. An old herbarium record (BLS data base). 59.]

P. tiliacea (*Parmelia tiliacea*) Appuldurcombe Park 1971*.

Parmotrema chinense (*Parmelia perlata*) On well-illuminated neutral to acid bark; locally frequent. All squares.

P. reticulatum (*Parmelia reticulata*) On oak and ash in old woods and parkland; widespread but local. A southern species. All except 38, 39.

Peltigera canina This nationally scarce species was recorded from 38 1971*, but most records refer to *P. membranacea* (see below). St. Helen's Duver CP 1984. 38, 68.

[*P. collina* Extinct. An old woodland species recorded from Shanklin (TS) and Apes Down (HHK).]

P. didactyla On recently disturbed ground; there are few records. Old records from Shanklin Down, Newbarn Down and near Ryde. One modern record from Bleak Down 1971*. 58.

[*P. horizontalis* Extinct. An old woodland species, known only from old records. Apes Down (JFR); Westover Down (HML); Newbarn Down (HML, BON).]

P. hymenina (*P. lactucifolia*) On mossy ground and tree banks; occasional. Shanklin Chine BJC 1973; Lynn Gravel Pit, Downend 1980; Whitecroft 1987; St George's Down and by the Medina at Scotland Farm, Godshill 1993 (all CP). 48, 57, 58.

P. membranacea On mossy ground and tree trunks; occasional. Bonchurch Landslip 1971*; Northpark Copse FR, CP 1981;

grassland on Brading Marshes CP 1984; Gatcombe Withy Bed CP 1995. 48, 57, 58, 68, 69.

P. neckeri On mosses on soil; rare or overlooked. Bembridge old railway track L&SS 2002. 68.

P. praetextata On stumps; local. Bonchurch Landslip 1971*; Walter's Copse, Newtown FR 1977; Gatcombe Withy Bed 1990; Briddlesford Copse 1992; Corf Heath 1993 (all CP). 49, 57, 58, 59.

P. rufescens On basic soils; local. Headon Warren BJC 1973; Cliff Copse, Shanklin CP 1977. 38, 58.

Pertusaria albescens On old oak or ash; local. Northpark Copse, Mudless Copse, Apes Down, Briddlesford Copse, etc. On stone in Calbourne churchyard BJC 1973. 47, 48, 58, 59. Var. *corallina* is more frequent, on nutrient-rich bark and old walls. All except 38, 57.

P. amara On bark, especially on trees in woods; sometimes on acid stone in churchyards, old buildings and greensand outcrops. Common and widespread. All squares.

P. coccodes On bark, generally on oak or ash; widespread but local. Occasionally on stone, e.g. Brighstone churchyard 1971*; top of sandstone table-tomb in Brading churchyard and west wall of Carisbrooke Castle, both BJC 1973; Godshill churchyard OLG 1990; Bonchurch New Church CP 1993. All except 38, 39, 47, 57.

P. flavida On wayside trees; very rare. On a sycamore and also a chest tomb in Brighstone churchyard, 1971*; at Carisbrooke Castle FR 1995; on poplar at Hurst Copse, Wootton PJW 1991. 48, 59.

P. hemisphaerica On rough, well-illuminated bark; scarce. Borthwood Copse, Nunwell Park, Quarr (all 1971*); Cliff Copse, Shanklin CP 1977; Briddlesford Copse, Lady Wood, Swainston, both FR 1980; Parkhurst Forest FR 1988. Saxicolous on Whitwell churchyard wall, the first saxicolous record from a churchyard in England, FR 1995. All except 38, 39, 47, 69.

P. hymenea On bark often in rather shaded sites; common. All except 38, 39.

P. leioplaca On smooth bark of ash, hazel and oak; widespread and frequent. All except 38, 39, 47, 69.

P. multipuncta On smooth bark; rather scarce. Parkhurst Forest and Mudless Copse FR 1976; Northpark Copse FR, CP 1980; Briddlesford Copse CP 1980; Burnt Wood CP 1986. 48, 49, 59.

P. pertusa On bark; common and widespread. Occasional on stone, e.g. Christchurch, Sandown CP 1983; greensand ledges at Cliff Copse, Shanklin BLS 1982 and Gatcliff CP 1989; St Edmund's Church, Wootton FR 1995. All squares.

P. pseudocorallina Rare on chert. Landslip, Bonchurch 1971*; St. Catherine's Point BLS 1982; Gatcliff CP 1989. 47, 57, 58.

[*P. velata* Extinct. An old woodland species. This threatened species has its national headquarters in the New Forest. Nineteenth century records from Quarr Wood (BM) and Shanklin (TS).]

Petractis clausa On calcareous rocks; scarce. Rocky outcrops at Totland Bay FR 1980; Niton Church wall VJG 1982; Tennyson Down FR 1977; Afton Down OLG 1990. 38, 57

Phaeographis dendritica On smooth bark in old woodlands; occasional. Parkhurst Forest CP 1976; Northpark Copse FR, CP 1980; Briddlesford Copse FR, CP 1982; Barton Wood, Osborne CP 1983; Bouldnor NS 1990; Swanpond Copse, Ryde CP 1990; Quarr Wood CP 1991; America Wood L&SS, CH 2002. 39, 48, 49, 57, 58, 59.

P. inusta An old woodland species; very rare. Parkhurst Forest, on oak FR 1988. 49.

P. lyellii An old woodland species; rare. Recorded from Parkhurst Forest on sweet chestnut FR 1988. Extinct east of the Isle of Wight and the New Forest. 49.

[*P. smithii* Presumed extinct. On smooth-barked woodland trees. America Wood 1857 (BM and K). An old record from Bonchurch (HPR) probably refers to this species.]

Phaeophyscia orbicularis On rocks and walls and on nutrient-enriched bark of elder, ash and formerly elm. All squares.

[*Phlyctis agelaea* Extinct. On trees; formerly recorded from Carisbrooke (BM). A species which is rapidly declining in England.]

P. argena On wayside trees, particularly oak and ash; frequent. On churchwall at Brading BJC 1973 and Freshwater All Saints Church FR 1995. All except 38.

Physcia adscendens On nutrient-enriched bark, wood and horizontal stone; frequent, occasionally fertile. All squares.

P. aipolia On nutrient-enriched bark of ash, elm and elder. All except 49, 69.

P. caesia On nutrient-enriched horizontal stone, particularly on limestone gravestones, also occasionally on natural calcareous outcrops. 38, 48, 57, 58, 68.

[*P. clementei* Extinct. On nutrient-rich bark. A rare and decreasing extreme southern species in this country. On elms between Brading and Alverstone 1971* Recorded in the last century from Ryde (BM). 58.]

P. leptalea (*P. semipinnata*) On exposed bark; very rare. St. Catherine's and Bembridge Down, both BJC 1973; Flowers Brook, Ventnor NS 1992. 47, 57, 68.

P. stellaris On branches and twigs. Recorded from sycamore twigs at Alum Bay BJC 1973. 38.

P. tenella On nutrient-enriched bark; common, although less frequent than *P. adscendens*. Sometimes fertile. All squares.

P. tribacia On nutrient-rich bark; rare, with no recent records. Abundant on elms SW of Brading and SW of Seaview; on elm and ash at Nunwell Park (all 1971*). Now lost from elms. 58, 69.

[*P. tribacioides* Extinct. On nutrient-rich bark. An endangered species with a pronounced south-western distribution in this country. Only recorded from elms: locally frequent between Brading and Alverstone, and north of Brading (both 1971*). 58, 68.]

Physconia distorta On nutrient-enriched bark; frequent. All except 38.

P. grisea On nutrient-enriched bark; frequent, until recently common on elm. Also on limestone. All except 38, 69.

P. perisidiosa On basic bark; very rare. Swainston Park 1971*, still present 1999. 48.

Placynthiella icmalea Various substrates, especially lignum; under recorded, but probably frequent. Not recognised before 1984. 57, 59, 68.

P. uliginosa On peaty soils; widespread and locally common, but earlier records are not separated from *P. icmalea*. All squares.

Placynthium nigrum Particularly found on calcareous gravestones; frequent, although not in all churchyards. All except 39, 49, 69.

[*P. tantaleum* Extinct. On basic rocks; one old record of this rare species, Brighstone Down 1907 (HML, BON).]

Platismatia glauca, Usually on horizontal branches of large trees; very local. Abundant on the roof tiles of a coach-house at Gatcombe House CP 1982. On peat at Headon Warren CP 1976. All except 47, 68, 69.

Polyblastia albida On hard, calcareous pebbles; very rare. Tennyson Down OLG 1990. 38.

Polysporina simplex On rough granite headstones in churchyards; occasional. Brading BJC 1973; Bonchurch New Church CP 1993; Brighstone and Chale Churches CP 1994; Freshwater All Saints' Church and St Edmund's, Wootton FR 1995. All except 39, 49, 58, 69.

Porina aenea On smooth bark; very local or overlooked. Appuldurcombe Park on elm 1971*; Shanklin Chine on elm BJC 1973; Long Copse, Gatcombe on field maple and Mudless Copse on hazel, both FR 1977; Wilmingham on elm CP 1979; St Lawrence Woods and Player's Woods, Ryde on ash trees, both CP 1983 (det BJC). 38, 48, 57, 58, 59.

P. leptalea On smooth bark; very rare. America Wood on hazel and Borthwood Copse on oak 1971*; Hurst Copse, Wootton on hornbeam PWJ 1991. 58, 59.

P. linearis On chalk and limestone; rare or overlooked. Bonchurch Landslip 1971*; Steephill Cove CP 1980. 57.

Porpidia cinereoatra On acid rock; very rare. Newchurch churchyard FR 1976 (hb. FR). 58.

P. crustulata. Sterile on a rock at St. Catherine's Point 1982 (BLS). Probably overlooked. Recorded from flints on Apes Down early 20th century (HHK). 47.

P. macrocarpa On acid stones; rare. On pebbles on Headon Warren 1982 (BLS) and Luccombe Down CP 1989. 38, 57.

P. platycarpoides Natural greensand outcrop, Gatcliff CP 1990 (det. OLG). 58.

P. tuberculosa Generally on headstones in churchyards; fairly widespread. All except 39, 59.

[*Protoblastenia calva* Extinct. Known only from a 19th century unlocalised record 'on chalk hills'. However, the specimen is on Carboniferous limestone (ABl, BM).]

P. rupestris On chalk and limestone; frequent, occurs on many of our Medieval churches. All except 39, 68.

[*Protopannaria pezizoides* (*Pannaria pezizoides*) Extinct. Amongst moss and rocks at Bonchurch (ABl, BM).]

[*Pseudocyphellaria aurata* Extinct. On trees. This splendid lichen was recorded from the tops of trees in Shanklin Great Wood by Rev. Bloxam in 1857 (BM). J.M. Crombie collected fine material from near Ryde (BM). There are no other records. Critically endangered in Britain today.]

Psilolechia lucida On vertical acid stone or brick, also on natural rock at Shanklin Chine. 47, 48, 57, 58, 68.

Punctelia borreri (*Parmelia borreri*) On well-illuminated, moderately nutrient-enriched bark; widespread but scarce. A species with a southern distribution in this country. 47, 48, 57, 58, 68.

P. reddenda (*Parmelia reddenda*) An old woodland species; rare. Borthwood Copse and Appuldurcombe Park 1971*; Parkhurst Forest on oak and beech FR 1989; Stenbury on oak NS 1992; Beech Copse, Godshill on fallen beech CP 1994. 49, 57, 58.

P. subrudecta s. lat. (*Parmelia subrudecta*) On bark; common and widespread. Saxicolous at Christchurch, Sandown CP 1983; terricolous on Afton Down OLG 1990. All squares.

Pycnothelia papillaria On heathland; very rare. Bonchurch Down NS 1992. 57.

Pyrenocollema halodytes On intertidal calcareous rocks and acorn barnacle shells. Under-recorded. 28, 47, 57, 68.

P. subarenisedum Occasional on base-rich cliffs at Bouldnor VJG & OLG 2002. 39.

Pyrenula chlorospila On smooth bark in woods; locally frequent. All except 38, 49, 59, 69.

P. macrospora With the last species and more frequent. Particularly abundant on ash between St. Lawrence and Niton CP 1983. All except 49, 59, 69.

Pyrrhospora quernea On smooth bark, generally on oak or ash; common. Material collected from St. Blasius Church wall, Shanklin, resembled this species but had a coloured hypothecium BLS 1982. All squares.

Ramalina calicaris On bark and twigs; local. Headon Warren; Bembridge Down; St. Catherine's (where it is still locally frequent) (all BJC 1973). 38, 47, 68.

R. canariensis On wayside trees and sheltered church walls; local. All except 57, 59, 69.

[*R. cuspidata* There is an old record from St. Lawrence 1910 (H. F. Parsons, BON) where it may still occur.]

R. farinacea On bark and twigs; common. All squares.

R. fastigiata On bark and twigs; fairly common and widespread. Rare on church walls at Kingston and Christ Church, Sandown CP 1983. All squares.

R. fraxinea Common in the last century; despite declining in luxuriance and abundance, it is still frequent on exposed trees on south and west coasts. All except 39, 49, 59, 69.

R. lacera On bark of wayside trees; occasional. Very rare on old buildings. Brighstone Church 1971*; Carisbrooke Church BLS 1982; Shalfleet Church KAS 1998. All except 49, 68, 69.

R. siliquosa Abundant on many medieval churches on the south of the Island (Chale, Kingston, Niton, Godshill, Shanklin, etc.). On the gatehouse of Carisbrooke Castle. Fertile on rocks at St. Catherine's Point. Luxuriant on the Upper Greensand cliff at Niton with thalli in excess of 12 cm. CP 1971. 47, 48, 57, 58.

R. subfarinacea A largely maritime species, found here on exposed ancient buildings. Godshill Church 1971*; Carisbrooke Castle FR 1995; Chale Church FR 1995. 47, 48, 58.

Rhizocarpon distinctum On acid stones; very rare. Recorded from the west wall of Carisbrooke Castle BJC 1973; not refound here by FR in 1995. 48.

R. petraeum (*R. concentricum*) On chert and flint; scarce. Stones near Combley Great Wood and Quarr Abbey ruins CP 1987; Luccombe Down CP 1989. 57, 58, 59.

R. reductum (*R. obscuratum*) On acid stones, especially flints exposed on heaths and clay soils; frequent. All except 39, 49, 68, 69.

[*R. richardii* Extinct? On acid pebbles. Known only from old records: St. Boniface Down (BM) and flints on St. George's Down (HHK).]

Rinodina atrocinerea Known only from a few thalli on a single granite headstone in Freshwater All Saints churchyard BLS 1982. 38.

R. confragosa On coastal acid rocks. Recorded on sandstone of old wall just south of St. Catherine's Oratory BJC 1973 (E). 47.

R. efflorescens On acid bark. Only recorded from oak in Borthwood Copse 1971* (BM). 58.

R. gennarii On walls; local but widespread throughout the Island. All except 39, 59.

(*R. occulta* On acid rock. A single 19th century record from Shanklin (TS).)

R. oleae (*R. exigua*) On nutrient-rich bark; very rare, formerly scarce on elm. On elms south west of Brading and just west of Niton 1971*; St Catherine's CP 1992. 39, 47, 57, 58.

R. roboris On rough bark of large ash and oak in parks and woodland, formerly on elm; frequent. All except 38 and 47.

R. sophodes On smooth bark; rare or overlooked. On wayside ash above Bonchurch Landslip and on elms north of Brading, both 1971*; on sycamore at Alum Bay and sycamores at St. Catherine's, both BJC 1973; Hurst Copse, Wootton PWJ 1991. 38, 47, 57, 59, 68.

R. teichophila Rare on stone. Brighstone churchyard 1971* (BM); west wall of Carisbrooke Castle BJC 1973; Godshill churchyard OLG 1990; Newtown Church L&SS, CH 2002. 48, 49, 58.

Roccella phycopsis On medieval churches; a rare south-west maritime species. Early Island records are reported by Salwey (1860). A.G. More found a small patch on the ruins of St. Helen's Old Church in 1859 (BM). It is still present here (2002) in small quantity. In 1860, More discovered it 'in the greatest abundance all over the north side and tower of the church at Godshill' (HML, BON). It remains abundant on the tower and walls. A single specimen, which was collected, was found at Shanklin Church (1860 HHy), where it no longer occurs. Found in small quantity on the wall of the present St. Helen's Church, CP 2000. Also frequent on a single ancient oak at Nunwell Park CP 1991. 58, 68, 69.

[*Sarcogyne privigna* Extinct? On acid stone. A single old record from Shanklin of this rare species (HBH, BM).]

S. regularis On exposed chalk, limestone and mortar; local. Ruins of the old Quarr Abbey 1971*; Bembridge windmill BJC 1973; Tennyson Down FR 1977; Lakeside, Wootton on hardcore PWJ 1991; Brading churchyard FR, KAS 1993. 38, 59, 68.

Sarcopyrenia gibba On basic stone; rare or overlooked. Table top tomb in Brading churchyard FR, KAS 1993; St Edmund's Church, Wootton FR 1995; Newtown Church L&SS, CH 2002. 49, 68, 59.

Schismatomma cretaceum On old wayside trees; very rare. Borthwood Copse 1971*; Newbridge and Winstone Farm, Wroxall CP 1989; near Bembridge windmill CP 1996; America Wood CH 2002. 48, 58, 68.

S. decolorans On dry shaded bark of old trees, especially oak; locally frequent. All except 38, 39, 47.

S. niveum An old woodland species on old oaks; rare. Quarr 1971*; Mark's Corner, Parkhurst FR 1976; Briddlesford Copse FR, CP 1980; Northpark Copse FR, CP 1981; Bouldnor NS 1990; Nunwell Park CP 1991. 39, 48, 49, 58, 59.

S. quercicola An old woodland species; very rare. Parkhurst Forest on oak FR 1989. 49.

S. umbrinum On nutrient-enriched graves; may be overlooked. Freshwater All Saints' Church FR 1995. 38.

Scoliciosporum chlorococcum On twigs and stems of shrubs; probably overlooked. Borthwood Copse 1971* (BM); on sycamore and elders at St. Catherine's Point and on sycamore and heather at Headon Warren, all BJC 1973 (E). 38, 47, 58.

S. pruinosum On rough oak bark; very rare. Borthwood Copse 1971*; Parkhurst Forest FR 1989. 49, 58.

S. umbrinum On nutrient-enriched acid stone; widespread, but somewhat overlooked. All except 39, 49, 69.

Solenopsora candicans On calcareous rock; locally frequent in the south of the Island. Abundant on coastal rocks from St. Catherine's Point east to Old Park and sporadically further east to Ventnor. On Carisbrooke Castle walls BLS 1982 and several churches. Binstead Church and sea wall at Seaview CP 1999. All except 39, 49, 68.

S. holophaea Rare maritime species. North-east wall of Whitwell Church BJC 1973 (BM) but not refound 1995, FR; Carisbrooke Castle, on outer walls exposed to south-west winds FR 1995. 48, 57.

S. vulturiensis Maritime localities on rock and soil; rare. Whitwell Church and St. Catherine's Point BJC 1973; Freshwater All Saints' Church and Carisbrooke Castle FR 1995. 38, 47, 48, 57.

Sphinctrina turbinata A parasymbiont parasite on the thalli of *Pertusaria* spp. Rare but probably under-recorded. Wayside oak at Haylands, Ryde CP 1991. 57, 59.

Squamarina cartilaginea On dry chalk; very local. A co-dominant component of the *Fulgensia* community on Tennyson Down OLG 1976, and sparingly further westwards to the Needles headland CP 1999. 28, 38.

Staurothele hymenogonia Recorded from a quarry on the east side of Brading Down 1971* (BM). 68.

S. rupifraga On hard calcareous pebbles; very rare. Afton Down FR 1989 (det. PWJ); Tennyson Down OLG 1990. 38.

Steinia geophana Occasional on acid cliffs at Bouldnor VJG & OLG 2002. There is a record from 1923 from soil in a field below Ashey Down (HHK, BM). 39.

Stenocybe septata On holly bark in ancient woodlands; very local. Mark's Corner, Parkhurst FR 1976; America Wood and Borthwood Copse CP 1998. 49, 58.

[*Sticta limbata* Extinct. An old woodland species. Old records from Quarr (BM), Shanklin, Appuldurcombe (ABl) and Apes Down (PHM).]

[*S. sylvatica* Extinct. An old woodland species. Appuldurcombe and Shanklin Great Wood (ABl).]

[*Teloschistes chrysophthalmus* Extinct. On nutrient-enriched twigs. Recorded in the 19th century from hawthorn twigs near Ryde (BM). Possibly extinct in Britain today.]

[*T. flavicans* Extinct. On exposed twigs and bark. This striking species was reported to be 'very fine' on trees in Appuldurcombe Park and Wood by Rev. Bloxam in 1860. Also recorded around the same time from Ventnor, Ryde and Shanklin (BM).]

Tephromela atra On walls and headstones in churches; frequent, also common on natural rock outcrops at St. Catherine's and on sycamore. All except 38, 59, 69.

Thelidium minutulum On chalk pebbles; rare or overlooked. Tennyson Down OLG 1990. 38, 57.

T. zwackhii On sheltered chalk; rare or overlooked. Quarry on the east side of Brading Down 1971* (BM); calcareous cliffs at Bouldnor VJG & OLG 2002. 39, 68.

Thelopsis rubella An old woodland species on oak; rare. Briddlesford Copse FR, CP 1980; Hurst Copse, Wootton CP 1991. 59.

Thelotrema lepadinum An old woodland species on oak; rare. Briddlesford Copse CP 1980; Parkhurst Forest FR 1989. 49, 59.

Toninia aromatica On limestone and mortar; widespread and frequent. All except 39.

T. episema A commensal on large thalli of *Aspicilia calcarea*. Mottistone churchyard chest tomb FR 1998. 48.

T. sedifolia On dry chalk; rare. Tennyson Down, near the cliff edge OLG 1976; old trackway on Afton Down CP 1982; Steephill Cove CP 1984. 38, 57.

Trapelia coarctata On old brick and acid stone; occasional. On pebbles on Headon Warren BLS 1982. 38, 47, 48, 57, 58.

T. involuta The only record is from red-brick wall coping at Cowes BJC 1973. 49.

T. placodiodes On acid stonework; rare or overlooked. Godshill Church OLG 1990. 58.

Trapeliopsis flexuosa On lignum; occasional but under-recorded. Lakeside, Wootton PWJ 1991; Foreland CP 1991; Bonchurch Down NS 1992. 57, 59, 68.

T. granulosa On old wood; on peat at Headon Warren CP 1973. Frequent. All except 39, 57. Some records are probably *T. flexuosa*.

T. pseudogranulosa Rare or overlooked. Parkhurst Forest on trackside bank FR 1989. 49.

Tuckermannopsis chlorophylla (*Cetraria chlorophylla*) On acid bark; rare, perhaps overlooked. In canopy of large fallen ash, Cliff Copse, Shanklin CP 1991. 58.

Usnea articulata Found rarely in tree canopies. Formerly recorded from Appuldurcombe (BM). Occurs in Cliff Copse, Shanklin and sparingly westwards along the scarp towards Wroxall, with thalli up to 12cm. CP 1977; Great Wood Copse, Shanklin CP 1987; on oak and ash in Mudless Copse CP 1976; on a single oak at Northpark Copse CP 1980; Gatcombe Withy Bed FR 1977; Westridge Copse CP 1987. 48, 58.

U. ceratina On oak or ash; local and infrequent, but sometimes luxuriant. Also on wild cherry in Mudless Copse. All except 39, 47, 57, 68.

U. cornuta Occasional in woodland. Mudless Copse and Gatcombe Withy Bed, both FR 1977; Northpark Copse FR 1981; Bouldnor NS 1990; Parkhurst Forest FR 1988; Lake Common NS 1992. 39, 48, 49, 58.

U. flammea On exposed twigs and branches; very rare. On Headon Warren, on heather CP 1976 and honeysuckle BLS 1982, but in small quantity. Formerly at Blackgang Chine (HML, BON). 38.

U. florida In tree tops; very rare, perhaps overlooked. On large fallen oak in Parkhurst Forest CP 1990. 49.

U. rubicunda On oak; rare. America Wood 1971*; Walter's Copse, Newtown FR 1977; Rushcroft Copse, Swainston FR 1981. 48, 49, 58.

U. subfloridana On bark and twigs; widespread and frequent. Terricolous on St George's Down CP 1993. All squares.

Verrucaria amphibia On intertidal rocks. Hard chalk stack in Freshwater Bay, growing with *Lichina pygmaea*, the first record from chalk CP 1992 (det. A. Fletcher); luxuriant on rocks at New Ditch Point, Freshwater Cliffs 1992 (CP). A western species at the edge of its range here. 38.

V. aquatilis On damp rocks in Bonchurch Landslip 1971*. 57.

V. baldensis On dry, exposed limestone; frequent. 38, 47, 48, 57, 58, 59.

V. dolosa On flints and chert, sometimes in woodlands; under recorded. Headon Warren BJC 1973; Appley, Ryde 1984; Eaglehead Copse, Ashey 1984; Afton Down CP 1988 (det BJC); Tennyson Down OLG 1990. 38. 58, 69.

V. glaucina On limestone; widespread and fairly frequent. 38, 47, 48, 58, 68.

V. hochstetteri On sheltered limestone and mortar; frequent. All except 39.

V. hydrela On pebbles in clean streams; very local. Winkle Street BJC 1973; Bloodstone Well, Ashey Down CP 1995. 48, 58.

V. macrostoma On limestone; occasional, perhaps overlooked. Most records are likely to refer to f. *furfuracea* (see also *V. viridula* below). Brading Church FR, KAS 1993; Carisbrooke Castle FR 1995; St Edmund's Church, Wootton FR 1995. 48, 59, 68.

V. maura On seashores. Widespread around the coast on rocks and concrete sea-walls.

V. muralis On soft calcareous rocks; common. All squares.

V. murina On chalk pebbles; rare or overlooked. Tennyson Down OLG 1990. 38.

V. nigrescens On limestone; widespread and frequent. All except 39, 47.

V. pinguicula On limestone; very rare. Table top tomb in Godshill churchyard OLG 1990. 58.

V. rheitrophila On submerged flints in clean water; rare. Bloodstone Well, Ashey Down CP 1978; St Lawrence Well CP 1995. 57, 58.

V. simplex On chalk pebbles; rare. Tennyson Down OLG 1990. 38.

(*V. striatula* On rocky shores. Known only from a 19th century record from Luccombe Chine (BM).)

V. viridula on limestone and concrete; frequent. Some records may refer to *V. macrostoma* f. *furfuracea*. All except 39, 49.

Wadeana dendrographa On old trees; very rare. Recorded in the 19th century from Whitefield Woods (EMH). On a single huge ash at Long Copse, Gatcombe FR 1977 (hb FR), still present 1990 but becoming overgrown by ivy; on a large plane tree at Wootton Bridge PWJ 1991. 48, 59.

Xanthoparmelia mougeotii (*Parmelia mougeotii*) On acid stone and tiles; scarce but increasing. Roof of the Hare and Hounds public house BJC 1973; Godshill churchyard OLG 1990; Freshwater All Saints' Church and St Edmund's Church, Wootton FR 1995; Newtown Church C&SS, CH 2002. 38, 49, 58, 59.

Xanthoria calcicola On nutrient-enriched stone; rather frequent. All except 39.

X. candelaria s. *lat.* On trees with nutrient-enriched bark and on stonework; frequent. 48, 57, 58, 59.

X. parietina On nutrient-enriched stone, slate and bark; common. All squares. The f. *ectanoides* occurs on rocks around St. Catherine's Point, east to Old Park. 47, 57.

X. polycarpa On nutrient-enriched twigs, occasionally on lignum; occasional. All except 39, 69.

X. ulophyllodes On nutrient-enriched bark; very rare. An eastern species. BJC, 1973. 48.

REFERENCES

Bloxam, A. (1860) Lichens. In *A Guide to the Isle of Wight*, E. Venables. Lond.

Coppins, B. J. (1978) H. M. Livens lichen collection at Bolton Museum: notes on some interesting specimens. *Naturalist* 103, 105-107.

Coppins, B. J. (2002) *Checklist of Lichens of Great Britain and Ireland*. British Lichen Society, Huddersfield.

Gilbert, O. L. (1978) Fulgensia in the British Isles. *Lichenologist* 10, 33-45.

Gilbert, O.L. (2000) *Lichens*. New Naturalist Library. Harper Collins.

James, P.W & Coppins, B.J. (1979) Key to British sterile crustose lichens with *Trentepohlia* as phycobiont. *Lichenologist* 11 (3): 253-262.

Knight, H. H. (1932) The lichens of the Isle of Wight. *Proc. Isle Wight nat. Hist. archaeol. Soc.* 2, 221-232.

Pope, C. R. (1983) A lichen flora of the Isle of Wight. *Proc. Isle Wight nat. Hist. Archaeol. Soc.* VII (VIII), 577-599.

Pope, C. R. (1984) Field meeting on the Isle of Wight. *Lichenologist* 16, 59-62.

Purvis, O. W., Coppins, B. J., Hawksworth, D. L., James, P. W. & Moore, D. M. (1992) *The Lichen Flora of the Great Britain and Ireland*. The British Lichen Society.

Salwey, T. (1860) *Roccella tinctoria* and *R. phycopsis* in the Isle of Wight. *Phytologist* n.s. 4, 267-268.

Seaward, M. R. D. (1978) H. M. Livens lichen collection at Bolton Museum. *Naturalist* 103: 15-16.

Smith, J. E. & Sowerby, J. (1806) *English Botany*, Vol. 24. London: J.Sowerby.

Wheldon, J. A. (1909) Lichens. In: *A Guide to the Natural History of the Isle of Wight.* (Morey, F. ed.). 89-102. Isle of Wight County Press, Newport.

Stoneworts

NICK STEWART

The Charophytes, or Stoneworts, are a small and ancient group of freshwater green algae. Because of their large size, complex structure and tendency to become dominant in clear water, they were studied alongside vascular aquatic plants by Victorian botanists and acquired an honorary position alongside them in floras. The English name refers to the deposit of marl or calcium carbonate, which frequently encrusts their surfaces. Stoneworts are often characterised by a distinctive unpleasant smell.

The paucity of historic and modern records is probably a reflection of the generally uncommon occurrence of these plants on the Island. There is a scarcity of clean, clear bodies of water in the countryside today.

Nomenclature follows Bryant, Stewart & Stace (2002)

(*Chara pedunculata* (*Chara aculeolata*)
HEDGEHOG STONEWORT
Records of this species are errors for *Chara hispida* (q.v.))

Chara aspera
ROUGH STONEWORT
Rare. The recent, and perhaps also the old site, is from slightly brackish water.

sz68 Harbour Farm Lagoons, Bembridge 17/7/1999, A. Martin det. NFS (field).

sz38 Between Freshwater and Norton, 1892, S.A. Steuart (Groves & Groves 1895)

Chara globularis
FRAGILE STONEWORT
Rare. Apparently less common than formerly. The records are mostly restricted to the north-western part of the Island. Seen recently in three sites: Concrete trough in Prospect Quarry, north-west of Shalcombe, sz385866,

17/9/1992, BS det. JAB (field); pond in exposed situation near the coast, near Newtown, sz4392, 20/9/2000 CP, det. NFS (field); and small pond in clearing, Parkhurst Forest, sz471914, 23/5/1982, CP, det JAB (BM).

There are also the following historic records:

sz38 Golden's Common, Freshwater, pre-1871, F. Stratton (More 1871) [As *C. fragilis* which, at this date, could refer to either this species or *C. virgata*]; pool in brickfield, Thorley, 1921, J.F. Rayner (Rayner 1929); near Ningwood, 1922, J.F. Rayner (Rayner 1929).

sz39 or sz49 Pool, Hamstead, 1923, J.F. Rayner (Rayner 1929).

sz59 Pool, Whippingham, 1924, J. Groves (Rayner 1929).

[*Chara hispida*
BRISTLY STONEWORT]
Extinct. All records were from the western end of the Island but last seen in 1918. sz38 Golden's Common, Freshwater, 16/9/1867, R. Tucker, det NFS (OXF). Golden's Common referred to land in the vicinity of Golden Hill fort. Also Freshwater, 1871, A.G. More (CGE). These records are given in More (1871) under "*C. hispida* var. *pseudocrinita* and *C. polyacantha*" both of which are synonyms for *C. aculeolata*. It is probable that the Tucker specimen was not seen by either of the Groves brothers until it was donated to OXF in 1916. This is labelled as "*C. hispida* approaching subsp. *horrida*" with a note by J. Groves saying "the plant referred to by H. and J. Groves to be var. *horrida* in [Groves & Groves 1880] but not so extreme as the Baltic plant". Groves & Bullock-Webster (1924) goes further in saying "the Isle of Wight plant has remarkably long posterior bract-cells, but is not otherwise so extreme as that from the Baltic, and we think it better referred

to *C. hispida* proper."]; Easton Marsh, Freshwater, 1918, J.F. Rayner (Rayner 1929); old clay pit by Lee Copse, Thorley, 4/9/1918, J. Groves, det NFS (E, Rayner 1929)

Chara virgata
DELICATE STONEWORT
Occasional in the western half of the Island. Now known from three sites: old flooded pit within a disused Bembridge limestone quarry, Prospect Quarry near Shalcombe, sz385866, 1996, CP, det NFS (field); irrigation reservoir, Atherfield, sz452807, 2001, CP, det NFS (field); and artificial pond in unimproved pasture, by coast, Porchfield Rifle Ranges, sz4490, 2000, CP, det. NFS (field).

There are also the following historic records:

sz38 Small heathland pool between Freshwater and Yarmouth, 26/7/1922, J. Groves (BM), most probably the site referred to in Rayner (1929).

sz49 Pit in brickfield on the shores of the creek running into Newtown Bay, 10/10/1888, C. Bailey, det NFS (originally as *C. fragilis* = *C. globularis* s.l.) (MANCH).

Chara vulgaris
COMMON STONEWORT
Occasional in small pools, ditches and particularly in seepages and pools on slumped cliffs. It is the most frequent species on the Island with recent records from the following eleven sites:

var. **vulgaris** Masses in recently dredged water course, Freshwater Marshes sz344867, 8/1980, BS, det JAB (BM); freshwater pool on slippage clay at foot of cliffs, Shippard's Chine, Compton Bay, sz377841, 14/7/1982, CP, det JAB (field); stagnant pool on undercliffs, Sudmoor, sz393827, 12/8/1982, WL, det JAB (field); completely filling the upper and lower ponds, probably brackish, Watershoot Bay, sz494755, 30/8/1980, BS, det JAB (BM); impermanent pond on slipping coastal greensand at Chilton Chine, sz411821, 21/10/1984, CP, det JAB (field); pool on landslip, Reeth Bay, sz510754, 17/3/1996, RVL (field); ditch near cement mills at Brading Marsh, sz6187, 1981, CP (field).

near var. *papillata* Flooded pit in disused Bembridge limestone quarry, Prospect Quarry near Shalcombe, sz385866, 17/4/1984, CP, det JAB (field); shallow pool on undercliffs, Compton, sz375844, 11/8/1982, WL, det JAB (field); small pond just above High Water Mark, affected by salt spray, Rocken End, Watershoot Bay, sz493756, 4/1985, W.F. Farnham, det JAB (field); dew pond on south coast south-east of Brighstone, sz437811, 22/10/1996, RVL (field); plentiful in a drainage ditch, Brading Marsh, sz615873, 6/1980, CP, det JAB (field).

near var. *refracta* Old mill pond, restored in 1979, Mottistone Mill, Brighstone, sz423837, 24/6/1982, OHF, det JAB (BM);

There are also the following historic records:

sz38 var. *vulgaris* Yarmouth, pre-1871, A.G. More (More 1871)

var. *longibracteata* Near West Hill, Yarmouth, 10/1920, J. Groves, det GOA (BM)

var. *vulgaris* Freshwater, pre-1871, A.G. More (More 1871)

var. *refracta* Pond on cliffs above Compton Bay, 16/9/1919, H. Drabble & J. Groves (BM)

sz47 var. *vulgaris* Gault pond at the base of Gore Cliff landslide, 15/5/1932, C.V.B. Marquand (BM) - gault pond, Blackgang landslide, 14/4/1930, C.V.B. Marquand (BM)

sz48 var. *longibracteata* Small pond, near Shalfleet, 10/1880, G. Nicholson (BM, Groves & Groves 1881)

vars. *papillata* and *vulgaris* Pit in brickfield on the shores of the creek running into Newtown Bay, 10/10/1888, C. Bailey, det J. Groves, NFS (BM, MANCH);

sz58 or sz59 vars. *vulgaris* and *refracta* Pool in marshes, River Medina between Mill Hill and Newport, 7/7/1921, J. Groves, det. GOA (BM)

sz68 var. *vulgaris* Brading Marshes, 8/1853, A.G. More (CGE); ditches near Brading Harbour, pre-1871, A.G. More (More 1871)

var. *longibracteata* Brading, pre-1904, A.G. More (Townsend 1904)

sz68 var. *vulgaris* Sandown Marshes, pre-1871, A.G. More (More 1871)

Lamprothamnium papulosum
FOXTAIL STONEWORT

Rare in brackish lagoons. This nationally rare species is 'near threatened' in the stonewort Red List. In England it is restricted to five sites on the South Coast between The Fleet and Chichester, but has been lost from several others due to habitat loss, salinity changes and eutro-phication. It was regularly reported from the salt pans at Newtown until the 1880s, but it did not survive when they fell into disrepair following their abandonment.

It was rediscovered in 1980 at the other end of the Island, in the Harbour Farm Lagoons, Bembridge, which are large brackish ponds near the sea sz638882, 8/1980, BS, det JAB (BM). There was a good population in one lagoon but vast amounts of algae; also a floating fragment in another lagoon, Bembridge, 9/1996, T. Pankhurst *et al.* (RVL pers. com.). More recently, it has become much rarer there and appears sporadic in its appearance. This is probably due to algal blooms produced as a result of nutrient enrichment and exacerbated by the presence of a substantial population of carp (which are herbivorous and stir up the bottom sediment). A record of a scrap floating at brackish lagoon at Seaview sz623917, 9/1987 (Sheader & Sheader 1987) has never been confirmed despite several subsequent searches, and perhaps it was a bird-carried fragment.

The following historic records refer to its occurrence at Newtown (sz49): 'Found by me [A.G. More] in 1862, covering the bottom of the shallow brine-pans at the west mouth of Newtown Creek, close to the boiling houses' (More 1871) [Also Saltpans, Yarmouth, 8/1862, A.G. More (BM) belongs here and is the origin of the record from sz38 in Stewart & Church (1992). Although Yarmouth is some distance from Newtown, it is clear that there was only ever one 19th century locality].

Newtown, 9/1863, A.G. More, det NFS (BM, DBN) '. . . and again in 1863, in the pits or reservoirs on the east side of the Creek, close to the village of Newtown, growing in salt water from eight inches to two feet deep' (More 1871).

Salterns (reservoir) Newtown, 15/9/1869, Trimen & F. Stratton (E, BM); 'this plant is not now to be found in the 'salt pans' properly so called at Newtown, Isle of Wight. It grows only in the large reservoir into which the sea flows, and from which the water is admitted or pumped into the 'pans'. The 'pans' are completely dried up during the early spring and in the late autumn and from their nature I do not think that the [*Lampro-thamnium*] could grow in them' F. Stratton in Syme (1870).

'I also exhibit another interesting Isle of Wight plant, which had been thought to have become extinct, as it has not been reported to have been found for several years past, in the only British station where it is known to occur. This plant is *Lychnothamnus alopecuroides*, Braun, one of the Characeae, and was collected at Newtown, 14th July 1881. On the north coast of the Isle of Wight, between Cowes and Yarmouth, is a narrow creek with numerous arms, on whose shores many years ago a considerable trade was done in salt, produced by evaporation in the numerous salterns on each side of the creek. The newer method of obtaining salt from the brine formed by the beds of rock salt in Cheshire and other districts, has led to the abandonment of the salt pans at Newtown, and very few are now left. Some have been absorbed by the formation of oyster-parks on both sides of the creek, as well as by brick-works, and similar destruction threatens the remainder. The particular locality in which the *Lychnothamnus* occurred lies a few hundred yards north of the Coast Guard station at Newtown, and contiguous to the shores of the creek. At this point are the remains of three old salterns containing water as salt (to the taste) as that of the sea, and filled with *Ruppia spiralis* Hartm. The saltern nearest the creek is the only one of the three which contained the *Lychnothamnus* and from it I obtained, by wading, a number of very fine plants, many of whose stems would be 18 to 24 inches in length. They were greatly infested with a confervoid growth, which rendered it difficult to secure good-sized plants. I spent some time in a subsequent visit, exploring some of the ramifications of the creek, but failed to find a trace of the plant in any other station. One of the large oyster parks on the western side of the creek was carefully, but unsuccessfully, searched during this second visit, its nearly empty condition providing every facility. The most likely station for its being found on the western side of the creek is at a spot where there are two old salterns, both of which had been emptied of their vegetation a few days previously, apparently to adapt them for duck-ponds.' (Bailey 1882, BM, E, BEL, DBN, SLBI).

'In July 1881 the plant was again discovered by Mr. Charles Bailey of Manchester who kindly sent me specimens but in 1887 I again searched unsuccessfully for it. There is, however, much hope that the plant still lingers there, and the spore may be lying, waiting for a favourable opportunity of germinating. Search should always be made for it.' (Stratton 1897). There are no further records.

Nitella opaca
DARK STONEWORT

Rare, with only two recent records from the Brighstone area: shallow pool in arable field, south side of Military Road, Thorncross, alt. 35m, SZ436810, 26/5/1996, CDP & TDD, det NFS (CGE); and irrigation reservoir, Atherfield, SZ452807, 2001, CP, det. NFS (field).

There are also the following historic records:

SZ39 or SZ49 Shaded pool near Hamstead, 27/5/1921, J. Groves (BM, Rayner 1929).

SZ48 Pond near the Long Stone, Mottistone, 29/7/1919, J. Groves (BM); pond near the Long Stone, Mottistone, 1926 or 1927, J. Groves (Rayner 1929);

SZ49 Pool, near Gurnard Bay, pre-1871, F. Stratton (More 1871).

[*Tolypella glomerata*
CLUSTERED STONEWORT

Extinct. Only once recorded in 1919. Nationally scarce. SZ49 In fair quantity in some shallow pits near Elmsworth brick-works, just to the east of the mouth of the Newtown River, 13/5/1919, JG, det GOA (BM, Groves 1919, Rayner 1929).]

REFERENCES

Bailey, C. (1882) On the Isle of Wight station for *Lychnothamnus alopecuroides*, Braun. *Proceedings of the Manchester Literary and Philosophical Society*, **21**: 74.

Bryant, J.A., Stewart, N.F. & Stace, C.A. (2002) A checklist of the Characeae of the British Isles. *Watsonia*, **24**: 203-208.

Druce, G.C. (1919) New county and other records. *The Botanical Society and Exchange Club of the British Isles, Report for 1918*, **5**:365-412.

Groves, H. & Groves, J. (1880) A review of the British Characeae. *Journal of Botany, British and Foreign*, **18**: 97-103, 129-135, 161-167.

Groves, H. & Groves, J. (1881) Notes on British Characeae. *Journal of Botany, British and Foreign*, **19**: 353-356.

Groves, H & Groves, J (1895) Notes on the British Characeae, 1890-1894. *Journal of Botany, British and Foreign*, **33**: 289-292.

Groves, J. (1919) *Tolypella glomerata* Leonh. in the Isle of Wight. *Journal of Botany, British and Foreign*, **17**: 197.

Groves, J. & Bullock-Webster, G.R. (1920, 1924) *The British Charophyta*. 2 volumes. Ray Society, London.

More, A.G. (1871) A supplement to the 'Flora Vectensis'. *Journal of Botany, British and Foreign*, **9**: 202-211.

[Rayner, J.F. (1929) *A supplement to Frederick Townsend's Flora of Hampshire and the Isle of Wight*. Buncle, Arbroath.

Sheader, M. & Sheader, A. (1987) *A lagoon survey of the Isle of Wight, September 1987*. Nature Conservancy Council CSD report no. 820.

Stewart, N.F. & Church, J. (1992) *Red data books for Britain and Ireland: Stoneworts*. Joint Nature Conservation Committee, Peterborough.

Stratton, F. (1897) Botany: Isle of Wight in *The Hants and Dorset Court guide and County blue book: a fashionable record, professional register and general survey of the counties*. C.W. Deacon and Co.

Syme, J.B. (1870) Report for the year 1869. *The Botanical Exchange Club, Report of the Curator for 1869 and list of desiderata for 1870*: 7-17. [The same information is requoted in Syme, J.B. (1870) Report of the Curator of the Botanical Exchange Club for the year 1869. *Journal of Botany, British and Foreign*, **8**: 265.]

Townsend, F. (1883) *Flora of Hampshire including the Isle of Wight*. London.

Townsend, F. (1904) *Flora of Hampshire including the Isle of Wight*. Second edition.

Appendices

1. RECORDERS

AB	Butler, Andy	EVW	Watson, E.V.	L&SS	Street, Les & Sheila
ABl	Bloxam, Rev. A.	FM	Morey, Frank	LS	Snow, Lorna
ABo	Boucher, Ann	FMM	Minns, Fanny M.	MAW	Walton, Mike
ABr	Brewis, Lady Anne	FR	Rose, Francis	MB	Burnhill, Margaret
AC	Campbell, Ann	FS	Stratton, F.	MJ	Jennings, Malcolm
ACC	Crundwell, Alan C.	GAS	Stanley, Gretel A.	MK	Keens, Mary
AEW	Wade, A.E.	GB	Bullock, Gladys	ML	Lawley, Mark
AG	Gregory, Anthony	GE	Een, G.	MMS	Seabroke, Mercia M.
AJ	Jones, A.W.	GK	Kitchener, G.	MOH	Hill, Mark O.
AJES	Smith, A.J.E.	GOA	Allen, G.O.	MP	Pool, M.
AJG	Gray, A.J.	GPR	Rothero, Gordon	MR	Rowe, Mike
AL	Loydell, A.	GS	Stevens, Gillian	MS	Smith, Martin
AM	Marston, Anne	GT	Toone, Geoff	MSa	Saunders, M.
AMu	Murray, Anne	HBH	Hall, H.B.	MSh	Sheader, Martin
AN	Newton, Angela	HCN	Napier, H.C.	MW	Whitaker, Maureen
AR	Redfearn, Anthony	HH	Higgins, Hilary	NFS	Stewart, Nick F.
ARP	Perry, A.R.	HHK	Knight, H.H.	NH	Hodgetts, Nick
AWW	Westrup, Alick W.	HHy	Hyndman, H.	NS	Sanderson, Neil
BA	Angell, Barry	HLKW	Whitehouse, H.L.K.	OHF	Frazer, Oliver F.
BBS	British Bryological Society	HND	Dixon, H.N.	OLG	Gilbert, Oliver L.
BG	Goater, Barry	HML	Livens, H.M.	PDG	Goriup, P.D.
BJC	Coppins, Brian J.	HPR	Reader, H.P.	PE	Edwards, Peter
BM	Middleton, Brian	HM	Matcham, Howard	PHM	Milledge, P.H.
BS	Shepard, Bill	IB	Boyd, Ian	PJ	Jepson, Peter
BW	Warne, Brian	IR	Ralphs, Ian	PJS	Selby, Pete J.
CC	Chatters, Clive	JA	Appleyard, Joan	PJW	Wilson, Phil J.
CCT	Townsend, Cliff C.	JAB	Bryant, Jenny	PL	Long, P.
CDP	Preston, Chris D.	JAP	Paton, Jean A.	PM	Martin, Peter.
CH	Hitch, Chris	JB	Bevis, Jim	PS	Stanley, Paul
CHB	Binstead, C.H.	JC	Cox, Jonathan	PSi	Sivell, Paul
CL	Lipscombe, Chris	JDS	Sleath, Jonathan	PW	Wanstall, Peter
CP	Pope, Colin	JEG	Gaffney, Jean E.	PWJ	James, Peter W.
DB	Biggs, David	JFR	Rayner, J.F.	RAF	Finch, R.A.
DD	Dupree, David	JG	Groves, J.E.	RC	Carpenter, Roy
DDa	Dana, Dave	JH	Higgins, Father J.	RCS	Stern, Rod C.
DEA	Allen, David E.	JHB	Bloom, J.H.	RD	Day, Robin
DF	Frazer, Dorothy	JJG	Graham, James	RDF	Fitzgerald, R.D.
DFC	Chamberlain, D.F.	JKN	Norledge, John K.	RDP	Porley, Ron D.
DGL	Long, David	JM	Mott, John	RF	Fisk, Richard
DTH	Holyoak, David T.	JMB	Blackburn, John M.	RG	Grogan, Richard
EAP	Pratt, Ted	JMC	Morrison, J.C.	RHW	Woodall, Rob H.
EC	Clement, Eric	JO	Ounstead, John	RJC	Chancellor, R.J.
ECW	Wallace, Ted	JP	Pope, Jillie	RK	Kettell, Reg
EFW	Warburg, E.F.	JR	Ralfs, J.	RL	Lightbown, Richard
EHW	White, Ernest H.	JRL	Laundon, Jack R.	RPA	Attrill, Robin P.
EL	Laidlaw, Eric	JRM	Moon, John R.	RPB	Bowman, R. Paul
EMH	Holmes, E.M.	JS	Simons, Jonathan	RSC	Cropper, R.S.
EN	English Nature	JT	Turner, Johnny	RT	Tangney, Ray
EV	Venables, E.	KS	Sandell, Ken	RVL	Lansdown, R.V.

RW	Walls, Robin	TH	Holtzer, Tim	VG	Gwynn, Val
SB	Blackwell, Sue	TL	Laflin, Tom	VNG	Giavarini, Vince N.
SC	Colenutt, Simon	TLB	Blockeel, Tom L.	VS	Scott, Vera
ST	Telfer, Sue	TR	Ransom, Tom	WHW	Wilkinson, W.H.
SY	Young, Simon	TS	Salwey, T.	WL	Lutley, W.
TCR	Rich, Tim C.	TT	Tutton, Tony	WRS	Sherrin, W.R.
TDD	Dines, Trevor D.	TW	White, Thelma	WW	Wilson, W.
TF	Feben, Tina	UKD	Duncan, Ursula K.		

II. EXTINCTIONS

Documented evidence suggests that, in the period from 1850 to 2000, vascular plants have been lost from the Island at a rate of approximately one species every two years. This is the same figure that has been calculated for many larger counties and is actually quite favourable. Smaller areas tend to have higher extinction rates because the area of habitat available for a species is reduced.

According to this information, the rate of extinction was least in the final quarter of the nineteenth century and highest in the third quarter of the twentieth century, a period of widespread hedgerow removal and ploughing of old pastures. There is no evidence for an accelerated rate of decline towards the end of the twentieth century.

Species of wet and dry acid heathland habitats account for some 30% of the losses. The next greatest losses are plants of coastal and arable habitats (each 15%).

This table is based upon last recorded dates of probable native species and archaeophytes. Microspecies have not been included. Many of the species, especially for the earlier records, certainly survived beyond these dates. Undoubtedly, some of these plants are likely to be rediscovered in the future.

Silene conica 1715
Lathyrus japonicus pre 1800
Parnassia palustris pre 1800
Otanthus maritimus pre 1823
Cystopteris fragilis 1839
Carex elata 1840
Persicaria minor 1847

Thalictrum flavum 1850
Petrorhagia nanteuillii 1854
Chenopodium vulvaria 1856
Viola tricolor 1856
Ajuga chamaepitys 1856
Calamagrostis canescens 1857
Rhynchospora alba 1858
Carex acuta 1858
Carex punctata 1862
Carex hostiana 1862
Fumaria densiflora 1865
Carex diandra 1870
Polystichum aculeatum 1871
Eleocharis quinqueflora 1871
Euphorbia peplis 1872
Gentianella campestris 1879
Utricularia minor 1879
Hordeum marinum 1879
Dianthus armeria 1892
Lycopodium clavatum 1900
Vaccinium oxycoccus 1900
Sparganium natans 1900
Lithospermum arvense 1904
Minuartia hybrida 1905
Thymus pulegium 1905
Eriophorum latifolium 1908
Wahlenbergia hederacea 1909
Pulicaria vulgaris 1909
Botrychium lunaria 1913
Filago lutescens 1913
Persicaria bistorta 1917
Adonis annua 1918
Carex vulpina 1920
Carex montana 1920
Cerastium arvense 1923
Pedicularis palustris 1927
Cuscuta europaea 1928
Sagina nodosa 1931
Hypericum montanum 1931
Hypericum elodes 1931
Gladiolus illyricus 1931
Bupleurum rotundifolium 1932
Galium tricornutum 1932
Himantoglossum hircinum 1933

Agrostemma githago 1943
Filago pyramidata 1947
Vicia sylvatica 1951
Erodium maritimum 1955
Torilis arvensis 1958
Epipactis phyllanthes 1959
Trifolium squamosum 1965
Scandix pecten-veneris 1967
Potamogeton perfoliatus 1967
Lotus angustissimus 1968
Utricularia australis 1970
Butomus umbellatus (native) 1972
Pyrola rotundifolia 1973
Platanthera bifolia 1973
Epilobium lanceolatum 1974
Oenanthe silaifolia 1974
Centaurium tenuiflorum 1974 ?
Veronica scutellata 1974
Tephroseris integrifolia 1974
Galeopsis angustifolia 1975
Fumaria vaillantii 1978
Fallopia dumetorum 1978
Spirodela polyrhiza 1982
Ophrys insectifera 1982
Aceras anthropophorum 1983
Genista anglica 1984
Orchis ustulata 1991
Thymus pulegioides 1992
Juncus squarrosus 1992
Eleogiton fluitans 1994
Radiola linoides 1995

NUMBERS OF EXTINCT SPECIES IN TIME PERIODS

Pre 1800	3
1800 – 1825	1
1826- 1850	4
1851 – 1875	14
1876 – 1900	7
1901 – 1925	13
1926 – 1950	11
1951 – 1975	18
1976 – 2000	11

Index to Species

Accepted Latin names are in **bold italics**, synonyms in *light italics* and English names in Roman. Page numbers refer to principal entries in the text and to illustrations in **colour** and black and white.

Abraham-Isaac-Jacob 141
Abutilon theophrasti 91
Acaena novae-zelandiae 112
Acanthus mollis 153
Acarospora fuscata 219
   smaragdula 219
Acaulon muticum 208
   triquetrum 208, 9
Acer campestre 129
   cappadocicum 129
   platanoides 129
   pseudoplatanus 129
Aceras anthropophorum 196
Achillea millefolium 165
   ptarmica 165
Aconite, Winter 71
Aconitum napellus 71
Acrocordia conoidea 219
   gemmata 219
   salweyi 219
Adder's-tongue 66, **8**
Adiantum capillus-veneris 67
Adonis annua 73
Adoxa moschatellina 157
Aegopodium podagraria 133
Aesculus hippocastanum 129
Aethusa cynapium 134
Agonimia allobata 219
   gelatinosa 219
   octospora 219
   tristicula 219
Agrimonia eupatoria 112
   procera 112
Agrimony 112
   Fragrant 112
Agrostemma githago 85
Agrostis canina 185
   capillaris 184
   curtsii 184
   gigantea 184
   stolonifera 184
   vinealis 185
Ailanthus altissima 129

Aira caryophyllea 184
   praecox 184
Ajuga chamaepitys 144
   reptans 144
Alchemilla mollis 112
   vulgaris 112
Alder 79
Alexanders 133
Alisma plantago-aquatica 168
Alison, Golden 97
   Sweet 97
Alkanet, Green 141
Alliaria petiolata 95
Allium moly 191
   neapolitanum 191
   nigrum 191
   oleraceum 191
   paradoxum 191
   roseum 191
   subhirsutum 191
   triquetrum 191
   ursinum 191
   vineale 191
Allseed 129
   Four-leaved 84
Alnus glutinosa 79
Aloina aloides 207
   ambigua 207
   rigida 207
Alopecurus aequalis 186
   bulbosus 185
   geniculatus 185
   myosuroides 186
   pratensis 185
Althaea hirsuta 91
   officinalis 91
Alyssum saxatile 97
Amandinea punctata 219
Amaranth, Common 81
   Green 82
Amaranthus albus 82
   hybridus 82
   retroflexus 81
Amblystegium compactum 214
   riparium 214
   serpens 213
   tenax 213
   varium 213
Ambrosia artemisiifolia 167
Amelanchier lamarckii 116
Ammi majus 135

Ammophila arenaria 185
Anacamptis pyramidalis 195
Anagallis arvensis 103
   minima 103
   tenella 102
Anaphalis margaritacea 163
Anaptychia ciliaris 219
   runcinata 219
Anchusa arvensis 141
   azurea 141
Anchusa, Garden 141
Anemone apennina 71
   blanda 71
   nemorosa 71
Anemone, Balkan 71
   Blue 71
   Wood 71
Aneura pinguis 202
Angelica sylvestris 135
Angelica, Wild 135
Anisantha diandra 187
   madritensis 187
   sterilis 187
   tectorum 187
Anisomeridium biforme 219
   nyssaegenum 219
   polypori 219
   ranunculosporum 219
Anomodon viticulosus 213
Anthemis arvensis 165
   cotula 166
   punctata 165
   tinctoria 166
Anthoceros agrestis 203
   punctatus 203
Anthoxanthum odoratum 184
Anthriscus caucalis 132
   sylvestris 132
Anthyllis vulneraria 117
Antirrhinum majus 148
Antitrichia curtipendula 212
Apera spica-venti 185
Aphanes arvensis 112
   australis 112
Apium graveolens 134
   inundatum 135
   nodiflorum 135
Apple 114
   Crab 114
Apple-mint 145
Apple-of-Peru 137
Aptenia cordifolia 79

Aquilegia vulgaris 73
Arabidopsis thaliana 95
Arabis glabra 97
   hirsuta 97
   turrita 97
Arbutus unedo 101
Archangel, Yellow 142
Archidium alternifolium 204
Arctium lappa 158
   minus 158
Arenaria balearica 82
   serpyllifolia 82
Armeria maritima 89, **3**
Armoracia rusticana 97
Arrhenatherum elatius 183
Arrowgrass, Marsh 169
   Sea 169
Artemisia absinthium 165
   verlotiorum 165
   vulgaris 165
Arthonia cinnabarina 219
   didyma 219
   elegans 219
   impolita 219
   lapidicola 219
   pruinata 219
   punctiformis 219
   radiata 219
   spadicea 219
   vinosa 219
Arthopyrenia analepta 219
   lapponina 219
   punctiformis 219
   ranunculospora 219
Arum italicum 171, **9**
   maculatum 171
Arum, Dragon 171
Ash 147
Asparagus officinalis 192
Asparagus, Garden 192
Aspen 93
Asperula arvensis 155
   cynanchica 154
Asphodel, Bog 190
Aspicilia caesiocinerea 219
   calcarea 220
   contorta 220
   radiosa 220
   subcircinata 220
   tuberculosa 220
Asplenium adiantum-nigrum 68

marinum 68, 3
ruta-muraria 68
trichomanes 68
Aster novi-belgii 164
Astragalus danicus 116
glycyphyllos 116
Athyrium filix-femina 68
Atrichum undulatum 204
Atriplex glabriuscula 80
halimus 81
hortensis 80
laciniata 80
littoralis 80
longipes 80
patula 80
portulacoides 81
prostrata 80
x *gustafsoniana* 80
Atropa belladona 138
Aubretia 97
Aubretia deltoidea 97
Aucuba japonica 125
Aulacomnium androgynum 211
palustre 211
Aunt-eliza 193
Avena fatua 183
sativa 183
sterilis 183
strigosa 183
Avens, Water 112
Wood 112
Azolla filiculoides 69

Bacidia arceutina 220
bagliettoana 220
biatorina 220
epixanthoides 220
friesiana 220
incompta 220
laurocerasi 220
naegelii 224
phacodes 220
rubella 220
sabuletorum 226
vezdae 223
viridifarinosa 220
Baeomyces rufus 220
Baldellia ranunculoides 168
Ballota nigra 142
Balm 144
Balm-of-gilead 94
Balsam, Indian 132
Orange 131
Small 132
Balsam-poplar, Western 94
Bamboo, Arrow 179
Broad-leaved 179
Barbarea intermedia 96
verna 96
vulgaris 96
Barberry 73
Barbula acutus 207

convoluta 207
cylindrica 207
fallax 207
hornschuchiana 206
recurvirostra 206
revoluta 206
rigidula 207
sinuosa 207
tophacea 207
trifaria 207
unguiculata 207
vinealis 207
Barley, Foxtail 187
Meadow 187
Sea 187
Two-rowed 187
Wall 187
Bartramia pomiformis 211
Bartsia, Red 151
Yellow 151
Basil, Wild 145
Bastard-toadflax 125
Bay 70
Beak-sedge, White 175
Beard-grass, Annual 185
Bear's-breech 153
Bedstraw, Fen 155
Heath 155
Hedge 155
Lady's 155
Beech 77
Beet, Sea 81
Bellflower, Adria 153
Clustered 153
Creeping 154
Ivy-leaved 154
Nettle-leaved 154, 1
Spreading 153
Trailing 153
Bellis perennis 164
Belonia nidarosiensis 220
Bent, Black 184
Bristle 184
Brown 185
Common 184
Creeping 184
Velvet 185
Water 185
Berberis vulgaris 73
Bergenia crassifolia 104
Bermuda-grass 188
Berula erecta 133
Beta vulgaris 81
Betony 142
Betula pendula 78
pubescens 79
x *aurata* 78
Biatora epixanthoides 220
Bidens cernua 168
tripartita 168
Bilberry 101
Bindweed, Field 138
Hairy 139

Hedge 139
Large 139
Sea 138
Birch, Downy 79
Silver 78
Bird-in-a-bush 74
Bird's-foot 117
Bird's-foot-trefoil, Common 117
Greater 117
Narrow-leaved 117
Slender 117
Bird's-nest, Yellow 102
Bistort, Amphibious 87
Common 86
Red 86
Bitter-cress, Hairy 97
Wavy 97
Bittersweet 138
Bitter-vetch 118
Black-bindweed 88
Black-grass 186
Black-poplar 94
Hybrid 94
Blackstonia perfoliata 136
Blackthorn 114
Bladder-fern, Brittle 69
Bladder-senna 116
Bladderwort 153, 11
Lesser 153
Blasia pusilla 202
Blechnum spicant 69
Blinks 82
Bluebell 190, 6
Spanish 191
Blue-eyed-grass 193
Blue-sow-thistle, Common 161
Bogbean 139
Bog-myrtle 77
Bolboschoenus maritimus 174
Borage 141
Borago officinalis 141
Botrychium lunaria 66
Box 126
Brachypodium pinnatum 187
sylvaticum 187
Brachythecium albicans 214
glareosum 214
mildeanum 214
populeum 214
rivulare 214
rutabulum 214
velutinum 214
Bracken 67
Bramble 105
White-stemmed 105
Brassica elongata 99
juncea 99
napus 99
nigra 99
oleracea 99
rapa 99
Bridewort 105

Confused 105
Bristle-grass, Foxtail 189
Green 189
Knotroot 189
Rough 189
Briza maxima 181
media 181
minor 181
Broadleaf, New Zealand 125
Brome, Barren 187
Compact 187
Drooping 187
False 187
Field 186
Great 187
Hungarian 187
Meadow 186
Rescue 187
Rye 186
Smooth 186
Soft 186
Upright 187
Bromopsis erecta 187
inermis 187
ramosa 187
Bromus arvensis 186
commutatus 186
hordaceus 186
lepidus 186
racemosus 186
secalinus 186
x *pseudothominei* 186
Brooklime 149
Brookweed 103
Broom 122
Montpellier 122
Spanish 122
Broomrape, Common 153
Greater 152
Ivy 152
Oxtongue 152, 2
Yarrow 152, 8
Brunnera macrophylla 141
Bryoerythrophyllum recurvirostrum 206
Bryonia dioica 93
Bryony, Black 194
White 93
Bryum algovicum 210
alpinum 211
argenteum 210
bicolor 211
bornholmense 211
caespiticium 210
canariense 210
capillare 210
donianum 210
dunense 211
gemmiferum 210
imbricatum 210
inclinatum 210
intermedium 210
klinggraeffii 211

microerythrocarpum 211
pallens 210
pseudotriquetrum 210
radiculosum 211
rubens 211
ruderale 211
sp. 211
subapiculatum 211
torquescens 210
violaceum 211
Buckler-fern, Broad 69
  Narrow 69
Buckthorn 128
  Alder 128
Buckthorn, Mediterranean 128
  Sea 123
Buckwheat 87
  Green 87
Buddleja davidii 147
Buellia aethalea 220
  griseovirens 220
  ocellata 220
  punctata 219
  stellulata 220
Buffalo-bur 138
Bugle 144
Bugloss 141
Bullace 114
Bullwort 135
Bulrush 189
  Lesser 190
Bupleurum rotundifolium 134
  subovatum 134
  tenuissimum 134
Burdock, Greater 158
  Lesser 158
Bur-marigold, Nodding 168
  Trifid 168
Burnet, Fodder 112
  Salad 112
Burnet-saxifrage 133, 6
Bur-parsley, Greater 136
  Small 136
Bur-reed, Branched 189
  Least 189
  Unbranched 189
Butcher's-broom 192
Butomus umbellatus 168
Butterbur, Giant 167
Buttercup, Bulbous 71
  Celery-leaved 72
  Corn 72
  Creeping 71
  Goldilocks 72
  Hairy 71
  Meadow 71
  Small-flowered 72
Butterfly-bush 147
Butterfly-orchid, Greater 195
  Lesser 195
Butterwort, Pale 153
Buxus sempervirens 126
Byssoloma leucoblepharum 220

Cabbage, Bastard 100
  Wild 99
Cabbage-palm 194
Caduus nutans 158
Cakile maritima 100
Calamagrostis canescens 185
  epigejos 185
Calamint, Common 144
  Wood 144, 43, 1
Calendula officinalis 167
Calicium glaucellum 220
  salicinum 220
  viride 220
Calliergon cordifolium 214
  cuspidatum 214
  stramineum 214
Calliergonella cuspidata 214
Callitriche hamulata 146
  obtusangula 146
  platycarpa 146
  stagnalis 146
Calluna vulgaris 101
Caloplaca atroflava 220
  aurantia 220
  ceracea 220
  cerina 220
  cerinella 220
  chlorina 220
  citrina 220
  crenularia 220
  dalmatica 220
  flavescens 221
  flavovirescens 221
  holocarpa 221
  isidiigera 220
  lactea 221
  luteoalba 221
  marina 221
  obscurella 221
  saxicola 221
  teicholyta 221
  ulcerosa 221
  variabilis 221
Caltha palustris 70
Calypogeia arguta 200
  fissa 200
  muelleriana 200
Calystegia pulchra 139
  sepium 139
  silvatica 139
  soldanella 138
  x *lucana* 139
Camelina sativa 98
Campanula glomerata 153
  medium 153
  patula 153
  portenschlagiana 153
  poscharskyana 153
  rapunculoides 154
  rotundifolia 154
  trachelium 154, 1
Campion, Bladder 85

  Red 86
  Rose 85
  Sea 85, 3
  White 86
Campyliadelphus
  chrysophyllus 213
Campylium calcareum 213
  chrysophyllum 213
  stellatum 213
Campylophyllum calcareum 213
Campylopus brevipilus 205
  flexuosus 205
  fragilis 205
  introflexus 205
  paradoxus 205
  pyriformis 205
Canary-grass 184
  Awned 184
  Bulbous 184
  Lesser 184
  Reed 184
Candelaria concolor 221
Candelariella aurella 221
  medians 221
  reflexa 221
  vitellina 221
  xanthostigma 221
Candytuft, Garden 98
Cannabis sativa 76
Canterbury-bells 153
Capsella bursa-pastoris 98
  rubella 98
Caraway 135
Cardamine bulbifera 97
  flexuosa 97
  hirsuta 97
  pratensis 97
Carduus crispus 158
  tenuiflorus 158
Carex acuta 179
  acutiformis 176
  arenaria 175
  binervis 177
  caryophyllea 178
  curta 176
  diandra 175
  distans 178
  disticha 176
  divisa 176
  divulsa 175
  echinata 176
  elata 179
  extensa 178
  flacca 177
  hirta 176
  hostiana 178
  laevigata 177
  montana 178
  muricata 175
  nigra 179
  otrubae 175
  ovalis 176

  pallescens 178
  panicea 177
  paniculata 175
  pendula 177
  pilulifera 178
  pseudocyperus 177
  pulicaris 179
  punctata 178
  remota 176
  riparia 177
  rostrata 177
  spicata 175
  strigosa 177
  sylvatica 177
  viridula 178
  vulpina 175
  x *boenninghausiana* 175
  x *pseudoaxillaris* 175
Carlina vulgaris 158
Carpinus betulus 79
Carpobrotus edulis 79
Carrot, Sea 136
  Wild 136
Carthamus tinctorius 160
Carum carvi 135
Castanea sativa 77
Catabrosa aquatica 182
Catapodium marinum 182
  rigidum 182
Catapyrenium rufescens 221
  squamulosum 221
Catchfly, Forked 86
  Italian 85
  Night-flowering 85
  Nottingham 85
  Sand 86
  Small-flowered 86
Catillaria aphana 221
  atropurpurea 221
  chalybeia 221
  lenticularis 221
Catinaria atropurpurea 221
Cat-mint 144
  Garden 144
Cat's-ear 160
  Smooth 160
Cat's-tail, Sand 186
  Smaller 186
Caucalis platycarpos 136
Celandine, Greater 74
  Lesser 72
Celery, Wild 134
Celothelium ischnobelum 221
Centaurea calcitrapa 160
  cyanus 159, 14
  melitensis 160
  montana 159
  nigra 160
  scabiosa 159
  solstitialis 160
Centaurium erythraea 136
  pulchellum 136
  tenuiflorum 136

Centaury, Common 136
   Lesser 136
   Slender 136
Centranthus ruber 157, 2
Cephalanthera damasonium 194
Cephalozia bicuspidata 200
   connivens 200
   macrostachya 200
Cephaloziella baumgartneri 200
   divaricata 200
   hampeana 200
   sp. 200
   stellulifera 200
   turneri 200
Cerastium arvense 83
   diffusum 83
   fontanum 83
   glomeratum 83
   pumilum 83
   semidecandrum 83
   tomentosum 83
Ceratocapnos claviculata 75
Ceratochloa cathartica 187
Ceratodon purpureus 204
Ceratophyllum demersum 70
Ceterach officinarum 68
Cetraria aculeata 221
   chlorophylla 232
Chaenorhinum minus 148
Chaenotheca ferruginea 221
   furfuracea 221
Chaerophyllum temulum 132
Chaffweed 103
Chamaemelum nobile 165
Chamerion angustifolium 125
Chamomile 165
   Corn 165
   Sicilian 165
   Stinking 166
   Yellow 166
Chara aculeolata 234
   aspera 234
   globularis 234
   hispida 234
   pedunculata 234
   virgata 234
   vulgaris 234
Charlock 99
Chelidonium majus 74
Chenia leptophylla 208, 15
Chenopodium album 80
   bonus-hendricus 79
   dessicatum 80
   ficifolium 80
   glaucum 79
   murale 80
   polyspermum 80
   rubrum 79
   urbicum 80
   vulvaria 80
Cherry, Bird 114
   Dwarf 114
   Wild 114
Chervil, Bur 132
   Rough 132
Chestnut, Sweet 77
Chickweed, Common 82
   Greater 83
   Lesser 82
   Upright 83
   Water 83
Chicory 160
Chiloscyphus pallescens 201
   polyanthos 201
Chromatochlamys muscorum 221
Chrysanthemum segetum 166
Chrysosplenium oppositifolium 105
Chrysothrix candelaris 221
   chrysophthalma 221
Cicerbita macrophylla 161
Cichorium intybus 160
Cinclidotus fontinaloides 209
   mucronatus 206
Cinquefoil, Creeping 111
   Hybrid 111
   Marsh 111
   Sulphur 111
   Ternate-leaved 111
Circaea lutetiana 125
Cirriphyllum crassinervium 215
   piliferum 214
Cirsium acaule 159
   arvense 159
   dissectum 159
   eriophorum 158
   palustre 159
   vulgare 159
   x *celakovskianum* 159
   x *forsteri* 159
Cladium mariscus 175
Cladonia arbuscula 221
   caespiticia 221
   cariosa 222
   cervicornis 222
   chlorophaea 222
   ciliata 222
   coccifera 222
   coniocraea 222
   crispata 222
   digitata 222
   fimbriata 222
   floerkeana 222
   foliacea 222
   furcata 222
   glauca 222
   gracilis 222
   macilenta 222
   ochrochlora 222
   parasitica 222
   pocillum 222
   polydactyla 222
   portentosa 222
   pyxidata 222
   ramulosa 222
   rangiformis 222
   scabriuscula 222
   squamosa 222
   subcervicornis 222
   subrangiformis 222
   subulata 222
   uncialis 222
Cladopodiella fluitans 200
Clary, Meadow 146
   Wild 146
Clauzadea immersa 222
   metzleri 222
   monticola 222
Claytonia perfoliata 82
   sibirica 82
Cleavers 155
   Corn 155
Clematis flammula 71
   vitalba 71
Clinopodium acinos 145
   ascendens 144
   menthifolium 144, 43, 1
   vulgare 145
Cliostomum griffithii 222
Clover, Alsike 120
   Bird's-foot 120
   Clustered 120
   Crimson 121
   Hare's-foot 121
   Knotted 121
   Red 121
   Reversed 120
   Rough 121
   Sea 121
   Strawberry 120
   Subterranean 121
   Suffocated 120
   Sulphur 121
   White 120
   Woolly 121
   Zigzag 121
Clubmoss, Krauss's 66
   Stag's-horn 66
Club-rush, Bristle 174
   Floating 174
   Grey 174
   Sea 174
   Slender 174
   Wood 174
Cochlearia anglica 98
   danica 98
   officinalis 98
Cocklebur, Rough 167
   Spiny 167
Cock's-eggs 138
Cock's-foot 182
Cockspur 189
Coelocaulon aculeatum 221
   muricata 221
   muricatum 221
Coeloglossum viride 195
Colchicum autumnale 190
Collema auriforme 222
   crispum 222
   cristatum 223
   flaccidum 223
   furfuraceum 223
   fuscovirens 223
   limosum 223
   nigrescens 223
   tenax 223
Cololejeunea minutissima 202
   rossettiana 202
Colt's-foot 167
Columbine 73
Colutea arborescens 116
Comfrey, Caucasian 141
   Common 140
   Creeping 141
   Hidcote 140
   Russian 140
   White 141
Conardia compacta 214
Coniocybe furfuracea 221
Conium maculatum 134
Conocephalum conicum 202
Conopodium majus 133
Conringia orientalis 99
Consolida ajacis 71
Convolvulus arvensis 138
Conyza bilbaoana 164
   canadensis 164
   sumatrensis 164
Copse-bindweed 88
Coralroot 97
Cord-grass, Common 189
   Small 188, 4
   Townsend's 188
Cordyline australis 194
Coriander 133
Coriandrum sativum 133
Corncockle 85
Cornflower 159, 14
   Perennial 159
Cornsalad, Broad-fruited 157
   Common 157
   Hairy-fruited 157
   Keeled-fruited 157
   Narrow-fruited 157
Cornus sanguinea 125
Coronilla scorpioides 117
Coronopus didymus 99
   squamatus 99
Cortaderia selloana 188
Corydalis cava 74
   solida 74
Corydalis, Climbing 75
   Pale 74
   Yellow 74
Corylus avellana 79
Cotoneaster horizontalis 116
   bullatus 116
   dielsianus 116

integrifolius 116
simonsii 116
Cotoneaster, Diel's 116
  Entire-leaved 116
  Himalayan 116
  Hollyberry 116
  Wall 116
Cottongrass, Broad-leaved 173
  Common 173
Cottonweed 165
Couch, Common 187
  Onion 183
  Sand 187
  Sea 187
Cowherb 86
Cowslip 102
Cow-wheat, Common 150
  Field 150, 9
Crack-willow 94
  Hybrid 94
Crambe maritima 100
Cranberry 101
Crane's-bill, Bloody 130
  Cut-leaved 130
  Dove's-foot 131
  Druce's 130
  French 130
  Hedgerow 131
  Himalayan 130
  Long-stalked 130
  Meadow 130
  Pencilled 130
  Purple 130
  Round-leaved 130
  Shining 131
  Small-flowered 131
Crassula helmsii 103
Crataegus laevigata 116
  monogyna 116
Cratoneuron commutatum 213
  filicinum 213
Creeping-jenny 102
Crepis biennis 162
  capillaris 162
  nicaeensis 162
  setosa 162
  vesicaria 162
Cresponea premnea 223
Cress, Garden 98
  Hoary 99
  Thale 95
  Tower 97
Crithmum maritimum 133
Crocosmia paniculata 193
  x *crocosmiiflora* 193
Crocus tommasinianus 193
  vernus 193
Crocus, Early 193
  Spring 193
Crosswort 155
  Caucasian 154
Crowfoot, Ivy-leaved 72

Round-leaved 72, 10
Cruciata laevipes 155
Cryphaea heteromalla 212
Cryptolechia carneolutea 223
Ctenidium molluscum 216
Cuckooflower 97
Cudweed, Broad-leaved 163
  Common 162
  Heath 163
  Marsh 163
  Red-tipped 163
  Small 163
Cupressus macrocarpa 70
Currant, Black 103
  Flowering 103
  Red 103
Cuscuta epithymum 139
  europaea 139
Cyclamen hederifolium 102
Cymbalaria muralis 148
Cynodon dactylon 188
Cynoglossum officinale 141
Cynosurus cristatus 180
  echinatus 180
Cyperus eragrostis 175
  longus 174
Cyphelium inquinans 223
  sessile 223
Cypress, Monterey 70
Cyrtidula quercus 223
Cyrtomium falcatum 69
Cystopteris fragilis 69
Cytisus scoparius 122

Dactylis glomerata 182
Dactylorhiza fuchsii 196, 8
  maculata 196
  praetermissa 196
  x *grandis* 196
  x *hallii* 196
Daffodil, Garden 192
  Wild 192, 12
Daisy 165
  Oxeye 166, 8
  Seaside 164
  Shasta 166
Dame's-violet 96
Dandelion, Lesser 161
Dandelions 161
Danthonia decumbens 188
Daphne laureola 124
  mezereum 124
Darmera peltata 104
Darnel 180
Datura stramonium 138
Daucus carota 136
Day-lily, Orange 190
Dead-nettle, Cut-leaved 143
  Henbit 143
  Red 143
  Spotted 143
  White 142
Degelia plumbea 223

Deschampsia cespitosa 184
  flexuosa 184
Descurainia sophia 95
Desmatodon convolutus 208
Dewberry 111, 1
Dewplant, Purple 79
Dialytrichia mucronata 206
Dianthus armeria 86
  deltoides 86
  plumarius 86
Dichodontium pellucidum 204
Dicranella cerviculata 204
  heteromalla 204
  schreberiana 204
  staphylina 204
  varia 204
Dicranoweisia cirrata 204
Dicranum bonjeanii 204
  majus 205
  montanum 205
  scoparium 205
Didymodon acutus 207
  fallax 207
  insulanus 207
  luridus 207
  rigidulus 207
  sinuosus 207
  tophaceus 207
  umbrosus 207
  vinealis 207
Digitalis purpurea 148
Digitaria sanguinalis 189
Dimerella lutea 223
  pineti 223
Diploicia canescens 223
Diplophyllum albicans 201
Diploschistes muscorum 223
  scruposus 223
Diplotaxis muralis 99
  tenuifolia 99
Diplotomma alboatrum 223
  chlorophaeum 223
Dipsacus fullonum 158
Dirina massiliensis 223
Disphyma crassifolium 79
Ditrichum cylindricum 204
  flexicaule 204
  gracile 204
  heteromallum 204
Dock, Aegean 89
  Broad-leaved 89
  Clustered 88
  Curled 88
  Fiddle 89
  Golden 89
  Water 88
  Willow-leaved 88
  Wood 89
Dodder 139
  Greater 139
Dog-rose 113
  Hairy 113

Round-leaved 113
Dog's-tail, Crested 180
  Rough 180
Dog-violet, Common 92
  Early 92
  Heath 92
  Pale 93
Dogwood 125
Doronicum pardalianches 167
Downy-rose, Harsh 113
  Sherard's 113
  Soft 113
Dranunculus vulgaris 171
Drepanocladus aduncus 214
Dropwort 105
Drosera intermedia 92
  rotundifolia 92
Dryopteris affinis 69
  carthusiana 69
  dilatata 69
  filix-mas 69
  x *deweveri* 69
Duchesnea indica 112
Duckweed, Common 171
  Fat 171
  Greater 171
  Ivy-leaved 171
  Least 172

Echinochloa crus-galli 189
Echium pininana 140
  vulgare 140
Eelgrass 170
  Dwarf 171
  Narrow-leaved 171
Elaeagnus umbellata 123
Elder 156
  Dwarf 156
Elecampane 163
Eleocharis multicaulis 173
  palustris 173
  quinqueflora 174
  uniglumis 173
Eleogiton fluitans 174
Elephant-ears 104
Elm, Cornish 76
  Dutch 76
  English 76
  Jersey 76
  Small-leaved 76
  Wych 78
Elodea canadensis 169
  nuttallii 169
Elytrigia atherica 187
  juncea 187
  repens 187
  x *obtusiuscula* 187
Encalypta streptocarpa 206
  vulgaris 206
Enchanter's-nightshade 125
Endocarpon pusillum 223
Enterographa crassa 223
  elaborata 223

Entodon concinnus 215
Entosthodon fascicularis 209
　obtusus 209
Ephemerum recurvifolium 209
　serratum 209
　sessile 209
Epilobium ciliatum 124
　hirsutum 124
　lanceolatum 124
　montanum 124
　montanum x *ciliatum* 124
　obscurum 124
　palustre 124
　parviflorum 124
　parviflorum x *ciliatum* 124
　roseum 124
　tetragonum 124
　x *limosum* 124
　x *subhirsutum* 124
Epipactis helleborine 194
　palustris 194, 2
　phyllanthes 194
　purpurata 194
Epipterygium tozeri 210
Equisetum arvense 66
　fluviatile 66
　palustre 66
　sylvaticum 66, 12
　telmateia 66
　x *littorale* 66
Eranthis hyemalis 71
Erica cinerea 101
　tetralix 101
Erigeron acer 164
　glaucus 164
　karvinskianus 164
Eriophorum angustifolium 173
　latifolium 173
Erodium cicutarium 131
　maritimum 131
　moschatum 131
Erophila glabrescens 97
　verna 97
Eruca vesicaria 100
Eryngium maritimum 132, 5
Erysimum cheiranthoides 96
　cheiri 96
Escallonia 103
Escallonia macrantha 103
Eschscholzia californica 74
Eucladium verticillatum 206
Euonymus europaeus 126
　japonicus 126
Eupatorium cannabinum 168
Euphorbia amygdaloides 127
　characias 127
　cyparissias 127
　exigua 127
　helioscopia 127
　lathyris 127
　oblongata 126
　paralias 127

　peplis 126
　peplus 127
　platyphyllos 126
　portlandica 127
　x *pseudovirgata* 127
Euphrasia anglica 151
　arctica 151
　confusa 151
　nemorosa 151
　nemorosa x *confusa* 151
　nemorosa x *pseudokerneri* 151
　pseudokerneri 151
　tetraquetra 151
　tetraquetra x *pseudokerneri* 151
Eurhynchium crassinervium 215
　hians 215
　praelongum 215
　pumilum 215
　schleicheri 215
　speciosum 215
　striatulum 215
　striatum 215
　swartzii 215
Evening-primrose, Fragrant 125
　Large-flowered 125
　Small-flowered 125
Everlasting, Pearly 163
Everlasting-pea, Broad-leaved 119
　Narrow-leaved 119
Evernia prunastri 223
Eyebrights 151

Fagopyrum esculentum 87
　tataricum 87
Fagus sylvatica 77
Fallopia baldschuanica 88
　convolvulus 88
　dumetorum 88
　japonica 87
　sachalinensis 88
False-acacia 116
Fat-hen 80
Fellhaneropsis vezdae 223
Fennel 134
Fen-sedge, Great 175
Fenugreek 120
　Sickle-fruited 119
Fern, Chain 69
　Lady 68
　Lemon-scented 67
　Limestone 68
　Maidenhair 67
　Marsh 67
　Royal 67, 3
　Water 69
Fern-grass 182
　Sea 182
Fescue, Bearded 180

　Dune 180
　Giant 179
　Meadow 179
　Rat's-tail 180
　Red 179
　Squirreltail 180
　Tall 179
Festuca arundinacea 179
　arundinacea x *multiflorum* 180
　filiformis 179
　gigantea 179
　ovina 179
　pratensis 179
　rubra 179
Feverfew 165
Ficus carica 76
Field-rose, Short-styled 113
Field-speedwell, Common 149
　Green 149
　Grey 149
Fig 76
Figwort, Common 147
　Water 147
Filago lutescens 163
　minima 163
　pyramidata 163
　vulgaris 162
Filipendula ulmaria 105
　vulgaris 105
Finger-grass, Hairy 189
Fissidens adianthoides 206
　bryoides 205
　celticus 205
　crassipes 205
　cristatus 205
　dubius 205
　exilis 205
　gracilifolius 205
　incurvus 205
　pusillus 205
　taxifolius 205
　viridulus 205
Flavoparmelia caperata 223
　soredians 223
Flax 128
　Fairy 128
　New Zealand 194
　Pale 128
Fleabane, Blue 164
　Canadian 164
　Common 164
　Guernsey 164
　Mexican 164
　Small 164
Fleawort, Field 167
Flixweed 95
Flowering-rush 168
Fluellen, Round-leaved 148
　Sharp-leaved 148
Foeniculum vulgare 134
Fontinalis antipyretica 212
Fool's-water-cress 135

Forget-me-not, Changing 141
　Creeping 141
　Early 141
　Field 141
　Great 141
　Tufted 141
　Water 141
　Wood 141
Forsythia 147
Forsythia x *intermedia* 147
Fossombronia pusilla 202
　wondraczekii 202
Fox-and-cubs 162
Foxglove 148
Fox-sedge, False 175
　True 175
Foxtail, Bulbous 185
　Marsh 185
　Meadow 185
　Orange 186
Fragaria vesca 111
　x *ananassa* 111
Frangula alnus 128
Frankenia laevis 93
Fraxinus excelsior 147
Fringecups 105
Frogbit 168
Frullania dilatata 201
　tamarisci 201
Fuchsia 125
Fuchsia magellanica 125
Fulgensia fulgens 224, 7
Fumaria bastardii 75
　capreolata 75
　densiflora 75
　muralis 75
　officinalis 75
　parviflora 75
　purpurea 75
　reuteri 75, 14
　vaillantii 75
Fumitory, Common 75
　Dense-flowered 75
　Few-flowered 75
　Fine-leaved 75
Funaria hygrometrica 209
　obtusa 209
Fuscidea lightfootii 224

Galanthus nivalis 192
Galega officinalis 116
Galeopsis angustifolia 143
　bifida 143
　speciosa 143
　tetrahit 143
Galingale 174
　Pale 175
Galinsoga parviflora 167
　quadriradiata 168
Galium aparine 155
　mollugo 155
　odoratum 155
　palustre 155

saxatile 155
tricornutum 155
uliginosum 155
verum 155
x *pomeranicum* 155
Gallant-soldier 167
Garlic, Few-flowered 191
  Field 191
  Hairy 191
  Neapolitan 191
  Rosy 191
  Three-cornered 191
  Yellow 191
Gastridium ventricosum 185
Gaudinia fragilis 183
Gazania rigens 162
Genista anglica 122
  monspessulana 122
  tinctoria 122, 8
Gentian, Autumn 137
  Early 137, 6
  Field 136, 11
Gentianella amarella 137
  anglica 137, 6
  campestris 136, 11
  x *davidiana* 137
Geranium columbinum 130
  dissectum 130
  endressi 130
  himalayense 130
  lucidum 131
  molle 131
  pratense 130
  purpureum 131
  pyrenaicum 131
  robertianum 131
  rotundifolium 130
  sanguineum 130
  versicolor 130
  x *magnificum* 130
  x *oxonianum* 130
Germander, Wall 144
Geum rivale 112
  urbanum 112
Giant-rhubarb 123
Gladiolus communis 193
  illyricus 193, 53
Gladiolus, Eastern 193
  Wild 193, 53
Glasswort, Common 81
  Long-spiked 81
  One-flowered 81
  Perennial 81
Glaucium flavum 74
Glaux maritima 103
Glechoma hederacea 144
Glyceria declinata 183
  fluitans 182
  maxima 182
  notata 183
  x *pedicellata* 182
Gnaphalium sylvaticum 163
  uliginosum 163

Goat's-beard 161
Goat's-rue 116
Goldenrod 164
  Canadian 164
Golden-saxifrage, Opposite-leaved 105
Gold-of-pleasure 98
Good-king-henry 79
Gooseberry 103
Goosefoot, Fig-leaved 80
  Many-seeded 80
  Nettle-leaved 80
  Oak-leaved 79
  Red 79
  Slim-leaf 80
  Stinking 80
  Upright 80
Gorse 122, 10
  Dwarf 122
Granium pusillum 131
Grape-hyacinth, Garden 191
Grape-vine 128
Graphina anguina 224
Graphis elegans 224
  scripta 224
Grass-of-Parnassus 105
Grass-poly 123
Greenweed, Dyer's 122, 8
Grimmia ovalis 209
  pulvinata 209
  trichophylla 209
Griselinia littoralis 125
Groenlandia densa 170
Gromwell, Common 140
  Field 140
Ground-elder 133
Ground-ivy 144
Ground-pine 144
Groundsel 167
  Heath 167
  Sticky 167
Guelder-rose 156
Guizotia abyssinica 167
Gunnera tinctoria 123
Gyalecta flotowii 224
  truncigena 224
Gymnadenia conopsea 195
Gymnocarpium robertianum 68
Gymnocolea inflata 200
Gymnostomum calcareum 207
  viridulum 207
Gypsywort 145
Gyroweisia tenuis 207

Haematomma caesium 226
  elatinum 226
  ochroleucum 224
Hair-grass, Crested 184
  Early 184
  Silver 184
  Tufted 184

Wavy 184
Hairy-brome 187
Hard-fern 69
Hard-grass 182
  Curved 182
Harebell 154
Hare's-ear, Slender 134
Hart's-tongue 67
Hawkbit, Autumn 160
  Lesser 160
  Rough 160
Hawk's-beard, Beaked 162
  Bristly 162
  French 162
  Rough 162
  Smooth 162
Hawkweeds 162
Hawthorn 116
  Midland 116
Hazel 79
Heath, Cross-leaved 101
Heather 101
  Bell 101
Heath-grass 188
Hebe salicifolia 149
  x *franciscana* 150
Hedera helix 132
Hedge-parsley, Knotted 135
  Spreading 135
  Upright 135
Helianthemum nummularia 92
Helianthus annuus 167
Helictotrichon pratense 183
  pubescens 183
Heliotrope, Winter 167
Hellebore, Green 71, 13
  Stinking 70
Helleborine, Broad-leaved 194
  Green-flowered 194
  Marsh 194, 2
  Violet 194
  White 194
Helleborus foetidus 70
  viridis 71, 13
Hemerocallis fulva 190
Hemlock 134
Hemp 76
Hemp-agrimony 168
Hemp-nettle, Bifid 143
  Common 143
  Large-flowered 143
  Red 143
Henbane 138
Hennediella heimii 208
  macrophylla 208
Heracleum mantegazzianum 135
  sphondylium 135
Herb-robert 131
Herniaria glabra 84
  hirsuta 84
Herzogiella seligeri 215

Hesperis matronalis 96
Hibiscus trionum 91
Hieracium eboracense 162
  exotericum 162
  sabaudum 162
  trichocaulon 162
  umbellatum 162
Himantoglossum hircinum 197
Hippocrepis comosa 117, 7
Hippophae rhamnoides 123
Hippuris vulgaris 146
Hirschfeldia incana 100
Hogweed 135
  Giant 135
Holcus lanatus 184
  mollis 184
Hollowroot 74
Holly 126
Holly-fern, House 69
Homalia trichomanoides 213
Homalothecium lutescens 214
  sericeum 214
Honckenya peploides 82
Honesty 97
Honeybells 191
Honeysuckle 156
  Henry's 156
  Himalayan 156
  Japanese 156
  Wilson's 156
Hookeria lucens 213, 11
Hop 76
Hordeum distichon 187
  jubatum 187
  marinum 187
  murinum 187
  secalinum 187
Horehound, Black 142
  White 143
Hornbeam 79
Horned-poppy, Yellow 74
Hornwort, Rigid 70
Horse-chestnut 129
Horse-radish 97
Horsetail, Field 66
  Great 66
  Marsh 66
  Shore 66
  Water 66
  Wood 66, 12
Hottentot-fig 79
Hottonia palustris 102
Hound's-tongue 141
House-leek 104
Humulus lupulus 76
Hyacinthoides hispanica 191
  non-scripta 190, 6
  non-scripta x *hispanica* 191
Hydrocharis morsus-ranae 168
Hydrocotyle ranunculoides 132

vulgaris 132
Hylocomium splendens 216
Hymenelia prevostii 224
Hyoscyamus niger 138
Hypericum androsaemum 90
  calycinum 90
  elodes 90
  hircinum 90
  hirsutum 90
  humifusum 90
  maculatum 90
  montanum 90
  perforatum 90
  pulchrum 90
  tetrapterum 90
  x *inodorum* 90
Hyperphyscia adglutinata 224
Hypnum andoi 216
  cupressiforme 215
  jutlandicum 216
  lacunosum 216
  lindbergii 216
  mammillatum 216
  resupinatum 216
Hypocenomyce scalaris 224
Hypochaeris glabra 160
  radicata 160
Hypogymnia physodes 224
  tubulosa 224
Hypotrachyna revoluta 224

Iberis umbellata 98
Ice-plant, Heart-leaf 79
Ilex aquifolium 126
Impatiens capensis 131
  glandulifera 132
  parviflora 132
Indian-rhubarb 104
Inula conyzae 163
  crithmoides 163
  helenium 163
Iris foetidissima 193
  germanica 193
  orientalis 193
  pseudacorus 193
Iris, Bearded 193
  Stinking 193
  Turkish 193
  Yellow 193
Isatis tinctoria 95
Isolepis cernua 174
  setacea 174
Isopterygium elegans 215
Isothecium alopecuroides 214
  myosuroides 214
  myurum 214
  striatulum 215
Ivy, Atlantic 132
  Common 132

Jacob's-ladder 139
Jasione montana 154
Johnson-grass 189

Juglans regia 77, 28
Juncus acutiflorus 172
  ambiguus 172
  articulatus 172
  bufonius 172
  bulbosus 172
  conglomeratus 172
  effusus 172
  foliosus 172
  gerardii 172
  inflexus 172
  maritimus 172
  squarrosus 172
  subnodulosus 172
  x *diffusus* 172
Juneberry 116
Jungermannia caespiticia 201
  gracillima 200
  hyalina 201
Juniper, Common 70
Juniperus communis 70

Kangaroo-apple 138
Ketmia, Bladder 91
Kickxia elatine 148
  spuria 148
Knapweed, Common 160
  Greater 159
Knautia arvensis 158
Knawel, Annual 84
Kniphofia uvaria 190
Knotgrass 87
  Cornfield 87
  Equal-leaved 87
  Ray's 87
  Sea 87, 4
Knotweed, Giant 88
  Himalayan 86
  Japanese 87
Koeleria macrantha 184
Koromiko 149
Kurzia pauciflora 200

Laburnum 122
Laburnum anagyroides 122
Lactuca serriola 161
  virosa 161
Lady's-mantle 112
  Garden 112
Lady's-tresses, Autumn 195, 6
Lagarosiphon major 169
Lamb's-ear 142
Lamiastrum galeobdolon 142
Lamium album 142
  amplexicaule 143
  hybridum 143
  maculatum 143
  purpureum 143
Lamprothamnium papulosum 235
Lapsana communis 160
Larkspur 71
Lathraea squamaria 152

Lathyrus annuus 119
  aphaca 119
  hirsutus 119
  japonicus 118
  latifolius 119
  linifolius 118
  nissolia 119
  palustris 119
  pratensis 119
  sativus 119
  sylvestris 119
  tuberosus 119
Laurel, Cherry 114
  Portugal 114
  Spotted 125
Lauristinus 156
Laurus nobilis 70
Lavandula x *intermedia* 146
Lavatera arborea 91
Lavender, Garden 146
Lecanactis abietina 224
  lyncea 224
  premnea 223
  subabietina 224
Lecania chlorotiza 224
  cyrtella 224
  erysibe 224
  naegelii 224
  turicensis 224
Lecanographa lyncea 224
Lecanora actophila 224
  albella 224
  albescens 224
  argentata 224
  campestris 225
  carpinea 225
  chlarotera 225
  conferta 225
  confusa 225
  conizaeoides 225
  crenulata 225
  dispersa 225
  expallens 225
  fugiens 225
  gangaleoides 225
  helicopsis 225
  muralis 225
  orosthea 225
  pallida 224
  polytropa 225
  pulicaris 225
  rupicola 225
  sulphurea 225
  symmicta 225
  varia 225
Lecidea fuscoatra 225
  lichenicola 225
  turgidula 225
Lecidella elaeochroma 225
  scabra 225
  stigmatea 225
Leek, Broad-leaved 191
Legousia hybrida 154

Leiocolea turbinata 200
Lejeunea lamacerina 202
  ulcina 202
Lemna gibba 171
  minor 171
  minuta 172
  trisulca 171
Lempholemma chalazanum 225
Leontodon autumnalis 160
  hispidus 160
  saxatilis 160
Leopard's-bane 167
Lepidium campestre 98
  draba 99
  heterophyllum 98
  perfoliatum 99
  ruderale 99
  sativum 98
  virginicum 98
Lepidozia reptans 200
Lepraria incana 225
Leproloma vouauxii 225
Leptobarbula berica 207
Leptobryum pyriforme 209
Leptodictyum riparium 214
Leptodon smithii 213
Leptodontium gemmascens 206
Leptogium biatorinum 225
  gelatinosum 225
  lichenoides 225
  plicatile 225
  schraderi 225
  teretiusculum 225
  turgidum 225
Leskea polycarpa 213
Lettuce, Great 161
  Prickly 161
  Wall 161
Leucanthemum vulgare 166, 8
  x *superbum* 166
Leucobryum glaucum 205
Leucodon sciuroides 212
Leucojum aestivum 192
Leycesteria formosa 156
Leymus arenarius 187
Lichina pygmaea 225
Ligustrum ovalifolium 147
  vulgare 147
Lilac 147
Lilium martagon 190
Lily, Martagon 190
Lime 90
  Small-leaved 90, 25
Limnanthes douglasii 131
Limonium binervosum 89
  humile 89
  vulgare 89, 4
  x *neumanii* 89
Linaria maroccana 148
  purpurea 148
  repens 148

supina 148
vulgaris 148
x *sepium* 148
Linum bienne 128
  catharticum 128
  usitatissimum 128
Liquorice, Wild 116
Listera ovata 195
Lithospermum arvense 140
  officinale 140
Little-robin 131
Lobaria amplissima 226
  pulmonaria 226, 13
  scrobiculata 226
  virens 226
Lobelia erinus 154
Lobelia, Garden 154
Lobularia maritima 97
Lolium multiflorum 180
  perenne 180
  remotum 180
  temulentum 180
  x *boucheanum* 180
Lombardy-poplar 94
London-rocket 95
  False 95
Lonicera henryi 156
  japonica 156
  nitida 156
  periclymenum 156
Loosestrife, Dotted 102
  Purple 123
  Yellow 102
Lophocolea bidentata 201
  fragrans 201
  heterophylla 201
  semiteres 201
Lophozia bicrenata 200
  excisa 200
  ventricosa 200
Lords-and-ladies 171
  Italian 171, 9
Lotus angustissimus 117
  corniculatus 117
  glaber 117
  pedunculatus 117
Lousewort 151
  Marsh 151
Love-in-a-mist 71
Loxospora elatina 226
Lucerne 120
Lunaria annua 97
Lungwort 140
  Narrow-leaved 140, 16
Lunularia cruciata 202
Lupin, Narrow-leaved 122
  Tree 121
Lupinus angustifolius 122
  arboreus 121
Luzula campestris 173
  forsteri 172
  multiflora 173
  pilosa 173

sylvatica 173
x *borreri* 173
Lychium barbaratum 137
  chinense 137
Lychnis coronaria 85
  flos-cuculi 85
Lycopersicon esculentum 138
Lycopodium clavatum 66
Lycopus europaeus 145
Lyme-grass 187
Lysimachia nemorum 102
  nummularia 102
  punctata 102
  vulgaris 102
Lythrum hyssopifolium 123
  portula 123
  salicaria 123

Madder, Field 154
  Wild 156
Mahonia aquifolium 73
Male-fern 69
  Scaly 69
Mallow, Common 91
  Dwarf 91
  Least 91
  Small 91
Malus domestica 114
  sylvestris 114
Malva moschata 91
  neglecta 91
  parviflora 91
  pusilla 91
  sylvestris 91
Maple, Cappadocian 129
  Field 129
  Norway 129
Marchantia polymorpha 202
Marchesinia mackaii 202
Mare's-tail 146
Marigold, Corn 166
  Pot 167
Marjoram, Wild 145
Marram 185
Marrubium vulgare 143
Marsh-bedstraw, Common 155
Marsh-mallow 91
  Rough 91
Marsh-marigold 70
Marsh-orchid, Southern 196
Marshwort, Lesser 135
Mat-grass 179
Matricaria discoidea 166
  recutita 166
Matthiola incana 96, 47, 9
Mayweed, Scented 166
  Scentless 166
  Sea 166
Meadow-foam 131
Meadow-grass, Annual 181
  Bulbous 182
  Early 181, 15

Flattened 182
Narrow-leaved 181
Rough 181
Smooth 181
Spreading 181
Wood 182
Meadow-rue, Common 73
Meadowsweet 105
Meconopsis cambrica 74
Medicago arabica 120
  lupulina 120
  polymorpha 120
  sativa 120
Medick, Black 120
  Spotted 120
  Toothed 120
Medlar 116
Megalaria grossa 226
Megaspora verrucosa 226
Melampyrum arvense 150, 9
  pratense 150
Melanelia elegantula 226
  exasperata 226
  fuliginosa 226
  laciniatula 226
  subaurifera 226
Melaspilea ochrothalamia 226
Melica uniflora 183
Melick, Wood 183
Melilot, Furrowed 119
  Ribbed 119
  Small 119
  Tall 119
  White 119
Melilotus albus 119
  altissimus 119
  indicus 119
  officinalis 119
  sulcatus 119
Melissa officinalis 144
Mentha aquatica 145
  arvensis 145
  pulegium 145
  requienii 145
  spicata 145
  suaveolens 145
  x *piperita* 145
  x *smithiana* 145
  x *verticillata* 145
  x *villosa* 145
Menyanthes trifoliata 139
Mercurialis annua 126
  perennis 126
Mercury, Annual 126
  Dog's 126
Mespilus germanica 116
Metzgeria fruticulosa 202
  furcata 202
  temperata 202
Mezereon 124
Micarea bauschiana 226
  denigrata 226
  erratica 226

melaena 226
nitschkeana 226
peliocarpa 226
prasina 226
Michaelmas-daisies 164
Microbryum curvicolle 208, 7
  davallianum 208
  floerkeanum 208
  rectum 208
  starckeanum 208
Microlejeunea ulicina 202
Mignonette, White 100
  Wild 100
Milium effusum 179
Milk-vetch, Purple 116
Milkwort, Common 129
  Heath 129
Millet, Common 189
  Wood 179
Mimulus x *robertsii* 147
Mind-your-own-business 77
Mint, Corn 145
  Corsican 145
  Round-leaved 145
  Spear 145
  Tall 145
  Water 145
  Whorled 145
Minuartia hybrida 82
Misopates orontium 148
Mistletoe 126
Mnium hornum 211
  stellare 211
Moehringia trinervia 82
Moelleropsis nebulosa 226
Moenchia erecta 83
Molinia caerulea 188
Monkeyflower, Hybrid 147
Monk's-hood 71
Monotropa hypopitys 102
Montbretia 193
Montia fontana 82
Moonwort 66
Moor-grass, Purple 188
Moschatel 157
Mouse-ear, Common 83
  Dwarf 83
  Field 83
  Little 83
  Sea 83
  Sticky 83
Mouse-ear-hawkweed 162
  Shaggy 162, 2
Mousetail 73
Muehlenbeckia complexa 88
Mugwort 165
  Chinese 165
Mullein, Dark 147
  Great 147
  Hoary 147
  Moth 147
  Orange 147
  Twiggy 147

Muscari armeniacum 191
Musk-mallow 91
Mustard, Ball 98
   Black 99
   Chinese 99
   Garlic 95
   Hare's-ear 99
   Hedge 95
   Hoary 100
   Tower 97
   White 100
Mycelis muralis 161
Mycoblastus caesius 226
   fucatus 226
   sterilis 226
Mycoporum quercus 223
Mylia anomala 200
Myosotis arvensis 141
   discolor 141
   laxa 141
   ramosissima 141
   scorpioides 141
   secunda 141
   sylvatica 141
Myosoton aquaticum 83
Myosurus minimus 73
Myrica gale 77
Myriophyllum alterniflorum 123
   aquaticum 123
   spicatum 123
Myxobilimbia lobulata 226
   sabuletorum 226

Narcissus agg. 192
   pseudonarcissus 192, 12
   x *medioluteus* 192
Nardia scalaris 201
Nardus stricta 179
Narthecium ossifragum 190
Navelwort 103
Neckera complanata 213
   crispa 213
   pumila 213
Neofuscelia verruculifera 226
Neottia nidus-avis 194
Nepeta cataria 144
   x *faassenii* 144
Neslia paniculata 98
Nettle, Common 76
   Small 77
Nicandra physalodes 137
Nicotinia rustica 138
   x *sanderae* 138
Nigella damascena 71
Niger 167
Nightshade, Black 138
   Deadly 138
   Green 138
   Leafy-fruited 138
Nipplewort 160
Nitella opaca 236
Nit-grass 185

Normandina pulchella 226
Nothoscordum borbonicum 191
Nuphar lutea 70
Nymphaea alba 70
Nymphoides peltata 139

Oak, Evergreen 78, 7
   Pedunculate 78
   Red 78
   Sessile 78
   Turkey 77
Oat 183
   Bristle 183
Oat-grass, Downy 183
   False 183
   French 183
   Meadow 183
   Yellow 184
Ochrolechia androgyna 226
   parella 226
   subviridis 227
   turneri 227
Odontites vernus 151
Oenanthe crocata 134
   fistulosa 133
   lachenalii 134
   pimpinelloides 134, 8
   silaifolia 133
Oenothera biennis 125
   cambrica 125
   glazioviana 125
   rubricaulis 125
   stricta 125
Oleaster, Spreading 123
Onion, Wild 191
Onobrychis viciifolia 117
Ononis repens 119
Onopordium acanthium 159
Opegrapha atra 227
   calcarea 227
   chevallieri 227
   conferta 227
   corticola 227
   gyrocarpa 227
   herbarum 227
   mougeotii 227
   multipunctata 227
   ochrocheila 227
   parasitica 227
   prosodea 227
   saxatilis 227
   sorediifera 227
   varia 227
   vermicellifera 227
   vulgata 227
Ophioglossum vulgatum 66, 8
Ophrys apifera 197
   insectifera 197
   sphegodes 197
Orache, Babington's 80
   Common 80

   Frosted 80
   Garden 80
   Grass-leaved 80
   Kattegat 80
   Long-stalked 80
   Shrubby 81
   Spear-leaved 80
Orchid, Bee 197
   Bird's-nest 194
   Burnt 196
   Early-purple 196, 13
   Fly 197
   Fragrant 195
   Frog 195
   Green-winged 196
   Lizard 197
   Man 196
   Pyramidal 195
Orchis mascula 196, 13
   morio 196
   ustulata 196
Oregon-grape 73
Oreopteris limbosperma 67
Origanum vulgare 145
Ornithogalum angustifolium 190
Ornithopus perpusillus 117
Orobanche artemisae-campestris 152, 2
   hederae 152
   minor 153
   purpurea 152, 8
   rapum-genistae 152
Orpine 104
Orthodontium lineare 209
Orthotrichum affine 212
   anomalum 212
   diaphanum 212
   lyellii 212
   pulchellum 212
   stramineum 212
   striatum 212
   tenellum 212
Osier 95
   Silky-leaved 95
Osmunda regalis 67, 3
Otanthus maritimus 165
Oxalis acetosella 130
   articulata 130
   corniculata 129
   debilis 130
   incarnata 130
   latifolia 130
   stricta 130
   valdiviensis 129
Oxlip, False 102
Oxtongue, Bristly 160
   Hawkweed 160

Pachyphiale carneola 227
Paeonia officinalis 89
Pallavicinia lyellii 202
Palustriella commutata 213

Pampas-grass 188
Panicum miliaceum 189
Pannaria pezizoides 229
   rubiginosa 227
Pansy, Field 93
   Garden 93
   Wild 93
Papaver argemone 74
   atlanticum 74
   hybridum 74
   rhoeas 74
   somniferum 74
   dubium 74
Parapholis incurva 182
   strigosa 182
Parentucellia viscosa 151
Parietaria judaica 77
Parmelia borreri 229
   caperata 223
   elegantula 226
   exasperata 226
   glabratula 226
   laciniatula 226
   mougeotii 232
   pastillifera 227
   perlata 227
   quercina 227
   reddenda 229
   reticulata 227
   revoluta 224
   saxatilis 227
   soredians 223
   subaurifera 226
   subrudecta 229
   sulcata 227
   tiliacea 227
   verruculifera 226
Parmelina pastillifera 227
   quercina 227
   tiliacea 227
Parmotrema chinense 227
   reticulatum 227
Parnassia palustris 105
Parrot's-feather 123
Parsley, Corn 135
   Cow 132
   Fool's 134
   Garden 135
   Stone 135
Parsley-piert 112
   Slender 112
Parsnip, Wild 135
Parthenocissus inserta 128
   quinquefolia 128
Pastinaca sativa 135
Pea, Fodder 119
   Indian 119
   Marsh 119
   Sea 118
   Tuberous 119
Pear 114
Pearlwort, Annual 84
   Heath 84

Knotted 84
Procumbent 84
Sea 84
Pedicularis palustris 151
  sylvatica 151
Pellia endiviifolia 202
  epiphylla 202
Pellitory-of-the-wall 77
Peltigera canina 227
  collina 228
  didactyla 228
  horizontalis 228
  hymenina 228
  lactucifolia 228
  membranacea 227
  neckeri 228
  praetextata 228
  rufescens 228
Penny-cress, Field 98
Pennyroyal 145
Pennywort, Floating 132
  Marsh 132
Pentaglottis sempervirens 141
Peony, Garden 89
Peppermint 145
Pepperwort, Field 98
  Least 98
  Narrow-leaved 99
  Perfoliate 99
  Smith's 98
Periwinkle, Greater 137
  Lesser 137
Persicaria amphibia 87
  amplexicaulis 86
  bistorta 86
  hydropiper 87
  lapathifolia 87
  maculosa 87
  minor 87
  mitis 87
  wallichii 86
Persicaria, Pale 87
Pertusaria albescens 228
  amara 228
  coccodes 228
  flavida 228
  hemisphaerica 228
  hymenea 228
  leioplaca 228
  multipuncta 228
  pertusa 228
  pseudocorallina 228
  velata 228
Petasites fragrans 167
  japonicus 167
Petractis clausa 228
Petrorhagia nanteuilii 86, 51
Petroselinum segetum 135
  crispum 135
Phacelia 139
Phacelia tanacetifolia 139
Phaeoceros laevis 203
Phaeographis dendritica 228

inusta 228
lyellii 228
smithii 228
Phaeophyscia orbicularis 228
Phalaris aquatica 184
  arundinacea 184
  canariensis 184
  minor 184
  paradoxa 184
Phascum curvicolle 208
  cuspidatum 208
  floerkeanum 208
Pheasant's-eye 73
Philonotis arnellii 212
  caespitosa 212
  fontana 212
  marchica 212
  rigida 211
Phleum arenarium 186
  bertolonii 186
  pratense 186
Phlyctis agelaea 228
  argena 228
Phormium tenax 194
Phragmites australis 188
Phuopsis stylosa 154
Phyllitis scolopendrium 67
Physcia adscendens 228
  aipolia 228
  caesia 228
  clementei 228
  leptalea 228
  semipinnata 228
  stellaris 228
  tenella 229
  tribacia 229
  tribacioides 229
Physcomitrium pyiforme 209
Physconia distorta 229
  grisea 229
  perisidiosa 229
Picris echioides 160
  hieracioides 160
Pigmyweed, New Zealand 103
Pignut 133
Pigweed, White 82
Pillwort 67
Pilosella aurantiaca 162
  officinarum 162
  peleteriana 162, 2
Pilularia globulifera 67
Pimpernel, Bog 102
  Scarlet 103
  Yellow 102
Pimpinella saxifraga 133, 6
Pine, Scots 70
Pineappleweed 166
Pinguicula lusitanica 153
Pink 86
  Childing 86, 51
  Deptford 86
  Maiden 86
Pink-sorrel 130

Garden 130
Large-flowered 130
Pale 130
Pinus sylvestris 70
Pirri-pirri-bur 112
Placynthiella icmalea 229
  uliginosa 229
Placynthium nigrum 229
  tantaleum 229
Plagiochila asplenioides 201
  porelloides 201
Plagiomnium affine 211
  elatum 211
  rostratum 211
  undulatum 211
Plagiothecium curvifolium 215
  denticulatum 215
  latebricola 215
  nemorale 215
  ruthei 215
  succulentum 215
  undulatum 215
Plantago afra 147
  arenaria 147
  coronopus 146
  lanceolata 146
  major 146
  maritima 146
  media 146
Plantain, Branched 147
  Buck's-horn 146
  Glandular 147
  Greater 146
  Hoary 146
  Ribwort 146
  Sea 146
Platanthera bifolia 195
  chlorantha 195
Platismatia glauca 229
Platygyrium repens 215
Pleuridium acuminatum 204
  subulatum 204
Pleurochaete squarrosa 206
Pleurozium schreberi 215
Ploughman's-spikenard 163
Plum, Cherry 114
  Wild 114
Poa angustifolia 181
  annua 181
  bulbosa 182
  compressa 182
  humilis 181
  infirma 181, 15
  nemoralis 182
  pratensis 181
  trivialis 181
Pogonatum aloides 203
  nanum 203
  urnigerum 203
Pohlia annotina 210
  camptotrachela 210
  carnea 210

  lescuriana 210
  lutescens 210
  melanodon 210
  nutans 210
  wahlenbergii 210
Polemonium caeruleum 139
Polyblastia albida 229
  gelatinosa 219
Polycarpon tetraphyllum 84
Polygala serpyllifolia 129
  vulgaris 129
Polygonatum x *hybridum* 190
Polygonum arenastrum 87
  aviculare 87
  maritimum 87, 4
  oxyspermum 87
  rurivagum 87
Polypodium aculeatum 69
  cambricum 67
  interjectum 67
  vulgare 67
  x *bicknellii* 69
Polypody 67
Polypogon monspeliensis 185
  viridis 185
Polysporina simplex 229
Polystichum setiferum 69
Polytrichum commune 204
  formosum 204
  juniperinum 204
  longisetum 203
  piliferum 204
Pond-sedge, Greater 177
  Lesser 176
Pondweed, Blunt-leaved 170
  Bog 169
  Broad-leaved 169
  Curled 170
  Fennel 170
  Horned 170
  Lesser 170
  Opposite-leaved 170
  Perfoliate 170
  Shining 169
  Small 170
Poplar, Grey 93
  White 93
Poppy, Atlantic 74
  Californian 74
  Common 74
  Long-headed 74
  Opium 74
  Prickly 74
  Rough 74
  Welsh 74
  Yellow-juiced 74
Populus alba 93
  nigra 94
  tremula 93
  trichocarpa 94
  x *canadensis* 94
  x *canescens* 93
  x *jackii* 94

Porella arboris-vitae 201
  obtusata 201
  platyphylla 201
Porina aenea 229
  leptalea 229
  linearis 229
Porpidia cinereoatra 229
  crustulata 229
  macrocarpa 229
  platycarpoides 229
  tuberculosa 229
Potamogeton berchtoldii 170
  crispus 170
  lucens 169
  natans 169
  obtusifolius 170
  pectinatus 170
  perfoliatus 170
  polygonifolius 169
  pusillus 170
Potato 138
Potentilla anglica 111
  anserina 111
  erecta 111
  norvegica 111
  palustris 111
  recta 111
  reptans 111
  sterilis 111
  x *mixta* 111
Pottia bryoides 208
  crinita 208
  heimii 208
  intermedia 208
  lanceolata 208
  recta 208
  starckeana 208
  truncata 208
Primrose 102
Primrose-peerless 192
Primula veris 102
  vulgaris 102
  x *polyantha* 102
Privet, Garden 147
  Wild 147
Protoblastenia calva 229
  rupestris 229
Protopannaria pezizoides 229
Prunella vulgaris 144
Prunus avium 114
  cerasifera 114
  cerasus 114
  domestica 114
  laurocerasus 114
  lusitanica 114
  padus 114
  spinosa 114
Pseudephemerum nitidum 204
Pseudocrossidium
  hornschuchianum 206
  revolutum 206
Pseudocyphellaria aurata 229
Pseudofumaria alba 74

  lutea 74
Pseudosasa japonica 179
Pseudoscleropodium purum 214
Pseudotaxiphyllum elegans 215
Psilolechia lucida 229
Pteridium aquilinum 67
Pterocarya fraxinifolia 77
Pterogonium gracile 212
Pterygoneurum ovatum 207
Ptilidium pulcherrimum 201
Puccinellia distans 180
  fasciculata 181
  maritima 180
  rupestris 181
Pulicaria dysenterica 164
  vulgaris 164
Pulmonaria longifolia 140, 16
  officinalis 140
Punctelia borreri 229
  reddenda 229
  subrudecta 229
Purslane, Pink 82
Pycnothelia papillaria 229
Pyrenocollema halodytes 229
  subarenisedum 229
Pyrenula chlorospila 230
  macrospora 230
Pyrola rotundifolia 101
Pyrrhospora quernea 230
Pyrus communis 114

Quaking-grass 181
  Greater 181
  Lesser 181
Quercus cerris 77
  ilex 78, 7
  petraea 78
  robur 78
  rubra 78
  x *rosacea* 78

Radiola linoides 129
Radish, Garden 100
  Sea 100
  Wild 100
Radula complanata 201
Ragged-robin 85
Ragweed 167
Ragwort, Common 166
  Hoary 166
  Marsh 166
  Oxford 166
  Silver 166
Ramalina calicaris 230
  canariensis 230
  cuspidata 230
  farinacea 230
  fastigiata 230
  fraxinea 230
  lacera 230
  siliquosa 230

  subfarinacea 230
Ramping-fumitory, Common 75
  Martin's 75, 14
  Purple 75
  Tall 75
  White 75
Ramsons 191
Ranunculus acris 71
  aquatilis 73
  arvensis 72
  auricomus 72
  baudotii 72
  bulbosus 71
  ficaria 72
  flammula 72
  hederaceus 72
  lingua 72
  omiophyllus 72, 10
  parviflorus 72
  peltatus 73
  penicillatus 73
  repens 71
  sardous 71
  sceleratus 72
  trichophyllus 73
  x *segretii* 73
Rape 99
  Long-stalked 99
Raphanus raphanistrum 100
  sativus 100
Rapistrum rugosum 100
Raspberry 105
Reboulia hemisphaerica 202
Red-hot-poker 190
Redshank 87
Reed, Common 188
Reseda alba 100
  lutea 100
  luteola 100
Restharrow, Common 119
Rhamnus alaternus 128
  cathartica 128
Rhinanthus minor 151
Rhizocarpon distinctum 230
  concentricum 230
  obscuratum 230
  petraeum 230
  reductum 230
  richardii 230
Rhizomnium pseudopunctatum 211
  punctatum 211
Rhodobryum roseum 211
Rhododendron 100
Rhododendron ponticum 100
Rhus hirta 129
Rhynchospora alba 175
Rhynchostegiella curviseta 215
  litorea 215
  tenella 215
Rhynchostegium confertum 215

  megapolitanum 215
  murale 214
  riparioides 214
Rhytidiadelphus squarrosus 216
  triquetrus 216
Ribes nigrum 103
  rubrum 103
  sanguineum 103
  uva-crispa 103
Riccardia chamedryfolia 202
  latifrons 202
  multifida 202
Riccia fluitans 202
  glauca 203
  sorocarpa 203
Rinodina atrocinerea 230
  confragosa 230
  efflorescens 230
  exigua 230
  gennarii 230
  occulta 230
  oleae 230
  roboris 230
  sophodes 230
  teichophila 230
Robinia pseudoacacia 116
Roccella phycopsis 230, 15
Rock-cress, Hairy 97
Rocket, Sea 100
  Eastern 95
  Garden 100
  Tall 95
Rock-rose, Common 92
Rorippa microphylla 96
  nasturtium-aquaticum 96
  palustris 97
  sylvestris 97
  x *sterilis* 96
Rosa arvensis 112
  caesia 113
  canina 113
  'Hollandica' 113
  micrantha 113
  mollis 113
  multiflora 112
  obtusifolia 113
  pimpinellifolia 112
  rubiginosa 113
  rugosa 113
  sherardii 113
  stylosa 113
  tomentosa 113
  x *andegavensis* 113
  x *coronata* 113
  x *dumetorum* 113
  x *nitidula* 113
  x *pseudorusticana* 112
  x *sabinii* 113
  x *verticillacantha* 112
Rose, Burnet 112
  Dutch 113
  Field 112

Japanese 113
  Many-flowered 112
Rosemary 146
Rose-of-Sharon 90
Rosmarinus officinalis 146
Rowan 114
Rubia peregrina 156
Rubus adscitus 108
  aequalidens 108
  albionis 106
  altiarcuatus 107
  amplificatus 107
  angusticuspis 110
  armeniacus 108
  armipotens 108
  bloxamii 109
  boudiccae 107
  boulayi 106
  caesius 111, 1
  cardiophyllus 107
  cinerosus 109
  cissburiensis 107
  cockburnianus 105
  conjungens 110
  cornubiensis 107
  curvispinosus 107
  dasyphyllus 110
  dentatifolius 109
  divaricatus 106
  dumnoniensis 107
  echinatus 109
  effrenatus 109
  elegantispinosus 108
  errabundus 106
  fissus 106
  flexuosus 109
  formidabilis 109
  fruticosus 105
  'H1056' 111
  'H107' 110
  'H165' 111
  'H252' 111
  'H375' 111
  'H388' 111
  'H863' 111
  halsteadensis 110
  hantonensis 109
  hindii 110
  hylophilus 108
  idaeus 105
  insectifolius 109
  lamburnensis 108
  largificus 109
  leightonii 109
  leucandriformis 106
  leyanus 109
  lindleianus 106
  micans 109
  mollissimus 106
  moylei 109
  mucronatiformis 108
  nemoralis 108
  nemorosus 110
  nessensis 106
  orbus 108
  oxyanchus 107
  phaeocarpus 110
  pictorum 110
  plicatus 106
  polyanthemus 108
  praetextus 110
  prolongatus 108
  pruinosus 110
  purbeckensis 107
  pyramidalis 107
  rilstonei 110
  riparius 107
  rudis 110
  rufescens 110
  salteri 107
  scaber 110
  sectiramus 110
  silvaticus 107
  sprengelii 108
  subinermoides 108
  sulcatus 106
  surrejanus 108
  transmarinus 110
  trichodes 109
  tuberculatus 110
  ulmifolius 108
  venetorum 110
  vestitus 108
  vigorosus 106
  winteri 108
Rumex acetosa 88
  acetosella 88
  conglomeratus 88
  crispus 88
  dentatus 89
  hydrolapathum 88
  maritimus 89
  obtusifolius 89
  pulcher 89
  salicifolius 88
  sanguineus 89
  x *muretii* 88
  x *abortivus* 88
  x *dufftii* 89
  x *mixtus* 89
  x *ogulinensis* 89
  x *pratensis* 88
  x *pseudopulcher* 88
  x *ruhmeri* 88
  x *sagorskii* 88
  x *schulzei* 88
Ruppia cirrhosa 170
  maritima 170
Rupturewort, Hairy 84
  Smooth 84
Ruscus aculeatus 192
Rush, Blunt-flowered 172
  Bulbous 172
  Compact 172
  Frog 172
  Hard 172
  Heath 172
  Jointed 172
  Leafy 172
  Saltmarsh 172
  Sea 172
  Sharp-flowered 172
  Soft 172
  Toad 172
Russian-vine 88
Rustyback 68
Rye-grass, Flaxfield 180
  Italian 180
  Perennial 180

Safflower 160
Saffron, Meadow 190
Sage, Wood 144
Sagina apetala 84
  maritima 84
  nodosa 84
  procumbens 84
  subulata 84
Sainfoin 117
Salicornia europaea 81
  procumbens 81
  pusilla 81
Salix alba 94
  aurita 95
  caprea 95
  cinerea 95
  daphnoides 94
  fragilis 94
  purpurea 94
  repens 95
  triandra 94
  viminalis 95
  x *calodendron* 95
  x *multinervis* 95
  x *reichardtii* 95
  x *rubens* 94
  x *rubra* 94
  x *smithiana* 95
Salpichroa origanifolia 138
Salsify 161
  Slender 161
Salsola kali 81
Saltmarsh-grass, Borrer's 181
  Common 180
  Reflexed 180
  Stiff 181
Saltwort, Prickly 81
Salvia pratensis 146
  verbenaca 146
Sambucus ebulus 156
  nigra 156
Samolus valerandi 103
Samphire, Golden 163
  Rock 133
Sandwort, Fine-leaved 82
  Mossy 82
  Sea 82
  Three-nerved 82
  Thyme-leaved 82
Sanguisorba minor 112
Sanicle 132
Sanicula europaea 132
Saponaria officinalis 86
Sarcocornia perennis 81
Sarcogyne privigna 230
  regularis 230
Sarcopyrenia gibba 231
Sasa palmata 179
Satureja montana 144
Savory, Winter 144
Saw-wort 159
Saxifraga granulata 105
  tridactylites 105, 15
  x *arendsii* 105
Saxifrage, Garden Mossy 105
  Meadow 105
  Pepper 134
  Rue-leaved 105, 15
Scabiosa columbaria 158, 6
Scabious, Devil's-bit 158
  Field 158
  Small 158, 6
Scandix pecten-veneris 133
Scapania aspera 201
  compacta 201
  irrigua 201
  nemorea 201
  undulata 201
Schismatomma cretaceum 231
  decolorans 231
  niveum 231
  quercicola 231
  umbrinum 231
Schistidium apocarpum 209
Schoenoplectus
  tabernaemontani 174
Scilla autumnalis 190, 5
  peruviana 190
Scirpus sylvaticus 174
Scleranthus anuus 84
Scleropodium cespitans 214
  purum 214
  tourettii 214
Scoliciosporum chlorococcum 231
  pruinosum 231
  umbrinum 231
Scorpion-vetch, Annual 117
Scorpiurium circinatum 214
Scrophularia auriculata 147
  nodosa 147
Scurvygrass, Common 98
  Danish 98
  English 98
Scutellaria galericulata 143
  minor 143
Sea-blite, Annual 81
Sea-heath 93
Sea-holly 132, 5
Sea-kale 100
Sea-lavender, Common 89, **4**
  Lax-flowered 89

Rock 89
Sea-milkwort 103
Sea-purslane 81
Sea-spurrey, Greater 85
   Lesser 85
   Rock 84
Securigera varia 117
Sedge, Bottle 177
   Brown 176
   Carnation 177
   Common 179
   Cyperus 177
   Distant 178
   Divided 176
   Dotted 178
   Flea 179
   Glaucous 177
   Green-ribbed 177
   Grey 175
   Hairy 176
   Long-bracted 178
   Oval 176
   Pale 178
   Pendulous 177
   Pill 178
   Prickly 175
   Remote 176
   Sand 175
   Smooth-stalked 177
   Soft-leaved 178
   Spiked 175
   Star 176
   Tawny 178
   White 176
Sedum acre 104
   album 104
   anglicum 104
   cepaea 104
   dasyphyllum 104
   forsterianum 104
   rupestre 104
   sexangulare 104
   spurium 104
   telephium 104
Selaginella kraussiana 66
Selfheal 144
Seligeria calcarea 209
   calycina 209
   paucifolia 209
   recurvata 209
Sempervivum tectorum 104
Senecio aquaticus 166
   cineraria 166
   erucifolius 166
   jacobaea 166
   squalidus 166
   sylvaticus 167
   viscosus 167
   vulgaris 167
   x *albescens* 166
   x *baxteri* 167
Serapias neglecta 197
Seriphidium maritimum 165

Serratula tinctoria 159
Service-tree, Wild 115, 12
Setaria italica 189
   pumila 189
   verticillata 189
   viridis 189
Shaggy-soldier 168
Sheep's-bit 154
Sheep's-fescue 179
   Fine-leaved 179
Shepherd's-needle 133
Shepherd's-purse 98
   Pink 98
Sherardia arvensis 154
Shield-fern, Hard 69
   Soft 69
Silaum silaus 134
Silene conica 86
   conoidea 86
   dichotoma 86
   gallica 86
   italica 85
   latifolia 86
   noctiflora 85
   nutans 85
   uniflora 85, 3
   vulgaris 85
   x *hampeana* 86
Silky-bent, Loose 185
Silverweed 111
Silybum marianum 159
Sinapis alba 100
   arvensis 99
Sison amomum 135
Sisymbrium altissimum 95
   irio 95
   loeselii 95
   officinale 95
   orientale 95
Sisyrinchium bermudiana 193
   striatum 193
Skullcap 143
   Lesser 143
Small-reed, Purple 185
   Wood 185
Smyrnium olusatrum 133
Snapdragon 148
Sneezewort 165
Snowberry 156
Snowdrop 192
Snowflake, Summer 192
Snow-in-summer 83
Soapwort 86
Soft-brome, Lesser 186
   Slender 186
Soft-grass, Creeping 184
Solanum dulcamara 138
   laciniatum 138
   nigrum 138
   physalifolium 138
   rostratum 138
   sarachoides 138
   tuberosum 138

Soleirolia soleirolii 77
Solenopsora candicans 231
   holophaea 231
   vulturiensis 231
Solidago canadensis 164
   virgaurea 164
Solomon's-seal, Garden 190
Sonchus arvensis 161
   asper 161
   oleraceus 161
Sorbus aria 115
   aucuparia 114
   intermedia 115
   latifolia 115
   torminalis 115, 12
Sorghum halepense 189
Sorrel, Common 88
   Sheep's 88
Southbya nigrella 201
Sowbread 102
Sow-thistle, Perennial 161
   Prickly 161
   Smooth 161
Sparganium emersum 189
   erectum 189
   natans 189
Spartina anglica 189
   maritima 188, 4
   x *townsendii* 188
Spartium junceum 122
Spearwort, Greater 72
   Lesser 72
Speedwell, Germander 149
   Heath 149
   Ivy 149
   Marsh 149
   Slender 149
   Thyme-leaved 149
   Wall 149
   Wood 149
Spergula arvensis 84
Spergularia marina 85
   media 85
   rubra 85
   rupicola 84
Sphagnum auriculatum 203
   capillifolium 203
   compactum 203
   contortum 203
   cuspidatum 203, 10
   denticulatum 203
   fallax 203
   fimbriatum 203
   palustre 203
   papillosum 203
   recurvum 203
   squarrosum 203
   subnitens 203
Sphinctrina turbinata 231
Spider-orchid, Early 197
Spike-rush, Common 173
   Few-flowered 174
   Many-stalked 173

Slender 173
Spinach, New Zealand 79
Spindle 126
   Evergreen 126
Spiraea salicifolia 105
   x *pseudosalicifolia* 105
Spiranthes spiralis 195, 6
Spirodela polyrhiza 171
Spleenwort, Black 68
   Maidenhair 68
   Sea 68, 3
Spotted-orchid, Common 196, 8
   Heath 196
Springbeauty 82
Spring-sedge 178
Spurge, Balkan 126
   Broad-leaved 126
   Caper 127
   Cypress 127
   Dwarf 127
   Mediterranean 127
   Petty 127
   Portland 127
   Purple 126
   Sea 127
   Sun 127
   Twiggy 127
   Wood 127
Spurge-laurel 124
Spurrey, Corn 84
   Sand 85
Squamarina cartilaginea 231, 7
Squill, Autumn 190, 5
   Portuguese 190
Squinancywort 154
St John's-wort, Pale 90
   Hairy 90
   Imperforate 90
   Marsh 90
   Perforate 90
   Slender 90
   Square-stalked 90
   Trailing 90
Stachys annua 142
   arvensis 142
   byzantina 142
   officinalis 142
   palustris 142
   sylvatica 142
   x *ambigua* 142
Starflower, Spring 191
Star-of-bethlehem 190
Star-thistle, Maltese 160
   Red 160
   Yellow 160
Staurothele hymenogonia 231
   rupifraga 231
Steinia geophana 231
Stellaria dioica 86
   graminea 83
   holostea 83

media 85
neglecta 83
pallida 82
uliginosa 83
Stenocybe septata 231
Sticta limbata 231
  sylvatica 231
Stitchwort, Bog 83
  Greater 83
  Lesser 83
Stock, Hoary 96, 47, 9
Stonecrop, Biting 104
  Caucasian 104
  English 104
  Pink 104
  Reflexed 104
  Rock 104
  Tasteless 104
  Thick-leaved 104
  White 104
Stonewort, Bristly 234
  Clustered 236
  Common 234
  Dark 236
  Delicate 234
  Foxtail 235
  Fragile 234
  Hedgehog 234
  Rough 234
Stork's-bill, Common 131
  Musk 131
  Sea 131
Stratiotes aloides 169
Strawberry, Barren 111
  Garden 111
  Wild 111
  Yellow-flowered 112
Strawberry-tree 101
Suaeda maritima 81
Succisa pratensis 158
Sumach, Stag's-horn 129
Sundew, Oblong-leaved 92
  Round-leaved 92
Sunflower 167
Sweet-briar 113
  Small-flowered 113
Sweet-grass, Floating 182
  Plicate 183
  Reed 182
  Small 183
Swine-cress 99
  Lesser 99
Sycamore 129
Symphoricarpos albus 156
Symphytum caucasicum 141
  grandiflorum 141
  'Hidcote Blue' 140
  officinale 140
  orientale 141
  x *uplandicum* 140
Syntrichia intermedia 209
  laevipila 209
  latifolia 209

papillosa 209
ruraliformis 208
ruralis 208
Syringa vulgaris 147

Tamarisk 93
Tamarix gallica 93
Tamus communis 194
Tanacetum parthenium 165
  vulgare 165
Tansy 165
Taraxacum angustisquameum 162
  arenastrum 161
  atactum 161
  cophocentrum 162
  corynodes 161
  dilaceratum 161
  exacutum 162
  fasciatum 162
  fulviforme 161
  fulvum 161
  hamatum 161
  insigne 161
  lacistophyllum 161
  laevigatum 161
  lepidum 161
  leptodon 162
  macranthoides 162
  nordstedtii 161
  oblongatum 162
  oxoniense 161
  pachymerum 162
  pallidipes 161
  pannucium 161
  polyodon 162
  pseudohamatum 161
  pulchrifolium 162
  retzii 161
  rubicundum 161
  sahlinianum 161
  Sect. *Erythrosperma* 161
  Sect. *Hamata* 161
  Sect. *Ruderalia* 161
  Sect.*Celtica* 161
  sellandii 161
  stenacrum 161
  subbracteatum 161
  subexpallidum 161
  subhamatum 161
  subxanthostigma 162
  undulatiflorum 161
  undulatum 161
  wallonicum 161
Tare, Hairy 118
  Slender 118
  Smooth 118
Tasselweed, Beaked 170
  Spiral 170
Taxiphyllum wissgrillii 215
Taxus baccata 70
Teaplant, Duke of Argyll's 137
Teasel, Wild 158

Tellima grandiflora 105
Teloschistes chrysophthalmus 231
  flavicans 231
Tephromela atra 231
Tephroseris integrifolia 167
Tetragonia tetragonioides 79
Tetraphis pellucida 204
Teucrium chamaedrys 144
  scorodonia 144
Thalictrum flavum 73
Thamnobryum alopecurum 213
Thelidium minutulum 231
  zwackhii 231
Thelopsis rubella 231
Thelotrema lepadinum 231
Thelypteris palustris 67
Thesium humifusum 125
Thistle, Carline 158
  Cotton 159
  Creeping 159
  Dwarf 159
  Marsh 159
  Meadow 159
  Milk 159
  Musk 158
  Slender 158
  Spear 159
  Welted 158
  Woolly 158
Thlaspi arvense 98
Thorn-apple 138
Thorow-wax 134
  False 134
Thrift 89, 3
Thuidium philibertii 213
  tamariscinum 213
Thyme, Basil 145
  Garden 145
  Large 145
  Lemon 145
  Wild 145
Thymus polytrichus 145
  pulegioides 145
  vulgaris 145
  x *citriodorus* 145
Tilia cordata 90, 25
  x *europaea* 90
Timothy 186
Toadflax, Annual 148
  Common 148
  Ivy-leaved 148
  Pale 148
  Prostrate 148
  Purple 148
  Small 148
Tobacco, Wild 138
Tolypella glomerata 236
Tomato 138
Tongue-orchid, Scarce 197
Toninia aromatica 231
  episema 231

lobulata 226
sedifolia 231
Toothwort 152
Tor-grass 187
Torilis arvensis 135
  japonica 135
  nodosa 135
Tormentil 111
  Trailing 111
Tortella flavovirens 206
  inflexa 206
Tortula acaulon 208
  atrovirens 208
  intermedia 209
  laevipila 209
  lanceola 208
  latifolia 209
  marginata 207
  modica 208
  muralis 208
  papillosa 209
  protobryoides 208
  rhizophylla 208
  ruralis 208
  subulata 207
  truncata 208
  viridifolia 208
Trachystemon orientalis 141
Tragopogon hybridus 161
  porrifolius 161
  pratensis 161
  x *mirabilis* 161
Trapelia coarctata 231
  involuta 231
  placodioides 232
Trapeliopsis flexuosa 232
  granulosa 232
  pseudogranulosa 232
Traveller's-joy 71
Treacle-mustard 96
Treasureflower 162
Tree-mallow 91
Tree-of-heaven 129
Trefoil, Hop 121
  Lesser 121
  Slender 121
Trichostomopsis umbrosa 207
Trichostomum brachydontium 206
  crispulum 206, 7
Trifolium arvense 121
  campestre 121
  dubium 121
  fragiferum 120
  glomeratum 120
  hybridum 120
  incarnatum 121
  medium 121
  micranthum 121
  ochroleucon 121
  ornithopodioides 120
  pratense 121
  repens 120

resupinatum 120
scabrum 121
squamosum 121
striatum 121
subterraneum 121
suffocatum 120
tomentosum 121
Triglochin maritimum 169
  palustre 169
Trigonella corniculata 119
  foenum-graecum 120
Tripleurospermum inodorum 166
  maritimum 166
Trisetum flavescens 184
Tristagma uniflorum 191
Triticum aestivum 188
Tritomaria exsectiformis 200
Tuckermannopsis chlorophylla 232
Tufted-sedge 179
  Slender 179
Tulip, Wild 190
Tulipa sylvestris 190
Turgenia latifolia 136
Turnip 99
Tussilago farfara 167
Tussock-sedge, Greater 175
  Lesser 175
Tutsan 90
  Stinking 90
  Tall 90
Twayblade, Common 195
Typha angustifolia 190
  latifolia 189
  x *glauca* 190

Ulex europaeus 122, 10
  minor 122
Ulmus glabra 76
  minor 76
  procera 76
  x *hollandica* 76
Ulota bruchii 212
  crispa 212
  phyllantha 212
Umbilicus rupestris 103
Urtica dioica 76
  urens 77
Usnea articulata 232
  ceratina 232
  cornuta 232
  flammea 232
  florida 232
  rubicunda 232
  subfloridana 232
Utricularia australis 153, 11
  minor 153

Vaccaria hispanica 86
Vaccinium myrtillus 101
  oxycoccus 101
Valerian, Common 157

Marsh 157
Red 157, 2
Valeriana dioica 157
  officinalis 157
Valerianella carinata 157
  dentata 157
  eriocarpa 157
  locusta 157
  rimosa 157
Velvetleaf 91
Venus's-looking-glass 154
Verbascum blattaria 147
  nigrum 147
  phlomoides 147
  pulverulentum 147
  thapsus 147
  virgatum 147
Verbena bonariensis 142
  officinalis 142
Vernal-grass, Sweet 184
Veronica agrestis 149
  anagallis-aquatica 149
  arvensis 149
  beccabunga 149
  chamaedrys 149
  filiformis 149
  hederifolia 149
  montana 149
  officinalis 149
  persica 149
  polita 149
  scutellata 149
  serpyllifolia 149
Veronica, Hedge 150
Verrucaria amphibia 232
  aquatilis 232
  baldensis 232
  dolosa 232
  glaucina 232
  hochstetteri 232
  hydrela 232
  macrostoma 232
  maura 232
  muralis 232
  murina 232
  nigrescens 232
  pinguicula 232
  rheitrophila 232
  simplex 232
  striatula 232
  viridula 232
Vervain 142
  Argentinian 142
Vetch, Bithynian 118
  Bush 118
  Common 118
  Crown 117
  Fine-leaved 118
  Fodder 118
  Horseshoe 117, 7
  Hungarian 118
  Kidney 117
  Spring 118

Tufted 118
Wood 118
Vetchling, Grass 119
  Hairy 119
  Meadow 119
  Yellow 119
Viburnum lantana 156
  opulus 156
  tinus 156
Vicia bithynica 118
  cracca 118
  hirsuta 118
  hybrida 118
  lathyroides 118
  lutea 118
  pannonica 118
  parviflora 118
  sativa 118
  sepium 118
  sylvatica 118
  tenuifolia 118
  tetrasperma 118
  villosa 118
Vinca major 137
  minor 137
Viola arvensis 93
  canina 92
  hirta 92
  lactea 93
  odorata 92
  palustris 93
  reichenbachiana 92
  riviniana 92
  tricolor 93
  x *bavarica* 92
  x *scabra* 92
  x *wittrockiana* 93
Violet, Hairy 92
  Marsh 93
  Sweet 92
Violet-willow, European 94
Viper's-bugloss 140
  Giant 140
Virginia-creeper 128
  False 128
Virgin's-bower 71
Viscum album 126
Vitis vinifera 128
Vulpia bromoides 180
  ciliata 180
  fasciculata 180
  myuros 180

Wadeana dendrographa 232
Wahlenbergia hederacea 154
Wallflower 96
Wall-rocket, Annual 99
  Perennial 99
Wall-rue 68
Walnut 77, 28
Water-cress 96
  Hybrid 96
  Narrow-fruited 96

Water-crowfoot, Brackish 72
  Common 73
  Pond 73
  Stream 73
  Thread-leaved 73
Water-dropwort, Corky-fruited 134, 8
  Hemlock 134
  Narrow-leaved 133
  Parsley 134
  Tubular 133
Water-lily, Fringed 139
  White 70
  Yellow 70
Water-milfoil, Alternate 123
  Spiked 123
Water-parsnip, Lesser 133
Water-pepper 87
  Small 87
  Tasteless 87
Water-plantain 168
  Lesser 168
Water-purslane 123
Water-soldier 169
Water-Speedwell, Blue 149
Water-starwort, Blunt-fruited 146
  Common 146
  Intermediate 146
  Various-leaved 146
Water-violet 102
Waterweed, Canadian 169
  Curly 169
  Nuttall's 169
Wayfaring-tree 156
Weasel's-snout 148
Weissia brachycarpa 206
  condensa 206
  controversa 206
  longifolia 206
  microstoma 206
  rutilans 206
  sp. 206
  squarrosa 206
  sterilis 206
  tortilis 206
Weld 100
Wheat, Bread 188
Whin, Petty 122
Whitebeam 115
  Broad-leaved 115
  Swedish 115
Whitlowgrass, Common 97
  Glabrous 97
Whorl-grass 182
Wild-oat 183
  Winter 183
Willow, Almond 94
  Creeping 95
  Eared 95
  Goat 95
  Green-leaved 94
  Grey 95

Holme 95
Purple 94
White 94
Willowherb, American 124
Broad-leaved 124
Great 124
Hoary 124
Marsh 124
Pale 124
Rosebay 125
Short-fruited 124
Spear-leaved 124
Wingnut, Caucasian 77
Winter-cress 96
American 96
Medium-flowered 96
Round-leaved 101
Wireplant 88
Woad 95

Woodruff 155
Blue 155
Wood-rush, Field 173
Great 173
Hairy 173
Heath 173
Southern 172
Wood-sedge 177
Thin-spiked 177
Wood-sorrel 130
Wood-spurge, Leathery 127
Woodwardia radicans 69
Wormwood 165
Sea 165
Woundwort, Field 142
Hedge 142
Hybrid 142
Marsh 142

x *Festulolium loliaceum* 180
x *F. brinkmannii* 180
x *Festuca braunii* 180
Xanthium spinosum 167
strumarium 167
Xanthoparmelia mougeotii 232
Xanthoria calcicola 232
candelaria 232
parietina 232
polycarpa 233
ulophyllodes 233

Yarrow 165
Yellow-cress, Creeping 97
Marsh 97
Yellow-eyed-grass, Pale 193
Yellow-rattle 151
Yellow-sedge, Common 178

Yellow-sorrel, Chilean 129
Procumbent 129
Upright 130
Yellow-vetch 118
Hairy 118
Yellow-wort 136
Yellow-woundwort, Annual 142
Yew 70
Yorkshire Fog 184

Zannichellia palustris 170
Zostera angustifolia 171
marina 170
noltei 171
Zygodon baumgartneri 212
conoideus 212
rupestris 212, 13
viridissimus 212